COMPLEXOMETRIC
TITRATIONS

COMPLEXOMETRIC TITRATIONS

GEROLD SCHWARZENBACH

*Professor of Chemistry at the
Eidg. Technische Hoschschule, Zürich*

AND

HERMANN FLASCHKA

*Professor of Chemistry at the
Georgia Institute of Technology,
Atlanta, Georgia, U.S.A.*

SECOND ENGLISH EDITION
Translated and revised
in collaboration with the authors by

H. M. N. H. IRVING

M.A., D.PHIL., D.SC., F.R.I.C.

*Professor of Inorganic and Structural Chemistry
at the University of Leeds and sometime
Vice-Principal of St. Edmund Hall, Oxford*

METHUEN & CO LTD
11 NEW FETTER LANE LONDON EC4

Die komplexometrische Titration
was first published by
Ferdinand Enke Verlag, Stuttgart in 1955
First English edition was published in 1957
This second English edition,
translated from the Fifth German Edition of 1965 and
incorporating later revisions by the author,
was first published in 1969
This translation © Methuen and Co. Ltd., 1969
Printed in Great Britain
by Richard Clay (The Chaucer Press), Ltd.,
Bungay, Suffolk

SBN 416 19290 4

Distributed in the United States of America
by Barnes and Noble, Inc.

Preface to the Fifth German Edition

It was in a lecture to the Swiss Chemical Society in the spring of 1945 [45-3] that it was first pointed out that metal cations – regarded as Lewis acids – could be titrated with alkali salts of nitrilotriacetic acid and ethylenediaminetetra-acetic acid. The end-point was detected by the change in pH. This discovery was the unexpected fruit of systematic studies on complex formation by aminocarboxylic acids. Uramildiacetic acid [46-1], which was synthesized from aminobarbituric acid, was also of interest in this context, but under certain conditions the reaction mixture turned red on standing, due to small amounts of murexide formed by atmospheric oxidation. The change in colour produced by calcium ions was observed, quite by chance, when washing up the reaction vessels in hard tap-water, and an investigation of this phenomenon led to the discovery of the idea of metal-indicators [49-4]. Murexide was naturally introduced at once for the titration of calcium, and a search was instituted for indicators for other metals. Such indicators were found among certain commercial mordant dyestuffs. Equilibrium studies on several members of the Eriochrome black group [48-4] showed that Eriochrome black T was the most suitable indicator for titrations, and the first publication on the complexometric determination of the hardness of water using Erio T appeared in January 1948 [48-5].

It can thus be said that complexometric titrations date from 1945. Appreciation of this new analytical procedure began to spread internationally from 1950 onwards, at first for determinations of the hardness of water [50-4 to 50-9]. However, shortly afterwards metals other than the alkaline earths were determined complexometrically, and the introduction of masking agents made it possible to deal with mixtures of metals.

With the first edition of this book in 1955 as a monograph of some 100 pages in length there came the first comprehensive treatment of complexometry, just ten years after its discovery. This book showed how the complicated equilibria with which we have to deal in complexometric titrations can be handled with quite simple mathematics by introducing the concept of effective or conditional stability constants (referred to at that time as 'apparent' constants) and by calculations with distribution coefficients.

In quick succession the book passed through three further editions (1956, 1957, 1960) which were only slightly expanded. For the present Fifth Edition, which appears exactly twenty years after the discovery of complexometric titrations, the manuscript has had to be completely rewritten to cope with the enormous growth of new material. Whereas the bibliography of the first edition referred to 181 papers, the present list comprises 1,437 items, even though papers which contain nothing essentially new have been omitted. The literature has been surveyed to the end of 1963. The former plan of indexing it alphabetically under author's names has been abandoned to facilitate the inclusion of new publications without the need for renumbering. However, the references have been grouped according to the year in which the papers were published, and the number of papers quoted in successive years presents a good picture of the development of complexometry.

Complexometry has now become a fully developed analytical technique capable of dealing with the majority of the known elements (cf. Table 13). A whole range of chelating agents are now available as titrants, and some 120 colour indicators as well as several instrumental procedures serve to detect the end-points. A large number of masking agents enable selective titrations to be carried out in the presence of a variety of other components. It is improbable that new titrating agents and indicators will be discovered in the future that will be a substantial improvement on those already known. Theories covering the different types of complexometric titrations have been fully worked out, and the factors which determine the accuracy of the various determinations are well understood. Admittedly there are still too few data on stability constants of complexes to apply exact theory to every case. Indeed, equilibrium studies have so far been carried out with only 16 of the 100 or so metallochromic indicators that have been introduced (Table 12), and even then only with a few cations. Our knowledge of the stability of metal complexes formed by the commonly used masking agents is also scanty. To obtain such data demands time-consuming researches, but they would be rewarding; for it is only from a knowledge of the exact values of the stabilities of different complexes that the potentialities of complexometry can be fully realized. However, many selective titrations of great importance in analytical practice can be shown to be possible by a combination of the proper chelating agent in the titrant with a suitable masking agent and indicator. Thus zinc can be determined in the presence of cadmium, or cadmium in the presence of zinc, despite the chemical similarity of the two cations (cf. p. 122). The most favourable combinations can only be discovered on the basis of the stability constants of the relevant complexes.

In Part II we thought it inappropriate merely to report a series of recipes for titrations, since the procedures actually used must generally be adapted to particular circumstances. Therefore for every element that can be determined complexometrically we have reported all the possible methods of titration and the various interfering factors so as to make it possible for the reader to find out for himself the most favourable practical procedure for his particular problem, after consulting the original literature should this prove necessary. The numerous procedures that are also given are rather in the nature of practical examples.

GEROLD SCHWARZENBACH
HERMANN FLASCHKA

Translator's Preface

My first task is to express my deep gratitude to Professor Schwarzenbach and to Professor Flaschka for granting me the privilege of writing an English translation of their latest monograph *Complexometric Titrations* and thus playing some small part in bringing this fine work to the attention of a wider public.

When I prepared the first English translation of Schwarzenbach's original monograph in 1957 I took the opportunity of incorporating material which he was then preparing for the second German edition, and the bibliography was extended from 181 to 305 entries, making it complete up to August 1956. The second, third, and fourth German editions differed but slightly in content from the English translation, but the situation was completely altered when in 1965, in collaboration with Professor Flaschka, the entire work was recast and very greatly expanded to take into account the astonishing growth in the literature dealing with the theory and practice of complexometry, which had by then become one of the most widely practised and most fully understood techniques of the analytical chemist. Something of the rate of development can be gleaned from the number of papers published over equivalent successive periods, viz. 10 from 1945–47, 25 from 1948–50, 133 from 1951–53, 348 from 1954–56, 410 from 1957–59, 541 from 1960–62, and more than 400 from 1963–65.

Although the basic theory of complexometry is now very well established, new procedures and applications are constantly being found, and I was reluctant to embark upon a translation which, in view of the inevitable delays in preparation and publishing, might contain no reference to important work carried out during the previous four years or so. The generous co-operation of Professor Flaschka made it possible to complete the bibliographies to the middle of 1967, and all the relevant literature references were incorporated in the new translation together with a considerable number of additions to the original German text necessitated by this new material.

I have regrouped Part I of the German text into seven chapters, but equations are numbered consecutively to facilitate cross-referencing. To save space I have in most cases followed the German authors in using symbols as abbreviations for the full names of the elements. Special attention should be drawn to certain major changes in nomenclature. The German authors use a variety of symbols α_H, α_H^F, α_H^{MF}, β_A, β_B, β_A^*, β_B^* for the 'distribution coefficients' which play such an important role in the theoretical treatment, and the symbol K for an overall (gross or cumulative) stability constant. Since nowadays the symbol β is almost universally adopted for overall stability constants (cf. *Chem. Soc. Special Publication No. 17* 'Stability Constants'), this could cause confusion to English readers, and I have therefore replaced K by β wherever this occurs. The consequential changes necessary in the symbolism for distribution coefficients has enabled me to introduce the very simple and systematic nomenclature for 'α-coefficients' used by Anders Ringbom in his important recent book *Complexation in Analytical Chemistry* [63-64], which many practising analysts will doubtless wish to use in conjunction with the present work. An annotated list of the symbols used in the text immediately precedes Chapter One.

The Department of Inorganic and Structural Chemistry,
The University of Leeds, England

July, 1967.

ADDENDUM

Early in 1968 I was told that the manuscript of the translation had been lost in transit to the printers. Every effort to find it – including the offer of a handsome reward – proved unsuccessful. Unfortunately the carbon copies sent to Professor Flaschka in the U.S.A. had, by that time, also been destroyed, Professor Flaschka thinking, quite reasonably, that the book was more advanced in the printing process than was actually the case.

A number of other commitments prevented me from returning immediately to the problems created by this loss, but at the instigation both of the publishers (who had ascertained that no new German edition would appear for some years) and of many English-speaking colleagues who had been looking forward to the appearance of the new translation, I set about reconstructing the material from my notes. Part One can be regarded as incorporating *all* the new material sent to me by Professor Flaschka together with additional material on indicators (Chapter Three) based on his notes. Part Two is a full translation of the experimental sections of the German text, and it includes some, but alas not all, of Flaschka's additional material. However, I have been at pains to extend the Bibliographies for 1964 onwards and have included as many references to papers published in 1968 as I could. I have resisted the temptation to omit some of the voluminous literature on the grounds that papers were largely repetitive of earlier work or were published in journals that were not too generally accessible. In this way I have left the reader with the last word and hope that he will find this book *Complexometric Titrations* as up-to-date as circumstances permit.

Leeds, November 1968.   H. M. N. H. IRVING

Contents

Preface to the Fifth German Edition   *page* vii
Preface   xi
List of Symbols   xix

PART I. TITRATING AGENTS, INDICATORS,
         AND INSTRUMENTAL METHODS
         OF DETERMINING END-POINTS

1 **Titration with Complexing Agents**   3
  1.1 Introduction   3
  1.2 Polyamines   4
  1.3 Complexones   8

2 **Basic Theory of Complexometric Titrations**   14
  2.1 Formation and stability of complexes   14
  2.2 Titration curves in complexometry   21

3 **Colour and Fluorescence Indicators in Complexometry**   31
  3.1 Introduction   31
  3.2 Weak colour indicators   31
  3.3 Metallochromic indicators   34
  3.4 Fluorescence indicators   63
  3.5 Redox indicators   67
  3.6 Source, purity, and nomenclature of indicators   71

4 **Colour Changes in Metallochromic Indicators**   77
  4.1 pH and pM indicators   77
  4.2 The dependence of indicator constants upon the pH   78
  4.3 Indicator constants   81
  4.4 Transition curves   84
  4.5 Visibility of the colour change   86

5 **Instrumental Indication of the End-Point**   90
  5.1 Introduction   90
  5.2 Photometric titrations   90

5.3 Potentiometric titrations 98
5.4 Conductimetric titrations 104
5.5 High-frequency titrations 105
5.6 Amperometric titrations 106
5.7 Coulometric titrations 109
5.8 Chronopotentiometric titrations 110
5.9 Thermometric titrations 110
5.10 Radiometric titrations 111

6 **Different Types of Titrations and the Accuracy Attainable** 113
6.1 Direct titration 113
6.2 Back-titration 115
6.3 Substitution titrations 117
6.4 Collective and selective titrations 119
6.5 Complexometry combined with precipitation reactions 123
6.6 Cyanide methods 123
6.7 Amalgam methods 124
6.8 Alkalimetric–complexometric titrations 127

7 **Selectivity** 129
7.1 Introduction 129
7.2 Separations 130
7.3 Masking 131
7.4 De-masking 136
7.5 Consecutive titrations 137
7.6 Indirect analysis 141

PART II. TITRATION PROCEDURES
AND WORKING INSTRUCTIONS

1 Volumetric solutions 146
2 Preparation of indicators 151
3 Buffer mixtures 152
4 Additional chemicals 154
5 Lithium 154
6 Sodium 154
7 Potassium (rubidium and caesium) 156
8 Beryllium 158
9 Magnesium and calcium 159
10 Barium and strontium 179
11 Boron 184
12 Aluminium 184

13 Scandium 193
14 Yttrium and the rare earths (lanthanons) 194
15 Plutonium 198
16 Titanium 199
17 Zirconium and hafnium 203
18 Thorium 209
19 Vanadium 213
20 Niobium and tantalum 216
21 Chromium 218
22 Molybdenum 223
23 Tungsten 226
24 Uranium 227
25 Manganese 230
26 Rhenium 234
27 Iron 235
28 Cobalt 242
29 Nickel 245
30 Platinum Metals 250
31 Copper 252
32 Silver 258
33 Gold 260
34 Zinc 260
35 Cadmium 268
36 Mercury 271
37 Gallium 273
38 Indium 275
39 Thallium 277
40 Carbon (carbonates, cyanides, organic compounds) 281
41 Silicon (silicic acid) 291
42 Germanium 292
43 Tin 292
44 Lead 295
45 Nitrogen derivatives 300
46 Phosphorus 301
47 Arsenic 306
48 Antimony 308
49 Bismuth 309
50 Oxygen 313
51 Sulphur 313
52 Selenium 319
53 Fluorine 319

54 Halogens (excluding fluorine) 322

Bibliography 325
Author Index 453
Subject Index 480

LIST OF SYMBOLS USED IN TEXT

(with references to the page or equation where they are first introduced)

ABBREVIATIONS

$M^{\nu+}$ A metal cation carrying a charge of $+\nu$ (p. 14).

M^* Another metal cation (when it is necessary to distinguish between two in the same solution) (p. 25).

$A^{\lambda-}$ A monodentate complexing agent, of charge $-\lambda$, e.g. NH_3, OH^-, acetate ion, and also other components of buffers and auxiliary complexing agents, e.g. tartrate, citrate, etc.

$Z^{\lambda-}$ The general symbol for any chelating agent, e.g. a molecule of a polyamine or the anions of EDTA, NTA, DCTA, EGTA, DTPA.

The following special symbols are also used:

tren   Tri-(2-aminoethyl)amine (p. 4).

penten   *NNN'N'*-tetra-(2-aminoethyl)ethylenediamine (p. 4).

EDTA   Ethylenediaminetetra-acetic acid or its anions (p. 8).

Y^{4-} The anion of EDTA (p. 9).

NTA   Nitrilotriacetic acid or its anions (p. 9).

X^{3-} The anion of NTA

DCTA   *Trans*-1,2-diaminocyclohexanetetra-acetic acid or its anions (p. 9).

EGTA   Bis-(2-aminoethyl)ethyleneglycol-*NNN'N'*-tetraacetic acid or its anions (p. 9).

DTPA   Diethylenetriaminepenta-acetic acid or its anions (p. 9).

$D^{\lambda-}$ The anion of a dyestuff used as a metal indicator (p. 44).

$In^{\lambda-}$ The anion of a metal indicator which is not itself a dyestuff (p. 41).

Isopr   Isopropyl (p. 39).

gl   Carboxymethylaminomethyl (p. 39).   } Groups in

mgl   *N*-Methyl-*N*-carboxymethylaminomethyl   dyestuff
       (p. 42).   mole-

mim   *N,N*-Bis-(carboxymethyl)aminomethyl (p. 42).   cules.

STOICHEIOMETRY

n The number of ligands in the complex $MA_n^{\nu-n\lambda}$. (p. 19) eqn. (20).

N The maximum possible value of n for a particular metal (p. 5).

j The number of ionizable hydrogen atoms in the species $H_jZ^{j-\lambda}$ (p. 12) eqn. (9).

J The maximum possible value of j in strongly acid solution (p. 15) eqn. (18).

CONCENTRATIONS

μ The ionic strength of the solution.

[ ]   This symbol denotes the concentration, in moles per litre, of the species enclosed in the square brackets. Thus [M] denotes the concentration of free, completely solvated metal ions. The charge on the particle is omitted on grounds of generality.

pM $= -\log_{10}$ [M] (pp. 6, 21).

$[\ ]_t$ The total concentration of a species as determined by analysis. Thus $[M]_t$ represents the total concentration of metal in solution and includes the concentration of free ions and of all complexes containing M (p. 22) eqn. (31).

Similarly $[Y]_t$ represents the total concentration of EDTA and includes that of the free ions Y^{4-}, the hydrogen complexes HY^{3-}, H_2Y^{2-}, H_3Y^-, and H_4Y, and all the possible metal complexes which may be present.

pM_t $= -\log_{10} [M]_t$ (p. 113).

$[\ ]'$ This symbol distinguishes the total concentration of a particular element or group. For example:

$[M]'$ $= \sum\limits_{n=0}^{N} [MA_n]$, and denotes the total concentration of metal present as the simple ion, M, or combined with the monodentate partner, A. It does *not* include the concentration of any metal combined with a polydentate chelating agent such as EDTA or NTA (p. 19) eqn. (22).

$[Z]'$ $= \sum\limits_{j=0}^{J} [H_jZ]$, and denotes the total concentration of a chelating agent such as EDTA, which is *not* bound to a metal (p. 15) eqn. (17).

$[D]'$ $= \sum\limits_{j=0}^{J} [H_jD]$ and denotes the total concentration of a dyestuff which is *not* bound to a metal (p. 78) eqn. (43).

$[MD]'$ $= \sum\limits_{j=0}^{J} [MH_jD]$ and denotes the total concentration of the various protonated forms of a 1 : 1 complex of dyestuff and metal (p. 80) eqn. (48).

DIMENSIONLESS COEFFICIENTS

a This denotes the number of moles of titrant per mole of the constituent being titrated which have been added at any stage in a titration, e.g. the number of moles of EDTA per mole of metal, or the number of moles of NaOH per mole of acid. At the equivalent point (end-point), a is therefore a whole number; usually $a = 1$.

$\alpha_{Z(H)}$ The value of the quotient $[Z]'/[Z]$ at any given hydrogen-ion concentration. Since

$$\alpha_{Z(H)} = (\Sigma[H_jZ])/[Z] = (\Sigma\beta_{H_jz}[H]^j[Z])/[Z] = \Sigma\beta_{H_jz}[H]^j$$

its numerical value is a function only of the pH and of the stability constants of the proton complexes, H_jZ (cf. Fig. 2) and is not directly dependent upon the actual concentration of the complexing agent (p. 15) eqn. (18).

$\alpha_{D(H)}$ The value of the distribution (or alpha) coefficient $[D]'/[D]$ at a given hydrogen-ion concentration (p. 79) eqn. (46).

$\alpha_{MD(H)}$ The value of the distribution (or alpha) coefficient $[MD]'/[MD]$ at a particular hydrogen-ion concentration (p. 80) eqn. (50).

α_M $= 1 + ([MZ]/[Z])$.

$\alpha_{M(A)}$ The value of the quotient $[M]'/[M]$ produced by an auxiliary complexing agent when the concentration of its free ligand is $[A]$. Since $\alpha_{M(A)} = \Sigma\beta_{MA_n}[A]^n$, it is a function only of $[A]$ and the stability constants of the relevant metal complexes (p. 19) eqn. (23).

$\alpha_{M(B)}$ The value of the distribution (or alpha) coefficient $[M]'/[M]$ due to some other auxiliary complexing agent B (p. 20) eqn. (27).

$\alpha_{M^*(A)}$, $\alpha_{M^*(B)}$ Corresponding alpha coefficients for the metal M*.

EQUILIBRIUM CONSTANTS

K, β Concentration (stoicheiometric) equilibrium constants governing the equations of mass-action. They are valid only for a particular, specified ionic strength, μ, and for a particular temperature. All the data in this book refer to $\mu = 0.1$ and to 20°C (p. 5) eqn. (5, 6).

K Denotes the equilibrium constants for individual stages in a stepwise reaction. More specifically:

K_{H_jZ} $= [H_jZ]/[H][H_{j-1}Z]$ is the individual or stepwise formation (or stability) constant for the species H_jZ eqn. (9).

pK_j = $\log K_{H_jZ}$. This positive logarithm of the association constant is identical with the negative logarithm of the dissociation constant, and therefore identical with the conventional pK value of the acid or proton donor H_jZ. The pK values of dyestuffs are distinguished as pK_j^D or pK_{H_jD}.

K_{MA_n} = $[MA_n]/[A][MA_{n-1}]$ is the individual formation (or stability) constant for the species MA_n. It is the equilibrium constant for the process:

$$A^{-\lambda} + MA_n^{\nu-(n-1)\lambda} \longrightarrow MA_n^{\nu-n\lambda}$$

and will often be abbreviated to K_n (p. 5) eqn. (5).

K_{MZ} = $[MZ]/[M][Z]$ is the formation constant of the chelate complex, MZ (p. 11).

K_{MD} = $[MD]/[M][D]$ is the formation constant of the metal-dyestuff complex (p. 84).

β Denotes the gross, or overall constant for a series of step reactions. For example:

β_{H_jZ} = $[H_jZ]/[Z][H]^j = K_{HZ} . K_{H_2Z} . K_{H_3Z} \ldots K_{H_jZ}$

$= \prod_{j=1}^{j} K_{H_jZ}$; $\log \beta_{H_jZ} = \sum_{j=1}^{J} pK_j$.

β_{MA_n} = $[MA_n]/[M][A]^n = K_{MA} . K_{MA_2} . K_{MA_3} \ldots K_{MA_n}$

$= \prod_{n=1} K_{MA_n}$.

$K_{MZ'}^{\text{eff.}}$ = $[MZ]/[M][Z]'$ (p. 15) eqn. (17).

$K_{M'Z}^{\text{eff.}}$ = $[MZ]/[M]'[Z]$.

$K_{M'Z'}^{\text{eff.}}$ = $[MZ]/[M]'[Z]'$ (p. 20) eqn. (25).

These give effective ('conditional') formation constants for the chelated complex MZ valid for a specific hydrogen-ion concentration [H], or for a specific ligand concentration [A], or for the case where both [H] and [A] are specified. Effective stability constants for dyestuffs, viz. $K_{MD'}^{\text{eff.}}$, $K_{M'D}^{\text{eff.}}$, or $K_{M'D'}^{\text{eff.}}$ are also used and will frequently be represented simply as $K_{MD}^{\text{eff.}}$.

Part One

Titrating Agents, Indicators, and Instrumental
Methods of Determining End-points

CHAPTER ONE

Titration with Complexing Agents

1.1. Introduction

A reaction can only serve as a basis for a titration procedure if it fulfils the following conditions:

(i) the reaction must be rapid;
(ii) it must proceed stoicheiometrically;
(iii) the change in free energy must be sufficiently large.

It is then generally possible to find some means of detecting the end-point, e.g. by the change in colour of an indicator, by the response of an indicator electrode, or through measurement of conductivity or the diffusion current at a dropping mercury electrode.

The three conditions mentioned above are not always fulfilled in the formation of metal complexes. On the one hand, we may have to deal with reactions that proceed slowly, as for example, in the formation of inert ('robust') complexes such as those formed by Co(III), Cr(III), and the platinum metals in particular. On the other hand, the innumerable labile complexes that are formed practically instantaneously from their components are generally weakly bonded so that the third condition is not fulfilled. Hg^{2+}, Ag^+, and Ni^{2+} form exceptions to this rule for their complexes, especially those with cyanide ions, have long since formed the basis for a limited number of titration procedures [47-3; 63-64].

Generally speaking, the second condition, too, is not fulfilled in the formation of ordinary complexes. The replacement by other ligands such as NH_3 or CN^- of water molecules from the hydration sheath of the metal ions present in aqueous solution usually takes place in steps [A-3]. In consequence, the reaction presents points of similarity to the neutralization of a polybasic acid which also proceeds in stages. Thus on successive additions of the complexing agent there results a whole series of different complexes of which, as a rule since the stages overlap, several are present simultaneously in the solution in appreciable concentration. Complex formation is therefore incomplete until after a certain excess of the complexing agent has been added. Accordingly, the end-point is characterized neither by a rapid decrease in the logarithm of the concentration of the metal ion nor by a rapid increase in the concentration of the free ligand (complexing agent): it is drawn out just like the end of the neutralization of a very weak polybasic acid such as a polyphenol [49-6].

1.2. Polyamines

One can achieve more favourable conditions if the individual ligands required by the metal cation are all linked together. For example, in place of several discrete molecules of ammonia we can present to the cation an organic polyamine whose basic nitrogen atoms are located in a carbon skeleton in such a way that 5- or 6-membered chelate rings result on complex formation [52-17; 55-118].

The tetramine and hexamine with the following constitutions exemplify this: we shall refer to them by the abbreviations 'tren' and 'penten' [50-3; 53-2]

$$N \begin{cases} CH_2.CH_2.NH_2 \\ CH_2.CH_2.NH_2 \\ CH_2.CH_2.NH_2 \end{cases}$$

'tren'; pK's 10·29, 9·69, 8·56

$$\begin{matrix} H_2N.CH_2.CH_2 \\ H_2N.CH_2.CH_2 \end{matrix} N.CH_2.CH_2.N \begin{matrix} CH_2.CH_2.NH_2 \\ CH_2.CH_2.NH_2 \end{matrix}$$

'penten'; pK's 10·20, 9·70, 9·14, 8·56

By using a complexing agent of this kind [55-118] the stoicheiometry of complex formation is simplified, and in consequence intermediate stages are eliminated, for the metal cation no longer requires four or six discrete molecules of ammonia,

$$Ni^{2+} \longrightarrow Ni(NH_3)^{2+} \longrightarrow Ni(NH_3)_2^{2+} \ldots \longrightarrow Ni(NH_3)_6^{2+} \quad (1)$$

but only a single molecule of polyamine

$$Ni^{2+} + penten \longrightarrow Ni(penten)^{2+} \quad (2)$$

At the same time we obtain all the advantages of the chelate effect [52-50; 54-55] as it is called. By this is meant the increase in the stability of the chelate complex in comparison with that of the corresponding complex with unidentate ligands. Specifically this equals the logarithm of the ratio of the equilibrium constants of the two reactions 1 and 2:

$$\beta_6 = \frac{[Ni(NH_3)_6^{2+}]}{[Ni^{2+}][NH_3]^6} = 3 \times 10^8; \quad K_{MZ} = \frac{[Ni(penten)^{2+}]}{[Ni^{2+}][penten]} = 2 \times 10^{19} \quad (3)$$

so that for this particular case the chelate effect amounts to

$$\text{Chel.} = \log K_{MZ} - \log \beta_6 = 10 \cdot 8 \qquad (4)$$

As shown by the figures for this example, the chelate effect can attain an astonishingly high value. It is not confined to nickel, for the effect is quite general, as shown in Table 1. This universality in the occurrence of the chelate effect derives from the fact that it is an entropy effect and has nothing to do with the strength of the co-ordinate bond [52-50; 54-9; 54-63; 63-80] formed by the relevant cation. Columns 2–7 in Table 1 give the formation constants of individual amine complexes, $M(NH_3)_j^{y+}$ of the chosen cation M as defined by:

$$K_j = \frac{[M(NH_3)_j]}{[M(NH_3)_{j-1}][NH_3]} \qquad (5)$$

where j can take any value from 1 to the maximum coordination number N. The individual constants give a picture of the stepwise formation of the complexes with ammonia. The product of the individual constants is identical with the so-called gross or overall stability constants, β_j, of the relevant complex:

$$\beta_j = \frac{[M(NH_3)_j]}{[M][NH_3]^j} = K_1 . K_2 . K_3 \ldots K_j = \prod_{j=1}^{j} K_j \qquad (6)$$

which is also shown in columns 8 and 9 of Table 1 for the cases $j = 4$ and $j = 6$. These constants β_4 and β_5 must then be compared with the formation constants $K_{M(tren)}$ and $K_{M(penten)}$ of complexes with the organic tetramine and hexamine when we wish to calculate the chelate effect.

TABLE 1

Formation constants for ammine complexes, valid for 20°C and ionic strength $\mu = 0 \cdot 1M$ [57-138; 59-140]

1 Cation	2 $\log K_1$	3 $\log K_2$	4 $\log K_3$	5 $\log K_4$	6 $\log K_5$	7 $\log K_6$	8 $\log \beta_4$	9 $\log \beta_6$	10 $\log K_M$ (tren)	11 $\log K_M$ (penten)
Co^{2+}	2·1	1·6	1·1	0·8	0·2	0·6	5·6	5·2	12·8	15·8
Ni^{2+}	2·8	2·2	1·7	1·2	0·7	0·0	7·8	8·5	14·8	19·3
Cu^{2+}	4·1	3·5	2·9	2·1	—	—	12·6	—	18·8	22·4
Zn^{2+}	2·3	2·3	2·4	2·1	—	—	9·1	—	14·7	16·2
Cd^{2+}	2·6	2·1	1·4	0·9	−0·3	−1·7	6·9	4·9	12·3	16·8
Hg^{2+}	8·8	8·7	1·0	0·8	—	—	19·3	—	—	—
Ag^+	3·2	3·8	—	—	—	—	—	—	7·8	—

Figure 1 demonstrates the effect of chelation in a titration. Nickel is again chosen as the example of its titration, on the one hand, with ammonia and, on the other, with the hexamine 'penten'. The number of atoms of amino-nitrogen per mole of Ni^{2+} is shown along the abscissa, whereas the ordinate gives the values of $pNi = -\log [Ni^{2+}]$, which corresponds to the pH in an alkimetric titration.

Naturally, on adding the basic amine to an unbuffered solution of the metallic salt not only the concentration of nickel but also the pH

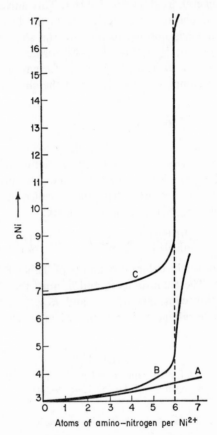

Figure 1. The titration of Ni^{2+} with amines.

CURVE A. Titration of Ni^{2+} with ammonia in the presence of $0.1M$ NH_4Cl.
CURVE B. Titration of Ni^{2+} with 'penten' in the presence of $0.1M$ NH_4Cl.
CURVE C. Titration of Ni^{2+} with 'penten' in the presence of $0.1M$ NH_4Cl and $0.1M$ NH_4OH. In each case $[Ni]_t = 10^{-3}$. The curves have been calculated from the constants given in Table 1, using equations analogous to those developed in Section 2.2.

changes, and this ultimately leads to the precipitation of nickel hydroxide. Such precipitation can be avoided if the titration is conducted in the presence of a substantial excess of ammonium chloride. These considerations lead to the conditions specified in the legends appended to Fig. 1.

The cation to be titrated is always present in a total concentration $[Ni]_t = 10^{-3}$M. Curve A illustrates the titration of this amount of nickel with ammonia in the presence of 0·1M NH_4Cl. At the equivalent point ($a = 6$ mole NH_3 per mole Ni) the concentration of free nickel ion has only fallen to one-fifth of its original value, and this point is not therefore characterized by an abrupt change in pNi. Complex formation is not really complete until a substantial excess of ammonia has been added.

The situation is quite different in curve B, where the titration, also in the presence of 0·1M NH_4Cl, is carried out with the hexamine. We now get quite a respectable jump in pNi, but this does not occur exactly at the point of equivalence but rather too late, since as soon as all the nickel has been complexed the pH value begins to rise steeply, whereupon the stability of the complex with 'penten' is also increased.

During the titration with a complexing agent it is far better to maintain the pH constant and to choose a high value for it. This is shown in curve C, which represents a titration with 'penten' in the presence of both ammonium chloride (0·1M) and ammonia (0·1M), i.e. in a buffer of pH about 9·3. The addition of ammonia, of course, changes the nickel into its ammine complex, with the result that pNi is increased so that the titration curve no longer begins at pNi = 3 but at about 6·9. During the titration the NH_3-complex is gradually transformed into the 'penten' complex, and finally there is a jump in pNi which now occurs exactly at the stoicheiometric end-point: this jump in pNi amounts to fully six units, i.e. it is even somewhat larger than the jump in pH obtained when HCl is titrated with NaOH.*

The specific example of Fig. 1 demonstrates that titration can actually be carried out with organic polyamines. However, the procedure would always remain restricted to those metals that show a marked preference for basic nitrogen as donor atoms, i.e. pre-eminently to Cu, Ni, Co, Zn, Cd, and Hg. The polyamines have yet a further disadvantage, since their complexes are very sensitive to acid because the complexing agent is able to take up a number of protons. For example, even at pH 7 the hexamine 'penten' exists as the tetra-

* The curves in Fig. 1 have been calculated with the stability constants given in Table 1 by using equations analogous to those given in Section 2.2.

ammonium ion $H_4penten^{4+}$, so that four protons must be displaced by the metal cation on complex formation:

$$Ni^{2+} + H_4penten^{4+} \longrightarrow Ni(penten)^{2+} + 4H^+ \qquad (7)$$

At pH = 7 therefore we have to deal with the process (7) rather than with reaction (2), and instead of the constant K_{MZ} of eqn. (3) we obtain the very much smaller 'conditional constant' of the complex which takes on the following value at pH 7:

$$K_{Ni(penten')} = \frac{[Ni(penten)^{2+}]}{[Ni^{2+}][penten']} = 4 \cdot 9 \times 10^9$$

Here the term [penten] denotes the total concentration of complexing agent that is not actually bound to nickel; thus

$$[penten'] = \sum_{j=0}^{4} [H_j penten]$$

Despite these drawbacks, the polyamines remain interesting substances with which a number of selective complexometric titrations can be carried out. Up till now unfortunately they are not available completely pure, so that their uses as titrants have still only been investigated to a limited extent [57-9; 57-87; 57-88; 59-113].

1.3. Complexones*

The complexing agents used almost exclusively nowadays in complexometry are aminopolycarboxylic acids (also known as complexones), of which ethylenediaminetetra-acetic acid (also termed diaminoethanetetra-acetic acid and ethylenedinitriletetra-acetic acid: EDTA) is by far the most important. Their anions have 4, 6 and even 8 atoms which can be used as donor atoms for the formation of 5-membered chelate rings with a metal atom. This special property derives from the fact that most of these ligand atoms are carboxylate oxygens. Indeed, the anions of NTA and EDTA can be derived formally from the bases 'tren' and 'penten' by replacing the terminal $-CH_2 \cdot NH_2$ groups by carboxylate groups.

The ability of $-COO^-$ to function as a ligand for a metal ion can be assessed from the behaviour of the acetate ion [57-148]. To be sure,

---

* The term *complexone* (plural *complexones*) is used throughout this book in accordance with established scientific usage to denote a complexing agent of the type of a polyaminocarboxylic acid, or a related salt. The word 'Complexone' is also registered as a trade-mark and reserved for certain commercial products marketed by the Uetikon Chemical Company, Switzerland. In America various complexones are marketed under the trade name of 'Versenes', and similar trade products appear in Europe under such names as 'Nervanaid', 'Sequestric acid', 'Tritriplex', 'Trilon', etc.

acetato-complexes are not particularly stable, but they are formed by practically all cations with multiple charges. When this general tendency towards complex formation is intensified by the chelate effect, as is the case with the anions of the aminopolycarboxylic acids, strong complexing agents result of low selectivity which can react with almost every cation. While ammonia and the polyamines form complexes only with metals of the transition series and B-group (d^{10} cations), the aminopolycarboxylic acids are able to form complexes with cations of the A-group (d° cations), e.g. with aluminium, the rare earths, the alkaline-earth metals, and even to a slight extent with Li^+ and Na^+.

We shall denote the anion of EDTA as Y^{4-} and use the general symbol $Z^{\lambda-}$ for any other chelating agent, e.g. for a polyamine ($\lambda = 0$) or the anions of NTA ($\lambda = 3$), DCTA ($\lambda = 4$), DTPA ($\lambda = 5$), and EGTA ($\lambda = 4$).

EDTA is used in well over 95% of all complexometric titrations. NTA has unique advantages for a few alkalimetric complexometric procedures [46-2; 48-1; 48-2] which are hardly ever used nowadays, and can be used as an auxiliary complexing agent. The somewhat more expensive reagent *trans*-diaminocyclohexanetetra-acetic acid (DCTA) generally forms somewhat stabler complexes than EDTA does (cf. Table 2) and they are also more 'robust', i.e. they are formed more slowly and they are dissociated more slowly, with consequential advantages in certain cases. Diethylenetriaminepenta-acetic acid (DTPA) binds metal ions which exhibit a coordination number of 8 better than EDTA does, and therefore has advantages in titrations of the larger cations, especially those of the lanthanides and actinides. However, only a few procedures have been tried out. Bis-(aminoethyl)-glycolether-*NNN'N'*-tetra-acetic acid (EGTA) is of particular interest in view of the big difference that exists between the

TABLE 2

Logarithms of the stability constants log K_{MZ} of EDTA and DCTA complexes: valid for 20°C and $\mu = 0.1M$

[Literature references 57-15; 57-130; 57-138; 59–126; 59-140; 60-107; 60-172; 60-180; 60-193; 61-39; 61-53; 61-165; 62-100; 62-156; 63-51; 63-77; 64-38; 64-57; 64-80; 65-35; 65-39; 63-123]

Cation	EDTA	DCTA	Cation	EDTA	DCTA
H^+ $j = 1$	10·26	12·35	Hf^{4+}	19·1 (?)	
2	6·16	6·12	Ce^{4+}	24·2	
3	2·67	3·52	V^{2+}	12·7	
4	2·0	2·4	V^{3+}	25·9	
5	1·6		VO^{2+}	18·8	20·1
6	0·9		VO_2^+	15·6	
Be^{2+}	~9		Mn^{2+}	13·8	17·4
Mg^{2+}	8·7	11·0	Fe^{2+}	14·3	—
Ca^{2+}	10·7	13·2	Fe^{3+}	25·1	29·3
Sr^{2+}	8·6	10·5	Co^{2+}	16·3	19·6
Ba^{2+}	7·8	8·6	Ni^{2+}	18·6	19·4
Ra^{2+}	7·1		Pd^{2+}	18·5	—
			Cu^{2+}	18·8	22·0
Al^{3+}	16·1	18·3			
Sc^{3+}	23·1	—	Ag^+	7·3	—
Ce^{3+}	15·9	16·9			
Y^{3+}	18·1	19·8	Zn^{2+}	16·5	19·3
La^{3+}	15·5	16·9	Cd^{2+}	16·5	19·9
Eu^{3+}	17·0	19·3	Hg^{2+}	21·8	25·0
Eu^{2+}	7·7	—			
Lu^{3+}	19·8	22·2	Ga^{3+}	20·3	23·6
			In^{3+}	24·9	
UO_2^{2+}	~10		Tl^+	5·3	
U^{4+}	25·5		Tl^{3+}	21·5	
Pu^{3+}	18·1				
Am^{3+}	18·2		Sn^{2+}	22·1 (?)	
			Pb^{2+}	18·0	20·3
Ti^3	21·3				
TiO^{2+}	17·3		Bi^{3+}	27·9	31·2
Zr^{4+}	29·5				

stability of its magnesium complex and those of the heavier alkaline-earth metals Ca, Sr, and Ba (cf. Table 3).

The stability constants of 1 : 1 complexes of the five selected complexing agents are collected in Tables 2 and 3. These constants are defined as follows:

$$K_{MZ} = [MZ]/[M][Z] \qquad (8)$$

TABLE 3

Logarithms of the stability constants log K_{MZ} of NTA, DTPA, and EGTA; valid for 20°C and $\mu = 0.1M$

[Literature refs. 56-125; 57-138; 57-139; 58-115; 59-142; 60-198; 62-101]

Cation	NTA	DTPA	EGTA	Cation	NTA	DTPA	EGTA
$H^+ j = 1$	9·73	10·58	9·46	Mn^{2+}	7·4	15·6	12·3
2	2·49	8·60	8·85	Fe^{2+}	8·8	16·0	11·8
3	1·9	4·27	2·68	Fe^{3+}	15·9	27·9	19·9
4		2·64	2·0	Co^{2+}	10·4	19·3	12·3
5		1·5		Ni^{2+}	11·5	20·2	13·6
				Cu^{2+}	13·0	21·5	17·8
Mg^{2+}	5·4		5·2				
Ca^{2+}	6·4	10·9	11·0	Ag^+	5·4		
Sr^{2+}	5·0	9·7	8·5				
Ba^{2+}	4·8	8·6	8·4	Zn^{2+}	10·7	18·6	12·9
				Cd^{2+}	9·8	19·3	16·7
Y^{3+}	11·4	22·2	16·8	Hg^{2+}		26·7	23·2
La^{3+}	10·5	19·6	15·6				
Lu^{3+}	12·5	22·6	17·8	Pb^{2+}	11·4	18·9	14·8

It should be noted that concentrations of the reactants (square brackets) and not activities are used in the mass-action equation. Strictly speaking, the resulting concentration (or stoicheiometric) equilibrium constants are only valid for the specified medium of ionic strength $\mu = 0.1M$. Such constants are not only easier to determine than thermodynamic equilibrium constants but they are more useful in practical applications, for the solutions that are to be titrated always contain some electrolyte or another. In fact, the need to add buffering material in a complexometric titration generally produces an ionic strength in the neighbourhood of $0.1M$. Thermodynamic constants could only be used if activity coefficients were known at the same time, which is naturally never the case.

When using 'conditional' concentration constants one must not forget that values of K_{MZ} in solutions of ionic strength appreciably less than $0.1M$ will be somewhat larger than the values quoted above in the tables, whereas in solutions with $\mu > 0.1$ the complexes are correspondingly weaker. A titration procedure can therefore fail if the concentration of inert electrolyte is too high, as for example, if a

B

salt brine has to be titrated. Of course, the values of stability constants also change if water-like organic solvents such as alcohol, acetone, or dioxan are added; their presence increases the stability of the complex. Furthermore, the values of the constants given in Tables 2 and 3 are temperature-dependent, and with few exceptions (e.g. Mg–EDTA) they become smaller as the temperature increases [63-80]. However, since the enthalpy change in complexing reactions is small, this effect is not large, and it only amounts to about 0.01–0.02 units in log K_{MZ} per degree centigrade.

The hydrogen ion has been included among the cations in the above Tables. The anions $Z^{\lambda-}$ of each of the five complexones can bind several protons corresponding to the ionic species $HZ^{(\lambda-1)-}$, $H_2Z^{(\lambda-2)-}$, ... or in general $H_2Z^{(\lambda-j)-}$. The figures quoted are the logarithms of the formation constants of those 'proton complexes' which are identical with the customary pK values (since the logarithm of a formation constant is identical with the negative logarithm of the corresponding dissociation constant):

$$\log K_{H_jz} = \log \frac{[H_jZ]}{[H][H_{j-1}Z]} = pK_j \qquad (9)$$

One must take note, however, that j signifies here the number of protons in the ionization of the species $H_jZ^{(j-\lambda)}$, whereas the value of the charge on the species $H_{j-1}Z^{j-1-\lambda}$ is customarily used as the index for a pK value. Thus with EDTA the numerical value of log $K_{H_4z} =$ pK_4 is identical with that of the negative logarithm of the first dissociation constant of the free acid, H_4Y; log $K_{H_3z} =$ pK_3 refers to the second dissociation constant (cf. H_3Y^-), pK_2 to the third, and pK_1 to the fourth dissociation constant (of HY^{3-}). There should be no confusion so long as one takes count of the sequence of numerical values, for the highest pK always corresponds to the species with the smallest number, j, of protons.

The data show that log K_{HZ} for NTA and both log K_{HZ} and log K_{H_2z} in the case of the other four complexing agents cannot refer to the dissociation of a carboxylic proton, since the value of pK for carboxylic acids never exceeds 5. This suggests that in the stepwise protonation of the anion $Z^{\lambda-}$ the H^+ is placed on the nitrogen atom first, and not until these have been made into ammonium groups are the carboxylate ions protonated. Indeed, in the case of EDTA, infrared spectroscopy has demonstrated that the species HY^{3-} and H_2Y^{2-} possess one and two N–H bands respectively and that protonated carboxylate groups first occur with the species H_3Y^- and H_4Y [63-25; 63-26]. This is confirmed by n.m.r. studies on solutions of aminopolycarboxylic acids [63-23]. The formally uncharged molecules of complexones are therefore certainly betains, as shown in the above structural formulae, with positive ammonium and negative carboxylate ions present simultaneously. This also explains the small pK value of the COOH groups that are present, since these are made more strongly acidic by the positive charge on the ammonium groups. The situation appears to be different in free aminopolycarboxylic acids in the solid state [55-56].

When one considers the betain character of the uncharged molecules $H_\lambda Z$ it is understandable that these, like all aminoacids, should also show weakly basic properties. Thus in strongly acid solutions free EDTA takes up a 5th and a 6th proton on the two ionized carboxylate groups, and its solubility, which is a minimum at pH 1·8, increases not only when the pH is increased but also when it is diminished [60-193], whence pK values for the species H_5Y^+ and H_6Y^{2+} can be found (cf. Table 2).

CHAPTER TWO

Basic Theory of Complexometric Titrations

2.1. Formation and stability of complexes

The reaction that takes place when a solution of a metal cation and one of a complexing agent are brought together cannot as a rule be described simply by eqn. (10) and the corresponding mass-action expression (8)

$$M^{\nu+} + Z^{\lambda-} \longrightarrow MZ^{\nu-\lambda} \tag{10}$$

since the free ion $Z^{\lambda-}$ does not begin to turn up until above pH 10. Thus in a solution neutralized with alkali to pH 7 NTA exists as the dialkali salt (cf. the pK values), so that the reaction of a cation with such a solution will proceed as follows:

$$M^{\nu+} + HX^{2-} \longrightarrow MX^{\nu-3} + H^+ \tag{11}$$

Hydrogen ions are therefore liberated during the formation of the metal complex and the pH decreases. If the solutions are unbuffered the lowering of pH may amount to several units. The first types of complexometric titration procedures were based on such pH effects: they are essentially alkalimetric procedures, since the liberated acid is simply titrated with NaOH [46-2; 48-1; 48-2].

Precisely the same conditions apply to EDTA, and the following reactions occur over the appropriate pH ranges [47-2]:

$$\text{pH 7–9} \quad M^{\nu+} + HY^{3-} \longrightarrow MY^{\nu-4} + H^+ \tag{12}$$

$$\text{pH 4–5} \quad M^{\nu+} + H_2Y^{2-} \longrightarrow MY^{\nu-4} + 2H^+ \tag{13}$$

Complex formation can thus be described quite generally by eqn. (14):

$$M^{\nu+} + H_jZ^{j-\lambda} \longrightarrow MZ^{\nu-\lambda} + jH^+ \tag{14}$$

for which the mass action expression is:

$$K = \frac{[MZ][H]^j}{[M][H_jZ]} = K_{MZ}/\beta_{H_jZ} \tag{15}$$

where $\qquad \log \beta_{H_jZ} = pK_1 + pK_2 + \ldots + pK_j \tag{16}$

The equilibrium constants of reactions 11, 12, and 13 can be obtained from the figures in Tables 2 and 3 with the aid of eqn. (15).

Another result of the basicity of the ion $Z^{\lambda-}$ is that the free energy of complex formation depends on the pH. It is measured in terms of the 'effective' (conditional) formation constant

$$(K_{MZ'}^{eff.})_H = [MZ]/[M][Z]'$$

where
$$[Z]_H' = \left(\sum_{j=0}^{J} [H_j Z]\right)_H = \alpha_{Z(H)} \cdot [Z] \tag{17}$$

In eqn. (17), $[Z]_H'$ denotes the total concentration of the complexing agent (in all its forms) that is not bound to metal, measured at some particular hydrogen-ion concentration [H]. This is calculated with the aid of the coefficient $\alpha_{Z(H)}$, which is obtained from pK values as follows:

$$\alpha_{Z(H)} = 1 + [H] \cdot K_{HZ} + [H]^2 \cdot K_{HZ} \cdot K_{H2Z} +$$
$$[H]^3 \cdot K_{HZ} \cdot K_{H2Z} \cdot K_{H3Z} + \ldots = 1 + \sum_{j=1}^{J} \beta_{HjZ}[H]^j \tag{18}$$

Combination of (8) and (17) lead finally to the desired conditional formation constant:

$$(K_{MZ'}^{eff.})_H = K_{MZ}/\alpha_{Z(H)} \tag{19}$$

Only in the exceptional case of strongly alkaline solutions (pH $\geqslant$ 11) will the figures given for K_{MZ} in Tables 2 and 3 serve as measures of the stability of the complex MZ', designated with the superscript dash. Since we always work at appreciably lower pH values where $\alpha_{Z(H)} > 1$, the appropriate stability constant $K_{MZ'}^{eff.}$ is given by (19). The influence of pH on the stability of a complex is extremely important. Its extent can be realized from the values of $\alpha_{Z(H)}$ given in Table 4. Log $\alpha_{Z(H)}$ is shown graphically as a function of pH in Fig. 2.

TABLE 4
Values of log $\alpha_{Z(H)}$ for various pH values calculated by using eqn. (18)

pH	1	2	3	4	5	6	7	8	9	10	11	12
EDTA	18·0	13·5	10·6	8·4	6·5	4·7	3·3	2·3	1·3	0·45	0·07	0·01
DCTA	20·4	16·5	13·2	10·6	8·5	6·7	5·7	4·4	3·4	2·4	1·4	0·51
NTA	11·2	8·6	6·9	5·7	4·7	3·7	2·7	1·7	0·8	0·19	0·02	0·00
DTPA	22·7	18·3	14·6	11·7	9·6	7·2	5·2	3·3	1·7	0·70	0·14	0·01
EGTA	18·0	15·3	12·5	10·3	8·3	6·3	4·3	2·6	0·77	0·12	0·01	0·00

A few examples will demonstrate the influence of acidity. Above pH 10 the alkaline-earth metals are practically completely complexed when an equivalent amount of EDTA is added. At pH 5, however, the effective stability of the complexes

$$(K_{MgY'}^{eff.} = 10^{2·2}; \quad K_{CaY'}^{eff.} = 10^{4·2})$$

has become so small that they can no longer be determined volumetrically, since the effective stability constant for this process must be 10^6 at the very least. But the presence of alkaline-earth metals does not interfere in the least with the titration of divalent heavy metals at pH 5, since their complexes still possess quite substantial stabilities, even in these weakly acidic solutions

$$(K_{ZnY'}^{eff.} = 10^{10 \cdot 0}, \ K_{CuY'}^{eff.} = 10^{12 \cdot 3} \text{ at pH 5})$$

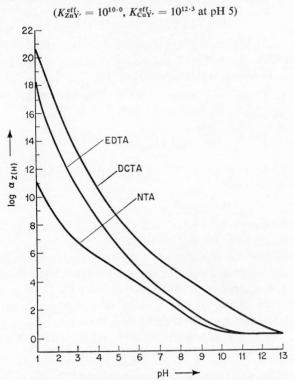

Figure 2. The decadic logarithm of the coefficient $\alpha_{Z(H)}$ for NTA, EDTA, and DCTA as a function of pH.

However, at pH 2 complexes of even the divalent heavy metals are formed to only a very limited extent

$$(K_{ZnY'}^{eff.} = 10^{3 \cdot 0}; \ K_{CuY'}^{eff.} = 10^{5 \cdot 3} \text{ at pH 2})$$

whereas the value for tervalent iron is $K_{FeY'}^{eff.} = 10^{11 \cdot 6}$ at pH 2; hence Fe(III) and several other ter- and tetra-valent cations can be determined very selectively in this strongly acid medium.

The figures given in Table 4 show that $\alpha_{Z(H)}$ is always greater for DCTA than for EDTA at any given pH value; this arises from the larger value of pK_1 for the homocyclic acid. In consequence, the effective stability of a complex formed with DCTA below pH 10 is scarcely any larger than that of EDTA. In this respect there is practically no advantage in using the more expensive cyclic reagent;

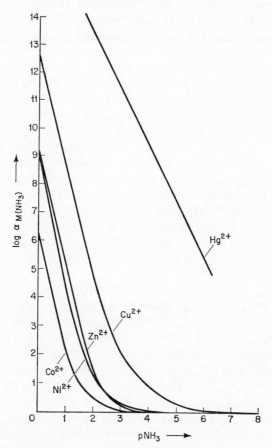

Figure 3. The decadic logarithm of the coefficient $\alpha_{M(NH_3)}$ for several cations shown as a function of the logarithm of the concentration of ammonia.

but if kinetic effects are exploited new analytical possibilities can be found which cannot be realized with EDTA.

Equation (19) does not exactly reproduce the influence of pH upon stability over the whole pH range. This is because complexes of aminopolycarboxylic acids are able to take up a proton in strongly acid solutions ($MZ^{\nu-\lambda} \longrightarrow MHZ^{\nu+1-\lambda}$) and to add on a hydroxyl ion in alkaline media ($MZ^{\nu-\lambda} \longrightarrow MZ(OH)^{\nu-\lambda-1}$); it seems that the extra proton becomes attached to a carboxylate ion that is no longer used as a ligand, and the hydroxyl ion either displaces a carboxylate ion from the coordination sphere or increases the coordination number to 7 or 8 [49-1; 61-177]. Hence the stability of the complex both at lower and at higher pH values is somewhat greater than that calculated by eqn. (19).

With complexes of divalent metals this effect does not become evident until below pH 3 or above pH 10, and even then does not amount to much. The same is true for complexes of the lanthanides and actinides. Admittedly the complexes of tervalent Al, V, Fe, Bi, Ga, In, Tl, and of Th(IV) are much more acidic [49-1], since the attachment of OH^- begins even in neutral solutions; however, these polyvalent cations whose complexes possess very large stability constants ($K_{MY} > 10^{20}$) are always titrated in a strongly acid medium so that any metal that is not chelated exists as free, unhydrolysed cation. In short, the pH range in which $MZ^{3-\lambda}$ is transformed into $MZ(OH)^{2-\lambda}$ is avoided in practice. This justifies our not seeking to complicate eqn. (19) by including coefficients analogous to α_H to take account quantitatively of the formation of hydrogen- and hydroxy-complexes [59-143; 63-64].

Equation (14) is also incomplete, in that the formation of a $1:2$ complex $MZ_2^{\nu-2\lambda}$ has been observed in exceptional cases along with that of the $1:1$ complex $MZ^{\nu-\lambda}$. This is particularly true for NTA, whose anion can only occupy four coordination positions. Large, highly charged cations can also form $1:2$ complexes with the sexadentate anion of EDTA; e.g. ThY_2^{4-}. The free energy of binding the second anion of the chelating agent (as measured by the equilibrium constant for the process $MZ + Z \rightleftharpoons MZ_2$) is, however, always very small compared with that for the first ($K_{MZ} \gg K_{MZ2}$; e.g. $K_{ThY} = 10^{23}$, whereas $K_{ThY2} = 10^{12}$ [52-41]), and thus there is no complication in any actual practical procedure. Triethylenetetramine-hexa-acetic acid can also form bimetallic (i.e. $2:1$) complexes [62-102; 63-27].

In complexometry the solution to be titrated often contains the metal to be determined not only as free cation $M^{\nu+}$ or chelated complex $MZ^{\nu-\lambda}$, but also in the form of other complexes, e.g. as ammine complexes if the solution has been made ammoniacal. Ammonium hydroxide is often added to produce a high pH; not only does it form a complex with the hydrogen ions but also with the cations present, whose hydroxide would otherwise ultimately precipitate. Other buffer components, such as acetate ions, can act as complexing agents for $M^{\nu+}$. Tartrate, citrate, and other substances are often added to prevent precipitation. We shall refer to all such substances as auxiliary complexing agents and denote them by the general symbol $A^{\lambda-}$. As shown already in eqn. (1) for the case $A = NH_3$, the reactions of a cation $M^{\nu+}$ with an auxiliary complexing agent take place in successive stages.

$$M^{\nu+} \underset{\longleftarrow}{\overset{A^{\lambda-}}{\longrightarrow}} MA^{\nu-\lambda} \underset{\longleftarrow}{\overset{A^{\lambda-}}{\longrightarrow}} MA_2^{\nu-2\lambda} \cdots \underset{\longleftarrow}{\overset{A^{\lambda-}}{\longrightarrow}} MA_n^{\nu-n\lambda} \cdots \underset{\longleftarrow}{\overset{A^{\lambda-}}{\longrightarrow}} MA_N^{\nu-N\lambda} \quad (20)$$

The equilibrium constants that characterize the successive steps will again be designated by K or β respectively with a subscript describing the composition of the complex. By analogy with (5) and (6) we define

$$K_{MAn} = \frac{[MA_n]}{[MA_{n-1}][A]}; \; \beta_{MAn} = \frac{[MA_n]}{[M][A]_n} = \prod_{7n=1}^{N} K_{MAn} \quad (21)$$

Just as, according to eqn. (17), the coefficient $\alpha_{Z(H)}$ measures the extent of the combination of protons with the chelating agent $Z^{\lambda-}$, so we can introduce a coefficient $\alpha_{M(A)}$ which describes the extent to which $M^{\nu+}$ is bound to the auxiliary complexing agent A. The following equation holds for the total concentration of metal that is not bound to the aminopolycarboxylic acid:

$$[M]' = [M] + [MA] + [MA_2] + \ldots = \sum_{n=0}^{N} [MA_n] = \alpha_{M(A)} \cdot [M] \quad (22)$$

and the value of $\alpha_{M(A)}$ can be calculated from the relevant stability constants by the equation:

$$\alpha_{M(A)} = 1 + [A]\beta_{MA} + [A]^2\beta_{MA2} + \ldots = 1 + \sum_{n=1}^{N} \beta_n[A]^n \quad (23)$$

Of course, the stabilities of the complexes MA_n formed with the auxiliary complexing agents will be lower than those formed by the complexones, $MZ^{\nu-\lambda}$, so that on introducing the chelating agents into a solution containing the metal and auxiliary complexing agent at some definite pH the reactions summarized by (24) take place between the various complexes MA_n and the various ionized species H_jZ derived from the aminopolycarboxylic acid:

$$MA_n^{\nu-n\lambda} + H_jZ^{j-\lambda} = MZ^{\nu-\lambda} + nA^{\lambda-} + jH^+ \quad (24)$$

This complicated state of affairs can, however, be described quite simply with the aid of the coefficients $\alpha_{Z(H)}$ and $\alpha_{M(A)}$ and by combining (8), (17), and (22) we obtain (25), the equilibrium constant for (24); we shall call this the effective stability constant $K_{M'Z'}^{eff.}$ of the complex MZ, since it provides a valid measure of the stability of the complex under the conditions in which it actually exists in the solution. Since total concentrations $[M]'$ and $[Z]'$ are involved in the equilibrium expression, this is made quite clear by the superscript dashes in $K^{eff.}$, thus:

$$K^{eff.}_{M'Z'} = \frac{[MZ]}{[M]'[Z]'} = K_{MZ}/\alpha_{Z(H)} \cdot \alpha_{M(A)} \tag{25}$$

Values of the coefficient $\alpha_{M(NH_3)}$ for NH_3 as an auxiliary complexing agent can easily be calculated from the stability constants for ammine complexes given in Table 1. Table 5 summarizes the results, which are also illustrated graphically in Fig. 3.

TABLE 5
Values of log $\alpha_{M(NH_3)}$ for various values of $pNH_3 = -\log [NH_3]$

$pNH_3 =$	0	1	2	3	4	5	6
Co^{2+}	6·1	2·2	0·5	0·05	0·01	0·00	0·00
Ni^{2+}	9·0	4·3	1·3	0·24	0·08	0·00	0·00
Cu^{2+}	12·6	8·6	4·9	2·0	0·94	0·26	0·03
Zn^{2+}	9·1	5·1	1·5	0·2	0·02	0·00	0·00
Cd^{2+}	7·2	3·5	1·1	0·1	0·01	0·00	0·00
Hg^{2+}	19·4	15·9	13·5	11·5	9·5	7·5	5·5
Aq^{+}	7·0	5·0	3·0	1·1	0·09	0·01	0·00

Since the stabilities of tartrate complexes are known [57-148], $\alpha_{M(Tart.)}$ can also be calculated by eqn. (23), and likewise $\alpha_{M(Citr.)}$ from the stability constants of citrate complexes so that the effective stabilities of complexonates $MZ^{\nu-\lambda}$ in the presence of tartrate and citrate can be obtained. Sometimes acetate complexes must also be taken into consideration, as when the titration is carried out in an acetate buffer. A few metals form complexes with chloride ions (Cd^{2+}, Pb^{2+}, Hg^{2+}) and the distribution coefficient $\alpha_{M(Cl)}$ can again be calculated from eqn. (23). A large number of such coefficients have been tabulated and displayed graphically in the new book by Anders Ringbom [63-64].

Two or more auxiliary complexing agents may be present simultaneously in the solution. $[M]'$, the total concentration of metal that is not bound to the complexone, is then given by the following equations:

$$[M]' = [M] + \sum_{n=1}^{N} [MA_n] + \sum_{j=n'}^{N'} [MB_{n'}] = \alpha_{M(A,B)} \cdot [M] \tag{26}$$

$$\alpha_{M(A,B)} = 1 + \sum_{n=1}^{N} \beta_{MAn}[A]^n + \sum_{n'=1}^{N'} \beta_{MBn'}[B]^{n'} = \alpha_{M(A)} + \alpha_{M(B)} - 1 \tag{27}$$

where the two auxiliary complexing agents are designated A and B respectively forming the complexes MA_n and $MB_{n'}$.

Complex formation with the hydroxyl ion can still occur, even when only a single auxiliary complexing agent has been added, so that strictly speaking according to (23) the term $(\alpha_{M(OH)} - 1)$ should be included in (27), where $\alpha_{M(OH)}$ can be calculated from (23) with $A = OH^-$ and the relevant stability constants of the hydroxy complexes, $\beta_{M(OH)}$. In almost every case that occurs in practice the concentration of hydroxy complexes in solution is sufficiently low for it to be possible to neglect $\alpha_{M(OH)}$ in comparison with $\alpha_{M(A)}$. Actually hydroxy complexes do not occur until the pH rises just before the precipitation of the metallic hydroxides begins, and the formation of a precipitate is avoided by the addition of an auxiliary complexing agent, which is only possible if $\alpha_{M(A)} \gg \alpha_{M(OH)}$.

One last application of a distribution coefficient is encountered in the titration of a metal M in the presence of a second metal which forms a much weaker complex with Z and which can be present in a relatively large concentration. For example, a heavy metal like Cu^{2+} can be titrated complexometrically in the presence of a great deal of calcium in a weakly acid medium (e.g. acetate buffer). The amount of the complexone Z that is not actually bound to the metal M now occurs either in protonated forms or as its calcium complex, and the distribution is described by $\alpha_{Z(H)}$ according to eqn. (18) and α_{Ca} according to (28):

$$\alpha_{Ca} = 1 + [Ca] \cdot K_{CaZ} \qquad (28)$$

The effective stability constant of the complex MZ in the presence of calcium and of the auxiliary complexing agent A (e.g. acetate ion) in a buffered solution at some specific pH value is given by eqn. (29):

$$K_{M'Z'}^{eff.} = K_{MZ}/(\alpha_{Z(H)} + \alpha_{Ca} - 1) \cdot \alpha_{M(A)} \qquad (29)$$

2.2. Titration curves in complexometry

The characteristic features of a titration procedure can best be appreciated by studying the appropriate titration curve. During a complexometric titration the value of pM changes and the sharpness of the jump in pM at the equivalent point provides a basis for estimating the various possibilities of detecting the end-point, whether by the use of a colour indicator or instrumentally; it also gives a clear indication of the accuracy that can be achieved.

2.2.1.

In the simplest possible case of the direct titration of a metal ion, M, we introduce from a burette into its buffered solution (which generally contains an auxiliary complexing agent, A, as well) a solution of the chelating agent, Z, in the form of an alkali-metal salt of the chosen aminopolycarboxylic acid. We now have to determine the concentration [M] and thence pM as a function of a, the number of moles of chelating agent added per mole of total metal $[M]_t$. We shall assume that $[M]_t$ and $[A]_t$ and that the pH and the ionic strength remain constant, i.e. that the small dilution produced by adding the standard solution of titrant can be neglected. The following equations are now valid:

Degree of titration $a = [Z]_t/[M]_t$ $\qquad$ (30)

Total metal concentration $[M]_t = [M]' + [MZ]$ $\qquad$ (31)

Total concentration of chelating agent $[Z]_t = a \cdot [M]_t = [Z]' + [MZ]$ (32)

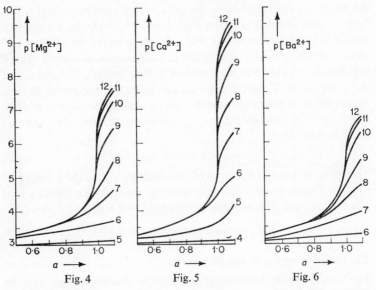

Fig. 4 $\qquad$ Fig. 5 $\qquad$ Fig. 6

Figures 4, 5, 6. Titration curves for magnesium (*Figure* 4), calcium (*Figure* 5), and barium (*Figure* 6) with ethylenediaminetetra-acetic acid at various pH values.

As before, [M]′ denotes the total concentration of metal that is not bound to the chelating agent and [Z]′ the total concentration of complexone that is not bound to the metal (eqns. (22) and (17)). From (31) and (32) with the aid of the effective complex formation constants (25) we can calculate values for the three concentrations [M]′, [Z]′, and [MZ] for every value of a, and with $\alpha_{M(A)}$ calculated by (22) we can also find [M].

The results of such calculations for titrations with EDTA are shown in Figs. 4–10. The auxiliary complexing agent was ammonia in every case, for with the ammonium chloride also present this can provide buffering between pH 8 and 11. The total concentration of NH_4^+ and NH_3 was kept constant at 0·1M, thus:

$$[A]_t = [NH_4^+] + [NH_3] = 0·1$$

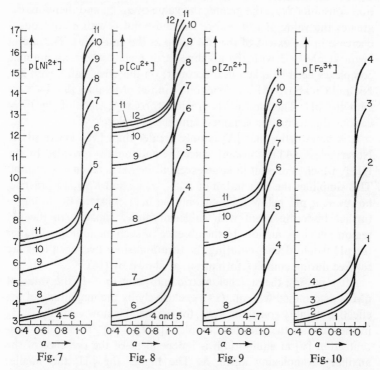

Figures 7, 8, 9, 10. Titration curves for nickel (*Figure* 7), copper (*Figure* 8), zinc (*Figure* 9), and iron (*Figure* 10) with ethylenediaminetetra-acetic acid at various pH values.

The concentration, $[NH_3]$, of free auxiliary complexing agent which has to be introduced into eqn. (23) to calculate $\alpha_{M(A)}$ varies therefore with the pH and can be obtained from the known basicity constant of NH_3, viz. $pK_{HA} = 9·3$:

pH =	4	5	6	7	8	9	10	11	12
$-\log [A] =$	6·3	5·3	4·3	3·3	2·32	1·48	1·08	1·01	1·00

The total concentration of the titrated metal was chosen to be $[M]_t = 10^{-3}M$, so that the titration curve begins at $a = 0$ with pM = 3, provided there has been no appreciable complex formation with the ammonia present. If, however, the metal does form a complex with ammonia, i.e. if $\alpha_{M(A)}$ from (23) is greater than 1, then $a = 0$ with $[M]' = 10^{-3}$ and $[M] = 10^{-3}/\alpha_{M(A)}$, so that pM > 3. The greater the concentration, $[A]$, of free ammonia and the formation constants β_{MA_n}, the greater the value of $\alpha_{M(A)}$, and therefore the greater the value of pM before the end-point – with a consequent decrease in the extent of the pM jump at the end-point. The curves begin at pM = 3 with the alkaline-earth metals where no ammine complexes need be taken into account; on the other hand, with Zn, Ni, and Cu above pH 7 we encounter initial pM values that increase with the pH. The increase is greatest for Cu^{2+}, since of the three cations this forms the strongest ammine complexes.

It is noteworthy that $[A]$ is only greater than $[M]_t$ above pH 8. Nevertheless, $[A]$ is constant, since the ammonia is provided by the NH_4^+, which is present in large excess if the pH is taken as constant. This simplifies the calculation of $\alpha_{M(A)}$, $\alpha_{Z(H)}$, and $K_{M'Z'}^{eff}$. In practice, however, a pH below 8 could only be held constant by adding a further buffering agent with sufficient buffer capacity in the pH region under discussion. In the absence of such an additional buffer the pH would decrease during the titration, since hydrogen ions are set free during complex formation, as shown by (14).

After passing the stoicheiometric end-point ($a = 1$) the value of pM is determined (cf. eqn. (19)) exclusively by the magnitude of the effective stability constants $K_{M'Z'}^{eff}$, for the metal is now virtually 100% in the form of the chelated complex ($[MZ] = 10^{-3}M$) and the concentration $[M]$ at equilibrium is independent of the presence of the auxiliary complexing agent A. The higher the pH, the smaller $\alpha_{Z(H)}$ becomes (cf. eqn. (18)) and the greater the value of K^{eff}. The pM curves therefore extend higher the greater the value of pH, although the effect of the increase above pH 10 admittedly does not make much difference, since at this pH $\alpha_{Z(H)}$ for EDTA is already as low as 2·8, and at higher alkalinities it rapidly approaches the constant limiting value of 1 (Table 4).

Graphs 7, 8, and 9 show very clearly that compromises have to be made in complexometric titrations. High values of pH are advantageous in order to produce the most stable complex during the titration and to secure jumps to the highest possible values of pM. Yet only with the alkaline-earth metals do we then achieve the greatest jump in pM (Figs. 4, 5, and 6). With the heavy metals the starting-point of the titration curve is displaced up the ordinate by reason of the added buffer component NH_3, which also combines with the metal that is being titrated: the jump in pM at the equivalence point (at $a = 1$) is consequently reduced. This effect also occurs when the pH is adjusted with basic substances other than ammonia; for any substance that can combine with a proton may also attach itself as a ligand to a metal ion, i.e. it can act as an auxiliary complexing agent and increase $\alpha_{M(A)}$. No appreciable improvement would be possible even with a hypothetical proton acceptor that showed no affinity for the metal to be titrated: for hydroxide ions would simply become attached to the metal as the pH is increased and, finally, when the solubility product is exceeded, the metallic hydroxide would be precipitated.

Theoretically speaking, the best buffer is one which will complex with the metal to the smallest extent over the amount needed to prevent the precipitation of its hydroxide. This ideal is approached by ammonia and various organic amines. Quite narrow limits are set to the possibilities of complexometry.

The graphs for Fe(III) (Fig. 10) were only calculated for low pH values, since in practice the titrations are always carried out in acid solution. Normally no special auxiliary complexing agent is added. The upwards displacement before the equivalent point of the titration curves at pH 3 and 4 as compared with those for pH 1 and 2 is a consequence of the formation of hydroxy complexes, and $\alpha_{M(OH)}$ was calculated (eqn. 23) from the relevant formation constants [51-1].

Figures 4–10 are only examples, and titration curves for many other metals can be found by the same kind of calculations. But where tartrate, citrate, or other less thoroughly studied substances are employed as auxiliary complexing agents, many of the formation constants of the relevant complexes needed for the calculation of $\alpha_{M(A)}$ are unknown. Some of these distribution coefficients are to be found in Ringbom's book [63-64].

2.2.2.

To assess the selectivity of complexometric titrations and to evaluate back-titrations, substitution titrations, and consecutive titrations, we must now consider those titration curves which result when two metals are present together. We shall designate these as M and M*. The calculation is only slightly more complicated than that for a single metal. We have simply to consider the concentrations [MZ] and [M*Z] of the two chelated complexes and the two concentrations [M]′ and [M*]′ of metals not bound to the complexing agent which are related to pM and pM* through their respective distribution coefficients $\alpha_{M(A)}$ and $\alpha_{M^*(A)}$; these species are in equilibrium with concentrations [Z]′ and [Z]. We thus have to evaluate five main concentration terms, viz. [M]′, [M*]′, [MZ], [M*Z], and [Z]′, from the five available eqns. (34)–(38). Each of these five concentrations can be calculated for every arbitrary value of a, the degree of titration. Note should be taken that a (plotted as abscissa) is defined as the number of moles of chelating agent added for each mole of total metal being titrated (eqn. (33)), so that $a = 1$ when the amount of complexone added is exactly equivalent to the total concentration of the two metals. We thus have:

$$\text{Degree of titration } a = [Z]_t/([M]_t + [M^*]_t) \tag{33}$$

$$\text{Total concentration of metal } [M]_t = [M]' + [MZ] \tag{34}$$

$$[M^*]_t = [M^*]' + [M^*Z] \tag{35}$$

$$\text{Total chelating agent } [Z]_t = [Z]' + [MZ] + [M^*Z] \tag{36}$$

$$\text{Effective stability constants } K^{\text{eff.}}_{MZ'} = [MZ]/[M]'[Z]' \tag{37}$$

$$K^{\text{eff.}}_{M^*Z'} = [M^*Z]/[M^*]'[Z]' \tag{38}$$

It is easy to envisage what the titration curve will look like if MZ and M*Z have very different stabilities, with $K^{\text{eff.}}_{MZ'} \gg K^{\text{eff.}}_{M^*Z'}$.

The most extreme case is exemplified by the titration of iron(III) in the presence of calcium at pH 3, where the appropriate curve of Fig. 10 will describe the change in pFe; for since $K^{\text{eff.}}_{CaY} \sim 1$, Ca^{2+} is virtually not complexed at all in such an acidic medium, so that $pCa = -\log [Ca]_t$ remains constant throughout the titration. In general, then, if the effective stability constant of the second metal is very small ($K^{\text{eff.}}_{M^*Z'} < 10^4$) the first metal can be titrated without any interference from the second.

A less extreme case is the titration of iron(III) at pH 3 in the presence of copper(II), whose effective stability constant ($K^{\text{eff.}}_{CuY'} = 10^{8 \cdot 2}$) is still considerable even in acid solution. The first portions of EDTA added naturally react with the iron ($K^{\text{eff.}}_{FeY'} = 10^{14 \cdot 5}$), and to start with, pFe changes just as if no copper were present (cf. Fig. 10). The jump in pFe which indicates that the complexing of the iron is complete is not so large, however, for it merges into a second flattened portion during which Cu^{2+} is changed into CuY^{2-} while $pFe = 9 \cdot 3$. Not until a quantity of EDTA equivalent to all the copper present has been added can pFe rise to the value (~ 13) of the upper end of the pM jump shown in Fig. 10. The entire curve for pFe thus exhibits two jumps (like the two lowest curves shown in Fig. 15). The value of $pCu = -\log [Cu]_t$ remains constant until all the iron has been complexed, and then there is a gradual increase which terminates with a small jump in pCu that marks the end-point: from Fig. 8 one can form an estimate of how the curve will run at pH 3. When, therefore, the stability constants for the two metals are very different, but the second is nevertheless sufficiently large ($K_{M*Z}^{eff.}> 10^6$), there will be two jumps, one for pM and one for pM*. If it is possible to detect the two jumps in pM, or the first in pM and that in pM* by a visual or instrumental method, both metals can be determined in a single titration (consecutive titration).

If the effective stability constants of MZ and M*Z are of much the same value the two metals will be complexed almost simultaneously when EDTA is added, and when an amount equivalent to all the M present has been added no jump in pM will occur. However, there will be a jump in pM* as well as in pM when $a = 1$, i.e. when the amount of titrant added is equivalent to the sum $[M]_t + [M*]_t$, and one or the other of these abrupt decreases in concentration can be used to indicate the end-point for the total metal content. This is of considerable importance in the titration of the alkaline earths as shown by the theoretical curves of Figs. 11–14.

In determining the hardness of water a mixture of calcium and magnesium has to be titrated. The curves shown in Fig. 11 illustrate the changes in pMg for the ratios Ca : Mg = 1 : 1, 10 : 1, and 100 : 1; the broken line shows pCa for the particular case where the ratio is 1 : 1. Changes in pMg are almost invariably used to indicate the end-point for the total concentration, because far better indicators are available for magnesium than for calcium (e.g. Erio T), and as Fig. 11 shows, the jump in pMg is very clear-cut, even in the presence of a large excess of calcium.

The same considerations that hold for the titration of Ca + Mg hold for the substitution titration of calcium alone. If the EDTA complex of magnesium, MgY^{2-}, is added to the solution of calcium that is to be determined there results a mixture of calcium and magnesium which in this case already contains an amount of the complexing agent exactly equivalent to the magnesium present: the amount of additional EDTA that has to be added until the indicator responds to the jump in pMg then corresponds to the amount of calcium present. The amount of calcium in the original solution has been substituted, so to speak, by an equivalent amount of magnesium, and then this has been titrated.

Strontium too, for which no good indicator is yet known, can be determined by such a substitution titration. Once again MgY^{2-} is added, which produces a solution containing both Sr and Mg: when this is titrated the jumps in pSr and pMg are additive, as shown in Fig. 12. Since the formation constants for SrY^{2-} and MgY^{2-} are nearly equal (Table 2), the curves for pSr and pMg practically coincide for equal amounts of the two metals.

That a substitution titration can be carried out even when no actual substitution takes place is shown in Fig. 13. Thus barium can be titrated with a magnesium indicator after adding magnesium com-

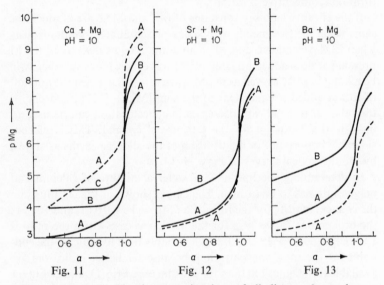

Figures 11, 12, 13. Simultaneous titrations of alkaline-earth metals at pH 10. Variations in pMg are shown as continuous curves.

Figure 11. In curves A, B, and C, the ratio of calcium to magnesium is 1:1, 10:1, and 100:1. Variations in pCa are shown by the broken curve.

Figure 12. In curves A and B the ratio of strontium to magnesium is 1:1 and 10:1. pSr shown by broken curve.

Figure 13. In curves A and B the ratio of barium to magnesium is 1:1 and 10:1. pBa shown by the broken curve.

plexonate, although $K_{BaY} < K_{MgY}$. Since the barium complex is quite appreciably less stable than that of magnesium, only a small percentage of the Ba^{2+} displaces magnesium from MgY^{2-}. Nevertheless, when $a = 1$ a jump in pBa is accompanied by a jump in pMg which can be used to detect the end-point.

The behaviour when the ratio $K_{M'Z}^{eff.}/K_{M^*Z'}^{eff.}$ is varied is beautifully illustrated in Fig. 15, which applies to the substitution titration of Ba by Zn. According to Table 2, the stability constant of ZnY^{2-} is greater than that of BaY^{2-} by a factor of $10^{8.7}$. But whereas ammonia does not coordinate to Ba^{2+}, zinc forms quite stable ammine complexes; therefore $\alpha_{M(A)}$ for zinc increases when NH_3 is added, and accordingly the effective stability constant for ZnY^{2-} decreases (eqn. (25)) until in a solution 1M (approx.) in ammonia ZnY^{2-} and BaY^{2-} have the same effective stability constants. Fig. 15 shows the

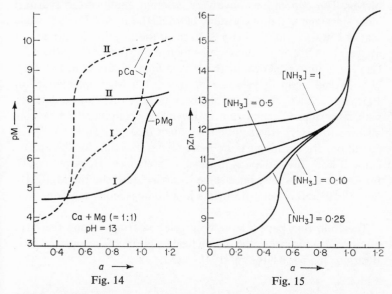

Fig. 14   Fig. 15

Figure 14. Simultaneous titration of Ca and Mg at pH 13. Full lines represent variations in pMg; broken lines, variations in pCa. The two curves labelled I are calculated on the assumption that no $Mg(OH)_2$ precipitates. Those labelled II are calculated for the case where $Mg(OH)_2$ precipitates, and equilibrium is established with this solid phase.

Figure 15. The curves represent the values of pZn during the simultaneous titration of Ba and Zn at pH 11 and various concentrations of ammonia. $[Zn]_t = [Ba]_t = 10^{-3}$. It should be noted that when $[NH_3] = 0.1$, the two metals react stepwise with EDTA, first the zinc and then the barium. At pH 10 the curves lie about 0.3 units lower.

effect of additions of ammonia on the changes in pZn during the total titration of zinc and barium. When $[NH_3] = 0.1$, $K_{Zn'Y'}^{eff.} \gg K_{Ba'Y'}^{eff.}$ but with increasing concentration of ammonia the two effective stability constants approach the same value, with the result that the jump in pZn disappears at $a = 0.5$. However, there is always a jump at $a = 1.0$, which indicates the completion of the complexing of both barium and zinc. If the complex ZnY^{2-} is added to a solution of Ba and the whole made ammoniacal until $[NH_3]$ is 0.5 to 1M and titration is carried out with EDTA using Erio T to indicate the jump in pZn, the amount of complexone required is a measure of the barium present.

Figure 14 shows clearly that not only can the sum of Ca + Mg be determined volumetrically (as in Fig. 11) but also Ca in the presence of Mg. The two cations have very different affinities for hydroxyl ions, since the stability constant of $MgOH^+$ is 380 [51-12], while that of $CaOH^+$ only 25 [53-59]. If, then, the titration is carried out above pH 13, $K^{eff.}$ is depressed much more for MgY^{2-} than for CaY^{2-}, so that, as shown by the broken lines of curve I in Fig. 14, a well-marked jump in pCa occurs at $a = 0.5$ when an amount of EDTA equivalent to the Ca has been added; this can be detected with murexide. If magnesium hydroxide is precipitated on making alkaline to pH 13 (the solubility product of $Mg(OH)_2$ is 10^{-10}) the conditions for titrating Ca are more favourable still, for the broken curve II now shows the way in which pCa changes. However, results for Ca are always low, since some is carried down by the $Mg(OH)_2$, and it is therefore better to titrate in a homogeneous solution.

The following papers have been selected from among the large number dealing with the theoretical aspects of complexometric titrations [65-64; 64-142; 65-104].

CHAPTER THREE

Colour and Fluorescence Indicators in Complexometry

3.1. Introduction

Changes in colour which can be used to indicate end-points in complexometric titrations are almost always due to complex formation between the indicator and the metal whose pM jump at the endpoint is to be detected. Two main cases can be distinguished. In the first the colour of the complex has its origin in the metal ion and the ligand bound directly to it, as in the thiocyanate complexes of Fe(III). Alternatively, the indicator may be an organic dyestuff that is able to form a metal complex: the ligand of the resulting complex is the chromophore in this case. A third group comprises some fluorescent substances capable of forming metal complexes, for these can be used as fluorescence indicators. Finally, there are redox indicators which can serve to detect the end-point in complexing reactions: here the metal whose pM jump is to be detected must be one component of a redox system. All told, there are four types of indicators which will be treated in separate sections. Those forming the second group, viz. dyestuffs that form metal complexes, are by far the most important.

The number of papers dealing with the theory, classification, and listing of complexometric indicators is now quite considerable: the following is a selection of references [56-116; 63-96; 63-97; 64-31; 64-34; 64-37].

3.2. Weak colour indicators

These are particularly suitable for the determination of iron(III), which gives colour reactions with a number of simple complexing agents.

Thiocyanate ions (I) give an orange-red coloration with Fe(III) [48-47; 60-161], but the resulting complexes $Fe(NCS)_j^{3-j}$ are only of low stability (cf. Table 6), so that alcohol or acetone is added to sharpen the change of colour. The pH should be between 2 and 3, because $\alpha_{M(OH)}$ for the iron must be kept small [57-4; 60-99]. Since the thiocyanate complexes are only weakly coloured, an excess of NH_4CNS must be used to ensure that, towards the end of the titration, as much as possible of the metal that has not been complexed by EDTA exists as a coloured complex. As the equivalence point is approached, the orange colour is slowly bleached, and the endpoint, at which the last trace of coloured thiocyanate complex should disappear, is difficult to establish accurately because of the yellow colour of the EDTA complex, FeY^-. If ether or amyl alcohol is added, the red $Fe(NCS)_3$ is extracted and the colour change is then observed in the organic phase [53-18].

Cobalt(II), which forms a blue thiocyanate complex, can also be titrated with SCN^- as indicator in a solution to which a good deal of alcohol or acetone has been added.

The azide ion, N_3^-, has also been used as a weak-colour indicator in the determination of Ga and In by back-titration with Fe(III) [66-17].

The so-called phenol reaction with ferric chloride is well known. Phenolic substances and enols which can chelate with the ferric ion give coloured complexes of high stability, and they can therefore serve as indicators, e.g. salicylic acid (II) [55-23], sulphosalicylic acid (III) [52-22; 60-162], hydroxynaphthoic acid (IV), [61-44], Tiron (V) [51-5], benzhydroxamic acid (VI) [61-103], N-phenyl-benzhydroxamic acid [62-95; 66-18], and cinnamohydroxamic acid [62-67; 63-105].

The formulae show that coordination through the two oxygen atoms of the respective anions can yield 6- and 5-membered chelate rings. The indicators are thus bidentate ligands and can form with

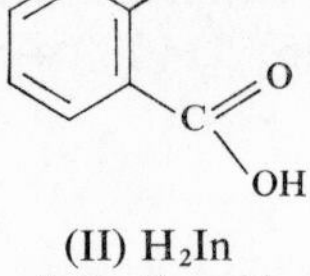

(II) H_2In
Salicylic acid

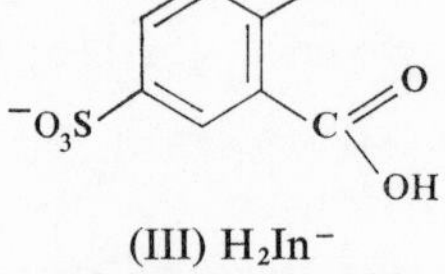

(III) H_2In^-
Sulphosalicylic acid

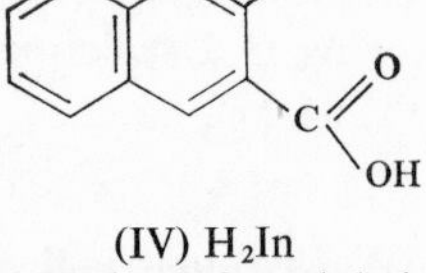

(IV) H_2In
3-Hydroxy-2-naphthoic
acid

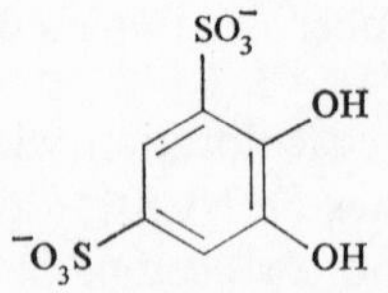

(V) Tiron; H_2In^{2-}

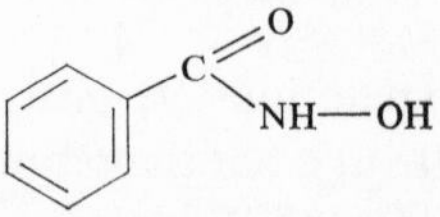

(VI) Benzhydroxamic acid; HIn

iron not only $1:1$ complexes but also $1:2$ and $1:3$ complexes. These colourless indicator ions will be designated by the symbol In; the protonated species HIn and H_2In occur in acid solutions.

TABLE 6
Logarithms of stability constants
$K_{H_jIn} = [H_jIn]/[H][H_{j-1}In]$ and $K_{FeIn_j} = [FeIn_j]/[FeIn_{j-1}][In]$

No.		$\log K_{HIn}$	$\log K_{H_2In}$	$\log K_{FeIn}$	$\log K_{FeIn_2}$	$\log K_{FeIn_3}$
(I)	SCN$^-$	—	—	2·3	1·6	−0·2
(II)	Salicylic acid	13·1	3·0	15·8	11·7	7·5
(III)	Sulphosalicylic acid	11·7	2·7	14·4	10·8	7·1
(V)	Tiron	12·6	7·7	20·7	15·2	11·0
(VI)	Benzhydroxamic acid	8·8	—	11·1	9·4	7·4

The salicylate and hydroxamate complexes of iron are brownish-red, while the $1:1$ complex with Tiron is green, the $1:2$ complex is violet, and the $1:3$ complex is yellowish-red. The $1:1$ complex with hydroxynaphthoic acid is blue.

As compared with metallochromic indicators (q.v.), the most significant feature of these indicators from the point of view of analysis is the low extinction coefficient ($\sim10^3$) of their complexes, so that concentrations of $10^{-5}-10^{-4}$ are necessary to produce a detectable colour. As with SCN$^-$, a high concentration of indicator must be used; this should be about ten times as large as that of the iron, so that at the end-point as much as possible of the iron that is still not chelated by EDTA should be present as the coloured indicator complex. The titration is only carried out under acid conditions (pH 1–4), because above pH 4 there is too little difference between the effective stability constants of the indicator and the EDTA complex. The changes in pFe can be taken from Fig. 10, which refers to a total metal concentration of 10^{-3}M. With a 10-fold excess of indicator the sum [HIn] and $[H_2In] = 10^{-2}$, and calculation with the data from Table 6 shows that in this pH and concentration range only FeIn and FeIn$_2$ occur in appreciable concentrations. The colour of the indicator complex disappears at the end-point when its concentration falls below the level of visibility at (approx.) 5×10^{-5}M; this corresponds to a value of pFe which lies well in the steep vertical portion of Fig. 10, i.e. at the exact end-point. However, owing to the feeble colour of the iron-indicator complex, the titration is only moderately accurate, because several per cent of the total of 10^{-3}M iron can still be present as the indicator complex when the colour is seen to disappear (or to appear in a back-titration). The titration error due to the feeble colour of the indicator complexes can be reduced by determining the end-point photometrically. Salicylic and sulphosalicylic acid have often been recommended for back-titration of excess EDTA with a ferric solution in the determination of Zr^{4+}, Hf^{4+}, Th^{4+}, Bi^{3+}, and several other highly charged cations [54-92; 60-160]. Iodide ions (VII) [48-7; 53-58; 54-6] and thiourea (VIII) [54-44] play a similar role as colourless indicators for bismuth(III) as thiocyanate does for iron and cobalt. The iodo- and thiourea-complexes are both moderately strongly yellow, so that the disappearance of this colour can mark the end-point when titrating bismuth with EDTA. However, the procedures are only of moderate accuracy.

3.3. Metallochromic indicators

Under this heading are included those organic colouring matters that undergo a change of colour when they form metal complexes. They are superior to the metal indicators discussed in Section 3.2, since the intensity of their colours is 10–100 times greater (molecular extinction coefficients 10^4–10^5); the metallochromic indicator need only be added in concentrations of 10^{-6}–10^5M to give a colour change that is still clearly detectable. The amount of metal bound to the indicator is therefore generally negligible (some 0·1% of the total metal to be determined). Generally speaking, not only the dyestuff but also its metal complex are coloured, so that the end of the titration is not characterized by the appearance (or disappearance) of a colour, but rather by a *change* in colour. Only the phthaleins (XVIII), (XIX), are monocolour indicators; here also the ligand is the site of the colour in the metal complex, since coordination of the metal ion simply induces colour through resonance in the ligand, and the intensity of colour is high compared with that of the weakly coloured indicators dealt with in Section 3.2.

The behaviour is completely analogous to that encountered in alkalimetric or acidimetric titrations with acid–base indicators.

3.3.1. GENERALITIES ON STRUCTURE

Like the titrants used in complexometry, the metallochromic indicators must also be chelating agents, for only these can form 5- or 6-membered chelate rings. When at least one of these donor atoms participates in the auxochromic or chromophoric groups of the dyestuff a change in colour will take place on complex formation. It so happens that coordination of a metal atom invariably has a very much smaller effect on the absorption spectrum than the attachment of a proton; nevertheless, when a metal complex is formed with the simultaneous loss of a proton the change in colour often achieves the sort of brilliance familiar in many acid–alkali indicators.

3.3.2. TRIPHENYLMETHANE DYESTUFFS

Representatives of this large class of dyestuffs that can be used as indicators are derived from benzaurin (IX). On absorption of light, which produces the red colour of the anion (IX), light energy is taken up by the electrons of the mesomeric system indicated by the arrows. This consists of the rings A and B attached to the central carbon atom and the oxygen atoms situated *para* to this, which serve as auxochromes.

The arrows show that by movement of the bonds the benzene ring A becomes quinonoid and the quinone ring B becomes benzenoid. The two canonical forms coexist, so that the negative charge is shared between the two oxygen atoms. The dyestuff (IX) is a pH indicator, for on acidification a proton becomes attached to one of the oxygen atoms, and this induces a colour change from red to yellow.

(IX) Benzaurin

The reason why benzaurin is not employed analytically is because its colour is slowly bleached both in strongly alkaline and in strongly acidic media, since a colourless carbinol base is formed by the attachment of OH^- to the central carbon atom.

This reaction can be prevented by appropriate substituents in the benzene ring C in positions 2″ or 6″ *ortho* to the central atom. Halogens and sulphonic groups are especially effective in this respect, and this substitution has no significant influence on the absorption spectrum, since this ring C is not involved in the mesomeric system.

In sulphonphthaleins, which are well-known pH indicators, a sulphonic acid group in position 2″ inhibits the formation of the colourless carbinol, and bicolour indicators result which are yellow in acids and turn red or blue as the pH is raised. In alkaline solutions of the phthaleins a carboxylate ion, $^-COO^-$, in position 2″ prevents the decolourization of the dyestuff through the formation of a carbinol. However, when the pH is lowered the colour disappears, because when the auxochromic oxygen atoms are protonated a lactone ring forms between the central carbon atom and the carboxylate group in the 2″ position, whereby the mesomeric system is disturbed. The phthaleins are therefore monocolour pH indicators which are colourless in acid, but red or blue in alkaline solution. The introduction of alkyl groups in positions 3,3′ and 5,5′ causes a bathochromic shift (to longer wavelengths) in both phthaleins and sulphonphthaleins, and the colour change occurs at a higher pH (cf. phenolphthalein with thymolphthalein, and phenol red with thymol blue). Halogen substituents in positions 3,3′ and 5,5′ also produce a bathochromic shift in the spectrum, but in contrast to alkyl groups, they increase the acidity so that the colour change takes place at a lower pH (cf. phenol red with bromophenol blue).

These phthalein and sulphonphthalein indicators have practically no ability to form complexes with metal ions. However, a metal indicator results if a substituent is introduced adjacent to the phenolic oxygen atoms (in 3,3′ or 5,5′ positions) which can act as a ligand for a metal in such a way that the auxochromic oxygen becomes co-ordinated with the simultaneous formation of a 5- or 6-membered ring. Since the electrons of the auxochromic oxygen form part of the electron system that determines the absorption spectrum, it is understandable why the coordination of this oxygen should result in a change of colour. In most indicators of this type both benzene rings A and B carry such a substituent capable of forming a metal complex. In principle, however, only one such substituent is necessary, and this could be introduced into either ring.

There are three substituents in particular, which, if introduced into positions 3,3′ or 5,5′ of a dyestuff of this class, are able to impart the characteristics of a chelating agent and a metal indicator: they are ^-OH, ^-COOH, and $^-CH_2 \cdot N(CH_2 \cdot COOH)_2$. Widely used and excellent complexometric indicators are illustrated by examples (X), (XI), and (XV) [53-43; 54-45; 54-75; 54-77; 54-78; 56-38; 56-39; 56-41; 56-84; 57-72; 57-79; 59-28; 61-2], and Table 7 includes further metal indicators of this type. With the exception of aluminon

(XII), it is noteworthy that in the phenyl residue C there is always a sulphonic acid group or a halogen adjacent to the central carbon atom which prevents the change to a colourless carbinol.

Double bonds are not shown in the structural formulae written for (X), (XI), and (XV), and they represent those forms of the dye-stuff that occur in strongly acid solution where the phenolic oxygen, the carboxylate group, and the amino groups are protonated. The sulphonic group cannot take up a proton even in strongly acid solutions, and it is always present as an ion. A positive charge resides on the mesomeric system comprising the central carbon atom, rings A and B, and the auxochromic oxygens, when these two latter are protonated. The zwitterions (represented in the formulae) which exist in strongly acid media are yellowish-red, and they lose a proton from one of the auxochromic oxygens even in a pH region in the neighbourhood of 1, whereby the colour changes to yellow. When the pH is raised further to 7 or 8 the second auxochromic oxygen releases its proton and the colour changes from yellow to purple, violet, or blue. The loss of protons from the carboxylate groups of (XI) and (XV) in the pH range 3–5 and the loss of a proton from the 3,3′ phenolic OH in (X) [56-38] or from the ammonium group of (XV) in a pH range above 8 produces no further marked change in

(X) Pyrocatechol violet, H_4In (XI) Eriochromecyanine R, H_4In

(XV) Xylenol orange, H_6In

colour [60-51]. The colour of metal complexes is always like that of the fully deprotonated dyestuff anion in strongly alkaline solution (violet, blue). An especially brilliant colour change takes place on co-ordination of a metal if the yellow protonated form of the dyestuff that occurs at pH <7 is transformed directly into the deeply coloured metal complex. This is illustrated by the absorption spectra of

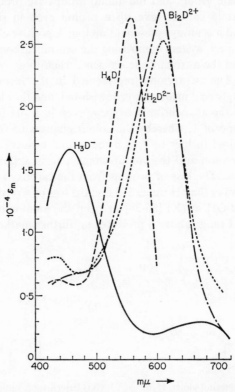

Figure 16. The absorption spectrum of pyrocatechol violet (X) in strongly acid solution (pH <0), in weakly acid solution (pH ~5), and in weakly alkaline solution (pH ~8), together with the spectrum of the bismuth complex, Bi_2D^{2+}. Wavelengths in millimicrons.

pyrocatechol violet shown in Fig. 16, where the molecule of struc-ture (X) is represented as H_4In corresponding to the four acidic hydrogens. In the anion H_3In^- a proton has been removed from one of the auxochromic oxygens and in H_2In^{2-} from both. Bi_2In^{2+} repre-sents the blue bismuth complex, in which each of the metal atoms is chelated to the oxygen atoms of the phenyl residues A or B respectively. It is formed via an intermediate violet-coloured 1 : 1 complex which has only a limited range of existence, but which can be observed just before the end-point in the complexometric titration of Bi with pyrocatechol violet as indicator.

It has been shown that in the titration of Hf, Y, and Pb with EDTA

TABLE 7

Metal indicators of structure (IX)

[The numbers denote the position of the substituents]

No.	Substituents	Name	Applications
(X)	3-OH, 3'-OH, 2''-SO$_3^-$	Pyrocatechol violet [54-75; 54-77; 54-78; 56-38; 56-41; 56-84]	Th, Ga, In, Bi
(Xa)	3-OH, 3'-OH, 2''-COO$^-$	Phthalein violet [65-30]	Ca, Mg, Ni, Co, Zn, Pb
(XI)	3-COOH, 5-CH$_3$, 3'-COOH, 5'-CH$_3$, 2''-SO$_3^-$	Eriochromecyanine R [53-43; 54-45; 57-72; 59-28]	Al, Zr, Th, Fe
(XII)	3-COOH, 3'-COOH, 3''-COO$^-$, 4''-OH	Aluminon [61-2]	Al
(XIII)	3-COOH, 5-CH$_3$, 3'-COOH, 5'-CH$_3$, 2''-Cl, 3''-SO$_3^-$, 6''-Cl	Eriochrome azurol S [55-26; 55-27; 55-29; 55-30; 55-31; 56-28; 59-14]	Ca, Sr, Ba, Al, Zr, Fe, Cu
(XIIIa)	3-CH$_3$, 5-COOH, 3'-CH$_3$, 5'-COOH, 2''-Cl, 3''-SO$_3^-$, 6''-Cl	Chromeoxane Pure Blue BLD [64-55]	Cu
(XIIIb)	Uncertain	Chromeoxane violet 5B [64-55]	Cu
(XIV)	3-COOH, 5-CH$_3$, 3'-COOH, 5'-CH$_3$, 3''-Az, 6''-Cl	Chromeoxane green GG [57-32]	Mg, Ca, Th, V
(XV)	5-mim, 3-CH$_3$, 5'-mim, 3'-CH$_3$, 2''-SO$_3^-$	Xylenol orange [56-39; 57-79; 57-137]	Al, Lanthanides, Th, Sc, Y, Zn, Cd, Hg, Th, Pb, Bi
(XVI)	5-mim, 3-CH$_3$, 3'-CH$_3$, 2''-SO$_3^-$	Semixylenol orange [62-18]	Zr
(XVII)	5-mim, 3-isopr, 5'-mim, 3'-isopr, 2''-SO$_3^-$, 6-CH$_3$, 6'-CH$_3$	Methylthymol blue [57-61; 58-34]	Pd, Pb, Sn, Sb, Th, Cd
(XVIII)	5-mim, 3-CH$_3$, 5'-mim, 3'-CH$_3$, 2''-COO$^-$	Metalphthalein [53-1; 57-50; 58-2]	Ca, Sr, Ba
(XIX)	5-mim, 3-isopr, 5'-mim, 3'-isopr, 2''-COO$^-$	Thymolphthalexone [59-46]	Ca, Sr, Ba
(XX)	5-gl, 3-isopr, 5'-gl, 3'-isopr, 2''-COO$^-$	Glycinethymol blue [58-50]	
(XXa)	$-$N(CH$_2$COOH)·C$_2$H$_4$·N: (CH$_2$COOH)$_2$ in place of OH in position 4, 2''-SO$_3^-$	Metallochrome violet A [63-95]	Cu, Cd, Co, Pb, Zn, Hg

mim = $-$CH$_2$·N(CH$_2$COOH)$_2$; gl = $-$CH$_2$·NH·CH$_2$COOH; isopr = $-$CH(CH$_3$)$_2$;

Az = $-$N$=$N$-$

the colour change of the xylenol orange indicator is also partly due to a change in the composition of the complex [64-40].

The indicators (Xa), (XVIII), (XIX), and (XX) of Table 7 have a carboxyl group in position 2″ : they are thus phhaleins derived from the well-known pH indicators phenol- and thymol-phthalein. It was with the metal phthalein (XVIII) that it was first demonstrated that a pH indicator could be transformed into a pM indicator by introducing a methyleneaminodiacetate group [53-1].

As previously mentioned, in contrast to the sulphonphthaleins ($-SO_3^-$ in position 2″), phenolphthaleins can form a lactone ring in which the ionic oxygen of the $-CO_2^-$ group in position 2″ becomes attached to the central carbon. This always happens when the auxochromic oxygen atoms in positions 4 and 4′ of rings A and B are protonated, i.e. in acid solution. Like the carbinols, the lactones are colourless, and thus the phthaleins (unlike the sulphonphthaleins) are monocolour indicators. They are colourless in acid solution, and not until the OH groups in positions 4 and 4′ are deprotonated by increase of pH does the lactone ring open and the indicator become red or blue.

(XVIII) Metalphthalein, H_6In

The various stages will now be considered in detail using metalphthalein (XVIII) as an example. The above constitutional formulae is that of the formally uncharged molecule (a zwitterion) that occurs in strongly acid solution: in view of the six potentially dissociable protons, it can be represented as H_6In. On raising the pH, hydrogen ions dissociate from the carboxyl groups even below pH 3 ($H_6In \longrightarrow H_4In^{2-}$), although the solution remains colourless. Between pH 7 and 8 the ion loses two more protons ($H_4In^{2-} \longrightarrow H_2In^{4-}$): this change is accompanied by the development of a pink coloration, since the ion H_2In^{4-} is pale red (absorption spectrum in Fig. 17, curve 6). The acidic hydrogens that still remain in the anion H_2In^{4-}

occur as hydrogen bridges between the auxochromic O and the N of the iminodiacetate groups. Complete deprotonation ($H_2In^{4-} \longrightarrow In^{6-}$) does not occur until between pH 10·5 and 13, and then the extinction coefficient increases to the high value typical of benzaurin derivatives (curves 7 and 1 of Fig. 17). The complexes of the indicator with Ca, Sr, and Ba are as intensely coloured as the fully dissociated species In^{6-} (Fig. 17); the Mg complex is much less strongly coloured, and complexes with the heavy metals are virtually colourless (cf.

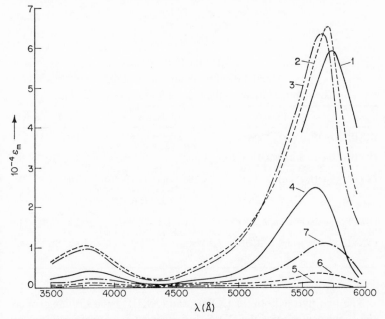

Figure 17. The absorption spectra of different ionized species and metal complexes derived from the metalphthalein (XVIII). In^{6-} (curve 1); Ba_2In^{2-} (curve 2); Ca_2In^{2-} (curve 3); Mg_2In^{2-} (curve 4); Zn_2In^{2-} (curve 5); H_2In^{4-} (curve 6); HIn^{5-} (curve 7); ε_m = molecular extinction coefficient. Wavelength (λ) in angstrom units.

spectra of Mg and Zn complexes in Fig. 17). Metalphthalein is therefore a good indicator only for the heavier alkaline earths. A pH between 10 and 11 is used where the solution of the indicator alone is pale pink and the addition of Ca^{2+}, Sr^{2+}, or Ba^{2+} produces a deep violet red.

A few indicators collected in Table 8 are classified as triphenylmethane dyestuffs and are derived from the parent structure (XXI), which only differs from that of (IX) by the oxygen bridge between positions 2 and 2′. A sulphonic or carboxylic acid group is located in ring C next to the central carbon atom.

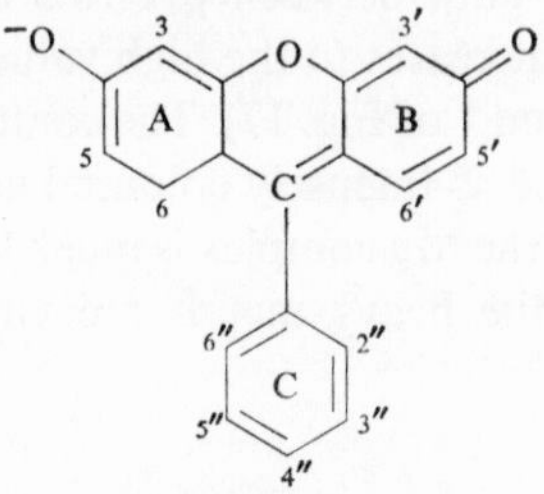

(XXI)

TABLE 8

Metal indicators of structure (XXI)
The numbers indicate the orientation of the substituents

No.	Substituents	Name	Indicator for
(XXII)	3-OH, 3′-OH, 2″-SO$_3^-$	Pyrogallol red [56-85; 56-88; 56-90; 56-92; 59-20]	Lanthanides, Th
(XXIII)	3-OH, 5-Br, 3′-OH, 5′-Br, 2″-SO$_3^-$	Bromopyrogallol red [56-90; 56-91; 59-20]	Co, Ni, Pb, Bi
(XXIV)	3-OH, 3′-OH, 2″-COO$^-$	Gallocyanine [55-79; 60-13; 61-30; 61-31; 61-32; 62-30; 62-31; 62-158; 63-102]	Ga, Bi, Pb, Th
(XXV)	3-mim, 5-mim, 2″-COO$^-$	Calcein [58-47; 58-52; 58-54; 63-104]	Ca, Sr, Ba, Mn
(XXVI)	3-mgl, 5-mgl, 2″-COO$^-$	Methylcalcein [59-11; 62-50]	Ca, Sr, Ba, Mn

mim = $-CH_2 \cdot N(CH_2 \cdot COOH)_2$; mgl = $-CH_2 \cdot N(CH_3)CH_2 \cdot COOH$

The two pyrogallol reds (XXII) and (XXIII) have similar properties to pyrocatechol violet (X). Solutions in acid are again yellow, and deprotonation of the OH groups, caused either by increase in pH or the formation of a metal complex, produces a change in colour to red or violet. The latter deeply coloured species also absorb at shorter wavelengths than the indicator (X) owing to the hypsochromic effect of the ether bridge between positions 2 and 2′. Many derivatives of tetrahydroxyfluorane (as (XXI) with 2″-COOH and 3-OH, 3′-OH or 5-OH, 5′-OH) have been synthesized by Itsuo Mori [65-33; 66-15] and studied as potential metal indicators. Substances (XXV) and (XXVI) correspond to metalphthalein (XVIII) itself, but in contrast calcein is coloured a yellowish-green, even in acidic solutions, since the carboxyl ion in position 2″ does not form a lactone with the central carbon atom, but is simply protonated when the solution is made acid. Over the appropriate pH range the addition of alkaline earths, in particular Ca^{2+}, changes the colour to dark yellow or brown, whereas the anion and the alkaline-earth complexes of (XVIII) are red. Thus difference in colour is again due to the hypsochromic effect of the 2–2' ether bridge. The precise orientation of the groups mim and mgl has not been completely established [59-144].

The parent substance of calcein which has no substituents in the 3,3' and 5,5' positions ((XXI) with $-COO^-$ in position 2'') is the well-known fluorescein, which shows an intense yellowish-green fluorescence in the medium pH range. Calcein and methylcalcein also fluoresce, but there is a great deal of difference between the intensity of the fluorescence of the indicator itself and that of its metal complex. The end-point in the titration or back-titration of Ca^{2+} can thus be detected by the quenching or appearance of fluorescence, especially if ultraviolet light is used (cf. Section 3.1 of this chapter). Unfortunately, most commercial samples of calcein (or fluorexon) contain rather a lot of fluorescein or of fluorescein that has been substituted with only a single methyliminodiacetate group; this profoundly affects the quality of the fluorescence end-point.

3.3.3. AZO DYES

In the simple azo dyes two aromatic nuclei are joined together by the chromophoric azo-group, $-N=N-$, and *ortho-* or *para-* to this are auxochromic substituents, almost always $-OH$ or amino groups, $-NH_2$, $-NHR$, $-NR_2$. All these azo dyes are pH indicators (the colour change with the products used industrially as dyes ought not to occur, save in a very acid or very alkaline region).

The change of colour when the pH is lowered is due to the attachment of a proton to the auxochromic groups (phenolate or amino group) or on one of the nitrogen atoms of the azo group ($R-N=N-R \rightarrow R-N=\overset{+}{N}H-R$), and by increasing the pH the process is reversed.

By introducing suitable ligands, azo dyes can also be made to act as metal indicators. Once again provision must be made for the formation of 5- or 6-membered chelate rings if the products are to be sufficiently stable. If the coordination of the metal involves an auxochromic group the absorption spectrum will be changed on complex formation, and an especially well-marked change in colour will result if the metal is bound to the azo group. This condition can be met if the substituents capable of coordination are introduced *ortho* to the azo group. The great majority of metallochromic azo indi-

C

cators carry −OH groups in this position (*o,o'*-dihydroxyazodyes), whereby these phenolic groups serve simultaneously as auxochromes and as ligands for the metal. In exceptional cases the carboxyl, −COOH, or arsonic acid group, $AsO(OH)_2$, are used as ligands *ortho* to the azo group.

As an example, Eriochromeblack T (Erio T for short) will be considered in greater detail. This indicator was referred to in the very first papers on complexometric titration [46-2, 48-4], and even now is one of the most frequently used metallochromic indicators. The structural formula (LIX) describes the red species, H_2D^-, that occurs in the solutions below pH 6; the ionized sulphonic acid group gives it a single negative charge and imparts solubility in water.

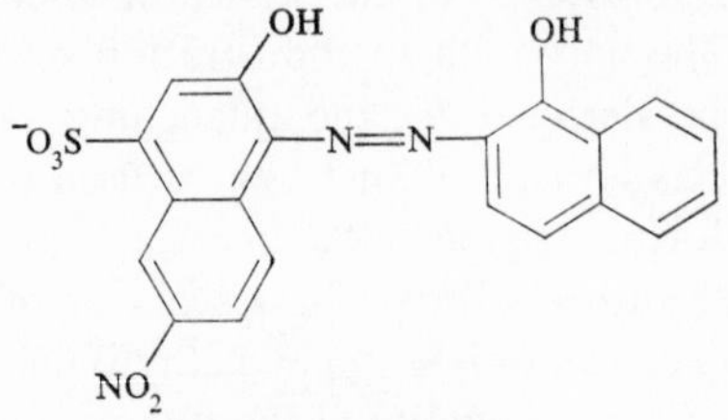

(LIX) Erio T (H_2D^-)

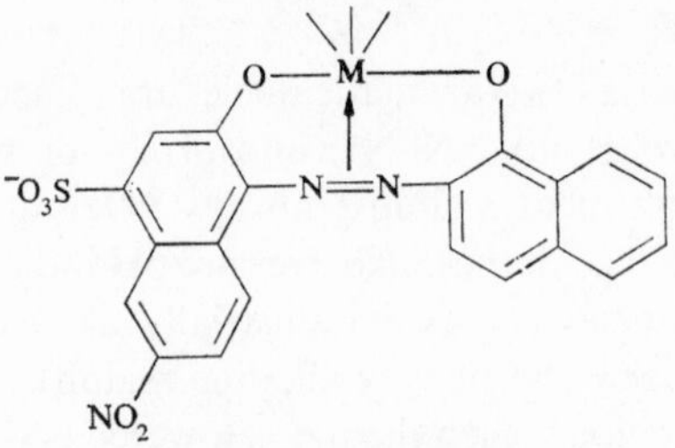

Metal complex, MD^-

In the abbreviated formulae, H_2D^-, the two hydrogens are those of the phenolic hydroxyl groups; on basification these can be removed, whereby red H_2D^- changes to blue HD^{2-} at pH 6·3, and finally into the fully deprotonated yellowish-orange form D^{3-} at pH 11·5. The metal complexes are red or violet, and the change in colour when they are formed is especially brilliant if the pH lies between 7 and 11, since this is the region in which the indicator exists as blue HD^{2-} and the colour changes to red on the addition of metal. Only the 1 : 1 complexes are important; for with the low concentrations of indicator and metal encountered under the conditions prevailing near the end-point of a complexometric titration the 1 : 2

complexes are formed in such low concentrations that there is no need to consider them. Figure 18 shows the absorption spectra of the ionic species H_2D^-, HD^{2-}, and D^{3-} of the metal-free dye and that of the magnesium complex MgD^-. There are very many azo dyes which form metal complexes; for they can be derivatives of benzene

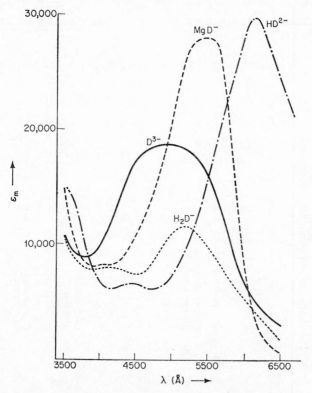

Figure 18. The absorption spectrum of Erio T at pH $\sim$5 (H_2D^-), pH $\sim$8 (HD^{2-}), and pH $\sim$13 (D^{3-}), together with that of the magnesium complex (MgD^-). $\varepsilon_m =$ molecular extinction coefficient. Wavelength (λ) in angstrom units.

or naphthalene, and the number of sites for substitution is quite large, so that although only a few substituents are used many variations are possible.

Up to now some 100 azo dyes have been successfully tried out as metal indicators. Those which have been described in the literature as having already been used for complexometric titrations are shown in Table 9. Most of them are mono-azo dyes derived from benzene or

TABLE 9

Metal indicators showing the structure of the basic skeleton with the numbering used to indicate the position of substituents. The charge on the group SO_3^- is omitted

No.	Structure Substituents Name	Name	Cations	Ref.
(XXVII)	2-OH, 3-SO₃, 5-Cl, 4'-OH, 6'-OH	Omega chrome garnet	Ni	[57-43]
(XXVIIa)	2-OH, 3-NO₂, 5-SO₃	—	Bi	[66-20]
(XXVIII)	6-COOH, 2'-OH, 5'-CH₃	Chrome Bordeaux B	Ca, Mg, Zn	[58-115]
(XXIX)	3-SO₃, 4-OH, 5-SO₃, 2'-OH, 3'-OH, 4'-OH	Chrome brown RR	Th	[60-55]
(XXX)	6-COOH, 2'-OH, 3'-SO₃, 6'-SO₃	Lacquer scarlet C	Ca	[57-109]
(XXXI)	3-Cl, 5-SO₃, 6-OH, 2'-OH	Magneson	Ca, In	[57-43]
(XXXII)	3-SO₃, 5-NO₂, 6-OH, 2'-OH	Alizarin chrome black R	Fe(III), Bi, Th	[57-43]
(XXXIII)	3-SO₃, 6-OH, 2'-OH	Palatine chrome violet	Mg, Ca, Sr, Ba, Mn, Zr	[58-91]
(XXXIV)	3-CH₃, 6-OH, 2'-OH, 4'-SO₃	Calmagite	Mg, Ca	[60-134; 66-14]
(XXXV)	2-OH, 4-OH, 2'-OH, 4'-SO₃	Sulphonnaphthylazoresorcinol	Zr	
(XXXVI)	4-NO₂, 3'-CH₂NH·CH₂·COOH, 4'-OH	Glycine naphthol violet	Mn, Ni, Cu, Co, Cd	[59-17]
(XXXVII)	4-NO₂, 3'-CH₂N(CH₂·COOH)₂, 4'-OH	Naphthol violet	Mg, Mn, Co, Cu, Zn, Cd, Bi	[57-81]
(XXXVIII)	6-AsO₃H₂, 2'-OH, 3'-SO₃, 6'-SO₃	Thorin	Bi, Th, U	[60-17]

	Substituents	Name	Metals	Refs
(XXXIX)	$1'$-OH, $3'$-SO$_3$, $6'$-SO$_3$, $8'$-OH	Chromotrope 2 R	Th	[61-37]
(XXXIXa)	2-OH, 5-Cl, $1'$-OH, $3'$-SO$_3$, $6'$-SO$_3$, $8'$-OH	—	Ca	[63-90]
(XL)	4-SO$_3$, $1'$-OH, $3'$-SO$_3$, $6'$-SO$_3$, $8'$-OH	SPADNS	Th, Zr	[55-92; 55-94]
(XLI)	6-COOH, $1'$-OH, $3'$-SO$_3$, $6'$-SO$_3$, $8'$-OH	Chromotrope 2 C	Fe(III), Th, Zr	[60-41]
(XLII)	6-OH, $1'$-OH, $3'$-SO$_3$, $6'$-SO$_3$, $8'$-OH	Acid chrome dark blue	Mg, Ca, Zn, Mn	[58-9; 60-87; 65-29]
(XLIIa)	6-CH(OH)COOH, $1'$-OH, $3'$-SO$_3$, $8'$-OH	Mandelic-azo-chromotropic acid	Th	[65-26]
(XLIII)	4-NO$_2$, 6-OH, $1'$-OH, $3'$-SO$_3$, $8'$-OH, $6'$-SO$_3$	Erio chrome green H	Ca, Ga	[57-43]
(XLIV)	3-SO$_3$, 6-OH, $1'$-OH, $3'$-SO$_3$, $6'$-SO$_3$, $8'$-OH	Acid chrome blue K	Mg, Ca, Mn, Cd, Pb	[58-9]
(XLV)	3-Cl, 6-OH, $1'$-OH, $3'$-SO$_3$, $6'$-SO$_3$, $8'$-OH	Erio SE	Mg, Ca, Mn, Ni, Cd, Zn, Pb	[58-8; 58-36; 61-120]
(XLVI)	3-NO$_2$, 5-Cl, 6-OH, $1'$-OH, $3'$-SO$_3$, $6'$-SO$_3$, $8'$-NH$_2$	Gallion	Ga	[60-191]
(XLVII)	3-Cl, 5-SO$_3$, 6-OH, $1'$-OH, $5'$-SO$_3$	Erio chrome violet	Mg, Ca, Mn, Ni, Zn, Cd, Hg, Pb	[61-126]
(XLVIII)	3-Cl, 6-OH, $1'$-OH, $4'$-SO$_3$	Acid chrome violet B	Fe(III)	[57-43]
(XLIX)	3-Cl, 5-SO$_3$, 6-OH, $1'$-OH	Chrome fast blue RLL	Ca, Sr, Ba, Ni, Cd	[57-43]
(L)	2-Cl, 3-Cl, 5-Cl, 6-OH, $1'$-OH, $4'$-SO$_3$	Solochromate fast violet B	Ca, Mn, Pb	[57-43]
(LI)	2-Cl,β-Cl, 5-Cl,6-OH, $1'$-OH, $5'$-SO$_3$, 7-SO$_3$, $8'$-NH$_2$	Omega chrome fast blue 2 G	Mg, Ca, Mn, Ni	[59-98]
(LII)	2-Cl, 3-Cl, 5-Cl, 6-OH, $1'$-OH, $5'$-SO$_3$, $8'$-NH-COCH$_3$	Eriochrome navy blue BRL	Mg, Mn, Zn, Cd	[59-100]
(LIII)	3-SO$_3$, 6-OH, $1'$-OH, $5'$-OH	Erio chrome black PV	Mg, Ca, Mn, Fe, Zn, Cd, Pb	[57-43]
(LIV)	6-AsO$_3$H$_2$, $1'$-OH, $3'$-SO$_3$, $6'$-SO$_3$, $8'$-OH	Arsenazo	Mg, Ca, Th, U, Pu, Lanthanons	[58-61; 61-89; 65-17; 65-22]

TABLE 9—continued

No.	Structure Substituents Name	Name	Cations	Ref.
(LV)	2-OH, 2'-OH, 4'-SO₃	Calcon	Mg, Ca, Mn, Zn, Cd	[56-61; 57-123; 59-102; 57-124; 59-112; 60-146]
(LVa)	2-OH, 3-COOH, 2'-OH, 4'-SO₃	Erio A	Ca	[64-48]
(LVI)	2-OH, 4-SO₃, 6-NO₂, 2'-OH	—	Mg, Mn, Ni, Cu, Zn, Pb	[48-4; 58-7]
(LVII)	2-OH, 4-SO₃, 2'-OH, 3'-COOH	Cal-red	Ca	[56-26]
(LVIII)	2-OH, 4-SO₃, 1'-OH	Erio B	Mg, Ca, Zn, Cd, Zr, U(IV)	[51-18; 60-178; 61-36]
(LIX)	2-OH, 4-SO₃, 6-NO₂, 1'-OH	Erio T (Solochrome black WDFA or T)	Mg, Mn, Zn, Cd, In, Pb, Zr, Lanthanons	[46-2; 48-4]
(LX)	2-OH, 8-SO₃, 1'-OH, 3'-SO₃, 7'-NH₂	Ponceau 3 R	Cu	[56-115]
(LXI)	4-SO₃, 1'-OH, 3'-SO₃, 6'-SO₃, 8'-OH	SNADNS	Zr, Th	[60-53]
(LXII)	5-SO₃, 1'-OH, 3'-SO₃, 6'-SO₃	SNADNS-5	Th	[60-192]
(LXIII)	3-SO₃, 6-SO₃, 8-OH, 1'-OH, 3'-SO₃, 6'-SO₃, 8'-OH	Beryllon II	Mg	[59-139]
(LXIV)	4-SO₃, 1'-OH, 3'-SO₃, 6'-SO₂, 7'-NO, 8'-OH	Nitroso-SNADNS	Zr, Th	[60-145]

(LXV) 2-OH, 1'-OH, 4'-SO_3, 2''-OH Acid alizarin black SE Ca, Mn [57-43]

(LXVI) 2-OH, 1'-OH, 4'-SO_3, 2''-OH, 6''-SO_3 Acid alizarin black SN Ca, Sr, Ba, Mn, Ni, Zn, Cd, Th [60-155]

(LXVIa) 2-COOH, 1'-OH, 3'-SO_3, 6'-SO_3, 8'-OH, 2''-COOH — Ba, Sr, Pb [65-34]

(LXVII) 4-SO_3, 1'-OH, 3'-SO_3, 6'-SO_3, 2''-OH Fast sulphon black F Cu [57-42]

TABLE 9—*continued*

No.	Structure	Substituents	Name	Cations	Ref.
(LXVIII)		4-SO₃, 8'-SO₃, 1''-OH, 3''-SO₃, 6''-NHC₆H₅	Brilliant Congo blue	Pb	[57-77]
(LXIX)			Stilbazo	Bi, Th	[61-147]
(LXIXa)		4-OH, 5-OH, 2'-SO₃	—	Sc	[64-35]
		1-OH, 3-SO₃, 6-SO₃, 8-OH, 6'-SO₃			
(LXX)			Calcichrome (structure uncertain and probably identical with Calcion)	Ca	[60-154; 63-103; 65-21]

	Substituents	Name	Metals	References
(LXXI)	2-OH, 4-OH	PAR (Pyridylazoresorcinol)	Sr, Ba, Al, Mn, Ni, Cu, Zn, Cd, Hg, In, Th, Bi, Lanthanides, Pb	[57-5; 59-67; 60-167; 61-86; 61-137; 62-11; 66-21]
(LXXIa)	2-OH, 4-OH, 5'-(2-N-methylpiperidine)	4-(2-N-methylanabasino-azo)-resorcinol	Ga, In	[64-41; 64-42; 64-43; 65-23]
(LXXII)	2-OH, 4-OH, 6-CH$_3$	PAO (Pyridylazo-orcinol)	Cu	[60-20]
(LXXIIa)	2-OH, 4-NHEt, 5-CH$_3$	—	In	[64-53]
(LXXIIb)	2-OEt, 4-NHMe, 5-CH$_3$	—	In	[66-16]
(LXXIII)	2-OH	PAN (Pyridylazonaphthol)	Mg, Ca, Sr, Ba, Al, Lanthanides, Mn, V, Fe, Co, Ni, Cu, Zn, Cd, Hg, Ga, In, Tl, Bi, Pb, Th, U	[55-18; 55-19; 55-20; 56-43; 56-44; 56-46; 58-64; 58-107; 59-82; 60-167; 61-60; 62-11]
(LXXIV)	2-OH, 3-OH, 6-SO$_3$	PADNS (Pyridylazodihydroxy-naphthalene sulphonic acid)	Cd, Cu, Zn, Pb	[60-20]
(LXXV)	1-OH, 3-SO$_3$, 6-SO$_3$, 8-OH	PACA (Pryidylazo chromotropic acid)	Cu	[58-28]
(LXXVI)	1-OH, 3-SO$_3$, 6-SO$_3$, 8-NH$_2$	PAHA (Pyridylazo-H-acid)	Cu	[58-28]

TABLE 9—continued

No.	Structure Substituents Name	Name	Cations	Ref.
(LXXVII)	5-SO$_3$, 8'-OH	α-Naphthylazoxine	Ga	[62-48]
(LXXVIII)	2-OH, 4-SO$_3$, 8'-OH	Hydroxysulphon-α-naphthylaz-oxine	Mg,	[57-122]
(LXXIX)	5'-SO$_3$, 8'-OH	α-Naphthylazoxine-sulphonic acid	Mn, Fe, Co, Ni, Cu, Zn, Cd, Th, Ga, Lanthanides	[57-122; 62-49]
(LXXX)	4-SO$_3$, 5'-SO$_3$, 8'-OH	SNAZOXS or NAS	Lanthanides, Fe, Co, Ni, Cu, Zn, Cd, Hg, Pb, Bi, In, Mn, Zr, Mo, Sn, Ti, Th	[60-67]
(LXXXI)	6-SO$_3$	Naphthylazoxine S	Ga	[61-112]
(LXXXII)	5-SO$_3$, 7-SO$_3$	Disulphononaphthylazo-8-hydr-oxyquinoline sulphonic acid or β-Naphthylazoxine 2 S	Tl(III)	[62-68]

	Pyridylazoxine	Tl(III)	
(LXXXIII)			[61-132]
(LXXXIIIa)	—	Cu	[64-38]
(LXXXIIIb)	—	Cu	[64-38]

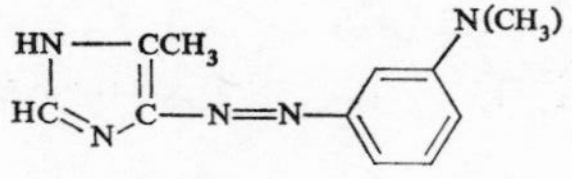

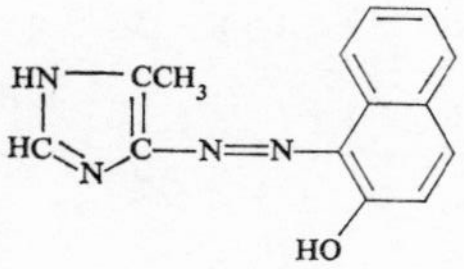

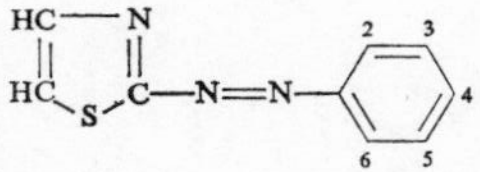

TAR (Thiazolylazoresorcinol) Mg, Ca, Mn, Co, Ni, Cu, Zn, Cd, Pb   [60-18; 60-50; 63-98; 64-45]
Ni   [64-45]
TAM   Co, Ni, Cu   [60-18; 60-50; 64-45]
TAC (Thiazolylazocresol)
TAO (Thiazolylazoorcinol)   Mg, Ca, Mn, Co, Ni, Cu, Zn, Cd, Pb   [60-18; 63-98]

(LXXXIV)   4-OH, 6-OH
(LXXXIVa)   4-OCH₃, 6-OH
(LXXXV)   4-CH₃, 6-OH
(LXXXVI)   2-CH₃, 4-OH, 6-OH

TABLE 9—*continued*

No.	Structure Substituents Name	Name	Cations	Ref.
(LXXXVII)	2-OH	TAN (Thiazolylazonaphthol)	Co, Ni, Cu, Zn, Pb	[60-18; 60-50; 63-98; 66-13]
(LXXXVIII)	2-OH, 6-SO$_3$	TAN-6-S (Thiazolylazonaphthol sulphonic acid)	Cu, Ni, Co, Zn, Pb, Lanthanides	[60-18; 60-50; 63-98]
(LXXXIX)		Eriochrome red B	Ca, Mn, Ni, Cu, Zn, Pb	[56-114; 57-2]
(LXXXIXa)	2-OH	—	Zn, Cd, Ca, Mg, Pb, Ni	[64-47]

naphthalene, the azo group in the latter case occurring in either the
α- or β-position. These are shown in the first five sections of Table 9,
which gives the structural formula of the parent dyestuff and the
orientation of the substituents. Of these forty-two indicators,
twenty-six are *o,o'*-dihydroxyazo dyes which bind a metal in the same
way as Erio T does, and three are *o*-hydroxy-*o'*-carboxyazo dyes in
which the metal is coordinated to the oxygen of the carboxyl group,
a nitrogen of the azo group, and the oxygen of the phenolic hydroxyl
group. In two cases, (XXXVIII) and (LIV), an arsonic acid group is
adjacent to the azo group; this is especially suitable for binding heavy
metal ions of high charge, such as Th^{4+}. The indicators (XXXVI) and
(XXXVII) include a methyleneaminoacetate or a methyleneiminodiacetate group respectively, which bind the metal simultaneously
with the auxochromic phenolic oxygen situated *para* to the azo
group. The remaining nine indicators are mono-*o*-hydroxyazodyes.
Of this whole series of forty-two indicators Erio SE (XLV), Erio A
(LVI), and above all Erio T (LIX) are used almost exclusively. An
excellent indicator which is coming increasingly into favour is
Calmagite (XXXIV), which has the advantage over Erio T (LIX)
that its aqueous solutions are more stable. Lindström has investigated sulphonated derivatives of this compound [66-14]. Calcon (LV)
is especially good for the titration of Ca in the presence of Mg above
pH 12. The indicator Cal-red (LVII) has been thoroughly investigated. Little practical use has been found for the bisazo dyes (LXV)
to (LXIX) and the trisazo dye (LXX).

Indicators (LXXI) to (LXXVI) are derivatives of pyridine with the
azo group adjacent to the pyridine nitrogen and to an *ortho* phenolic
OH in the second aromatic nucleus, so that on coordination to a
metal two 5-membered chelate rings are formed as shown in the
following structural formula. The coordination of two N and one O
atom introduces a high degree of selectivity in comparison to the
o,o'-dihydroxyazo dyes, in which one N atom and two O atoms are

Copper(II) complex of PAN

attached to the metal. Especially stable complexes are produced by those metals (in particular, Co, Ni, Cu, Zu, and Cd) that form ammine complexes; however, weaker adducts are formed by many other cations, e.g. Mn, Pb, and Bi. While the free dyestuffs are yellow in neutral or weakly acid solutions, the metal complexes are coloured red or violet. Both 1 : 1 and 1 : 2 complexes are formed. The copper complex is especially stable and strongly coloured.

This fact is exploited in back-titrations procedures, in which a known amount of EDTA is added and the excess titrated with a standard solution of $CuSO_4$ using PAN (LXXIII) or PAR (LXXI), as indicator. A whole series of metals can be determined in this way. Generally speaking, complexes formed by PAN are very sparingly soluble, and this leads to a long-drawn-out end-point in a direct titration, because the precipitate that is formed does not dissolve immediately when EDTA begins to appear in excess. The complexes with PAR are more soluble. The other pyridineazo dyes recorded in Table 9 are very little used.

The azoxines (LXXVII) to (LXXXIII) contain 8-hydroxyquinoline (oxine) as one component of the azo dye; this can coordinate to a metal through the quinoline nitrogen and the phenolic oxygen (cf. the sparingly soluble oxinates). The introduction of the oxine residue into an azo dye produces indicators which change from red to yellow when a metal is coordinated. The stabilities of most of these complexes are rather too low for them to be used in complexometry. The direct titration of Zn and Cu goes well, and so, too, does the back-titration of excess EDTA with Cu in the determination of a few other metals.

The thiazolylazophenols, (LXXXIV) to (LXXXVIII), have similar properties to the pyridineazo dyes and can be used like PAR and PAN. Below pH 7 the metal-free indicators are yellow, and change to red or violet on the addition of alkali when the phenol group in the *ortho* position loses its proton. The metal complexes are also red, blue, or violet. Bivalent Co, Ni, Cu, Zn, and Pb can be titrated directly within the pH range 4–7, and a number of other metals can be determined by back-titrating excess EDTA with Cu^{2+}. The sulphonated product TAN-6-S (LXXXVIII) is especially to be recommended because of the better solubility of its metal complexes.

The last member in the list of azo dyes is Eriochrome red B (LXXXIX), wherein a sulphonated β-naphthol is linked through an azo group to methylphenylpyrazolone, which, in its enolic form, has an acidic OH group adjacent to the azo group. Eriochrome red B is therefore related to the *o,o'*-dihydroxyazodyes. The indicator is yellow, and turns red on the addition of metals.

3.3.4. INDICATORS OF DIFFERENT STRUCTURE

The formazans comprise a special class of azodyes whose characteristic feature is that on complex formation not only are the protons from any acidic substituents set free but also the NH of the hydrazo-group is deprotonated whereupon the two possible limiting structures of the molecule become identical.

1,5-diphenylformazan

The following two compounds of this structure have been used as indicators in complexometry:

No.	Substituents	Name	Indicator for	Refs.
(XC)	X = phenyl; 3-SO$_3^-$, 6-OH, 2'-COOH	Zincon	Zn, Ca	[54-90; 55-2; 57-9; 58-4]
(XCI)	X = SH	Dithizone	Na, Al, Ni, Cd, Pb, Bi	[53-16; 55-98; 57-82; 58-16; 58-26; 60-27; 60-147; 61-61; 61-62]

In strongly acid solution zincon (XC) is reddish-violet, but the colour changes to yellow at pH 4·5 (H$_3$D$^-$ ⟶ H$_2$D^{2-}) and to reddish-orange at pH 8·3 (H$_2$D^{2-} ⟶ HD^{3-}); this abbreviated formula takes into account the protons from the substituents 6-OH and 2'-COOH as well as the NH proton. In complexometry only the blue zinc complex of composition ZnD^{2-} is used; in this the metal is coordinated with both the nitrogen atoms attached to the aryl groups and with an oxygen atom from each of the *ortho* substituents. Zinc can be titrated using zincon at pH values around 10. However, the brilliant colour change is used in particular for the determination of Ca^{2+} with EGTA in the presence of Mg^{2+} after the addition of a little zinc: the jump in pZn which occurs simultaneously with that in pCa is revealed by the zincon indicator.

Metal-free dithizone (XCI) is green, and its complexes are coloured yellow, red, or violet. Here the metal is bound to sulphur and to the nitrogen of the azo- or hydrazo-group. Owing to the insolubility of this indicator and its metal complexes in water, the titration must be carried out in a water–acetone or water–alcohol mixture. A two-phase system can also be used if some chloroform or carbon tetrachloride is added. Dithizone and its metal complexes dissolve in the immiscible organic phase, and it is there that the colour change is seen. The elements Ni, Zn, Cd, Pb, and Bi which form sulphides can be titrated in this way; Al is determined by back-titration with zinc solution, and Na is determined indirectly by titrating the zinc in sodium zinc uranyl acetate.

All hydroxyquinones are able to react with metals. For centuries past naturally occurring alizarin and cochineal have served to dye textile materials, and it was known that metals were involved. Even the simple hydroxybenzoquinones (XCII) and (XCIII) form metal complexes; however, the colour changes that occur when the hydroxynaphtha- and anthra-quinones form metal complexes are especially brilliant. The following compounds can serve as indicators in complexometric titrations.

Yellow rhodizonic acid forms reddish-brown precipitates with Sr^{2+} and Ba^{2+}, and their appearance can indicate the end-point when back-titrating an excess of EDTA. Chloranilic acid behaves similarly. Naphthol purple and the hydroxyanthraquinones are specially suited to the titration of the heavy and highly charged cations of the lanthanides and actinides in strongly acid solutions. Compound (XCVI) is often used; this is an alizarin dyestuff which has been sulphonated to render it water-soluble. In carminic acid (XCVIII), the colouring matter of the cochineal insect, it is the presence of a sugar residue that makes it soluble. Alizarincomplexone (XCVII) can be used for the complexometric titration of several metals in acid solution. It is especially noteworthy that its complex with cerium(III) can be used for the photometric determination of fluoride [60-194].

In all these complexes the metal is coordinated to the quinone oxygen and to the oxygen atom of the deprotonated phenolic OH situated in the peri-position. The hydroxyketones are thus pure oxygen donors, which accounts for their high affinity for the highly charged d° cations.

TABLE 10

No.	Structure, substituents	Name	Used for
(XCII)	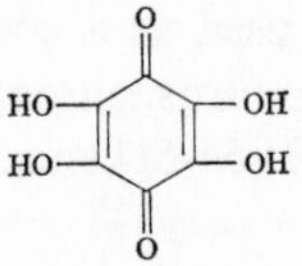	Rhodizonic acid [60-190]	Sr, Ba
(XCIII)	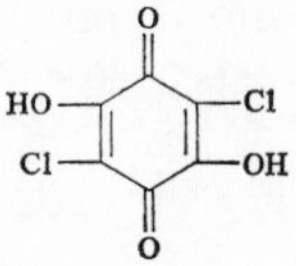	Chloranilic acid [54-45]	Zr
(XCIV)	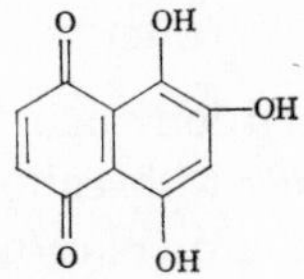	Naphthol purple [56-19]	Th
	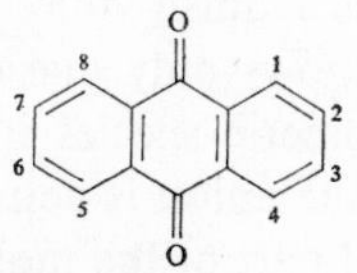	Anthraquinone	
(XCV)	1-OH, 4-OH	Quinizarin [60-106]	Th
(XCVI)	1-OH, 2-OH, 3-SO$_3^-$	Alizarin S [53-19; 66-21]	Al, Lanthanides, Th, Pu, Tl(III), Bi
(XCVa)	1-OH, 4-OH, 6-SO$_3^-$	Quinizarin sulphonic acid [64-44]	Th
(XCVII)	1-OH, 2-OH, 3-CH$_2$N(CH$_2$·COOH)$_2$	Alizarincomplexone [58-3; 60-102]	Pb, Zn, Co, In, Th, Ce
(XCVIII)	1-OH, 2-sugar, 3-OH, 5-COOH, 6-OH, 8-CH$_3$	Carminic acid [57-66; 58-79]	Lanthanides, Zr, Th

The following natural products which have also been used in complexometry are aromatic hydroxyketones too, with the structure:

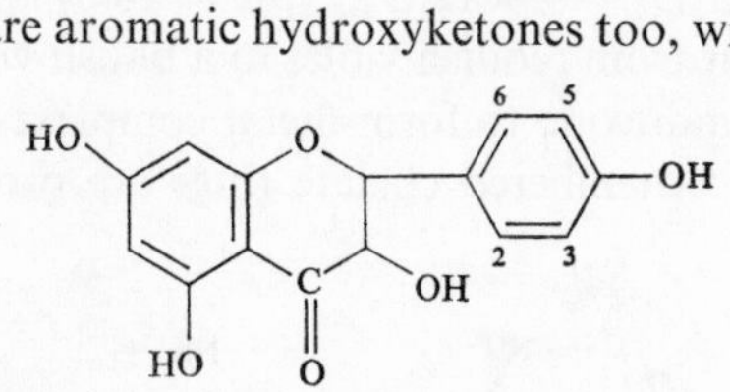

(XCIX) Quercitin [58-77] with a fourth OH-group in the 5- position. Used for the titration of U, Th, F$^-$, SO$_4^{2-}$.

(C) Morin [55-36] with a fourth OH-group in the 6- position. Used as a fluorescence indicator for Al, In, Ga.

(CI) Haematoxylin [55-8; 60-117; 61-40], the dyestuff from logwood. Used for the titration of Al, etc.

The ability of nitrosophenols to form metal complexes has been known for a long time (cf. the gravimetric determination of cobalt with α-nitroso-β-naphthol). The nitroso derivative of R-acid, whose anion has the structure (CII), has been used as a complexometric indicator for Zr [56-57; 60-115]. Mononitrosochromotropic acid (CIIa) has been used for Cu [61-19; 64-39; 64-51] and 2,7-dinitroso-chromotropic acid for Th [57-51].

(CII)   (CIIa)

Murexide is a dyestuff that was obtained long ago from animal material, but its ability to form complexes with cations has only recently been discovered [45-3]. Here we are concerned with the ammonium salt of purpuric acid, whose reddish-violet anion possesses the structure (CIII). This again shows only one of the four possible limiting structures of the mesomeric species indicated by arrows. In fact, the negative charge on the anion is actually spread over all four oxygen atoms in the central part of the molecule. Free purpuric acid, which is yellow in colour, is obtained only in the presence of strong mineral acids ($H_4D^- \longrightarrow H_5D$, pK ~ 0), and it is very unstable, in that it is almost immediately hydrolysed: in consequence, the violet solutions below pH 3 are bleached more and more rapidly as the acidity increases.

As the pH is increased, two more protons are removed from the imide groups, and since there are four such groups, the murexide anion is represented as H_4D^-. The change $H_4D^- \longrightarrow H_3D^{2-}$ (pK = 9·2) and $H_3D^{2-} \longrightarrow H_2D^{3-}$ (pK = 10·5) is accompanied by a change in colour from reddish violet to a bluish violet (cf. Fig. 19).

The ability of murexide to form metal complexes is easily understood, since two 5-membered chelate rings are produced by simul-

(CIII) Anion H_4D^- of murexide   Metal complex $MH_4D^{\nu-1}$

taneous coordination through the central nitrogen atom and two oxygen atoms (cf. the structural formula given for $MH_4D^{\nu-1}$). But it should be pointed out that the purpureate anion can even form a complex with calcium. Of course, all the transition metals and the d^{10} cations form corresponding complexes as well [49-5; 54-33; 54-35; 54-46; 55-49; 56-2; 56-16; 57-69; 57-104; 59-43; 60-82; 60-103], but

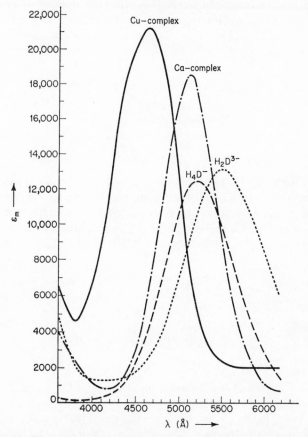

Figure 19. The absorption spectrum of murexide in neutral solution (H_4D^-) and in strongly alkaline solution (H_2D^{3-}) together with that of the calcium complex at pH 13 (CaH_2D^-) and that of the copper complex at pH 8 (CuH_2D^-). ε_m = molecular extinction coefficient; wavelength (λ) in angstrom units.

their stability constants are not very large. From studies of equilibria carried out to date it is clear that in addition to several metal-free dyestuff species, such as H_4D^-, H_3D^{2-}, H_2D^{3-}, a variety of 1 : 1

metal complexes $MH_4D^{\nu-4}$, $MH_3D^{\nu-2}$, $MH_2D^{\nu-3}$, etc, can coexist (cf. the spectra of CaH_2D^- and CuH_2D^- in Fig. 19). The range of existence of these species in the pM–pH diagram is discussed in Section 4.2.2.

The same mesomeric chromophore system found in murexide occurs also in other dyestuffs [49-5], but these have found hardly any applications in complexometry. However, the dyestuff (CIV) obtained by the ninhydrin reaction on albumin has been tried out [57-5].

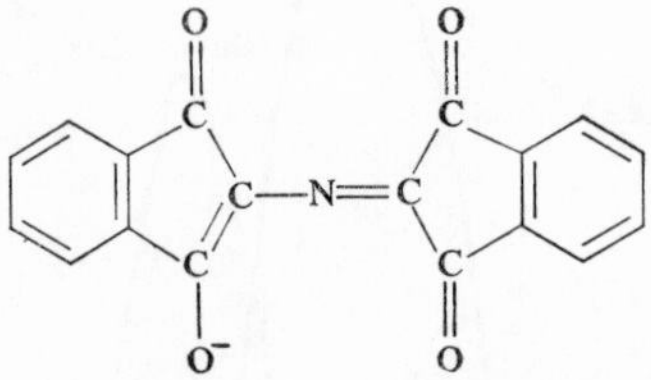

(CIV) Ninhydrin dyestuff used for Ca, Zn, Cu, Ni

Indicators (CV) and (CVI) are derivatives of oxalic acid. The compound N,N'-bis(2-hydroxyethyl)dithiooxamide can be obtained from rubeanic acid, which, as is well known, gives deeply coloured precipitates with Cu and Ni. With the disubstitution product (CV) the metal complexes can be kept in solution by adding a good deal of

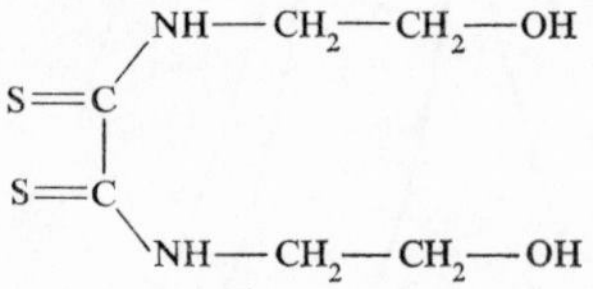

(CV)

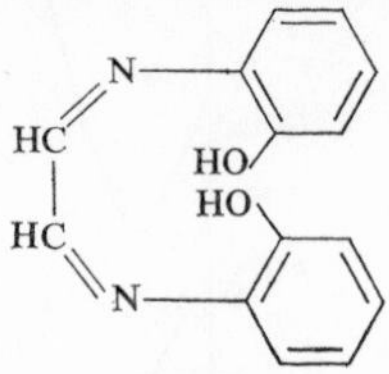

(CVI)

acetone (75%), and it is possible to titrate both Cu^{2+} and Ni^{2+} in such solutions, the end-point being indicated by a colour change from red to violet [62-4; 63-29]. Probably the metal forms a 5-membered chelate ring with the two sulphur atoms of (CV). Glyoxal-bis(2-hydroxyanil), (CVI), has been investigated by Bayer as a potential complexing agent [57-140]. Three 5-membered chelate rings are formed from the two nitrogen atoms and the oxygen atoms of the ionized phenolic groups. With this indicator Ca^{2+} can be titrated in strongly alkaline solution (approx. 0·1M NaOH), and there is then a colour change from pink to yellow [59-151; 61-142; 62-46]. Ferron (CVIIIa), the well-known absorptiometric reagent for Fe(III), has also been used for the complexometric determination of vanadium(V) [65-27].

$$O_2N-\langle\rangle-N{=}N-NH-\langle\rangle-N{=}N-\langle\rangle-SO_3^-$$

with substituent AsO_3H_2

(CIIIa) Sulpharsazen (plumbon)

The indicator sulpharsazen (plumbon) has been studied extensively [60-199; 62-159; 64-54; 65-25] and recommended for the titration of Pb, Zn, and Cd.

3.4. Fluorescence indicators

Compounds capable of fluorescing can also be furnished with several donor atoms so that they are then able to form metal complexes. The coordination of the cation is then usually accompanied by some change in fluorescent properties, e.g. the colour of the fluorescence changes or the fluorescence becomes intensified or partially or completely quenched by the presence of the metal. Such fluorescence transitions can, like the colour changes of ordinary indicators, serve to locate the end-point in complexometric titrations. Fluorescence indicators have also been called 'metalfluorechromic indicators' [59-12] or 'metalfluorescent indicators' [60-45]. They can be used to considerable advantage when coloured solutions have to be titrated. This is the situation with many transition metals whose EDTA-complexes are more strongly coloured than their aquo-complexes (e.g. the ion CoY^-), so that the determination of the end-point with an ordinary colour indicator is rendered harder by their presence. The fluorescence transition is best observed in a darkroom (dark box with a peephole) in which the titration vessel is exposed to the rays from an ultraviolet lamp (quartz lamp) [59-149]. The fluorescence transitions are often very brilliant, and with their help the end of a titration can be established very precisely. However, the observation of fluorescence phenomena is far more tiring to the eyes than watching for changes in colour.

For decades past aluminium has been identified by the yellowish-green fluorescence produced on adding morin (C). Gallium and indium show this phenomenon too, and morin has therefore been selected for indicating the end-point in the titrations of Ga [55-36] and In [53-38] with EDTA. It has also been known for a very long time that many metal oxinates fluoresce more strongly than metal-free 8-hydroxyquinoline (CVII; 'oxine') – or in a different way. By using oxine, gallium can be titrated at pH 2·5–3·5 (tartrate buffer) in the presence of NH_2OH, the end-point being detected by the sudden quenching of fluorescence [58-40]. 8-Hydroxyquinoline-5-sulphonic acid (CVIII), which produces water-soluble complexes and not precipitates with metals, has also been used as a fluorescence indicator. With zinc a strong yellow-green fluorescence appears round about pH 10 (ammoniacal buffer), and this is quenched at the end-point in the titration of this metal with EDTA [58-19]. Acidic solutions of 3-hydroxy-2-naphthoic acid (IV) fluoresce green: on adding aluminium there is a change in fluorescence colour to blue, and this has been exploited for the titration of aluminium. Fe(III) and Al can be determined successively with EDTA by using 3-hydroxy-2-naphthoic acid (IV), the iron at pH 2 to the disappearance of the blue colour of the ferric complex (cf. p. 33) and then the Al at pH 3 (glycine buffer) until the fluorescence change from blue to green [61-44].

(CVII)   Oxine
(CVIII) 5-SO_3^- ; oxine-5-sulphonic acid
(CVIIIa) 5-SO_3^-, 7-I; Ferron

The search for new fluorescence indicators was greatly accelerated by the discovery of calcein [56-72]. This condensation product of fluorescein, formaldehyde, and iminodiacetic acid has the structure (XXV) (Table 8), although the position of the grouping $-CH_2N(CH_2·COOH)_2$ has not been established beyond all doubt [59-144]. Its principal use is for the titration of calcium in a strongly alkaline solution. The point of equivalence was originally detected by the colour change from greenish-yellow to brown [56-72]. However, there is a simultaneous change in the fluorescence [56-72; 57-103]. Körbl showed that with a really pure product (called fluorexon [58-47]) this change in fluorescence is far more brilliant than the change in colour [58-54]. The fluorescence of fluorexon in a neutral, metal-free solution is practically the same as that of ordinary fluorescein, but, in contrast, its fluorescence vanishes on adding KOH when the concentration reaches about 0·025M. If an alkaline-earth metal, in particular Ca, Sr, or Ba, is now added to this strongly alkaline solution the fluorescence reappears, and it is quenched again when these cations are fully complexed by EDTA: in this way the end-point can be detected [60-6; 61-7]. When present in greater concentration sodium also produces fluorescence of appreciable intensity, so that when titrating alkaline earths the alkalinity must be achieved with KOH and not with NaOH [58-52]. Between pH 4 and 10 the fluorescence of metal-free calcein is quenched by cupric ions so that excess EDTA can be back-titrated with a standard solution of Cu^{2+}, the end-point being detected by the sudden disappearance of the fluorescence. By this means Cu, Co, and Cr(III) can be determined by a back-titration procedure [59-7; 59-12], and so can molybdenum [61-138]. The direct titration of Cu with EDTA or a polyamine can also be carried out [59-11]. Besides calcein (XXV), the condensation product of fluorescein with N-methylglycine, known as methyl-calcein (XXVI), is also a serviceable fluorescence indicator for calcium [60-65].

Experience with fluorexon has stimulated the synthesis of other fluorescing compounds with donor atoms capable of binding metals, especially compounds derived from the colourless coumarones and xanthones which fluoresce magnificently and show acid–base fluorescence transitions. An interesting metal-fluorescence indicator results from the condensation of 4-methylumbelliferone with formaldehyde and iminodiacetic acid [60-63; 60-64; 60-126]. This product, which has been named umbellicomplexone by one distributor and calcein blue by another, probably has the structure (CIX); the uncertainty concerns the position of the grouping

$$-CH_2N(CH_2 \cdot COOH)_2$$

which could also be situated between the phenolic OH and the oxygen of the hetero-ring.

(CIX) Umbellicomplexone or calcein blue, with R = $-CH_2 \cdot COOH$
(CX) Methylcalcein blue, with R = CH_3

This indicator is insoluble in water but dissolves on adding a weak base such as sodium acetate. Even at high dilution it shows an intense blue fluorescence in a pH range from 4 to 10 and maximum intensity at a wavelength of 445 mμ. The fluorescence is excited by ultraviolet light, and the best yield is produced by incident light of about 370 mμ, which practically corresponds with one of the two maxima of a mercury-vapour lamp. The fluorescence becomes weaker above pH 10 and is practically extinguished at pH 12. The cations Mn, Fe, Co, Ni, Cu, Hg, Pb, and Bi quench the fluorescence in the pH range 4–10, but the d° ions Mg^{2+}, Ca^{2+}, Sr^{2+}, and Ba^{2+} do not: Zn^{2+} and Cd^{2+} cause partial quenching. The addition of Ca^{2+}, Sr^{2+}, or Ba^{2+} to the non-fluorescing, strongly alkaline solution (pH $\sim$13) revives the emission of fluorescence at its full intensity; Mg^{2+}, Zn^{2+}, Cd^{2+}, and Pb^{2+} do this only partially, and the remaining cations have no effect at all. Procedures for detecting end-points with umbelli-complexone are based on these observations. Alkaline earths are titrated directly with EDTA in strongly alkaline solution to the point where the fluorescence is quenched. Heavy metals are usually deter-mined by a back-titration procedure in the pH range from 4 to 10 by back-titrating excess EDTA with Cu^{2+}, since the occurrence of cop-per in an uncomplexed form quenches the fluorescence completely.

Another fluorescence indicator is obtained by condensing methyl-umbelliferone with formaldehyde and N-methylglycine [60-65]. The product is called methylcalcein (CX), and its structure is shown above.

A further group of fluorescence indicators was discovered when searching for new redox indicators [60-153]: these are derivatives of benzidine of structures (CXI) to (CXVI).

(CXI) X = OCH$_3$, Y = H   (CXII) X = OC$_2$H$_5$, Y = H
(CXIII) X = COOH, Y = H   (CXIV) X = OCH$_3$, Y = SO$_3^-$
(CXV) X = OC$_2$H$_5$, Y = SO$_3^-$ (CXVI) X = OR,   Y = COOH

In the pH range from 4 to 10 *o*-dianisidinecomplexone (CXI) and *o*-phenetidinecomplexone (CXII) exhibit a blue fluorescence which is quenched by adding Cu^{2+}. Many cations can thus be determined with these indicators by using a back-titration procedure with a standard solution of a copper salt: this has been tried out with Bi, Cd, Ca, Ce, Fe(III), Fe(II), Hg, In, La, Ni, Pb, Th, Tl, Sn, and Zr. The same phenomena are shown by the compounds (CXIV) and (CXV), which have sulphonic acid groups in positions 3 and 3': they are used analytically in the same ways as compounds (CXI) and (CXII), but their fluorescence is more intense. The carboxylic acids (CXIII) and (CXVI) do not begin to fluoresce until above pH 8, and their emission is quenched by Ca^{2+}, so that calcium can be titrated directly using these indicators [60-153].

The stilbene derivatives (CXVII) and (CXVIII) behave in the same way as the benzidine derivatives [62-155]. Both compounds fluoresce blue at pH 4, and the emission is quenched by uncomplexed cupric ions, so that they can be used to indicate the end-point in back-titrations of a whole group of metals.

(CXVII) R = H   (CXVIII) R = SO₃⁻

3.5. Redox indicators
3.5.1.

If a solution containing a metal in two different oxidation states in reversible equilibrium is titrated with a complexing agent a sudden change in redox potential occurs at the end-point owing to the difference in the stabilities of the complexes of the metal in its two oxidation states. This jump in redox potential can be detected potentiometrically (cf. Section 5.2) or with a colour indicator.

The behaviour of iron has been investigated particularly thoroughly since the stability constants of the complexes of Fe^{2+} and Fe^{3+} with EDTA differ by a factor of about 10^{11}. Iron(III) can even be titrated in strongly acid solution, where iron(II) is virtually not complexed at all. The redox potential, $E = \pi + 0.059(\log [Fe^{3+}] - \log [Fe^{2+}])$, will follow the change in $[Fe^{3+}]$ in such a titration, for $[Fe^{2+}]$ remains practically constant: thus E changes linearly with pFe^{3+}. A jump in redox potential corresponding to the jump in pFe shown in Fig. 10 occurs as soon as an amount of EDTA equivalent to the trivalent iron has been run in [51-16]. The concentration of Fe^{2+} at this stage is unimportant, since it only affects the absolute value of E but not the changes in its value during the titration. The jump in redox potential occurs even when no Fe(II) is added to the solution of Fe(III) that is to be determined, for a very small, but nevertheless quite sufficient, concentration of Fe^{2+} is always present in equilibrium with Fe^{3+}.

Besides iron a number of other transition elements exist in different oxidation states. Despite the theoretical possibilities of their being used to indicate the end-point of a complexometric titration through a sudden jump in redox potential, practically no use can be made of them. The standard redox potentials, π, of the couples Ti(IV)/Ti(III), V(III)/V(II), and Cr(III)/Cr(II) are so low that solutions with a sufficient concentration of the lower oxidation state are strongly reducing, and therefore sensitive to atmospheric oxidation or even capable of evolving hydrogen. On the other hand, the system Mn(III)/Mn(II) is too strongly oxidizing, and in the case of the couple Co(III)/Co(II) complex formation in the cobaltic state is slow. However, complexometric titration with a redox indicator can be carried out with copper as well as with iron. Of the two oxidation states only Cu^{2+} forms a stable complex with EDTA, whereas the Cu^+ is practically uncomplexed, so that the redox potential $E = \pi + 0.059(\log [Cu^{2+}] - \log [Cu^+])$ follows the change in pCu shown in Fig. 8. Although the standard potential for the system Cu(II)/Cu(I) is only 0·17 volt as compared with 0·76 volt for the couple Fe(III)/Fe(II), the copper can be brought into the more favourable potential range of E by adding some thiocyanate to the solution: this complexes very strongly with univalent copper, so that $[Cu^+]$, the concentration of uncomplexed Cu(I), becomes very small, though nevertheless well defined. The same effect as produced by a small amount of SCN^- can also be achieved with larger concentrations of Cl^- [56-109]. Complex formation of vanadium(V) by EDTA can also be followed with a redox indicator [56-100; 57-15; 58-66; 63-1].

Among the many organic dyestuffs that can be reversibly reduced and reoxidized with a change in colour [56-126], only a few have proved to be suitable as redox indicators for complexometric titrations. Two derivatives of diphenylamine deserve special mention, for they can be oxidized to the following deeply coloured derivatives (CXIX) and (CXX) of indophenol and indamine:

(CXIX) Variamine blue
(CXX) Bindschedler's green

4-Methoxy-4'-aminodiphenylamine [53-36; 57-117; 60-89; 65-31], which is marketed under the name Variamine blue-B-base, has proved very suitable for the complexometric titration of iron(III). The colourless base is oxidized by uncomplexed Fe^{3+} in weakly acid solution to the deep-violet-coloured dyestuff (CXIX). The standard

$$H_2N-\!\!\!\bigcirc\!\!\!-NH-\!\!\!\bigcirc\!\!\!-OCH_3 \;+\; 2Fe^{3+} \longrightarrow$$

$$H_2\overset{+}{N}=\!\!\!\bigcirc\!\!\!=N-\!\!\!\bigcirc\!\!\!-OCH_3 \;+\; H^+ \;+\; 2Fe^{2+}$$

potential of the redox system dyestuff/leuco-base is given by $\pi = 0.60$ volt at pH 2 [53-36], which is just the right value for the indophenol to be reduced by Fe^{2+} as soon as all the Fe^{3+} has been complexed by EDTA: at the end-point of the complexometric titration the violet solution becomes colourless again. This method of determining iron works particularly well on the micro- and ultramicro-scale [54-60]. Variamine blue can also be used for the titration of copper by EDTA at pH 5–6, since Cu^{2+} is also able to oxidize the leuco-base to the indophenol, provided some SCN^- or much chloride is added: when all the Cu^{2+} has been complexed the reaction is reversed [55-17; 56-109] so that a change in colour is observed. In this way it is also possible to back-titrate an excess of EDTA with Cu^{2+}, and this procedure has been exploited for a back-titration of Al [56-106].

The dyestuff (CXX), which is marketed as its zinc double salt under the name Bindschedler's green, can be used in place of variamine blue. One remarkable feature is that the colour change is seen more easily if the commercial product is kept for some months before it is used without any steps being taken to protect it from the light. It then no longer dissolves in water with a green but with a brown colour, and on adding Fe^{3+} at pH values between 2.0 and 3.5 a deep green coloration is produced which vanishes again very abruptly when all the iron has been complexed with EDTA [56-108]. The zinc present in the indicator does not interfere, provided the titration is carried out below pH 3.5. The decomposition product of Bindschedler's green has also been used for determining excess EDTA in the back-titration of Cr(III) [56-111]. The appearance of the green colour is easily recognizable despite the intense colour of the violet chromium complex CrY^-.

The end-point in the titration of Fe(III) by EDTA can also be located with cacothelin (CXXI). This easily prepared nitration product [A-11] of unknown constitution from the alkaloid brucine gives a yellow solution which is turned amethyst by sufficiently strong reducing agents. Ferrous ions cannot effect this reduction so long as ferric ions are present, but the colour change takes place immediately all Fe^{3+} has been complexed. In this procedure it is necessary to add a ferrous salt, whose oxidation must be prevented by working in an atmosphere of CO_2 [56-27; 57-11].

It is noteworthy that vanadium(V) also forms a complex with EDTA [57-15; 58-66; 65-123], but its stability is so low that the vanadium can be displaced by most bivalent and tervalent cations. This gives rise to uncomplexed vanadium(V), which can oxidize colourless diphenyl-carbazide (CXXII) in weakly acid solution to a violet-coloured vanadium complex of diphenylcarbazone.

(CXXII) Diphenylcarbazide	Diphenylcarbazone

A number of titration procedures are based on this principle, which can be used for many cations, e.g. Mn, Pb, Zn, Cu, Ni, Ti, V, Zr, Al, Tl, Th, Cd, Co, Sn, Cr, and the rare earths [56-100]. In the direct titration some vanadium(V)–EDTA complex and diphenyl-carbazide or diphenylcarbazone is added to the solution buffered at pH 4·5–5·5, and titration is carried out with EDTA until the colour changes sharply. In the back-titration a known excess of EDTA is added together with some indicator, and the titration is carried out with vanadium(V) or also with Mn, Zn, Pb, etc, to the colour transition. When diphenylcarbazone is used the change of colour at the end-point is not due to any redox-process but to the formation of a complex of vanadium(V) with the indicator.

The end-point of a titration of EDTA with $Hg(NO_3)_2$ in acid solution can also be detected with diphenylcarbazide, for the first trace of Hg^{2+} in excess produces a violet colour. This has been used in a back-titration procedure for determining aluminium [60-150].

3.5.2.

In the presence of the redox couple $Fe(CN)_6^{3-}/Fe(CN)_6^{4-}$ there is an abrupt change in redox potential when zinc ions are introduced or removed, i.e. when pZn itself changes abruptly. This results from the fact that the ion Zn^{2+} prefers ferrocyanide to ferricyanide ion and binds it very strongly (complex formation or precipitation). If excess Zn^{2+} is present the solution is so strongly oxidizing that benzidine (CXXIII) or naphthidine (CXXIV) are oxidized to blue products. On this reaction are based procedures for the determination of a host of cations through back-titrations. The metal to be determined is treated with an excess of EDTA and ferro-ferricyanide, and benzidine [54-41; 54-42; 56-99] or naphthidine [53-34; 53-35; 55-67] is added and the pH adjusted to about 5: excess EDTA is then back-titrated with a standard solution of zinc until the blue or violet colour appears.

(CXXIII) Benzidine   (CXXIV) 3,3′-Dimethylnaphthidine

Variamine blue-B-base can be used in place of benzidine or naphthidine, and at the end-point this is oxidized to (CXIX) [59-59; 60-89].

3.6. Source, purity, and nomenclature of indicators

All the most important indicators for complexometry are obtainable from firms that sell special chemicals for analytical purposes. It is most important to realize that we are not always dealing with chemically homogeneous, pure compounds. With dyestuffs, and more particularly with their salts, even repeated recrystallizations and precipitations generally fail to lead to the pure compounds desired, but rather give mixtures of isomers whose components can only be separated by chromatography or by electrophoresis. Many of the dihydroxyazodyes that are marketed, e.g. Erio T, are not chemically homogeneous substances. Furthermore, the commercial indicators which contain the iminoacetate group for binding metals (e.g. (XV) to (XIX), (XXV), (CIX), and (CX)) are as a rule inhomogeneous, since the reaction mixture resulting from the condensation of the initial dyestuff with formaldehyde and iminodiacetic acid can only be resolved into its components by laborious chromatography. However, despite their inhomogeneity, the products listed as metallochromic indicators can be used for titrations without further purification – although it must be admitted that sharper colour and fluorescence transitions are obtainable with pure indicators. On the other hand, when conducting fundamental studies of equilibria designed to give information about the composition and stability of metal-indicator complexes commercial indicators should under no circumstances be used if one wishes to avoid the danger of getting erroneous results. Most of the azo dyestuffs listed in Table 9 which are not commercially available as indicators are technical mordant dyes, i.e. dyestuffs in which the colour is fixed on the fabric with the aid of metals (metallic mordants). Such products can be bought from firms that market the corresponding dyestuffs for the textile industry. But especial care must be taken in such cases in view of what has been said above about the inhomogeneity of dyestuffs in general, for this applies with particular force to commercial products. The metallizable dyes are often taken up directly by the fibre as pre-formed metal complexes, and the commercial products may be marketed in the form of such metal complexes without being labelled as such specifically. It will be obvious that dyestuffs that are already metallized cannot be used as indicators in analytical practice without further treatment.

Among the many indicators that have been proposed are some which are not yet obtainable commercially. Such compounds must be prepared *per se*, and the appropriate literature reference usually gives the procedure for their synthesis. But not all these working instructions are found to lead to a homogeneous product, and in some cases, too, justifiable doubts can be raised as to the correctness of the structural formulae ascribed by the author to the newly synthesized dyestuff (e.g. (LXX), calcichrome).

The nomenclature of indicators is in a confused state. The systematic names for organic compounds based on the established rules of nomenclature are too clumsy to use on account of their great length. Therefore, abbreviations of the scientific name are often used, e.g. HHSNN for 2-hydroxy-1-(2-hydroxy-4-sulphonic-1-naphthylazo)-3-naphthoic acid. Alternatively, trivial names are given which often refer to the metal for whose determination the particular indicator is used, e.g. calcein, magneson, thorin, zincon. With dyestuffs that are also marketed for the textile industry the description used by the firm which manufactures it has often gained general acceptance, e.g. Eriochrome black T, a mordant dye from the firm of I. R. Geigy and Co., which often uses the prefix 'Erio' for their products: an abbreviation of the Trade Name is also used, e.g. Erio T. Unfortunately it often happens that more than one trivial name is extant for one and the same substance, for each manufacturer selects his own particular

TABLE 11
Alphabetical list of indicators
The Roman numerals refer to the structural formulae shown in
Tables 6–10 and in the text in Sections 3.2–3.5

Acid alizarin black R (XXXII)
Acid alizarin black SE (LXV)
Acid alizarin black SN (LXVI)
Acid alizarin red B (XXX)
Acid alizarin violet (XXXIII)
Acid chrome black SNA (LXVI
Acid chrome blue K (XLIV)
Acid chrome blue 2 R (XLIX)
Acid chrome blue RRA (XLVII)
Acid chrome blue black (LV)
Acid chrome dark blue (LII; XLII)
Acid chrome red B (XXX)
Acid chrome violet B (XLVIII)
Alizarin S (XCVI)
Alizarin chrome black R (XXXII)
Alizarin fluorine blue (XCVII)
Alizarincomplexone (XCVII)
Alizarin-methyliminodiacetic acid
  (XCVII)
Alizarin red S (XCVI)
Alizarinsulphonic acid (XCVI)
Aluminon (XII)
Ammonium purpureate (CIII)
Anisidinecomplexone (CXI)
Anisidinecomplexone-3,3′-dicarboxylic
  acid (CXVI)
Anisidinecomplexone-3-3′-disulphonic
  acid (CXIV)
Anthranol fast black SE (LXV)
APANS (XXXVIII)
Arsenazo (LIV)
Arsenazo I (LIV)
Aurinetricarboxylic acid (XII)
Azin-bis-indanedione (CIV)
Azindibarbituric acid (CIII)

Benzaurin (IX)
Benzhydroxamic acid (VI)
Benzhydroxylamide (VI)
Benzidine (CXXIII)
Benzidinecomplexone-2,2′-dicarboxylic
  acid (CXIII)
Benzo blue FBL (LXVIII)
Beryllon II (LXIII)
Bindschedler's green (CXX)
BT (LIX)
Black T (LIX)
Brilliant Congo blue (LXVIII)
Brilliant Congo blue BFL (LXVIII)
Bromopyrogallol red (XXIII)

Cacothelin (CXXI)
Calcein (XXV)
Calcein blue (CIX)
Calcein W (XXV)

Calcichrome (LXX)
Calcion (LXX)
Calcon (LV)
Calcochrome black PV (LIII)
Calmagite (XXXIV)
Calconcarboxylic acid (LVII)
Cal-red (LVII)
Carminic acid (CXVIII)
Catechol violet (X)
Chloranilic acid (XCIII)
Chromeazurol S (XIII)
Chrome black T (LIX)
Chrome black 2 R (XLVII)
Chrome blue black B (LVIII)
Chrome Bordeaux B (XXVIII)
Chrome brown BR (XXIX)
Chrome fast black A (LVI)
Chrome fast black PV (LVIII)
Chrome fast blue 2 B (XLV)
Chrome fast blue FB (LI)
Chrome fast blue RLL (XLIX)
Chrome fast navy blue (XLIV)
Chrome fast red B (LXXXIX)
Chrome fast violet (XXXIII)
Chrome red B (LXXXIX)
Chromogene black EAG (or EA) (LVI)
Chromogene black (LVIX)
Chromotrope 8 B (LXI)
Chromotrope 2 C (XLI)
Chromotrope 2 R (XXXIX)
Chromotrope red 2 R (XXXIX)
Chromoxane pure blue BLD (XIIIa)
Chromoxane green GG (XIV)
Chromoxane violet 5 B (XIIIb)
Corinth CA (XLV)
o-Cresolphthalein-bis-methyliminodi-
  acetic acid (XVIII)
o-Cresolphthaleincomplexone (XVIII)
o-Cresolphthalexone (XVIII)
o-Cresolsulphonphthalein-bis-
  methyliminodiacetic acid (XV)
o-Cresolsulphonphthalein-
  monomethyliminodiacetic acid (XVI)

Diamond alizarin black SN (LXVI)
Diamond chrome blue 8 RL (L)
Diamond fast blue FB (LI)
Dibromopyrogallolsulphonphthalein
  (XXIII)
Dichlorodihydroxybenzoquinone (XCIII)
1,4-Dihydroxyanthraquinone (XCV)
1,2-Dihydroxybenzene-4,6-disulphonic
  acid (V)
3,3′-Dihydroxyfluorescein (XXIV)
Dimethylnaphthidine (CXXIV)

Table 11—*continued*

Diphenylcarbazide (CXII)
Diphenylcarbazone (CXII)
Diphenylthiocarbazone (XCI)
Dithizone (XCI)
DSNADNS (LXIII)
Dye of Patton and Reeder (LVII)

EBT (LIX)
Erio A (LVI)
Erio B (LVIII)
Erio R (LV)
Erio SE (XLV)
Erio T (LIX)
Eriochromal blue 2 RL (XLIX)
Eriochromeazurol S (XIII)
Eriochrome blue SE (XLV)
Eriochrome blue black B (LVIII)
Eriochrome blue black R (LV)
Eriochrome cyanine R (XI)
Eriochrome black T (LIX)
Eriochrome blue black RSS (LV)
Eriochrome navy blue BRL (LII)
Eriochrome green H (XLIII)
Eriochrome red B (LXXXIX)
Eriochrome red PE (XXX)
Eriochrome black A (LVI)
Eriochrome black PV (LVIII)
Eriochrome black SR (XXXII)
Eriochrome violet (XLVII)
Eriochrome violet BA (XXXIII)

Fast mordant blue (XLII)
Fast navy blue 2 R (XLVII)
Fast sulphon black F (LXVII)
Ferron (CVIIIa)
Fluorescein-bis-methylalanine (XXVI)
Fluorescein-bis-methylimino-diacetic
    acid (XXV)
Fluoresceincomplexone (XXV)
Fluorexone (XXV)

Gallein (XXIV)
Gallion (XLVI)
Gallocyanine (XXIV)
Glycinenaphthol violet GNN (XXXVI)
Glycinethymol blue (XX)
Glyoxal-bis-(2-hydroxyanil) (CVI)

H-Indicator (LXX)
Haematoxylin (CI)
HHSNN (LVII)
Hydron I (LI)
2-Hydroxybenzoic acid (II)
8-Hydroxyquinoline (CVII)
8-Hydroxyquinoline-5-sulphonic acid
    (CVIII)
Hydroxyethyldithio-oxamide (CV)
3-Hydroxy-2-naphthoic acid (IV)
2-Hydroxy-5-sulphobenzoic acid (III)

Iodide ion (VII)

Lacquer scarlet C (XXX)
Logwood dye (CI)
Lumogallion (XXVII)

Magneson (XXXI)
Metachrome brilliant blue RL (L)
Mandelicazochromotropic acid (XLIIa)
Metallochrome violet A (XXa)
Metalphthalein (XVIII)
4-(2-N-Methylanabasinoazo)-resorcinol
    (LXXIa)
Methyl calcein (XXVIII)
Methyl calcein blue (CX)
Methylphthalexone (XVIII)
5-Methyl-4-(2-pyridylazo)-resorcinol
    (LXXII)
Methylthymol blue (XVII)
Metomega chrome blue 2 RL (XLIX)
Metomega chrome cyanine BLL (LII)
Metomega chrome navy BLS (XXXI)
Morin (C)
Murexide (CIII)

Naphtharson (XXXVIII)
Naphthidine (CXXIV)
Naphthol purple (CXIV)
Naphthol violet (XXXVII)
α-Naphthylazoxine (LXXVII)
β-Naphthylazoxine S (LXXXI)
β-Naphthylazoxine 2 S (LXXXII)
α-Naphthylazoxinesulphonic acid
    (LXXIX)
NAS (LXXX)
Neothorin (LIV)
Niagara brilliant blue BFL (LXVIII)
Ninhydrin dyestuff (CIV)
Nitroso-chromotrope 8 B (LXIV)
Nitroso-R-salt (CII)
Nitroso-SNADNS (LXIV)
N,N′-bis-(2-hydroxyethyl)-dithiooxamide
    (CV)
N,N′-Bis-hydroxyethylrubeanic acid
    (CV)

Omega chrome black P (LVI)
Omega chrome black T (LIX)
Omega chrome blue black B (LVIII)
Omega chrome blue GFS (XLVII)
Omega chrome dark violet D (XXXIII)
Omega chrome fast blue B (XLV)
Omega chrome fast blue 2 G (LI)
Omega chrome garnet (XXVII)
Omega chrome green F (XLIII)
Omega chrome red B (LXXXIX)
Omega chrome violet B (XLVIII)
Oxine (CVII)
Oxine sulphonic acid (CVIII)

PACA (LXXV)
PADNS (LXXIV)
PAHA (LXXVI)
Palatine chrome black S (LXVI)
Palatine chrome green (XLIII)
Palatine chrome violet (XXXIII)
PAN (LXXIII)
PAO (LXXII)
PAR (LXXI)
Patton and Reeder's dye (LVII)
Phenazo black SA (LXVII)
Phenetidinecomplexone (CXII)
Phenetidinecomplexone-3,3'-disulphonic
  acid (CXV)
Phenolsulphonphthaleine-3,3'-di-
  carboxylic acid (XI)
Phthalein complexone (XVIII)
Phthalein purple (XVIII)
Plasmocorinth B (XLV)
Plumbon (CIIIa)
Ponceau 3 R (LX)
Potting black PV (LIII)
Purpuric acid (CIII)
Pyridylazo-dihydroxynaphthalene
  sulphonic acid (LXXIV)
Pyridylazoorcinol (LXXII)
Pyridylazoxine (LXXXIII)
Pyrocatechol disulphonic acid (V)
Pyrocatecholsulphonphthalein (X)
Pyrocatechol violet (X)
Pyrogallolphthalein (XXIV)
Pyrogallol red (XXII)
Pyrogallolsulphonphthalein (XXII)
Pyridylazo-H-acid (LXXVI)
1-(2-Pyridylazo)-2-naphthol (LXXIII)
Pyridylazo-R-acid (LXXV)
4-(2-Pyridylazo)-resorcinol (LXXI)

Quercitin (CXIX)
Quinizarin (XCV)
Quinizarin sulphonic acid (XCVa)

Rhodizonic acid (XCII)

Salicylic acid (II)
Semixylenol orange (XVI)
Solochromate black RN (XXXII)
Solochromate fast blue B (LII)
Solochromate fast grey RA (XXXII)
Solochromate fast violet B (L)
Solochrome black AS (or A) (LVI)
Solochrome black 6 BN (LVIII)
Solochrome black WDFA (or T) (LIX)
Solochrome violet R (XXXIII)

SNADNS (LXI)
SNADNS-4 (LXI)
SNADNS-5 (LXII)
SNAZOXS (LXXX)
SPADNS (XL)
Sulphanilic acid azo-chromotropic acid
  (XL)
Sulpharsazen (CIIIa)
Sulpho-α-naphthylazo-hydroxyquinoline
  sulphonic acid (LXXX)
Sulpho-β-naphthylazo-hydroxyquinoline
  sulphonic acid (LXXXI)
Sulphonaphthylazoresorcinol (XXXV)
Sulphosalicylic acid (III)
Stilbazo (LXIX)
Stilbazo R (LXIX)

TAC (LXXXV)
TAN (LXXXVII)
TAM (LXXXIVa)
TAN-6-S (LXXXVIII)
TAO (LXXXVI)
TAR (LXXXIV)
Tetrahydroxybenzoquinone (XCII)
Thiazolylazo-β-naphthol (LXXXVII)
Thiazolylazocresol (LXXXV)
Thiazolylazoresorcinol (LXXXVI)
Thiazolylazo-β-naphthol sulphonic acid
  (LXXXVIII)
Thiazolylazo-orcinol (LXXXVI)
Thiocyanate ion (I)
Thiourea (VIII)
Thiosalicylic acid (III)
Thorin (XXXVIII)
Thoron (XXXVIII)
Thoronol (XXXVIII)
Thymolphthalein-bis-methylimino-
  diacetic acid (XVII)
Thymolphthalexone (XIX)
Thymolsulphonphthalein-bis-methyl-
  iminodiacetic acid (XVII)
Tiron (V)

Umbellicomplexone (CIX)
Umbelliferone-methyliminodiacetic
  acid (CIX)

Variamine blue-B-base (CXIX)
Variamine blue (CXIX)

Wool black N (XXXII)

Xylenol orange (XV)

Zincon (XC)

description. Thus (XXV) is variously described as calcein, fluoresceincomplexone, and fluorexone. To make it easier for the reader to discover the structure of any of the indicators referred to in the text, all the names have been collected together and arranged alphabetically in Table 11. Each name is followed by the Roman numeral by which it has been described in Sections 3.2–3.5 of this chapter.

CHAPTER FOUR

Colour Changes in Metallochromic Indicators

4.1. pH and pM indicators

Metallochromic indicators respond to the pM value of a solution in just the same way that acid–base indicators respond to the pH. This is expressed by the following equations:

Metallochromic indicators:

$$\text{pM} = \log K_{\text{MD}}^{\text{eff.}} - \log ([\text{MD}]'/[\text{D}]') \tag{39}$$

pH indicators:

$$\text{pH} = pK_{\text{HIn}} - \log ([\text{HIn}]/[\text{In}]) \tag{40}$$

Here $[\text{D}]'$ and $[\text{MD}]'$ signify the concentrations of the metal-free dyestuff and of its metal complex – which will be of different colours. These are analogous to $[\text{In}]$ for the pH indicator and $[\text{HIn}]$, its differently coloured conjugated acid. The close analogy between expressions (39) and (40), which describe the colour changes of pM and pH indicators respectively, is due to the fact that both when the metal ion and the proton form compounds, practically only 1 : 1 complexes need to be taken into consideration. Equations (39) and (40) result from the application of the law of mass action to reactions (41) and (42):

$$\text{M}^{\nu+} + \text{D}'^{\lambda} \rightleftharpoons \text{MD}'^{\nu-\lambda} \tag{41}$$

$$\text{H}^+ + \text{In}^{\lambda-} \rightleftharpoons \text{HIn}^{1-\lambda} \tag{42}$$

It is often possible for the 1 : 1 dyestuff complex to take up a second or even a third molecule of dye ($\text{MD} \longrightarrow \text{MD}_2 \longrightarrow \text{MD}_3$), although this cannot, of course, occur with the proton complex. The formation constant of the 1 : 1 complex is, however, almost always very much larger than that of the higher complexes, so that at the low concentrations of dyestuff ($\sim 10^{-5}\text{M}$) and uncomplexed metal ($< 10^{-5}\text{M}$) present in solution at the equivalence point of a complexometric titration, only the 1 : 1 complex need be considered. Bimetallic dyestuff complexes, M_2D, do exist, just as, by analogy, the basic form of a pH indicator may take up two protons (In $\longrightarrow$ HIn $\longrightarrow$ H_2In). However, the attachment of the two metal ions also takes place in two completely discrete stages. What is invariably observed during a colour change is therefore the transformation of the 1 : 1 complex into the metal-free dyestuff, and vice versa, so that the relevant process is (41) and eqn. (39) applies. For the hypothetical case where the bimetallic complex, M_2D, is changed to the metal-free form, D, without going through the intermediate stage of MD the only change that need be made to eqn. (39) is to multiply the term which represents the relative amounts of the two coloured forms (which will now read log ($[M_2D]'/[D]'$) by the factor 1/2.

4.2. The dependence of indicator constants upon the pH
4.2.1.
Metallochromic indicators are simultaneously pH indicators. Thus in the absence of polyvalent cations the well-known Erio T is red in acid solutions, but changes to blue at pH 6·3 and to orange at pH 11·5. The red species, whose structure is shown in Section 3.3.3 (formula (LIX)), can thus lose two protons: for this reason it is represented by the abbreviation H_2D^- (since the ionized sulphonic acid group is a monovalent anion), and it is transformed into HD^{2-} (blue) and D^{3-} (orange) when the pH is raised. Analogous behaviour is shown in practice by all the metal indicators. The species designated as D' in eqns. (39) and (41) generally occurs as a mixture of several protonated states:

$$[D]' = [D] + [HD] + [H_2D] = \ldots \tag{43}$$

The result of this is that the change in colour governed by reaction (41) varies with the pH of the solution, even when only one and the same indicator is involved. For example, Erio T can give a red magnesium complex MgD^-, whose formation is scarcely detectable below pH 6, for there H_2D^- (which is also red) changes to red MgD^-: however, between pH 7 and 11 there is a colour change from blue to red (HD^{2-} to MD^-) and above pH 12 from orange to red (D^{3-} to MD^-). Since in reaction (41) different numbers of hydrogen ions are set free, depending upon the pH, its equilibrium constant (which is the effective stability constant of the complex MD) will likewise be pH-dependent. This is shown by combining eqn. (44) with eqns. (46) and (47):

$$K_{MD}^{eff.} = \frac{[MD]}{[M][D]'} = K_{MD}/\alpha_{D(H)} \tag{44}$$

where K_{MD} is the pH-independent equilibrium constant of reaction (45)

$$M^{\nu+} + D^{\lambda-} \rightleftharpoons MD^{\nu-\lambda} \tag{45}$$

and $\alpha_{D(H)}$ is the distribution coefficient for the dyestuff with respect to its various protonated forms:

$$\alpha_{D(H)} = ([D]'/[D]) = 1 + [H] \cdot K_{HD} + [H]^2 \cdot K_{HD} \cdot K_{H_2D} + \dots \tag{46}$$

The logarithms of the constants K_{HD}, K_{H_2D}, etc, are the pK values of the particular proton donors indicated by the subscript numeral (with Erio T, p$K_1 = 11 \cdot 5$ and p$K_2 = 6 \cdot 3$):

$$pK_1 = \log K_{HD}; \; pK_2 = \log K_{H_2D}, \text{etc} \tag{47}$$

The formation of dyestuff complexes can thus be described just as precisely as that of the colourless complexonates discussed in Chapter Two, eqns. (17), (18), and (19). The logarithm of K_{MD}^{eff} from eqn. (44) is identical with the effective pK value of the metallochromic indicator used in eqn. (39), and in consequence it is pH-dependent.

The various relationships are illustrated in Fig. 20 for colour changes with Erio T and magnesium. Along the ordinate are plotted the values of pMg at which the indicator is at the mid-point of its colour transition (i.e. $[MD]' = [D]'$); according to eqn. (39), these are identical with the values of $\log K_{MD}^{eff}$.

On adding Mg to a solution of the indicator buffered at some pre-determined pH, pMg decreases, and we pass from a region where the metal-free indicator exists (either as red H_2D^-, blue HD^{2-}, or orange D^{3-}) into the region where the metal complex MgD^- (red) exists, and this produces the change in colour. The transition interval where the indicator shows a mixed colour is left white in Fig. 20, and it is through this region that the curve giving pM values for which $[MD]' = [D]'$ passes. When the product $[Mg][OH]^2$, the solubility product of the metal hydroxide, exceeds 10^{-10} solid $Mg(OH)_2$ must precipitate, and this region is also shown in Fig. 20.

4.2.2.

In the majority of cases, as with Erio T, only a single 1 : 1 metal–dyestuff complex is formed, and eqn. (44) correctly describes the dependence of the transition constant of eqn. (39) upon the pH. But it so happens that not only the metal-free dyestuff but also its 1 : 1 metal complex is a pH indicator and occurs in different protonated forms, depending upon the pH. The first known example of this [49-4] is provided by murexide (whose structural formula (CIII) is given on p. 60), where both the metal-free purpureate ion H_4D^- and also its metal complex MH_4D^{v-1} can lose some of the protons of the imido groups in sufficiently alkaline solution. In eqn. (39) not only the concentration term $[D]'$ stands for the sum of various individual concentration term (cf. eqn. 43) but the term $[MD]'$ is composite as well:

$$[MD]' = [MD] + [MHD] + [MH_2D] + \ldots \qquad (48)$$

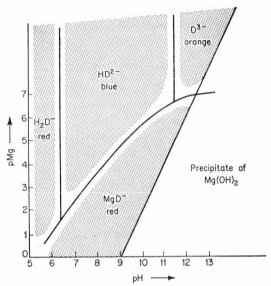

Figure 20. Erio T and Magnesium.

With the aid of the pK values of the various proton donors $MHD^{v+1-\lambda}$, $MH_2D^{v+2-\lambda}$, etc, it is again possible to calculate an α-value which takes into account the distribution of the complex MD' among the various species. Thus:

$$[MD]' = \alpha_{MD(H)} \cdot [MD] \qquad (49)$$

The equilibrium constant of reaction (41) thus takes the form:

$$K_{MD}^{\text{eff.}} = \frac{[MD]'}{[M] \cdot [D]'} = K_{MD} \cdot \frac{\alpha_{MD(H)}}{\alpha_{D(H)}} \qquad (50)$$

The colour change of murexide when used as an indicator for calcium provides a good example; this is illustrated by Fig. 21, where the metal-free forms H_4D^- (reddish-violet), H_3D^{2-} (violet), and H_2D^{3-} (blue-violet) are shown in equilibrium with the complexes CaH_4D^+ (orange-yellow), CaH_3D (reddish-orange), and CaH_2D^- (red).

A somewhat more complicated situation is provided by metal-phthalein (XVIII), and Fig. 22 depicts the colour changes with calcium in the pH range 6–14. Five metal-free species, viz. H_4D^{2-}, H_3D^{3-} (colourless), H_2D^{4-} (very pale pink), HD^{5-} (pink), and D^{6-} (deep red) are in equilibrium with the 1 : 1 complexes CaD^{4-} (deep red) and $CaHD^{3-}$ (pink). At high concentrations of calcium (low pCa) the bimetallic species, Ca_2D^{2-}, is formed; this also has a deep red colour. The structural formulae of these species is described in

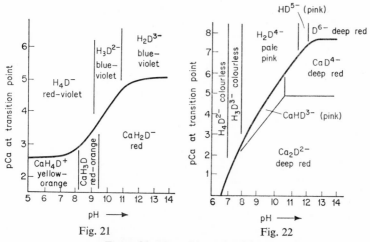

Figure 21. Murexide and calcium.
Figure 22. Metalphthalein and calcium.

Section 3.3.2. From Fig. 22 we see that only the restricted pH range from 10·6 to 12 is available for a really striking colour change from pale pink to deep red when pCa is reduced by adding calcium. Above pH 12 there is practically no detectable change, for the deep red complex CaD^{4-} is formed from D^{6-}, which also possesses a deep red colour (cf. Fig. 17). Below pH 10·6 we obtain from the pale pink H_2D^{4-} the species $CaHD^{3-}$, which is only a little pinker, and the deep red species Ca_2D^{2-} does not occur before a concentration of $[Ca] = 10^{-4}$. However, between pH 10·5 and 11·5 metalphthalein is a very good indicator for Ca^{2+} (and for Sr^{2+} and Ba^{2+} too), especially if the pink colour of the metal-free species HD^{5-} and H_2D^{4-} is masked by adding a green dye as an inner-filter (cf. Section 4.5).

4.3. Indicator constants

Only a few of the indicators described in Chapter Three have been studied sufficiently thoroughly to permit of the construction of diagrams of colour transitions of the type depicted in Figs. 20, 21, and 22. These demand a knowledge of the pK values of the metal-free dyestuff, the stability constants of its metal complexes, and, in many cases, their behaviour as proton acceptors or donors. The data so far available are summarized in Table 12. A few more pK values for metal indicators which have been measured experimentally by Ringbom appear in his recently published book [63-64].

In the first column of Table 12 the name of the indicator is ac-

TABLE 12

pK values for some indicators and the stability constants of their metal complexes ($\mu \sim 0.1$M)

Indicator	pK values for the metal-free dyestuff	Logarithm of the stability constants of the metal complexes	Ref.
(X) Pyrocatechol violet: H_4D	HD^{3-} 11·7; H_2D^{2-} 9·8 H_3D^- 7·8; H_4D 0·2	BiD^- 27·1; Bi_2D^{2+} 5·2 ThD 23·4; Th_2D^{4+} 4·4; GaD^- 22·2; Ga_2D^{2+} 4·6	[59-150] [56-37; 58-118]
(XIII) Eriochrome-azurol S: H_4D	HD^{3-} 11·5; H_2D^{2-} 4·9; H_3D^- 2·5		[61-185]
(XV) Xylenol orange: H_6D	HD^{5-} 12·3; H_2D^{4-} 10·5; H_3D^{3-} 6·4; H_4D^{2-} 3·2; H_5D^- 2·6	A few K_{MD}^{eff} values have been determined experimentally [63-64]	[60-51]
(XVII) Methylthymol blue: H_6D	HD^{5-} 13·4; H_2D^{4-} 11·2; H_3D^{3-} 7·2; H_4D^2 4·5	$MnHD^{3-}$ 9·2. A few values of K_{MD}^{eff} have been obtained experimentally [63-64]	[58-48]
(XVIII) Metal-phthalein: H_6D	HD^{5-} 12·0; H_2D^{4-} 11·4; H_3D^{3-} 7·8; H_4D^{2-} 7·0; H_5D^- <3; H_6D <2	MgD^{4-} 8·9; Mg_2D^{2-} 5·2; $MgHD^{3-}$ 7·5; CaD^{4-} 7·8; Ca_2D^{2-} 5·0; $CaHD^{3-}$ 6·4; BaD^{4-} 6·2; Ba_2D^{2-} 3·0; $BaHD^{3-}$ 4·8	[53-1]
(XXXIII) Palatine chrome violet: H_2D^-	HD^{2-} 13·0; H_2D^- 7·0	MgD^- 7·6; CaD^- 5·6; NiD^- 14·9; CuD^- 20·8; ZnD^- 12·5	[61-186; 62-157]
(XXXIV) Calmagite: H_2D^-	HD^{2-} 12·4; H_2D^- 8·1	MgD^- 8·1; CaD^- 6·1	[60-134]
(XLV) Erio SE: H_3D^{2-}	HD^{4-} 11·9; H_2D^{3-} 10·5; H_3D^{2-} 8·0	CaD^{3-} 4·3 at $\mu = 1$M; 5·0 at $\mu = 0.1$M	[61-120]
(LV) Calcon: H_2D^-	HD^{2-} 13·5; H_2D^- 7·3	MgD^- 7·6; CaD^- 5·3; ZnD^- 12·5 $$\log \frac{[ZnD(NH_3)]}{[Zn][D][NH_3]} = 16\cdot4$$	[48-4; 57-123; 59-114]
(LVI) Erio A: H_2D^-	HD^{2-} 13·0; H_2D^- 6·2	MgD^- 7·2; CaD^- 5·3	[48-4]
(LVIII) Erio B: H_2D^-	HD^{2-} 12·5; H_2D^- 6·2	MgD^- 7·4; CaD^- 5·7	[48-4]
(LIX) Erio T: H_2D^-	HD^{2-} 11·6; H_2D^- 6·3	MgD^- 7·0; CaD^- 5·4; BaD^- 3·0; MnD^- 9·6; ZnD^- 12·9	[48-4; 63-64]

TABLE 12—*continued*

Indicator	pK values for the metal-free dyestuff	Logarithm of the stability constants of the metal complexes	Ref.
(LXXI) PAR: H_3D^+	HD^- 12·4; H_2D 6·9; H_3D^+ 2·3 (valid for 50% dioxan)	$NiHD^+$ 13·2; $CoHD^+$ ~12; $Ni(HD)_2$ 12·8; $Mn(HD)_2$ 9·2; $Zn(HD)_2$ 11·1 (valid for 50% dioxan). CuD^+ 15·58, CuD_2 8·4 in 2–10% dioxan $\mu = 0·1$, 15°C	[62–11; 64–36]
(LXXIII) PAN: H_2D^+	HD 12·3; H_2D^+ ~2 (in 50% dioxan) HD 12·2; H_2D^+ 1·9 (in 20% dioxan)	NiD^+ 12·7; CoD^+ ~12; ZnD^+ 11·2; MnD^+ 8·5; NiD_2 12·6; ZnD_2 10·5; MnD_2 7·9 (in 50% dioxan); CuD^+ ~16 (in 20% dioxan); CuD^+ 17·19 at pH 6·5; CuD_2 8·4 at pH 10 in 2–10% dioxan $\mu = 0·1$, 15°C	[59–82; 62–11]
(XC) Zincon:* H_3D^-	HD^{3-} > 14; H_2D^{2-} 8·3; H_3D^- 4·5	$\dfrac{[ZnD].[H]}{[Zn].[HD]} = 0·1$	[58-4]
(CIII) Murexide: H_5D	H_3D^{2-} 10·9; H_4D^- 9·2; H_5D ~0	CaH_4D^+ 2·6; ZnH_4D^+ 3·1; CaH_3D 3·6; CdH_4D^+ 4·2; CaH_2D^- 5·0; CuH_2D^- ~15	[49-4]

* Zincon carries only two acid groups on the diphenylformazan residue, viz. a phenolic hydroxyl (pK 8·3) and a carboxyl group (pK 4·5). However, when complex formation takes place with Zn^{2+} the proton attached to the nitrogen of the hydrazine group is also replaced so that the indicator is represented by H_3D^-, the negative charge arising from the ionized 3-sulphonic acid group. However, the pK of H_3D^- cannot be measured, as it is certainly greater than 14, and therefore the stability of the complex ZnD^{2-} has to be reported in terms of the equilibrium constant for the exchange reaction

$$HD^{3-} + Zn^{2+} \rightleftharpoons ZnD^{2-} + H^+$$

companied by the Roman numeral by which it was designated in Chapter Three. The value of j in the general formula $H_jD^{\lambda-}$ shows the number of protons which the dyestuff takes up in strongly acid solution (pH = 0), i.e. the maximum number of acidic protons that need to be considered within the pH range from 0 to 14. In view of the almost ubiquitous presence of the sulphonic acid group which is introduced to solubilize the dyestuff, this is generally an anion, even in strongly acid solution, so that the charge, λ, is negative.

The second column of Table 12 gives the pK values of the metal-free dyestuff, so far as these are known. The number to the right of the formula H_jD thus denotes:

$$pK \text{ of } H_jD^{\lambda-} = \log([H_jD]/[H][H_{j-1}D]) = pK_j$$

To calculate the distribution coefficients $\alpha_{D(H)}$ it is necessary to use values of $K_{H_jD} = 10^{pK_j}$ in eqn. (46).

In the third column of Table 12 are the stability constants of the various metal complexes. The number that follows the formula is the decadic logarithm of the formation constant of the relevant complex.

For the 1 : 1 complex MH_jD ($j = 0$–4),

$$\log K_{MH_jD} = \log ([MH_jD]/[M][H_jD])$$

For the 1 : 2 complex $M(H_jD)_2$ ($j = 0$–1),

$$\log K_{M(H_jD)_2} = \log ([M(H_jD)_2]/[M(H_jD)][H_jD])$$

and for the 2 : 1 complex M_2D,

$$\log K_{M_2D} = \log ([M_2D]/[M][MD])$$

The acidity constants of the protonated complexes which we need to calculate $\alpha_{MD(H)}$ by means of eqn. (49) do not have to be tabulated, since they can be calculated from the figures given in Table 12. For example, the pK of MHD is given by

$$\log ([MHD]/[H][MD]) = \log K_{MHD} - \log K_{MD} + pK_{HD} \qquad (51)$$

4.4. Transition curves

It is instructive to express the colour transition as a percentage, i.e. to calculate for different values of pM the percentage of the dyestuff that exists in the metal-free form and as the metal complex. Thus

$$\text{Percentage colour transition} = 100[D]'/[D]_t \qquad (52)$$

where $[D]_t = [D]' + [MD]'$ signifies the total concentration of the indicator. On combining this with the mass-action expression for reaction (41) we find:

$$\text{Percentage transition} = 100/(1 + [M] \cdot K_{MD}^{eff.}) \qquad (53)$$

The transition curves for a few indicators have been calculated by eqn. (53) using data from Table 12, and the results are shown graphically in Figs. 23–30. From these we can appreciate how metal is abstracted from the metal–indicator complex as the concentration of metal in the solution is reduced (increase in pM achieved by adding EDTA), so that the different colour due to the metal-free form of the indicator results. Since according to eqns. (44) and (50) the effective stability constant, $K_{MD}^{eff.}$, is pH-dependent, a different transition curve is obtained for each pH value of the solution that is being titrated. The same assumptions were therefore made in these calculations as in those for the titration curves of Figs. 4–10, namely that the respective pH values were fixed by using NH_3–NH_4^+ buffers in which the total concentration $[NH_3] + [NH_4^+]$ was always 0·1M. Since the symbols $[D]'$ and $[MD]'$ represent the total concentration of a mixture of species (cf. eqns. (43) and (48)), not only the value of pM at which the colour transition takes place but also the phenomena visible around the end-point depend on the pH. For example, the colour changes from orange to red ($CuH_4D^+ \longrightarrow H_4D^-$) in the titration of copper at pH 5 using murexide as indicator, but from yellow to violet-blue ($CuH_2D^- \longrightarrow H_2D^{3-}$) at pH 11; this is clearly shown in Figs. 23–8.

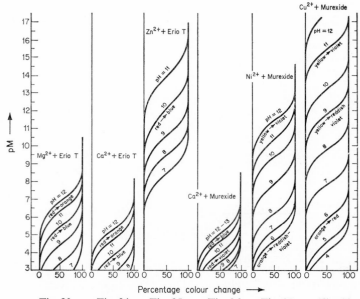

Fig. 23   Fig. 24   Fig. 25   Fig. 26   Fig. 27   Fig. 28

Figures 23–8. The effect of changes in pH on the colours of various combinations of metals and indicators.

Figures 23, 24, 25. Eriochrome T with Mg, Ca, and Zn respectively.

Figures 26, 27, and 28. Murexide with Ca, Ni, and Cu respectively.

With metalphthalein all the species have their maximum absorption at the same wavelength (Fig. 17): they are all red, but the intensity of absorption varies. If the intensity of coloration of the fully deprotonated anion D^{6-} is arbitrarily set at 100% the following figures hold for the other species, viz. HD^{5-} 20%; H_2D^{4-} 7%, H_3D^{3-} 0%, CaD^{4-} 100%, CaD_2^{2-} 100%, $CaHD^{3-}$ 20%, BaD^{4-} 100%, Ba_2D^{2-} 100%, $BaHD^{3-}$ 20%, Mg_2D^{2-} 20%, Zn_2D^{2-} 2%.

To take account of the special features of metalphthaleins the colour transition in Figs. 29 and 30 has not been calculated on the basis of eqn. (53), but the intensity of the colour is shown as a function of pCa and pBa. The fact that this indicator can only be used over a very narrow pH range is shown specially clearly in these diagrams. Thymolphthalexone (XIX) and glycine thymol blue (XX), which are also monocolour indicators, behave like metalphthalein.

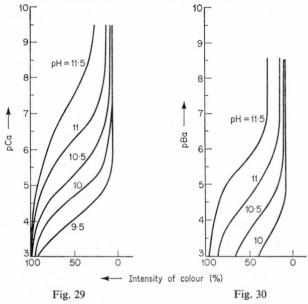

Fig. 29   Fig. 30

Figures 29 and 30. The effect of pH on the intensity of coloration of metalphthalein at different values of pCa (*Figure* 29) and pBa (*Figure* 30).

4.5. Visibility of the colour change

Curves of the type shown in Figs. 4–15, which present the change in pM as a function of *a*, the amount of titrant added, can be combined with curves of the type shown in Figs. 23–30 (pM against colour transition interval) to find out how 'sharp' the indicator transition is at the end of a titration. It will be obvious that the range of colour transition of the indicator (the flattened portion of the pM–transition curve) ought to lie in the steep portion of the titration curve (pM versus *a*) at *a* = 1. In good procedures the indicator will be completely transformed from the form MD to the form D between *a* = 0·99 and *a* = 1·01. If the transition range of the indicator coincides only approximately with the jump in pM in the titration curve a long-drawn-out ('blurred') end-point will be observed, since the changes in colour do not take place suddenly but begin too soon or finish too late. In Chapter Six we shall discuss the error which results when the correct value of pM at the end-point (i.e. the pM at the exact point of stoicheiometric equivalence) is not achieved. A long-drawn-out end-point may, however, have its origin in quite a different phenomenon, namely the slowness of the reaction that is taking place when the metal is transferred from the indicator to the EDTA. A trailing end-point of this sort due to kinetic factors can be improved by warming the titration solution or by adding a catalyst: no such palliative is possible if the reason is an unfavourable position of the reaction equilibrium.

An entirely different problem will now be discussed – the visibility of the colour change, a problem which is bound up with pyschosomatic processes in the retina and in the brain of the observer. It will be taken for granted that the illumination is satisfactory and that the titration is carried out against a bright background. The colour transition is then the more easily seen the more 'different' the limiting colours (of pure MD and pure D) appear to the eye. According to the theory of colour, these limiting colours ought to be as complementary as possible so that they neutralize each other to give a white. Complementary colours are, for example, red – bluish green, orange – blue, yellow – indigo, greenish yellow – violet, etc. An indicator which changes from one complementary colour to the other becomes intermediately colourless (or grey), as the two hues neutralize each other, and thus the neutral point is particularly easy to establish.

It is all too obvious that an indicator that has been selected because its transition interval occurs in the correct pM range will only rarely oblige us by changing into its complementary colour at the same time. But an approximation to this ideal behaviour can very frequently be achieved by modifying the colours by the addition of a suitable inert dyestuff (inert in the sense that it does not take part in the chemical reactions) which produces a background colour such that the limiting colours of the indicator before and after the end-point now become broadly complementary. Thus to an indicator that changes from yellow to red a particular shade of blue can be added in the correct proportion so that the colour change is from yellowish-green (mixed colour from yellow and blue) through neutral grey to violet (red plus blue). At the end of the titration the solution is often practically colourless, for the pale grey hue is barely detectable, and the equivalence point can be established far better than in the absence of the additional dyestuff, where the titration has to be carried out to an orange colour with a definite reddish hue. If the titration curve is almost vertical in a plot of transition interval against pM (a large pM jump) the colour of the solution being titrated changes abruptly from one complementary colour to the other and the end-point is extraordinarily 'sharp'.

Such 'screening dyes' were recommended quite early on in complexometric titrations as inner filters to improve the end-points. Empirical search for the appropriate dyestuff to add and to find its optimum concentration is naturally a very time-consuming and wearisome business, and we are indebted to Reilly, Flaschka, and Laurent [60-68] for showing how to reach the objective in a systematic way. Their procedure depends upon an extension of general colour theory to complementary tristimulus colorimetry, and it enables us to specify the visibility of the indicator transition by precise numerical values. The choice of the additional dyestuff can be made on the basis of the absorption spectra of the indicator in its limiting colours, and its own spectrum provides the information needed as to the concentration in which it should be added to make the colours before and after the end-point as complementary as possible. A simple scheme for such calculations and details of the practical operations has been reported for pH indicators by Flaschka [61-184]. This scheme can be used for metallochromic indicators without modification. But it must be realized that a mixed indicator developed in this way cannot be used unconditionally for all the metals that could otherwise be titrated with the particular metallochromic indicator alone. The complexes of a single dyestuff with different metals often have different colours (e.g. the complexes of Ca, Cu, etc, with murexide, cf. Fig. 19), so that for each metal the mixed indicator ought really to be individually composed to achieve maximum complementarity. However, a mixed indicator designed for one particular metal generally shows advantages over the simple indicator for several other cations. Thus the same murexide mixed indicator can be used for Cu and for Ni, although the composition must be changed entirely for the Ca titration. Solutions of mixed indicators generally keep as well as those of the unmodified indicators; but sometimes they are less stable, and it is then advantageous to use a solid diluent, such as common salt or sugar. It should not be overlooked that the tristimulus procedure presupposes an 'average observer'. However, small differences in sensitivity to different colours occur very frequently between different people – even when no actual colour-blindness is in question. For those observers who depart more markedly from the norm the composition of a mixed indicator must often be modified somewhat by trial and error to achieve the optimum effect.

The following are a few examples of mixed indicators. For the Ca titration, Knight [51-14] recommends a mixture of 0·2 g murexide with 0·5 g naptholgreen B ground up with 100 g NaCl: the colour change is from olive-green through a reddish-grey to pure blue. Another mixed indicator for Ca (described as Hydron (II)) is a 1 : 2 mixture of a 0·5% solution of eriochrome green H with a 0·25% solution of naphthol yellow S: the colour change is from red to green [60-108]. Murexide can also be screened with indigo carmine [64-30] or by adding 1–2 drops of a 0·1% solution of bromocresol blue [63-84]. The colour transition in the titration of Ca alone or of Mg $+$ Ca can be sharpened if 0·125 g of methyl yellow is mixed with 0·5 g Erio T: diethanolamine (100 ml) serves as the solvent. A mixture of alizarin S (45 ml of a 0·5% solution in ethanol) and methylene blue (15 ml of 0·1% aqueous solution) serves for the titration of lanthanides [56-95]. To improve the colour change of calmagite the addition of some picric acid has been proposed [63-93]. Such an addition should also prove valuable in the case of Erio T. A mixture of Alizarin S and xylene cyanole FF has been proposed for the titration of Th [57-121]. For the determination of Ba^{2+} and SO_4^{2-} a 1 : 5 mixture of solutions of sodium rhodizonate (0·05%) and methyl-amino-4-(p-tolyl)-anthraquinone (0·5% in 50% ethanol) has been recommended [59-83]. Grönquist [53-20] changed the colour transition of yellow to colourless in the titration of Bi^{3+} when using thiourea as indicator into the transition green to violet, simply by adding gentian violet. The residual pink colour in the titration of Ca using metalphthalein can be masked by adding diamine green and methyl red (0·05 g and 0·005 g per 0·1 g of metalphthalein) [53-1]. A solution of 0·3 g metalphthalein, 0·3 g naphthol green B, and 0·5 ml conc. ammonia in 100 ml water should be still better [56-68]. The transition of fluorescence indicators (Section 3.4) has also been improved by their admixture with other substances [61-85; 62-151]. An improved colour change for xylenol orange is obtained by using 1 ml of 0·02% indicator solution in combination with 2 drops of 0·02% aqueous Poirrier's Blue [65-15].

Another means of increasing the visibility of a colour change is to employ an extractive end-point. The theoretical aspects of this approach have been discussed by Still [65-16], and Betteridge [66-22].

CHAPTER FIVE

Instrumental Indication of the End-point

5.1. Introduction

In the practice of complexometry, as in most other volumetric procedures, visual observation of the end-point by means of a colour indicator is preferable on the grounds of speed and simplicity. However, instrumental methods were also investigated quite early on and have gained acceptance in practice. In comparing the relative merits of visual and instrumental indications of the end-point the complexity or length of the procedure and the apparatus required are not the sole criteria. For, as will be explained in detail in Chapter Seven, the reasons for preferring one method to another go much deeper and are intimately bound up with the problem of selectivity. Moreover, instrumental methods often lead to correct results when visual methods fail, as, for example, when dealing with turbid or strongly coloured solutions, or those that contain substances which can 'block' or (like oxidizing agents) actually destroy the indicator. In most cases automation of the titration procedure can only be achieved through an instrumental indication of the end-point.

Many of the methods that have been studied hitherto have been applied only to the titration of simple solutions and have still not gained acceptance in general analytical practice; but this does not decrease their potential value for future applications. In the following sections a brief survey will be given of various instrumental methods together with their practical applications, in so far as these have been reported: the number of references quoted ought not to be taken as indicating their relative importance. Details of practical operations will only be mentioned if they involve a significant departure from the usual procedure.

5.2. Photometric titrations

In photometric titrations the absorbance (optical density) of the solution being titrated is plotted against the volume of titrant used and the end-point is read off from the titration curve obtained in this way as the intersection of two straight lines (which may have had to be extrapolated) which meet at an angle. This procedure has been applied to every possible kind of titration, and the relevant literature is extremely voluminous. So far as the necessary apparatus is concerned, reference should be made to the excellent summary of the fundamental theory and practice of photometric titrations which has appeared in book form [58-22] as well as to the literature survey [58-23], which includes papers up to about 1957. Both publications are due to J. B. Headridge, and they cover all aspects of titrations and include complexometry. The construction of a simple but versatile instrument is described by Flaschka and Butcher [65-108]. The theory and practice of automatic titrations has been thoroughly treated in a book by J. P. Phillips [59-81].

In the following sections only the most important references will be cited: papers describing methods which depend upon some well-established principle and which differ only by using some special piece of equipment have been omitted.

5.2.1. TYPES OF INDICATION METHODS

It is appropriate to distinguish between different types of complexometric titration with photometric end-points which behave differently when viewed essentially from considerations of selectivity.

5.2.1.1. GROUP A. STEP INDICATION. In this group we are dealing with the same types of titrations that can be carried out with visual indicators and the human eye is simply replaced by the light detector of the apparatus. A characteristic feature is that an indicator is added whose concentration is small in comparison with that of the substance being titrated. Generally an organic dyestuff is involved ($\epsilon_{max.} = 10^3$–10^5) which can form a complex with the metal being titrated: at the end of the titration this complex is decomposed again, and thereby a sudden change in absorption spectrum is produced. In the ideal case the titration curve proceeds practically horizontally, rises (or falls) almost vertically at the end-point, and then continues to run horizontally; this behaviour has led to the description 'step indication' and 'step indicator' [61-49].

5.2.1.2. GROUP B. SLOPE INDICATION. (*Neigungsindikation.*) In this group too the end-point is revealed by a change in the slope of the titration curve, but here there is no double change as in the step method mentioned above. On the contrary, the titration curve in the ideal case consists of two straight lines inclined at an angle (of which one is generally more or less horizontal) whose intersection marks the end-point; hence the name 'slope indication' [61-49; 62-27]. It is convenient to distinguish three sub-groups.

(a) *Self-indicating systems.* This group includes systems to which it is unnecessary to add a special indicator, since changes in the optical absorption of one or more of the species taking part in the titration suffice to follow its course. This is especially easy to carry out with the coloured cations of the transition metals because the complexes present in solution before the titration starts (e.g. aquo- or ammine-complexes of copper(II)) possess a different absorption spectrum from that which appears during the titration (e.g. the EDTA complex of Cu(II)). It is therefore possible to find a wavelength at which the optical density will change during the titration, and indeed it will rise (or fall) linearly with the volume of titrant added right up to the end-point, after which it will remain constant. Since the molecular extinction coefficient of the coloured metal complexes is generally small ($\epsilon_{max.}$ $\sim$10–100), the most favourable range of optical densities is obtained with metal concentrations of from 10^{-2} to 10^{-3}.

The absorption due to the complexing agent in the titrating solution can also be used to detect the end-point. Thus the absorption due to the metal complexes of EDTA at $\sim$220 mμ is significantly lower than that due to the metal-free but protonated species H_2Y^{2-} and HY^{3-}, which begin to appear in solution after the end-point has been reached – provided the pH is not too high. Once again the end-point will be detected by a change in the slope of the titration curve.

(b) *Self-indicating systems after the addition of excess indicator.* In this system an indicator is added, but in a concentration which is at least equivalent to that of the metal to be determined: a factitious excess is generally used. Once again a complex is formed with the metal to be determined, though this usually has a smaller molecular extinction coefficient than the metal complexes formed with the metallochromic indicators of Group A (5.2.1.1). Since the whole of the metal has been transformed into its metal-indicator complex, the system now becomes 'self-indicating' and the titration curves become identical with those encountered in Section 5.2.1.2 (*a*). An example is the titration of Bi, which forms a yellow complex in the presence of excess thiourea. During the titration with EDTA the complex is gradually decomposed, so that the optical density at a suitable wavelength decreases linearly with the added titrant until the end-point is reached, after which it remains constant. A further example is the titration of copper, in which the ammine $Cu(NH_3)_4^{2+}$ is replaced by the EDTA complex, CuY^{2-}. This titration can be allocated to Section 5.2.1.2 (*a*) as described above, or to 5.2.1.2 (*b*), depending on whether one chooses to regard the ammonia as forming part of the solvent or as an indicator added in excess.

When dealing with metals in very low concentrations it is also feasible to add an excess of a metallochromic indicator and so produce a self-indicating system with a sufficiently large optical density. This procedure is especially valuable in the ultramicro region, and in this way submicrogram amounts of Mg can be determined in the presence of excess Erio T [62-27].

(c) *End-point indication produced by adding a self-indicating system.* A self-indicating system can also be added as an indicator if the metal to be determined is more strongly complexed than that in the indicator system. Thus copper can be introduced for the titration of Bi. When EDTA solution is run in, the Bi is complexed first, but this process is not accompanied by any change in optical density. But when all the Bi has been transformed into its EDTA complex the copper starts to form CuY^{2-} and the optical density rises with a concomitant change in the slope of the titration curve. The copper thus functions as a very special kind of 'photometric indicator' [54-50], and certainly in a very different way from the metallochromic indicator as envisaged in Section 5.2.1.1; in consequence, the titration curves look quite different. To avoid confusion with the more general concept of a photometric indicator it is advisable to speak of copper in this context as a 'slope indicator'.

The end-points for each metal can be read off from the titration curve of a solution that initially contains both Cu and Bi, for it begins with a horizontal portion (Bi), continues with a rising branch (Cu), and concludes with another horizontal portion. The mathematical treatment of such a curve is given in reference [61-113].

Systems that can be made self-indicating by adding an excess of a metallochromic indicator can be used for titrations on the ultramicro scale. Thus the determination of submicro amounts of Ca and Mg can be achieved in a single titration [62-27] if an excess of calmagite is added to their solution buffered at pH 10 (Erio T can also be used, but it is less stable). This produces the self-indicating system Mg–calmagite, which serves as a 'slope-indicator' for Ca, for calcium itself scarcely reacts with this dyestuff. Since the complex of EDTA with Ca is stronger than that with Mg, calcium is titrated first, and a titration curve results, consisting of three portions, horizontal–sloping–horizontal. The first break corresponds to the titration of Ca, the second to that of Mg. It is worth recalling that in the customary titration of a solution containing Ca and Mg with calmagite as a step-indicator (i.e. in an amount well below that corresponding to the amount of Mg) only gives the total concentration of the two metals.

5.2.2. THEORETICAL CONSIDERATIONS

A mathematical discussion of titrations of Group A (Section 5.2.1.1) has been given by several authors [53-45; 54-43; 54-69; 54-70; 55-12; 57-19; 59-122; 61-117; 65-109]. On the basis of such mathematical deductions Fortuin and his co-workers [54-43] arrive at the optimum conditions for a step-indication; these have been completely confirmed by experiment:

(*a*) The effective stability constant of the titration complex should be appreciably greater than that of the indicator complex – by a factor of at least 10^4.

(*b*) The effective stability constant of the indicator complex should be as high as possible, and at least 10^4–10^5.

(*c*) The concentration of indicator should be as small as possible.

(*d*) The concentration of the metal should be as large as possible.

(*e*) The difference between the molecular extinction coefficients of the metallized and metal-free forms of the indicator should be as large as possible.

If this last condition is modified to read 'as different as possible in colour' these conditions become practically identical with those required for a visual titration. The essential difference between photometric and visual titrations lies in the fact that deviations from ideal behaviour can be much greater in the former before unusable results are obtained.

Photometric titrations are often used especially for micro-determinations, where it is difficult to fulfil condition (*c*). (The possibility of working with excess indicator, and so providing a self-indicating system, must naturally be disregarded for visual titrations.) If, furthermore, it also happens that condition (*b*) is not exactly met, then non-linear titration curves are obtained which can nevertheless lead to the correct results, although the answer will depend on whether extrapolation is to the beginning or to the end of the step, or indeed whether the point of inflection is accepted as the end-point. If the formation constants of the indicator complexes are known it is possible to calculate the form of the titration curves theoretically, and it is then possible to specify exactly what must be done to obtain correct results. (As an example, cf. the titration of Ca using murexide, Fig. 37.)

Ringbom [62-118] has shown how it is possible to base the choice of experimental conditions on a mathematical analysis and to introduce a correction. The correction can be calculated from a knowledge of the stability constants of all the complexes present and the concentrations of the various species involved. Of course, effective stability constants have to be used, and these are not always known with sufficient accuracy for the relevant analytical conditions; indeed, it may often be quite impossible to calculate them at all. The procedure runs into difficulties, particularly where definitely limiting cases are under consideration, for here the results of calculation are particularly subject to uncertainty. In such cases it is far simpler in practice to carry out titrations under identical conditions with exactly known amounts of the materials in the test solution and to decide on the basis of the resulting titration curves which part of it is to be used for calculating the end-point.

Flaschka and his co-workers [61-48; 61-49] have given a mathematical treatment for Group B (Section 5.2.1.2). The mathematics for the special cases of Section 5.2.1.2 (c) has been worked out by Reilley [61-113]. The advantage of working without an indicator has been dealt with in previous sections, more particularly under the heading of selectivity. The distinction drawn between logarithmic and non-logarithmic titrations is naturally also valid for the special case of photometric end-point detection. A few cases will be discussed in the section on successive titrations where profitable use is made of this principle.

5.2.3. APPLICATIONS

Photometric determination of a step-indication has been proposed especially frequently for calcium. This is because the colour transition of murexide appears rather poorly defined to the human eye. To cite only a few of the relevant papers, the following call for special notice [54-10; 55-12; 55-69; 57-17; 58-84; 59-35; 60-175; 61-9; 64-5]. Practical use has been made of the procedure for the determination of calcium in serum [54-19; 58-45; 59-107], urine [54-19; 58-96], and spinal fluid [56-4]. Ca can be determined in the same way in the water-soluble constituents of gypsum [55-52], in natural waters [57-74], glass sands [61-9], steel [60-112], and cements [57-17]. The method has become widely used as a routine procedure in the analysis of silicates [59-112]. Other indicators can be used in place of murexide in the photometric determination of Ca, such as phthalein-complexone [55-7], Cu-PAN [62-119], and also calcein for small amounts of Ca in lithium salts [61-7], silicates [61-180], and blood serum [59-91].

The photometric determination of calcium in an aliquot portion of a test solution is often combined with the determination of the total concentration of Ca + Mg in a second aliquot portion to give Mg by difference. Naturally this procedure can also be called upon for the determination of Mg alone. Erio T has been used predominantly as the indicator [53-45; 54-69; 54-70; 56-48; 57-18; 58-84; 59-122; 61-8]. The method has been used for the analysis of glass sands [61-9], cement [57-17], steel [60-112], serum [59-129; 63-88], and spinal fluid [56-4]. Calmagite has been recommended for the determination of Ca + Mg [64-5], and pyrocatechol violet has been specially recommended for the analysis of serum [60-175]. The procedure has also been worked up for the analysis of silicates [59-112; 61-180].

Ca and Mg can both be determined in a single aliquot portion if the Ca is first titrated at pH 10 with EGTA, using murexide as indicator and then, after changing the wavelength, Mg is titrated with EDTA using Erio T [61-50].

The closely related problem of determining the hardness of water has also been put on a photometric basis by the methods discussed above.

Innumerable other metals have been determined photometrically, and the widest possible variety of indicators have been used. The following are a few examples. Th using chromeazurol S [55-71], pyrocatechol violet [59-76], alizarin red S [59-95], arsenazo I [62-19; 63-106], naphthol purple [56-19], or SNADNS [62-5]. Ba [56-10; 56-71], Zn [56-25; 63-47], and Cd [63-47] with Erio T; Sr with phthalein complexone [60-109]; rare earths with alizarin red S [59-95], and arsenazo I [61-54; 62-19; 63-106] and as a special case erbium with PAR [60-130]. Bi and Cu can be titrated using pyrocatechol violet [59-21], Ga [66-67], Bi and Pb with xylenol orange [60-47], Pb with PAR [64-126], Fe with sulphosalicylic acid [64-144] or nitrosalicylic acid [66-66], Ni with murexide (with the simultaneous masking of Co by nitroso-R salt [57-63]), and Ag [57-75] and Pd [55-1] after a replacement reaction with tetracyanonickelate.

The high accuracy of photometric titrations was employed in evaluating the suitability of cadmium complexonate as a primary standard: an azo-oxine was used as indicator [60-100]. The determination of iron proved to be largely free from interference by Al if it is titrated photometrically at pH 1–2 using salicylic acid [60-144]. Al can be determined using chrome azurol S [61-149]. After reduction with hydrazine sulphate in weakly acidic solution Mo can be directly titrated using alizarin complexone [60-83]. After oxidation with hydrogen peroxide in an alkaline medium Co can be determined by back-titrating a measured excess of EDTA with Bi in a strongly acid solution: the method is very selective, and pyrocatechol violet serves as the indicator [61-51]. Ni can be determined directly in the presence of Fe that has been masked with pyrophosphate. After the nickel titration excess EDTA can be added and the iron itself determined by back-titration with a standard solution of Ni [58-38]. Hydroxamic acid serves as the indicator in the determination of Zr when excess EDTA is back-titrated with a standard solution of Fe [55-80].

Self-indication is possible in the determination of Fe [53-37; 53-39], Cu [53-37; 53-39; 63-45], and Ni [53-37] in the visible region of the spectrum. The titration of traces of Ca [64-4] and of Ca, Mg, Zn, Cd, Ti, and Zr [54-14], Bi and Pb [55-97], Pb^{2+} in the presence of organo-lead [60-201], and Mo [59-62] is possible in the ultraviolet. On adding hydrogen peroxide, titanium is changed to a self-indicating system [57-20; 60-136]. The same is brought about for Co in 50% aqueous acetone by adding thiocyanate [60-171].

Slope-indication, using Cu as a photometric indicator, permits the determination of Bi [54-51], Ca [57-13; 57-14; 61-113], Mg [61-113], and a selective determination of Cd in the presence of Zn [62-26].

The increase in selectivity is most impressive in many cases of photometric titrations, as evidenced by the following examples. Fritz [59-96] titrated Sc in the presence of a 60-fold excess of rare earths by using Cu as a slope indicator. The successive titration of Fe–Cu [53-39] and Bi–Cu [54-51] works on the principle of self-indication. Step-indication in conjunction with changes of pH permit the titration of Th and rare earths: Th is first titrated at pH 2 using arsenazo, and then the pH is raised by adding urotropine and the rare earths determined with the same indicator [62-19]. The successive titration of Bi and Pb follows the same pattern using xylenol orange [60-47]. The possibility of determining Ca and Mg in one and the same portion of solution has already been discussed [61-50; 62-27].

Almost all the above methods use EDTA as the titrant. A few other examples will be given to show that the use of another complexing agent can sometimes lead to a further increase in selectivity. Thus the self-indicating titration of Cu with TRIEN (triethylenetetramine) in weakly acidic solution is almost specific [57-87; 57-88]. Cd can be titrated in the presence of Zn by using DTPA [61-113] or even better EGTA [62-26] with Cu as a slope indicator. EGTA permits the titration of Ca at pH 10 using murexide even in the presence of a 100-fold excess of Mg [61-50].

Automation can be achieved in two ways. An automatic burette can be coupled with a pen-recorder so that the complete titration curve is drawn. But when it is only a question of locating the break at the end-point it is often sufficient to plot the transmission against the volume of the titrant, a course which allows a considerable simplification of the apparatus. The second possibility is to arrange for the current produced in the phototitrator or the voltage change to operate a relay to stop the flow of titrant from the burette at the end-point. The signal can also be differentiated electronically once or twice, whereby the first or second derivative of the titration curve is obtained. Malmstadt, in particular, has used this development. Principles and applications have been treated exhaustively by Hadjiioannou [59-121]. Automatic titrations have been reported for Th [54-12], Cu and Zn [59-120], Fe [66-66], and Cu, Fe, and Co [60-78]. Use has been made of them in the determination of Ca and Mg in dolomite and limestone [58-110], serum [59-118; 63-88], carbonate minerals [55-87], and natural waters [59-119; 63-46]. The automatic delineation of the titration curve obtained in the presence of fluorescein complexone serves for the determination of serum calcium [59-61]. Micro-amounts of Th can be titrated photometrically with quercetin as indicator [58-70].

5.3. Potentiometric titrations

If a metal occurs in solution both in a higher oxidation state (ion $M_{ox.}^{\nu}$) and in a lower state (ion $M_{red.}^{\nu-n}$) corresponding to the addition of n electrons per atom (equilibrium $M_{ox.}^{\nu+} + n \cdot e^- \; \rightleftharpoons \; M_{red.}^{(\nu-n)+}$), the complexing of the metal will always be accompanied by a change in the redox potential, E, for the two cations will form complexes of different stability. This is easily seen by writing down the Nernst equation for the potential of an inert electrode. Thus for a solution which does not as yet contain any of the chelating agent (titrant Z)

$$E_i = \pi_0 + (s/n) \cdot \log\left([M_{ox.}]'/[M_{red.}]'\right) \tag{54}$$

where E_i is the initial potential, π_0 the standard electrode potential, and s a temperature-dependent numerical factor which is about $0 \cdot 06$ volt. Charges are omitted for simplicity's sake.

When both the higher and the lower oxidation states have been completely transformed into their complexes ($M_{ox.}^v \longrightarrow M_{ox.}Z$ and $M_{red.}^{v-n} \longrightarrow M_{red.}Z$) the potential of the electrode becomes

$$E_f = \pi_0 + \frac{s}{n} \log \frac{[M_{ox.}Z]}{[M_{red.}Z]} + \frac{s}{n} \log \frac{K_{red.}^{eff.}}{K_{ox.}^{eff.}} \tag{55}$$

where E_f is the final potential, and $K_{ox.}^{eff.}$ and $K_{red.}^{eff.}$ are the effective stability constants of the complexes $M_{ox.}Z$ and $M_{red.}Z$ defined by

$$K_{ox.}^{eff.} = \frac{[M_{ox.}Z]}{[M]'[Z]'} \, ; \; K_{red.}^{eff.} = \frac{[M_{red.}Z]}{[M]'[Z]'} \tag{56}$$

Equation (55) is quite simply obtained by substituting from eqn. (56) for the concentration terms $[M_{ox.}]'$ and $[M_{red.}]'$ in (54). When both the ions have been completely complexed the ratio $([M_{ox.}Z]/[M_{red.}Z])$ that occurs in eqn. (55) is exactly equal to the initial ratio $([M_{ox.}]'/[M_{red.}]')$, so that the change in potential, $(E_i - E_f)$, depends only on the ratio of the stability constants, $(K_{red.}/K_{ox.})$. This potential change can be quite considerable. Thus with Fe(III) and Fe(II) and EDTA as the complexing agent $K_{red.}/K_{ox.} = 10^{-11}$, so that E_f is some 660 mVs more negative than E_i.

Schwarzenbach and Heller [51-1] calculated the stability constant of the Fe(III)–EDTA complex on the basis of relationship (55) by measuring the standard potential of the Fe(III)/Fe(II) couple in the presence of EDTA and using the stability constant of the Fe(II)–EDTA complex that had already been measured. On introducing the complexing agent the decrease in potential takes place stepwise, for the cation that forms the more stable complex (generally $M_{ox.}^{v+}$) becomes complexed first. Thus if a mixture of Fe^{2+} and Fe^{3+} is titrated with EDTA the big jump in potential takes place as soon as an amount of complexing agent equivalent to the Fe(III) has been added. If the titration is carried out below pH 3 essentially only the first potential jump is realized, for the effective stability constant of $Fe(II)Y^{2-}$ has become so small that the Fe^{2+} present remains practically completely uncomplexed. The amount of Fe(II) in the solution therefore influences only the absolute magnitude of E_i and E_f, whereas the potential jump, $(E_i - E_f)$, is independent of it. It is even possible to titrate Fe(III) with EDTA using an inert indicator electrode without adding any Fe(II), for an extremely small amount of this must be present in every ferric solution (otherwise the potential according to eqn. (54) would be positive and infinite), and this small amount is sufficient to make the electrode potential responsive to the position of equilibrium. The theory of the potentiometric titration curve in complexometry is dealt with by Goldman and Meites [64-113].

Přibil *et al.* were the first to apply this practically to the determination of Fe(III). The iron solution is buffered with acetate to pH 5, a bright platinum electrode serves as the indicator and calomel as the reference electrode. On titrating with EDTA the potential at the end-point drops by 350 mV within 0·04 ml: steady potentials are established almost instantaneously. At lower pH values the potential jump coincides exactly with the point of equivalence, but it is not so large, and the time taken to establish a steady potential increases as the acidity increases. The titration curve looks almost the same as that for a redox titration, save that it bends round more rapidly after the potential jump to give an almost horizontal portion. According to Strafelda *et al.* [65-67], a Pt electrode may be used to give the indication in the titration of Co(II), Cu(II), and Tl(III) as well as Fe(III).

In addition to studying the redox behaviour of Fe(II)/Fe(III), Belcher *et al.* examined that of Cu(I)/Cu(II) in the presence of EDTA [55-104]. Cu(II) can be titrated in the same manner as Fe(III), but the potential jump at the end-point is considerably improved by adding some thiocyanate [55-105]. A potential jump at a platinum electrode is also obtained when Hg(II) is titrated with EDTA; this probably responds to the couple Hg(I)/Hg(II). A large number of metals can be determined by back-titration of excess EDTA with a standard ferric solution, namely: Al, Bi, Co, Ni, Cd, Pb, Zn [51-16], Ti [60-135], and mixtures of Al, Fe, and Cr [57-29].

Besides platinum, other metals have been used successfully as indicator electrodes: e.g. Ag [60-111; 62-17; 65-4], also automation [67-1], Hg (as a pool) [57-22], and a platinum wire coated electrolytically with PbO_2 [62-135]. In many cases where the combination of a platinum wire and an external calomel reference electrode is inapplicable success has been achieved with a bimetallic pair (Pt–W or Pt–Mo), both metals being immersed in the solution [55-6; 65-112]. In the titration of Pb with EDTA the electrode pair Pt–W gives a larger potential jump at the end-point [55-73], and it can also indicate the end-point in the titration of Cu, Ni, Cd, Zn, and even Mg [55-32; 62-16]. Since the potential jump is especially large with Pb, back-titrations with a standard lead solution have been tried out and found to be very good for the determination of Ba in nickel alloys [57-26]. The electrode pair C–Pt has also been studied [64-146] and has been used in the titration of iron. However, the reactions taking place at these electrodes, which are responsible for the observed potentials, have not been explained.

The response of a Pt electrode to the redox couple ferro-ferricyanide is more easily understandable. Here we are concerned with the back-titration of excess EDTA with a standard solution of zinc [54-41] after very small amounts of $K_4Fe(CN)_6$ and $K_3Fe(CN)_6$ have been added to the test solution. The first uncomplexed zinc ions to appear in the solution after the equivalence point combine with the ferrocyanide ions, $(Fe(II)(CN)_6^{4-})$, and in this way cause the potential to shift to a positive value. The determination of Al and Cr(III) in the same solution has been made the subject of a special paper [60-158]. Both cations are first bound up by adding an excess of complexone, and the excess of EDTA is then back-titrated with Zn. After the first end-point has been registered, NaF is added, whereby the aluminium is transformed into AlF_6^{3-}, and the equivalent amount of EDTA set free can then again be determined with the standard Zn solution: Cr is obtained by difference.

The stationary mercury-drop electrode is of special importance, and its theory and application have been treated comprehensively by Schmid [62-29]. The indicator electrode consists of a drop of mercury resting in a funnel-shaped enlargement of a glass tube bent into the form of the letter J: contact is made by a platinum wire passing through a seal in the lower bend, and this enables the potential to be measured. A calomel or a Ag/AgCl electrode serves as the external reference electrode, but an amalgamated gold or silver wire can also be used when absolute potential differences are not required and only the jump in potential at the end-point needs to be detected.

The solution containing the cation $M^{\nu+}$ that is to be titrated with such a mercury electrode is first treated with a few drops of a solution of the mercury–EDTA complex, whereupon the following equilibrium is set up:

$$M^{\nu+} + HgY^{2-} \rightleftharpoons MY^{\nu-4} + Hg^{2+} \tag{57}$$

Since HgY^{2-} with a stability constant of 10^{22} is appreciably more stable than the complex $MY^{\nu-4}$, this equilibrium usually lies well to the left. The concentration of the uncomplexed Hg^{2+} is very small in comparison with that of HgY^{2-}, but it is the former that determines the potential of the mercury-drop:

$$E_{Hg} = \pi_0 + 0.03 \log [Hg^{2+}] \tag{58}$$

where π_0 is the standard electrode potential of the Hg^{2+}/Hg couple.

Substituting in (58) for the concentration of mercuric ions by means of the mass-action constant for reaction (57), viz.:

$$\frac{[Hg][MY]}{[M][HgY]} = \frac{K_{MY}}{K_{HgY}} = K \tag{59}$$

we obtain the following expression for the potential:

$$E_{Hg} = \pi_0 + 0.03 \log (K_{MY}/K_{HgY}) +$$
$$0.03 \log [HgY] + 0.03 \log ([M]/[MY]) \tag{60}$$

When, as is often the case, the equilibrium in eqn. (57) lies far to the left the concentration of HgY^{2-} hardly changes at all during the titration, so that only the last term in (60) is a variable. At the beginning of a titration the concentration ratio ([M]/[MY]) will be very large, but after passing the equivalence point it will become very small, thereby producing the change in potential. There will be a marked jump in potential as E_{Hg} becomes negative at the equivalence point. As this is approached, the bulk of the metal M has already been transformed into its EDTA complex, so that the concentration term [MY] cannot increase much more: E_{Hg} therefore decreases linearly with pM. Thus in the neighbourhood of the end-point the indicator electrode displays the jump in pM that has been discussed theoretically in Section 2.2 (Figs. 4–10).

However, the potential of the mercury-drop can still indicate the end-point of a titration, even if the equilibrium (57) does not lie completely to the left; the potential jump will not be so large, however, for now although the last term in eqn. (60) becomes smaller as the EDTA is added, the penultimate term simultaneously increases. As Ringbom has pointed out [63-64], in practical analyses the effective stability constant of HgY^{2-} is often no larger than 10^{12}, since chloride ions and other species (NH_3, OH^-) may be present, and these form complexes with Hg^{2+} so that the distribution coefficient $\alpha_{M(A)}$ is larger for mercury(II) than for the cation $M^{\nu+}$ that is being titrated. The effect of all this is to displace the equilibrium of reaction (57) towards the right. Despite this, the potential change of the mercury-drop will indicate the end-point.

Equation (60) shows that the mercury electrode can also be used to measure the stability constant K_{MY} of a metal complex. If known concentrations of HgY^{2-}, $M^{\nu+}$, and $MY^{\nu-4}$ are present together in solution measurement of E_{Hg} leads to a value for the ratio of stability constants K_{MY}/K_{HgY}. This so-called 'pHg method' has been worked out by Schwarzenbach and Anderegg [54-28] and also by Reilley and Schmid with their co-workers [56-76; 58-11; 58-12; 58-14].

By using potentiometric end-point indication with the mercury electrode, complexometric titrations of many metals have been studied practically, especially by Reilley *et al.* Consecutive titrations are also possible. For example, all three metals in a mixture of Bi, Pb, and Ca can be determined by successive titrations at three different pH values. Reports on automation using this and other electrodes have been published [64-145; 65-110].

In an analogous fashion it is possible to utilize the Cu(I)Y/Cu(II)Y [65-37] and the Fe(II)Y/Fe(III)Y [65-111] systems for indication purposes, and this makes it possible to titrate a number of metal ions to which an electrode does not *per se* respond.

Back-titrations of excess EDTA can obviously be carried out with a standard solution of mercuric nitrate, and many metals can be determined in this way [58-57]. A wealth of possibilities is opened up if masking agents are included [58-62]. The following determinations have been reported: Th [58-58]; Al and Mn [58-59]; binary mixtures of alkaline earths or Mg with Zn or Cd [58-60]; Ba, Sr, or Mg with Pb, Co, Ni, or Cu [58-56; 58-60]; binary mixtures of Hg [60-163] with La [59-125], Cr [61-13], Ga or In [61-14]; Sc and Pd [62-120]; Bi [59-124] and Hg in ternary mixtures with two other metals [60-164]. After modification the method becomes quite suitable for ultramicro-titration, and permits, e.g, the determination of 2–4 μg of Ni through a back-titration with a Cu solution [66-9].

The mercury electrode also serves to indicate the end-point in the determination of Cu, Zn, Pb, and Ca with EDTA that has been liberated coulometrically (q.v.) [56-77].

The use of the mercury electrode is not restricted to titrations with EDTA. Other complexones can be employed, e.g. EGTA for the determination of Ca [63-65] and also the far more selective polyamines, such as triethylenetetramine or tetraethylenepentamine [59-113; 62-99]. As in the titration with EDTA, it is necessary to add a small amount of the mercury complex of the chelating agent that is subsequently to be used as the titrant.

Potentiometric titrations with EDTA have been extended to non-aqueous media [66-69]. Constant-current electrodes can also be used to detect the end-point in complexometric titrations; with their aid Cd [54-22] and Ca [59-73] have been titrated and Zn and Cu determined in serum with a polyamine [62-99]. Cd and Pb can be determined with a polarized mercury electrode [59-80]: even a.c. polarization has been described [60-32].

5.4. Conductimetric titrations

The changes in conductivity that occur during a complexometric titration can be used to detect the end-point by plotting the conductivity against the volume of titrant. If the titration is carried out in a buffered solution the break in the titration curve at the end-point is only poorly defined, since ions of specially high mobility are neither formed nor removed during the complexation reaction, and the electrolyte added for buffering is responsible for most of the conductivity. A greater effect is produced in unbuffered solutions, for then complex formation is accompanied by large changes in pH. However, buffering is generally essential in analytical procedures, as otherwise the decrease in pH might become too large, and thereby reduce the stability of the complex to an impractically low value. Titrations in unbuffered media are therefore restricted to very dilute solutions. A further limitation in the possible applications of these procedures stems from the need to keep the concentration of inert electrolyte as low as possible in order to avoid having to make measurements against a large background conductivity.

Hall and his collaborators [54-32] employed conductivity titrations to standardize metal salt solutions. The procedure is quicker and more accurate than other methods, but has little practical significance, for a known amount of EDTA is titrated with a solution of a metal salt according to the method of back-titration. The same authors [59-104] have also studied the influence of the degree of dilution and the concentration of the ammonia buffer on the shape of the titration curves.

Direct titrations with EDTA and DCTA, both as solutions of their di-sodium salts, have been thoroughly studied by Výdrá and Karlik [57-34]. The course of the titration curves is almost identical for either complexing agent, but DCTA forms complexonates of greater stability. The shape of the titration curves is markedly dependent on the experimental conditions. In unbuffered, weakly acidic solutions the curves rise steeply in consequence of the high mobility of the hydrogen ions set free. After the end-point has been passed only a slightly rising curve, a horizontal portion, or even a falling curve is observed. What actually happens depends on the pH of the solution, and this is determined by the buffering action of the titrant. As anticipated, good results are only obtainable in highly diluted solutions.

In solutions that have been buffered with ammonia the titration curves first rise gently and become abruptly much steeper after the equivalence point: this is due to the formation of mobile dissociation products from the excess of titrant. The use of borate buffers affords the greatest advantages: hydrogen ions set free initially are hereby trapped, and the titration curve starts out horizontally and the gradient increases after the end-point.

A number of metals (Ca, Sr, Ba, Mg, Zn, Cd, Cu, Ni, Co, and Mn) have been titrated under various conditions, and they all give practically identical curves. The titration of Ca and Mg in a borate buffer has been used in determining the total hardness of water. The procedure is useful in the presence of oxidizing agents (e.g. hypochlorite) which would destroy visual indicators like Erio T [57-35]. The determination of hardness conductimetrically has also been proposed by Pasovskaya [57-28]. The equivalent conductance of some EDTA salts has been measured by Hall *et al.* [64-147].

5.5. High-frequency titrations

'Classical' conductivity titrations employ audiofrequencies. High-frequency titrations employ far higher frequencies up to a range of several megacycles. The form of the titration curves and the various limitations are in general the same as for procedures at lower frequencies. The essential advantage of the high-frequency method is that the electrodes do not have to be immersed in the solution and the response is appreciably more sensitive, especially to hydrogen ions. This latter feature makes it abundantly clear why high-frequency titrations have been employed to clarify some of the fundamental problems of complexometry. Thus, Blaedel and Knight [54-2] have investigated the stoicheiometry of EDTA titrations of Cu, Zn, Ca, and Mg by a differential method and find that the molar ratio of 1 : 1 in complex formation is fulfilled within the limits of experimental accuracy. It is of interest that in a series of measurements made when determining magnesium sulphate the variations in the end-point were some two to four times larger than for titrations of magnesium chloride: no explanation can be given. The same authors investigated the use of EDTA (free acid and di-sodium salt) as primary standards, and recommend high-frequency titrations for the evaluation of the various commercial preparations [54-3].

Hara and West used high-frequency titrations to follow the formation of various metal complexes of EDTA both by direct titration and also by acid–base titration of solutions containing the metals and the complexing agent in different proportions. In addition to the usual cations, Ag and Tl were also studied and shown to form 1 : 1 complexes [54-47]. The uranyl ion forms a complex in the pH range from 3·5 to 5·0, and though this is weak, it nevertheless enables a titration to be carried out [55-42]. The determination of a variety of metals in unbuffered solutions and at high dilution (down to 10^{-4}M) has been carried out [55-41], including that of Th [55-43] and the rare earths [56-59]. Zapp used the method in studying complex formation [60-56]. Bellomo and Bruno determined the sensitivity of the titration for many metals: under the conditions they used this was (with the exception of thorium) from about 2 to 6 μg per ml for Hg, Pb, Zn, Cd, Ni, Co, Cu, Ce(III), and Al [58-18; 60-149]. Alkaline earths and magnesium could not be titrated satisfactorily. Dyatlova *et al.* have reported on the titration of Th, Ga, and Fe [63-110].

Procedures have been worked out for the determination of Al in Al–Mg alloys, and for Ni in Ni–Cr alloys [58-27] and for the analysis of natural waters [62-117].

5.6. Amperometric titrations

The advantage of amperometric titrations is that they can be carried out without a complex-forming indicator. They thus present conditions favourable from the point of view of selectivity: a number of consecutive titrations of practical importance are therefore possible. The procedures are also very sensitive and permit of determinations with very small amounts and at a very low concentration. For example, the determination of cadmium is still effective in 10^{-7}M solutions and can be carried out with astonishing accuracy [56-22]; cf. also [66-63]. Automation to allow of continuous titration has now been achieved [65-106; 66-64]. Masking by precipitation is often possible, since the presence of a precipitate frequently causes no interference.

Classification is possible from a variety of viewpoints. For example, procedures may be differentiated according to the material used for the electrodes. The majority of investigations describe the application of the dropping-mercury electrode in the classical way. Přibil *et al.* [51-15] were the first to use an amperometric titration for the determination of Bi and Cd as well as for a few other polaro-graphically active metals. Many authors have followed their example, as the undermentioned references will show. However, in addition to the classical method, square-wave polarography has also been introduced: thus Fe(III) has been titrated in very dilute solutions at pH 4·7–5·0 [61-35], and so has In [62-36]; in the latter case consecutive titrations In–Cd or In–Pb are possible. Martin and Reilley [59-148] employ two stationary polarized mercury-drop electrodes. The authors give a full theoretical treatment of the basic principles of the titration. The principle has been applied to the determination of Ca and Mg in serum and urine [64-8]. Others report on the use of the rotating dropping-mercury electrode [58-74].

Platinum provides an alternative electrode material which has been used either as a rotating micro-electrode (e.g. in the titration of Fe(III) [59-44], Sr, Cu, Ni, or Mo(VI) [64-26]) or in the determination of Al by back-titrating excess of EDTA with Fe(III) [60-151] or as a stationary foil, as in the titration of Zn in Co-baths [57-101]. A vibrating platinum electrode [60-189] can be used in the titration of vanadium when both V(III) or V(IV) can be present in the sample. If both species are present, a consecutive titration is possible, since two breaks occur, the first of which is due to V(III). Kies [58-44] describes titrations using two polarized Pt-electrodes (dead-stop end-point). With such a set-up, Th has been titrated correct to 0.4% in amounts of 0.3–15 mg [60-57], and it is also possible to titrate U [64-90]. Systems involving two polarized electrodes have been studied in detail and applied to the titration of various metal ions by Vorliček and Výdrá [65-64; 65-105]. Guerrin *et al.* [60-66] describe the use of a copper electrode. Uranium has been titrated at a graphite electrode [63-48] in concentrations of 10^{-3}–10^{-8} g in 25 ml after reduction to U(IV). A rotating tantalum electrode has been used [60-75] for the determination of Zr $(+1.1$ volt); mercury can also be titrated at a Ta-electrodes [60-74], as well as Mo [63-14] or Zr through Fe(III) [64-76].

Another classification can be based on the electrode reaction that serves to indicate the end-point. If one is dealing with a metal that can be reduced polarographically this is carried out quite simply at a plateau potential of the relevant polarographic wave as, e.g. with Ti [61-72], Bi [51-15; 60-42], Cd [51-15; 56-22; 56-94], Ga [59-130], or Fe [57-33; 58-41]. For inactive (i.e. non-reducible) ions there is always the possibility of a back-titration. Thus Al can be determined by back-titration of excess EDTA with Fe(III) [60-151] and Zr by back-titration with Bi [56-13; 60-71]. Goldstein *et al.* [63-103] recommend the ion VO^{2+} for general use as a reagent in back-titrations. Some inactive metals can be determined by way of the amperometry of a polarographically active metal obtained after a suitable displacement reaction. For example, an equivalent amount of Zn^{2+} can be set free by the action of Ca^{2+} on a strongly ammonia-

E

cal solution of ZnY^{2-} and titrated with EDTA [52-36; 54-88; 63-73]. Th(IV) can be titrated indirectly through Pb^{2+} displaced from PbY^{2-} [57-91]. The use of the 'complexone wave', as it has been called, presents a further possibility: this wave results neither from the reduction nor from the oxidation of the metal-free complexone but depends on its dissolving mercury from the electrode (anodic depolarization $Hg + H_jY^{j-4} \longrightarrow HgY^{2-} + jH^+ + 2e^-$). Since this oxidation takes place at different potentials, depending on the composition of the supporting electrolyte, a potential appropriate to the conditions must be applied. However, the potential is always very positive, which has the great advantage that oxygen does not interfere. The method using the anodic diffusion wave was described first by Michel *et al.* [53-51; 54-29] and developed by Reilley and his co-workers [56-78; 62-33] and worked up for consecutive titrations such as Zn–Mg, Bi–Pb, Cu–Zn–Ca, Fe–Mn, etc. The procedure is, of course, equally applicable to the determination of active (polarographically reducible) ions.

Finally, one more possible way of titrating inactive metals should be mentioned wherein an active metal is used as an amperometric indicator. The theory of such titrations is completely analogous to the procedures described in the section on photometric titrations under the heading 'slope-indication' (5.2.1.2). The essential condition is that the titrant should form a more stable complex with the metal that is to be determined than with the indicator metal. The latter causes a sharp change in the slope of the titration curve (diffusion current versus volume of titrant) when all the metal to be determined has been complexed. Thus Ca and Mg can be titrated in this way if Tl(I) is used as the indicator metal [62-64] and Th with Fe(II) as the indicator metal [62-32]. With Cu as indicator Cd can be determined in the presence of considerable quantities of Zn [63-9].

Israel *et al.* [66-65] studied applications of a polarized working electrode that has been short-circuited. This produces a spontaneous current whose strength is proportional to the concentration of the species reacting at the electrode. The principle has been applied to titration with EDTA.

Amperometric titrations in non-aqueous media have been studied by Arthur and Hunt [67-6].

A few examples of practical applications will now be given. Amperometric titration has been used for the determination of Ni and Cd in the active mass of alkaline accumulators [62-10], Zr in Nb-alloys [60-71], Al [60-151] and Bi [60-95] in alloys, Hg in organic compounds after combustion in an oxygen-filled flask [58-86], Ca via Zn-displacement [64-5; 64-7], especially in solutions containing large amounts of alkali metal salts [63-73], In in sphalerite [58-100] and Cd metal [56-15], Zn in cobalt-plating baths [57-101], alloys [56-3], and oils [53-48]. EDTA [55-86; 59-73; 60-88] and DCTA [60-88] can also be determined by an amperometric titration.

5.7. Coulometric titrations

Reilley and Porterfield [56-77] seized on the possibility of reducing mercury(II) from its complex with EDTA as a basis for coulometric titrations. An amount of EDTA equivalent to that of the mercury(II) reduced is liberated in this process, and so becomes available for forming complexes with other metals. 100% current efficiency is easily achieved when working with about 45 mA. The discharge of the mercuric ions takes place at a mercury electrode: platinum foil serves as the anode, and the anode and cathode compartments are separated by a sintered-glass frit. The progress of the titration is followed potentiometrically with a mercury-drop electrode. In an actual determination EDTA solution with a slight excess of mercury salt is first added and the pH adjusted to about 8·5 with an ammonia buffer. The excess of mercury is then back-titrated coulometrically. The sample solution is then introduced and the titration continued (by switching on the current) until a jump in potential again occurs. The result is calculated from the difference in the quantity of electricity used in the fore and the main titration. Preliminary removal of dissolved atmospheric oxygen is absolutely essential. Cu, Pb, Zn, and Ca can be titrated in this way. Schmid and Reilley titrate Fe(III) salts with FeY^{2-} obtained by the electrolytic reduction of a solution of $Fe(III)Y^-$: the end-point is determined potentiometrically with a bright platinum electrode [56-124].

This procedure has potentialities for determining very small amounts of material. The smallest currents can be maintained extremely constant by modern techniques, and the measurement of time can be carried out with great accuracy. Thus the 'titrating agent' (the electrons) can be 'diluted' as much as one wishes (by reducing the current) without any dilution of the analytical sample during the course of the titration.

Monk and Steed [62-25] exploited this possibility in determinations of Ca, Ni, Cu, Zn, Cd, Sr, Ba, and Pb coulometrically at pH 10·5, and Zn, Cd, Y, and Nd at pH 4·5. Only one millilitre of solution was used, and the end-point was detected by means of two polarized mercury electrodes. The coefficient of variation in three to five replicate analyses was only 0·5–1%. Christian *et al.* [65-1] report on the coulometric generation of EGTA and its use to titrate Ca in the presence of Mg.

5.8. Chronopotentiometric titrations

This kind of titration depends on the following principle. A current of constant strength is suddenly applied to an electrode dipping in the solution to be titrated and that is completely at rest: the transient change in potential is followed as a function of time. The solution contains a supporting electrolyte, so that the magnitude of the current is controlled (as in polarography) by the transport of reducible (or oxidizable) materials to the electrode by linear diffusion. Thus, on closing the circuit, the potential rises immediately, reaches the decomposition potential of the first component, and stays there for a short time and then rises to the decomposition of the next component that can react at the electrode, since in the immediate neighbourhood of the electrode the first substance is quickly exhausted. Experimental results which are in full agreement with theoretical predications show that the duration of the current (the 'transition time') at the particular decomposition potential is directly proportional to the square of the concentration of the corresponding species in solution. If therefore the square root of the transition time is plotted against the volume of titrant two lines of different slopes result, and their intersection marks the end-point. Reilley and Scribner employed this procedure for the titration of Cu with EDTA [55-75]. When account is taken of the considerable outlay for apparatus and the fact that this and other titrations can be carried out far more simply, it becomes obvious that this method has little chance of gaining a footing in analytical practice.

5.9. Thermometric titrations

This technique, known also as enthalpy titrations, is based on the exploitation of the heat effects that accompany complex formation. In the procedure described by Jordan [57-97] the concentrated standard solution of titrant (approx. molar) is allowed to flow from a burette at a constant rate into the sample solution, which is contained in a calorimeter. Temperature changes are measured with a thermistor that forms part of a Wheatstone's bridge and displayed with a pen-recorder. The titration curves (temperature against volume of titrant) show a break at the end-point. Determinations of Pb, Cu,

Zn, Cd, Ni, Co, Ca, and Mg were carried out by titrations down to about 0·5–2 mM per litre with an accuracy of 3%. Harmelin [62-44] reports on similar results with a large number of divalent cations.

The titration curves can also be used to calculate heats of reaction. The possibility of consecutive titrations is of interest when several metals are present together, even when the stability constants of their complexes are of comparable magnitude. In these titrations, besides the ratio of the stability constants, the difference between the heats of formation of the metal complexes is of paramount importance. The situation is particularly favourable for the determination of Ca and Mg with EDTA, since their respective enthalpies are not only very different but even of opposite sign.

Up to now the method has hardly been evaluated practically. Probably this will be speeded up by an apparatus devised by Priestley *et al.* [63-66; 63-67] by which the procedure is greatly simplified by automation. The volume of titrant can either be read off on a counter or the entire titration curve can be presented on a pen-recorder. Many types of titration (not all of them complexometric) have been studied in this way.

5.10. Radiometric titrations

The principle of this relatively new technique can best be explained by an example, viz. the complexometric titration of zinc. Some solid silver iodate containing radioactive silver is added to the zinc solution that is to be titrated. This gives a heterogeneous system, since this salt is only sparingly soluble. When EDTA solution is added the zinc ions are complexed first of all and not until this has been completed is any of the silver transformed into its EDTA complex – with a sudden concomitant increase in the solubility of the radio-active silver iodate. The end-point can therefore be established by following the solubility of the precipitate during the course of the titration. This can be done for each point on the titration curve by removing a test-portion of the solution by filtration, measuring the radioactivity of the filtrate, and then returning the test-portion to the reaction vessel. If the count is plotted against the volume of titrant two lines result whose intersection marks the end-point. The theo-retical assumptions in such titrations have been discussed by Braun and his co-workers, who also give some practical examples [58-32; 64-14; 64-15; 66-1]. The application of precipitation membranes has been described [64-13]. The determination of Ca, Sr, Mg [59-60], and of Cu and Zn [60-133] in pure solution has been investigated.

Applications to analytical practice would seem to be limited, since these metals can be determined far more easily by other methods. However, in some special cases the technique seems to offer advantages. For example, Müller reports on some titrations in which radioactive $^{110m}AgIO_3$ is employed as indicator [66-71]. Ishibashi *et al.* [66-70] report the following 'titrations' of Co. To a series of aliquot portions of the sample solution spiked with ^{60}Co are added increasing and known volumes of EDTA. The uncomplexed Co is separated from the CoY by means of a cation exchanger and the clear supernatant liquid is counted. The results are plotted to give a 'titration curve'. Solutions as dilute as $10^{-6}M$ in Co can be analysed.

CHAPTER SIX

Different Types of Titrations and the Accuracy Attainable

This chapter will deal with all the various possible types of titrations. The accuracy attainable in each case depends on the course of the titration curve, and it can be predicted if all the relevant equilibrium constants are known. It is also possible to specify the conditions under which one metal can be determined in the presence of others and what effects masking agents have upon these conditions.

6.1. Direct titration

In a direct titration the solution of the metal that is to be determined is adjusted to a definite pH, generally by the addition of buffering materials that as a rule form complexes with the metal. Then a solution of a chelating agent, generally a standard solution of EDTA, is run in from a burette until there is a visible change in colour, or until a sudden change in optical density, electrode potential, conductivity, diffusion current, or other indication of the end-point is detected instrumentally.

As already described in Chapter Two, the end-point is characterized by a jump in the value of pM which should be as large as possible. The precise value of pM at the equivalence point (which will be denoted by $pM_{eq.}$) can be calculated from eqn. (61):

$$pM_{eq.} = 0.5 \, (\log K_{M'Z'}^{\text{eff.}} + pM_t) + \log \alpha_{M(A)} \tag{61}$$

This expression is derived from eqns. (30), (31), and (32) together with eqn. (25) by setting $a = 1$ (equivalence point) and noting that $[M]'$ can be neglected in comparison with $[MZ]$, since at the end-point of the titration practically all the metal has been transformed into its complex. The term pM_t in (61) signified the negative logarithm of the total concentration of metal ($pM_t = -\log [M]_t$).

In actual titrations, however, the end-point located by the colour change of an indicator will never occur at the exact value $pM_{eq.}$ but at an end-point which will deviate more or less from this ideal value: we will denote this by $pM_{end.}$. The difference

$$\Delta pM = pM_{end.} - pM_{eq.} \tag{62}$$

is a measure of the titration error. It is easily shown* [63-64] that this error can be calculated from eqn. (63).

$$\text{Titration error as a percentage of } [M]_t = \frac{460 \cdot \Delta pM}{([M]_t \cdot K_{M'Z'}^{eff.})^{\frac{1}{2}}} \tag{63}$$

Equation (63) demonstrates what is already obvious from a purely qualitative consideration of the titration curve, namely that the discrepancy between the value of pM at the experimental end-point and at the true point of equivalence, has a smaller effect on the titration error the steeper the pM jump at the end-point. If $[M]_t$ is large the first branch of the curve lies far down on the pM axis and if $K_{M'Z'}^{eff.}$ is large the branch that follows the jump in pM will be high up; it thus follows that the pM jump will be longer and steeper the more the product in the denominator of eqn. (63) increases in value.

* At the true point of equivalence the concentrations of non-chelated metal and of complexing ligand are exactly equal, and practically all the metal is present in the form of the complex MZ:

$$[M]'_{eq.} = [Z]'_{eq.}; \; [MZ]_{eq.} = [M]_t \tag{1*}$$

When combined with equation (2*):

$$K_{M'Z'}^{eff.} = [MZ]/[M]'[Z]' \tag{2*}$$

eqn. (1*) gives rise to (61). If at the actual end of the titration the concentration of unchelated metal and of chelating agent amount to $[M]'_{end.}$ and $[Z]'_{end.}$ respectively the titration error expressed as a percentage of $[M]_t$ will be given by:

$$\text{Error} = 100([Z]'_{end.} - [M]'_{end.})/[M]_t \tag{3*}$$

Consider now the differences:

$$\Delta Z = [Z]'_{end.} - [Z]'_{eq.}; \; \Delta M = [M]'_{end.} - [M]'_{eq.} \tag{4*}$$

So long as we remain in the immediate neighbourhood of the end-point [MZ] is practically equal to $[M]_t$, and according to (2*) the product

$$([M]'_{eq.} + \Delta M)([Z]'_{eq.} + \Delta Z)$$

will be constant, and this leads to:

$$\Delta M = -\Delta Z \tag{5*}$$

Introducing (5*) and (1*) into equation (3*) we find:

$$\text{Error} = -100.2 \cdot \Delta M/[M]_t \tag{6*}$$

But since $d[M]/[M] = d \ln [M] = -2 \cdot 3 \cdot d(pM)$, the relationship

$$\Delta M/[M] = -2 \cdot 3 \cdot \Delta pM \tag{7*}$$

holds so long as we confine our attention to the linear portion of the pM jump. If the value of ΔM from (7*) is substituted in (6*) and [M] is replaced by $[M]'_{eq.}$ using eqns. (1*) and (2*) we obtain eqn. (63).

With a well-chosen indicator (log K_{MD}^{eff} should be as near to $pM_{eq.}$ as possible) it is feasible to arrive at the correct value of pM to about ± 0.2–0.5 units with the naked eye alone. Since in most cases the concentration of the metal to be titrated lies between 10^{-2} and $10^{-4}M$ (on average $[M]_t = 10^{-3}$), it becomes obvious that the effective stability constant of the chelated complex that is formed must be at least 10^7 if the titration error is to be less than 1%. This condition is easily met in the titration of Mg with EDTA at pH 10, whereas barium lies just on the permissible limits. By using instrumental endpoint detection the correct value of $pM_{eq.}$ can be located more precisely, and thus ΔpM and the resultant titration error can be kept smaller. However, in many complexometric titrations the jump in pM is often so large that very accurate results can be achieved even with visual observations. Thus according to (63) the error in determining cadmium in $0.1M$ NH_4OH at pH 10 ($K_{M'Z'}^{eff.} = 10^{12.5}$) with $[M]_t = 10^{-4}$ and $\Delta pM = 0.5$, is only 0.01%. Under these circumstances it is not at all essential to locate $pM_{eq.}$ very precisely, for at the endpoint the indicator will change very abruptly from one coloured form into the other.

It should be pointed out that in calculations with eqn. (63) no account is taken of the amount of metal that is still bound to the indicator at the end-point: however, this can be neglected provided the concentration of indicator is small in comparison with $[M]_t$.

6.2. Back-titration

Direct titration is naturally preferable to back-titration. However, the former cannot always be carried out, as, for example, when no indicator exists for the metal to be determined, or if it reacts too slowly with the complexing agent in the standard solution, or if it cannot be kept up in solution at the pH most favourable for the titration. Then one turns to a back-titration in which a small known excess of EDTA is added to the metal being determined and the excess is measured with a second cation M^*. If M reacts only slowly with the complexing agent it is necessary to warm the solution before back-titrating: alternatively, one must wait until M is fully complexed. If M would be thrown out of solution at the pH used, an excess of titrant is added in acid solution, and not until afterwards is the pH raised to the value appropriate for the back-titration. Naturally there has to be an indicator suitable for M^*, the cation with which the back-titration is to be carried out.

In back-titrations both the error in the strength of the complexone solution as well as that in the titration value of the solution of M* comes into play, and this is especially unfavourable when a large excess of EDTA has been taken and the final result has to be calculated from the difference between two numbers of similar magnitude. The error which results from the value of $pM_{eq.}$, which should prevail at the end-point, not being exactly equal to $pM_{end.}$ is no greater in the back-titration than in a direct titration using M*, and its magnitude can be calculated in the same way by eqn. (63). The theoretical value of $pM^*_{eq.}$ follows from eqn. (61) using the appropriate constants and the value of $\alpha_{M^*(A)}$ for the metal M*, which serves for the back-titration. The conditions prevailing at the end-point are, of course, different, depending on whether MZ is inert or whether, like M*Z, it quickly establishes equilibrium with uncomplexed cations and the ligand of the complexing agent. This leads us directly to the discussion in the following two sections.

6.2.1.

If the metal M that is to be determined reacts only slowly with the complexone, then MZ dissociates slowly and the curve for the back-titration is completely unaffected by the presence of MZ, since this behaves as a completely inert species. The resulting curve is therefore identical with that of the direct titration of M*, but followed out, of course, in the reverse direction.

Good examples are provided by the titration of Cr(III) and Co, in which the back-titration of Co is carried out after its oxidation to the kinetically inert Co(III) complex. The complexes CrY^- and CoY^- are so inert that the back-titration of excess EDTA can be carried out in strongly acid solution by a metal M* which forms a more stable complex than CrY^- or CoY^-, e.g. by Fe(III), Th, or Bi.

The complexonates of a few other elements whose reactions with EDTA are not outstandingly slow are nevertheless sufficiently inert to interfere remarkably little in back-titrations. At pH 3 the aluminium complex AlY^- has an effective constant of only $10^{5.5}$; nevertheless, back-titration of excess EDTA with Fe(III) presents no difficulties because the exchange reaction $AlY^- + Fe^{3+} \longrightarrow FeY^- + Al^{3+}$ is sufficiently slow. Similarly, in the determination of Ni the excess of EDTA can be back-titrated in acid solution even with Bi, provided the solution is cooled down with ice to slow down the exchange reaction $NiY^{2-} + Bi^{3+} \longrightarrow BiY^- + Ni^{2+}$.

6.2.2.

If MZ is not an inert complex the effective stability constant of M*Z must not be greater than that of MZ: otherwise after reaching the equivalence point of the back-titration more M* would react with MZ to give M*Z and M, and this must not occur if the indicator is to respond sharply to M. Because of this displacement reaction the end-point is impossible or at least ruined, since the decrease in pM* cannot take place sufficiently abruptly. This point can be taken care of in selecting the cation to be used for the back-titration. The effective stability constant of M*Z must be smaller than that of MZ, but it must be greater than 10^7:

$$K_{M'Z'}^{eff.} > K_{M*'Z'}^{eff.} > 10^7 \qquad (64)$$

If the conditions of eqn. (64) are fulfilled the back-titration again gives a titration curve (pM* against volume of titrant) identical with that of the direct titration of M* – but traced in the opposite direction. It may also be that the complex MZ takes no part in the re-actions in this instance, not for kinetic reasons but due to its thermo-dynamic stability. Here again $pM_{eq.}^*$ at the equivalence point can be estimated from eqn. (61) and the titration error due to $pM_{end.}^*$ not being identical with $pM_{eq.}^*$ from (63).

The conditions of eqn. (64) are almost always met if Mg is chosen for the back-titration, for most metals form more stable complexes than it does with EDTA. In practice, Mg is used especially often for back-titrations, and the fact that Erio T is a readily available and a good indicator for this cation is an additional advantage. However, since the stability constant of MgY^{2-} is not much above the reliable lower limit specified in eqn. (63), the end-points it gives are not al-ways particularly sharp. Generally speaking, the conditions of (63) are also fulfilled by Mn^{2+}, and in point of fact the end-point is sharper in back-titrations with this cation. If the complex MZ of the metal M being determined is very stable many other metals can fill the role of M* in back-titrations: examples are Cu and Zn, which can also be used in acid solution.

6.3. Substitution titrations

If a direct titration is impossible for the reasons given in Section 6.2 a substitution titration can also be carried out in place of a back-titration. The solution of the metal to be determined is then usually treated with the magnesium complex of EDTA, whereupon the following exchange takes place:

$$M^{\nu+} + MgY^{2-} \longrightarrow MY^{\nu-4} + Mg^{2+} \qquad (65)$$

provided the complex with M is more stable than MgY^{2-}, which is almost always the case. In solution, therefore, the cation $M^{\nu+}$ is replaced by an equivalent amount of Mg^{2+}, which is titrated in its place, since several good indicators are available for it. If the direct titration of $M^{\nu+}$ is impossible because it reacts too slowly with EDTA, then reaction (65) will also be slow, and time must be allowed for it to go to completion, heating if necessary to accelerate this process, before starting to titrate the liberated Mg. If there is any danger of M being thrown out of solution at the pH needed for the titration the magnesium complexonate is introduced into the acidic solution before this is subsequently made ammoniacal.

The metal to be titrated can be replaced by metals other than Mg, and of course the complexing ligand need not be EDTA. To name but one example, exchange may be effected with dissolved or suspended Cu diethyldithiocarbamate, $Cu(DITICA)_2$ [64-70]. See also the cyanide methods of Section 6.6. To generalize, the metal M to be determined reacts with M^*Z to give M^* and MZ, and subsequently the cation M^* is titrated in place of M. The displacement represented by

$$M + M^*Z \longrightarrow M^* + MZ \qquad (66)$$

will, of course, only proceed from left to right if the effective stability constant of MZ is greater than that of M^*Z. Furthermore, the effective stability constant of M^*Z must be at least 10^7 if the titration error is not to be too large, and the conditions imposed by eqn. (64) must equally be fulfilled in a substitution titration. The titration curve for a mixture of MZ and M^* (pM* against the volume of titrant) is practically the same as if MZ were not present, so that $pM^*_{eq.}$ can be obtained with eqn. (61) and the titration error from (63). Under certain conditions the replacement reaction between M and M^*Z can be forced to the right even when $K_{MZ} < K_{M^*Z}$ by adding selective auxiliary complexing agents which greatly increase $\alpha_{M^*(A)}$, the distribution coefficient for the ion M^*, so that $K^{eff.}_{M^*Z'}$ becomes greater than $K^{eff.}_{M^*Z'}$. Thus barium, for which there is no good indicator, can be replaced by Zn if ammonia is added to the solution until $[NH_3] \sim 1$, whereby $K^{eff.}_{Zn'Y'}$ is reduced to about $10^{7.4}$, i.e. to a value lower than that of K_{BaY}. The solution containing Ba to be determined is treated with the complex ZnY^{2-}, made ammoniacal, and the liberated Zn is then titrated with EDTA. The jump at the end-point at $pZn_{eq.} = 14.3$ (according to eqn. (61)) is still very well developed and can be detected using Erio T. Indeed, the sharpness of the colour transition is even better than might be expected from a combination of Figs. 15 and 25, because in the strongly ammoniacal solution the indicator has the composition $ZnD(NH_3)^-$ and not ZnD^-, and for this reason the effective indicator constant is greater than that calculated by eqn. (50) from the figures given in Table 12.

The exchange reaction (66) is also of significance in the direct titration of M if a second metal M* with its appropriate indicator is used to locate the end-point. In this case the equilibrium position of (66) is normally to the left, so that virtually no substitution of M by M* takes place and the concentration of the added M*Z is small compared with $[M]_t$. Nevertheless, when all the metal M has been complexed a jump in pM* takes place, as explainedi n Chapter Two, because there is an abrupt increase in log [Z] at this stage: it is this jump in pM* which, through the indicator which responds to M*, reveals the end of the titration of the metal M.

The end-point in the complexometric titration of innumerable metals can thus be detected with a mercury electrode if a small amount of the mercury complex HgY^{2-} is added to the solution. In (66) M* can now represent Hg^{2+}, and the indicator is the electrode which displays the jump in pHg. Further examples are afforded by titrations with a mixture of PAN and the copper complex of EDTA as indicator. The dyestuff PAN is a very selective and sharp indicator for Cu^{2+}. If, for example, Al is being titrated (M = Al^{3+} and M* = Cu^{2+} in eqn. (66)) practically no substitution of Al^{3+} by Cu^{2+} takes place when CuY^{2-} is added to Al^{3+} because the stability constant of AlY^- is smaller than that of CuY^{2-}. However, in consequence of equilibrium (66) the concentration of Cu^{2+} is nevertheless large enough for the Cu–PAN complex to be formed: when all the Al has been titrated there is an abrupt increase in the value of pCu, so that the Cu–PAN complex is broken down with a concomitant colour change which reveals the end-point.

6.4. Collective and selective titrations

If a solution contains two metals there is generally no problem in determining their total concentration. If the cation M forms the more stable and the cation M* the less stable complex, then, on introducing the chelating agent, first M and then M* will become bound, and the indicator chosen for the titration must be the one which responds to a jump in pM* which will take place when complexing of both metals has been completed. As discussed in Section 6.3 under substitution reactions, the presence of MZ resulting from the complexing of the first metal has practically no influence upon the titration curve (pM* against volume of titrant), so that the value of $pM^*_{eq.}$ can be calculated for eqn. (61), and once again the titration error can be estimated from $\Delta pM = pM^*_{end.} - pM^*_{eq.}$. The effective stability constant for M*Z must, of course, be used for both equations.

If the effective stability constants of the two complexes MZ and M*Z are sufficiently different, and if a sufficiently large jump in pM occurs upon the complete complexing of M which can be revealed by an indicator suitable for M, then the determination of M is possible in the presence of M*. M can thus be determined selectively in the presence of M*, and this result, taken together with the result of a titration for total concentration, makes it possible to determine each of the metals quantitatively. To assess the feasibility of carrying out such selective titrations and stepwise titrations the course of the pM curve must be considered, though in fact we need only study the region covering the first break in the curve, since we can calculate the pM value ($pM_{eq.}$) prevailing at the point of equivalence and discuss the error resulting from the fact that in practice we do not exactly hit on this pM value at the end-point, since the experimental value ($pM_{end.}$) deviates somewhat from the theoretical ($pM_{eq.}$).

Right up to the end of such a selective titration of M the second metal M* remains virtually entirely uncomplexed, and its concentration therefore remains constant, $[M^*]' = [M^*]_t$. However, one result of the presence of M* is that the jump in pM is not so great, because in the presence of M* the complex MZ has a smaller effective stability constant than if no second metal were there. The calculation can be carried out if in eqn. (67):

$$K^{eff.}_{M'Z'} = [MZ]/[M]'[Z]' \qquad (67)$$

the value of $[Z]'$ is taken as the total concentration of the complexing agent that is not combined with the metal M. Thus:

$$[Z]' = \Sigma[H_jZ] + [M^*Z] = [Z](\alpha_{Z(H)} + \alpha_{M^*} - 1) \qquad (68)$$

for we must now include in the customary summation for the forms of the ligand that are not combined with the metal M that amount, $[M^*Z]$, which is blocked by the second metal. The distribution coefficient $\alpha_{Z(H)}$ is calculated as usual with eqn. (18) and α_{M^*} is calculated with eqn. (69):

$$\alpha_{M^*} = 1 + \frac{[M^*Z]}{[M^*]} = 1 + [M^*]_t \cdot \frac{K_{M^*Z}}{\alpha_{M^*(A)}} \qquad (69)$$

where $\alpha_{M^*(A)}$ is the distribution coefficient according to eqn. (23) for the metal M*. (The coefficient α_M has already been written out in eqn. (28) for the case M* = Ca.) Of the terms in the brackets in (68), the figure 1 can be neglected in comparison with $\alpha_{Z(H)}$ or α_{M^*}, and only the first of these cases is of interest here, for if $\alpha_{M^*} < \alpha_{Z(H)}$, M*Z is such a weak complex that M* does not interfere at all in the titration of M. We thus obtain the following expression for the effective stability constant of the complex MZ in the presence of the second metal M*:

$$K_{M'Z'}^{\text{eff.}} = \frac{K_{MZ} \cdot \alpha_{M^*(A)}}{[M^*]_t \cdot K_{M^*Z} \cdot \alpha_{M(A)}} \tag{70}$$

This constant can be used in eqn. (67) to find the pM value at the equivalence point in the selective titration of M in the presence of M*. Thus:

$$pM_{\text{eq.}} = 0.5 \, (\log K_{MZ} - \log K_{M^*Z} + \\ \log \alpha_{M(A)} + \log \alpha_{M^*(A)} + pM_t + pM_t^*) \tag{71}$$

Furthermore, the constants from eqn. (70) can be introduced into eqn. (63) to evaluate the error resulting from a deviation ΔpM in the value of pM at the end-point from the theoretical value:

$$\text{Titration error as a \% of } [M]_t = 460 \cdot \Delta pM \cdot \left(\frac{[M^*]_t \cdot K_{M^*Z} \cdot \alpha_{M(A)}}{[M]_t \cdot K_{MZ} \cdot \alpha_{M^*(A)}} \right)^{\frac{1}{2}} \tag{72}$$

This equation shows that the error in the selective titration is proportional to the ratio of the effective stability constants of the complexes of the two metals. If ΔpM is set equal to 0.5 (as in a usual titration) and the total concentration of the two metals is taken as approximately the same ($[M^*]_t/[M]_t = 1$) we see from eqn. (72) that the ratio ($K_{M^*Z}^{\text{eff.}}/K_{MZ}^{\text{eff.}}$) must not exceed 10^{-4}–10^{-5} if the error is to remain less than 1%. The metal that is to be determined selectively must therefore form a complex more stable by a factor of 10^4–10^5 than the accompanying metal. For example, Ca cannot be determined in the presence of Mg with EDTA, although this is possible with EGTA (Tables 2 and 3). When end-points are detected instrumentally it is possible to set ΔpM less than 0.5 so that acceptable results can be obtained when there is a smaller difference between the stability of MZ and M*Z.

It is worth noting that the error does not depend upon the absolute concentration of the metal that is being titrated but on the ratio $[M^*]_t/[M]_t$; in consequence, it is independent of the dilution, a result which contrasts with the titration of a metal on its own, where the error increases as the solution becomes more dilute (eqn. (63)).

On the other hand, according to eqn. (71), $pM_{eq.}$ increases with the dilution. This opens up the possibility of choosing the correct concentration such that $pM_{eq.}$ coincides with the pM value of the indicator at its transition point. The apparently anomalous observation that the end-point in a selective titration becomes sharper as the solution is diluted has actually been made by several authors.

Equation (72) shows that the error depends not on the absolute values of the two stability constants but on their ratio. Under certain circumstances, therefore, it may pay to use a complexing agent that forms weaker complexes than EDTA, provided there is then a greater difference between the stabilities of MZ and M*Z. For example, in the titration of Ni in the presence of Mn, triethylenetetramine ($\log K_{Ni(tren)} = 14\cdot8$; $\log K_{Mn(tren)} = 5\cdot8$) is preferable to EDTA ($\log K_{NiY} = 18\cdot6$; $\log K_{MnY} = 13\cdot4$).

Selective titrations are often rendered possible by the use of complexing agents which increase the value of $\alpha_{M*(A)}$, the distribution coefficient of the metal in whose presence the metal is being determined. This is essentially a matter of masking, as it is called.

The complexes of La and Zn with EDTA have very similar stabilities, so that only the total concentration can be determined by titration. But if KCN is added only the Zn is complexed (the value of $\alpha_{Zn(CN)}$ increases to 10^{11} with $0\cdot1M$ KCN at pH 8), so that the ratio of the effective stability constants of LaY^- and ZnY^{2-} now reaches a value of about 10^{10}. Thus we have *masked* Zn with cyanide so that it no longer interferes in the titration of La.

Another example is the titration of Zn in the presence of Cd. The complexes of these two cations with EDTA have almost identical stabilities, and titration in an ammoniacal solution only gives the total concentration. But if potassium iodide is added to give a concentration of at least $0\cdot5M$, then virtually only the cadmium is complexed as CdI_4^{2-} and $\alpha_{Cd(I)}$ rises to about 10^5: the effective stability constants of CdY^{2-} and ZnY^{2-} now differ by a factor of about 10^5, and the selective titration of zinc is possible in the presence of Cd. Conversely, if the object is to determine Cd in the presence of Zn, then it is better to use EGTA as the titrant, for K_{CdZ} and K_{ZnZ} already differ by a factor of $10^{2\cdot2}$ (Table 3). If now an auxiliary complexing agent is added which binds the zinc preferentially, for example, the polyamine 'tren' (Table 1), the selective titration of Cd in the presence of Zn is rendered possible.

Further discussion of masking will appear in Chapter Seven.

6.5. Complexometry combined with precipitation reactions

6.5.1.

It is often possible to deal with one component that has to be determined in a solution by separating it in the form of a sparingly soluble precipitate that contains a metal that can be determined complexometrically. This precipitate is then separated, collected, washed, and after being redissolved the metal in it is determined complexometrically. The principle was first applied to the determination of sodium [52-25] by titrating the zinc in the precipitated $NaZn(UO_2)_3Ac_9$.aq., and the determination of phosphate by titrating the magnesium in precipitated $(NH_4)MgPO_4$.aq. Other examples are the determination of pyrophosphate through the titration of Zn in precipitated $Zn_2P_2O_7$ [55-11] or of Mn in $Mn_2P_2O_7$ [55-109]; similarly, Mo and W by titration of calcium in $CaMoO_4$ or $CaWO_4$ respectively [55-99]. The titration of lead can follow the selective precipitation of PbO_2 [55-74], and other recommended procedures are the titration of calcium after precipitation of $CaSO_3$ [54-5] and that of cobalt and nickel after precipitation as sulphide by thioacetamide [55-63; 55-77]. Pectin can be determined by precipitation with calcium and determination of the Ca in a solution of this precipitate. A whole group of organic compounds and pharmaceuticals can be determined using this principle (see Part Two).

6.5.2.

Many substances can also be determined by the device of adding an excess of some metal that can be determined complexometrically and which forms a precipitate with the component that is to be determined. But in this case, instead of isolating the precipitate and determining its metal content, it is the excess of metal in the filtrate that is titrated. One method for determining sulphate is based on this principle [53-1]: here after precipitating $BaSO_4$ with Ba the excess is back-titrated with EDTA. Chromate, oxalate [55-22], sulphide [55-13], and other anions can be determined by analogous methods.

6.6. Cyanide methods

The anion CN^- is distinguished among monodentate ligands by the high stability of the complexes it forms with certain metals. This is illustrated by the following stability constants:

$$Ni(CN)_4^{2-}, \beta_4 = 10^{28}; Ag(CN)_2^-, \beta_2 = 10^{22}; Hg(CN)_4^{2-}, \beta_4 = 10^{36}$$

Of all the ligands that can only occupy a single coordination site, the cyanide ion alone can compete with EDTA and displace this from a metal complex. This forms the basis for the complexometric determination of cyanide as well as for the titration of silver, palladium, gold, and the halogens.

6.6.1.

To determine cyanide, the sample is made ammoniacal and an excess of $NiSO_4$ is added. Excess free Ni^{2+} is then back-titrated with EDTA using murexide [52-32].

6.6.2.

Silver, palladium(II), and gold(III) react with the tetracyano-nickelate ion in the following manner:

$$Ni(CN)_4^{2-} + 2Ag^+ \longrightarrow Ni^{2+} + 2Ag(CN)^- \qquad (73)$$

$$Ni(CN)_4^{2-} + Pd^{2+} \longrightarrow Ni^{2+} + Pd(CN)_4^{2-} \qquad (74)$$

$$Ni(CN)_4^{2-} + Au^{3+} \longrightarrow Ni^{2+} + Au(CN)_4^- \qquad (75)$$

To determine the precious metals [52-26; 53-21; 55-1] an ammoniacal solution of pure, recrystallized $K_2Ni(CN)_4$ is prepared and added to the test solution: the nickel liberated is then titrated with EDTA. The ammoniacal solution of $K_2Ni(CN)_4$ will also dissolve the sparingly soluble silver precipitates of AgCl, AgBr, AgI, and AgSCN, so that, as shown in eqn. (73), a titratable amount of Ni is liberated exactly equivalent to the silver content, and hence to the amount of halogen combined with it. To determine halogens (with the exception of fluoride) or thiocyanate, silver nitrate is added and the precipitated silver salt is collected, washed, and introduced into ammoniacal $K_2Ni(CN)_4$ and the Ni liberated is titrated [52-26].

6.7. Amalgam methods

In a short communication Wehber [55-15] drew attention to the possibility of determining substances with oxidizing properties by allowing them to react with liquid amalgams and then carrying out a complexometric titration of the metal ions that go into solution. However, he has published nothing further in this field. Scribner and Reilley [58-13] followed up the suggestions and made a thorough study of the possibilities and obtained extremely interesting results. By choice of a suitable amalgam and by working at the correct pH and if necessary adding other reagents, selectivity in the determination of mixtures can be improved. Of especial significance, however, is the possibility – and it is really astonishing that it has as yet hardly been exploited – of carrying out an indirect complexometric determination of substances, and especially of organic compounds, that cannot directly be titrated complexometrically. A few nitro- and nitroso-compounds were analysed as examples of this procedure, and the results were extraordinarily satisfactory.

Four types of procedure can be distinguished in the applications of amalgam methods.

6.7.1. TYPE I

This is exchange of a metal which can be readily titrated or masked for another metal which does not show these properties. A mixture of Pb and Mn provides an example. An aliquot portion of the sample is used to obtain the total concentration of the two metals. In a second aliquot portion Pb is exchanged for Zn with zinc amalgam: zinc can easily be masked by cyanide, so that the determination of Mn presents no problems. The extension of this principle to ternary mixtures, e.g. Pb–Zn–Ca, presents no real difficulties if de-masking of zinc cyanide by formaldehyde is introduced as an additional stage.

6.7.2. TYPE II

This is exchange of a metal that is not readily titrated or masked for one that is. A mixture of Cu and Ni provides an example of this type, which is the exact converse of Type I. Once again the total concentration of the two metals is determined in one aliquot portion, and in another the mixture is shaken with lead amalgam, whereupon Cu is replaced by Pb but Ni remains unaffected. After adding cyanide to mask the nickel the Pb (which is equivalent to the original Cu) is titrated. Nickel is obtained by difference.

6.7.3. TYPE III

This procedure makes use of the difference between the equivalent weight factor for the reduction and the titration reaction. Thus the titration of a mixture of Pb and Bi with EDTA gives the total concentration, with each metal reacting in the ratio 1 : 1 with the titrant. In an aliquot portion Bi^{3+} is replaced by Pb^{2+} by shaking with lead amalgam, in which process three gram atoms of lead go into solution for every two gram atoms of bismuth. Subsequent titration with EDTA requires a greater consumption of reagent than in the first case. For the two titration values two simultaneous equations can be set up, and their solution leads to the concentration of each of the component metals.

6.7.4. TYPE IV

This group includes all substances for which a direct complexometric titration is impossible or presents considerable difficulties. For example, the ions Ag^+ or Tl^+ can readily be exchanged for Zn^{2+} (or Cd^{2+}) by shaking with zinc (or cadmium) amalgam. Many reducible organic substances, such as nitro-compounds, can also be determined volumetrically in this way, as can reducible anions, such as arsenate or chromate.

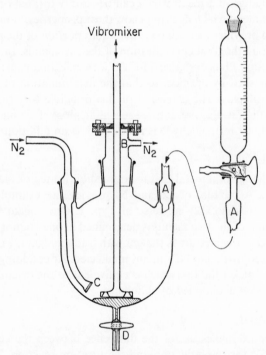

Figure 31. Apparatus for reduction with amalgams.

It is not difficult to see how combinations of several of these types of procedures carried out on a sufficient number of aliquot portions of the test solution can be adapted to the analysis of multi-component mixtures. Very many additional examples are to be found in the literature, and no great stretch of imagination is needed to add to their number. It should be pointed out that the examples quoted

above include mixtures which could also be analysed complexometrically in other ways and serve here only to illustrate the general
principles of amalgam methods.

Practical procedures must include the following manipulations:
(i) buffering the test solution to the desired pH; (ii) de-aeration of the
test solution so that the base metal in the amalgam that is to serve as
the reductant should not also be oxidized by atmospheric oxygen;
(iii) addition of amalgam and bringing this into intimate contact with
the test solution; (iv) removal of the amalgam after reduction is
complete; (v) titration with EDTA.

Scribner and Reilley describe a special shaking vessel which
facilitates all these operations. The apparatus shown in Fig. 31 has
also proved its worth. After introducing the buffered test solution
into the flask it can be de-aerated by leading in nitrogen through the
sintered-glass frit C. The desired amount of amalgam (about 5 ml) is
then introduced through A from the calibrated tube and the vibro-
mixer is set into operation, whereby the liquid amalgam is broken up
into a suspension of fine droplets. After reduction (5–10 minutes) the
amalgam is run off through the capillary tap D, indicator is added
through A, a burette is attached to the standard joint at A, and the
titration is carried out.

Scribner and Reilley thoroughly investigated the effects of pH, and
have published a good deal of information on the most suitable
amounts of amalgam to use, on their concentration, and on the rate
of reduction. In those cases where the reduction is carried out at low
pH it is possible for a small amount of the metal dissolved in the
amalgam to go into solution with evolution of hydrogen, so that
blanks must be determined with solutions of known composition and
taken into account in the calculation. Doležal and Zyká [64-67] have
extended the technique to many more organic and inorganic ions
and have given practical examples.

Aseni [61-154] used the method for actual analyses of Ag–Bi and
Ag–Cd mixtures.

6.8. Alkalimetric–complexometric titrations

In Section 6.3 of Chapter Three it was shown how Mg^{2+} can be
substituted for the metal that is to be determined, the determination
being completed by the volumetric estimation of Mg. It should be
noted that it is also possible to exchange the metal that is to be determined by an equivalent amount of hydrogen ion and to finish the
determination volumetrically by titration with NaOH [46-2]. This

can be brought about if in reaction (65) the proton complex of EDTA (i.e. HY^{3-} or H_2Y^{2-}) is used instead of the magnesium complex, MgY^{2-}, whereupon the reactions (12) and (13) displayed in Section 2.1 now take place.

To carry this out in practice two burettes are set up side by side containing $0.1M$ NaOH and $0.1M$ EDTA (as Na_2H_2Y). The solution containing the metal that is to be determined, which must not contain any buffering material, is treated with a substantial amount of NH_4Cl and a few drops of a pH indicator (methyl-red–bromocresol-green mixture) and titrated to pH 5, where the mixed indicator has a neutral grey shade. Then some EDTA is run in and the acid liberated is neutralized by NaOH from the second burette. These two operations are repeated successively until the addition of a portion of EDTA solution no longer produces any lowering in pH. The result is calculated from the amount of NaOH used: each gram ion of OH^- corresponds to $\frac{1}{2}$ gram atom of metal ion. The method suffers from the disadvantage that the solution containing the metal to be determined must be exactly neutralized, and this often presents difficulties due to the hydrolysis of the cation. The method is, however, to be recommended as a pre-titration stage if a back titration is to be carried out, for it does not require a special indicator appropriate to the metal being determined, and furthermore, by this method the addition of a large excess of EDTA can be avoided. The back-titration is commenced immediately after the alkalimetric titration is finished by adding buffer and the indicator appropriate to the metal that is going to be used for the back-titration.

The colour change of the pH indicator is only reasonably sharp if the added excess of EDTA is kept as small as possible, for metal-free EDTA acts as a buffer between pH 5 and 7 ($H_2Y^{2-} \longrightarrow HY^{3-}$). Since NTA has no buffer region between pH 3 and 10, the colour transition is sharper if the alkalimetric–complexometric titration is carried out with the di-alkali metal salt of NTA [48-1; 48-2]. The hydrogen ions resulting from the pH effect shown in reaction (11) of Section 2.1 can also be titrated iodometrically by adding iodide–iodate mixture and titrating the liberated iodine with thiosulphate. If Z^{3-} represents the anion of NTA we have:

$$M^{\nu+} + HZ^{2-} + \tfrac{1}{6}IO_3^- + \tfrac{5}{6}I^- \longrightarrow \tfrac{1}{2}H_2O + \tfrac{1}{2}I_2 + MZ^{\nu-3} \qquad (76)$$

CHAPTER SEVEN

Selectivity

7.1. Introduction

Aminopolycarboxylic acids of the type of EDTA form complexes of sufficient stability with nearly all the polyvalent cations on which to base titration procedures. The advantage of this almost universal potentiality for complex formation is that a large number of different elements can be titrated with a single volumetric solution; but it has the concomitant disadvantage of low selectivity.

Nevertheless, the formation constants of several EDTA complexes are sufficiently different for certain metals to be determined selectively in the presence of others. As shown in Section 6.4, this requires a difference between the two effective stability constants by a factor of at least 10^4–10^5. A glance at Table 2 shows that Al, the lanthanides, and the divalent cations of Co, Ni, Cu, Zn, and Cd ought to be capable of selective titration in the presence of the alkaline earths, and Fe^{3+}, Ga^{3+}, In^{3+}, Tl^+, Bi^{3+}, Zr^{4+}, and Th^{4+}, which form complexes with stability constants of over 10^{20}, must be specially capable of selective titration in the presence of practically all the divalent cations: all this is realizable in practice. In a practical procedure the pH is reduced to such an extent that those metals which are not to be included in the titration value are virtually not complexed at all, since due to the increase in $\alpha_{Z(H)}$ the effective stability constants of their complexes become very small indeed. The titration of Al and the divalent heavy metals in the presence of the alkaline-earth metals is carried out at pH $\sim$5 and in the selective titrations of trivalent cations in the presence of divalent cations at pH 3 or below. At the same time there is the added advantage of protecting the metal that is to be determined from the danger of precipitation of its hydroxide without having to add the auxiliary complexing agents which would be necessary at higher pH values. Any auxiliary complexing agent (e.g. NH_3) would lead to a decrease in the effective stability constant of the EDTA complex of the metal that is being titrated, and hence to a deterioration in the degree of selectivity.

In certain cases it is possible to carry out a selective titration that is impossible with EDTA by using some other complexone. Thus Ca can be determined complexometrically in the presence of Mg with EGTA but not with EDTA. The polyamines, which form complexes practically only with cations of transition metals and the univalent and divalent d^{10} cations (Table 1), should be more selective titrating agents than the polyaminocarboxylic acids. For this reason certain polyamines have already often been proposed for use as volumetric titrants. But this feature has only been applied to a very limited extent in analytical practice because the stability of complexes with polyamines decreases very rapidly with decrease in pH in consequence of their high pK_j values.

Since there are so few cases where the stability constants of the EDTA complexes of the metals in a particular mixture lie more than 4–5 logarithmic units apart, complications often arise when one wishes to determine one metal complexometrically in the presence of others. It is often necessary to precede the actual titration by some separation stage or stages. On the other hand, the addition of a masking agent can often achieve the desired effect, and there are even possibilities of procedures involving successive titrations by which several metals can be determined consecutively.

7.2. Separations

An immediate possibility consists of throwing down one component as a solid and filtering off the precipitate. The customary precipitating agents of gravimetric analysis are often used for this purpose. It must be stressed, however, that the requirements for a precipitate that is subsequently to be weighed are not the same as for one which is subsequently to be titrated. In the latter case it is essential to have a clean separation from all the other metals in the mixture, but it is not necessary for the precipitate to have a strictly stoicheiometric composition. For example, sulphides which are generally useless for gravimetric analysis provide good precipitates for separations and can be thrown out with advantage by using thioacetamide in place of hydrogen sulphide [55-62]: 'cementation' reactions [62-152] can also be used for separations.

The introduction of ion-exchangers [58-42; 59-6] is of great significance, as Wünsch was the first to point out [57-111]. The considerable dilution of the test-solution that generally results from passage through the ion-exchanger is not normally a drawback to the complexometric finish. The various possibilities are discussed in Samuelson's monograph [63-82].

Liquid–liquid extraction was used quite early on in conjunction with EDTA titrations [53-40; 54-38]. For further possibilities the book by Freiser and Morrison [57-141] should be consulted.

7.3. Masking

The term 'masking' or 'camouflaging' applied to one constituent in a solution is taken to mean that it is made to react chemically with some added substance (the masking agent) so that it can no longer take part in some subsequent chemical reaction. A metal can be masked by its precipitation from solution, or it can be made into a complex, or oxidized or reduced. Frequently it is necessary to employ more than one of these techniques at the same time. For example, Cu can be masked by reduction to Cu(I) with ascorbic acid and complexation with I^- [64-95]. Furthermore, of course, more than one masking agent may be added at the same time: cf. Patrovsky [65-114].

7.3.1. MASKING BY PRECIPITATION

If interfering metals can be precipitated in some form they are eliminated from the actual solution, and the distinction between masking and actual separation is simply that the precipitate is not removed by filtration before titration is carried out. It will be obvious that the concentration of the cation masked by a precipitation reaction (which can be calculated from the known solubility product of the precipitate) must be made so low that no reaction can take place with EDTA.

One such case of precipitation masking was employed early on in the titration of Ca in the presence of Mg in strongly alkaline solution (for the titration curve see Fig. 14), whereby $Mg(OH)_2$ is thrown down. But this procedure already pin-points the weak features of masking by precipitation, viz. the low results due to partial coprecipitation of the metal that is to be titrated: moreover, the adsorption of the indicator on the precipitate produces a trailing endpoint. Even when no adsorption takes place the presence of a precipitate makes it harder to detect the colour change.

Other metals, such as Fe(III), Ti(IV), Zr, or Sn(IV), can be masked as precipitated hydroxides, but again there is the difficulty of coprecipitation of the metal that is to be determined, for the hydroxides of practically all metals are sparingly soluble and tend to come out together. The fluoride ion is a rather more selective precipitant and is suitable for masking Ca and Mg [45-74; 61-81; 62-89] and Al [54-74] or Zr [64-47]. Under certain circumstances Fe(III) can be masked with F^- [61-109; 62-23]. Small amounts of Al in alkaline solution can be masked with sodium silicate [61-64], and phosphate removes Ti(IV) even from its complexonate – a result which can be exploited for the indirect determination of Ti, since the complexes FeY^- and AlY^- are not so attacked [54-42]. Sulphate ions are still more selective, and can be used to precipitate Pb when Bi is to be titrated [60-95] or for masking Ba. Sulphide ions are highly selective, but, in view of the dark colour of the resulting precipitates, are only useful for masking traces of metals. Thus traces of heavy metals (Fe, Cu) which may be present when determining the hardness of water are removed by the addition of a sulphide to prevent their blocking the Erio T used as indicator. Diethyldithiocarbamate can be used for the same purpose.

7.3.2. MASKING THROUGH COMPLEX FORMATION

The overwhelming majority of cases of masking depend upon the formation of stable, soluble complexes. The masking agent must therefore be a selective complexing agent which will react as little as possible with the metal M that is to be titrated but will combine with the interfering metal M^* as strongly as possible so that the ratio of the effective stability constants $(K_{MZ} \cdot \alpha_{M^*(A)}/K_{M^*Z} \cdot \alpha_{M(A)})$ is increased as much as possible by its presence. The addition of A (the auxiliary complexing agent which we identify in this instance with the masking agent) ought only to increase the distribution coefficient $\alpha_{M^*(A)}$ of the foreign metal and not $\alpha_{M(A)}$, the coefficient for the metal that is to be titrated, and the ratio of the stability constants ought to be at least 10^4–10^5 if the titration error (given by eqn. (72)) is to remain small. It is thus possible to distinguish between total and partial masking of the foreign metal. In the first case $K_{M^*Z}^{eff.}$ has become so small that M^* no longer reacts with the titrating solution, whereas with partial masking, after complete complexing of M by the complexing agent in the titration solution, M^* also reacts, when in this case a successive titration becomes possible.

The most valuable masking agents are those that form very stable complexes with the foreign (interfering) metals so that large values of $\alpha_{M^*(A)}$ are achieved with small concentrations of them; this implies that the stability constants in eqn. (23) must be large so that the concentration term, [A], can be kept small. If the complexes M^*A_j are only weak ones high concentrations of the masking agent would have to be used, and this in turn would produce an unfavourable increase in ionic strength. However, there are cases where even weak complexing agents can produce good results, as in the masking of Cd by iodide, which is added in amounts up to 50 g KI per 100 ml to permit the titration of Zn in the presence of a 3,000-fold excess of Cd [64-148; 65-113]. High concentrations of Cl^- are employed to mask Bi and Pb [64-144]. Large amounts of sulphate are also used to mask Th [59-110; 60-46].

The following sections will describe the most important masking agents, arranged in order of increasing selectivity, and following the order, oxygen donors and fluoride ion, nitrogen donors, sulphur donors, iodide and cyanide ions.

The hydroxide ion, OH^-, is mainly used as a precipitating agent, although it forms a soluble complex in masking Al as aluminate in the titration of Ca. On the whole, fluoride is a weakly selective precipitating agent; but it forms soluble complexes with (III)- and (IV)-valent metals, and F^- can be used to mask Sn(IV) in the titration of Sn(II) [59-25]. Almost all metal carbonates are sparingly soluble, but UO_2^{2+} gives a soluble carbonato-complex so that CO_3^{2-} can be used to mask U(VI) [52-25]. Fe(III) [56-98; 58-38] and to a small extent Al [56-98] can be masked with pyrophosphate $P_2O_7^{4-}$. Acetate is only a weak complexing agent, but it enables the titration of Ga to be carried out, possibly by preventing the formation of hydroxy-complexes [61-78]. Sulphosalicylic acid is a masking agent for Al in the titration of Fe(III) [58-62], and for Fe(III) as well as Al (in conjunction with F^- for Th) in the titration of the lanthanides [62-89]. Al can be masked by Tiron, which likewise is a derivative of phenol [55-68]. The anion of acetylacetone also coordinates through oxygen, and this enables this enolizable β-diketone to mask Fe, Al, Pd, and UO_2^{2+} [60-30]; it is especially useful for masking Al if Pb or Zn [63-14] or the rare earths [65-14] are to be titrated. Tartrate has been used as an auxiliary complexing agent in the titration of Pb in alkaline solution [52-24; 52-29] or to hold Mg in solution when Ca is titrated at pH 12 [61-88]. Tartrate can also be used in cooled solutions to mask small amounts of Fe and Al when Mn is being titrated [60-77] and to mask W [59-3] as well as Nb, Ta, and W in the titration of Co [59-2]. The increase in selectivity in titrations with EDTA that can be achieved by adding citrate has been demonstrated by Fritz and his co-workers [58-62]. Lactic acid will mask Ti(IV), Sn(IV), and Sb(IV) [65-46; 65-94]. Triethanolamine is also a slightly selective and weak complexing agent (since it coordinates through only one N^- but through three HO^- groups) which serves to mask the tervalent ions of Fe, Al, Mn [53-11], and Cr [61-94] in an alkaline medium. With Mn, atmospheric oxidation produces a green complex of Mn(III) whose colour is so intense that it swamps that of the indicator if large amounts of Mn are present.

The simplest complexing agent that coordinates through nitrogen is ammonia, which is so frequently employed as a buffering material in complexometric titrations. It prevents the precipitation of hydroxides by an ammine complexing agent by increasing the value of $\alpha_{M(NH_3)}$ for the metal. However, high concentrations of NH_3 may also mask the reactions of the metal with the indicator or even with EDTA: Cu, for example, can no longer be titrated using murexide. The polyamines provide unique complexing agents for Ni, Cu, Zn, and Cd. Tetraethylenepentamine is used in the titration of the alkaline earths or of Pb [59-114]. 1,10-Phenanthroline permits masking of the same metals as well as that of Mn in the titration of In [60-48].

Compounds that coordinate through sulphur are selective masking agents for those metals that form sulphides. Thiourea has often been used to mask Cu [57-107; 66-44], but if Fe(III) is to be titrated this must be protected from reduction by adding fluoride [61-95]. The masking by thiourea can readily be reversed by boiling with hydrogen peroxide [65-73]. Dimercaptopropanol or **BAL** (British Anti-Lewisite) [54-79] in conjunction with cyanide serves to resolve a variety of mixtures of metals: Unithiol [60-37; 60-38; 60-59; 60-60] and thioglycollic acid [61-96] behave similarly. Cd and Zn can be masked with cystein [61-114], and Cu, Zn, Cd, Ni, Co, and Hg with β-aminoethylmercaptan [63-75]. 3-Mercaptopropionic acid can be used to mask Pb [61-62; 63-76] and thiosemicarbazide to mask Hg [57-83] in the titration of Zn or Cd. Dimercaptosuccinic acid [64-119] at pH 6 masks Cu and at pH 10 it masks Ni, Pb, Co, and Cd as well. Dithiocarbaminoacetic acid has recently been studied as a masking agent [66-47]. Thiosulphate masks Cu [58-63; 59-49] and Hg [65-78].

The cyanide ion forms complexes with very much the same cations as NH_3, but the cyanide complexes are far more stable than the corresponding ammines. KCN is frequently used as a masking agent for Co, Ni, Cu, Zn, Cd, Hg, and the precious metals in the titration of the alkaline earths, lanthanides, Mn, and other metals. Under the right conditions Fe and Mn can also be masked [61-93]. Iodide provides a very selective reagent that is used in low concentrations for masking Hg(II), in particular [57-65]. As already mentioned, Cd and Pb can also be masked by high concentrations of iodide ions

[64-148; 65-113]. KI also masks Cu in the presence of ascorbic acid [64-95]. The fact that hydrogen peroxide find some use in masking Ti(IV) is of some interest. In acidic solution H_2O_2 prevents the titanium from forming hydroxy-complexes which only react slowly with EDTA, and thus enables the direct titration of this metal to be carried out [60-136]. Complete masking is effected in alkaline solution [57-3]. H_2O_2 is also a masking agent for V [64-27].

7.3.3. REDOX MASKING

In some cases the interfering metal can be masked by changing it into a different oxidation state.

In the case of Fe(III), masking can be effected by reduction [56-84] if hydroxylamine or ascorbic acid is added. The difference between the stability constants of Fe(II) and Fe(III) amounts to some 10 logarithmic units, so that in many selective titrations Fe(II) does not interfere where Fe(III) would. Masking of iron with cyanide also involves reduction, for $Fe(CN)_6^{4-}$ interferes less than the ion $Fe(CN)_6^{3-}$, which is strongly coloured and a moderately strong oxidizing agent. Přibil [61-93] has reported an interesting redox masking of Fe and Mn with KCN. In the titration of Zn [58-63; 59-49] or Ni [58-16; 61-108; 62-45] Cu(II) can be masked with $S_2O_3^{2-}$, whereby reduction to a thiosulphato-complex of Cu(I) takes place. As a rule, thallium does not interfere in the titration of other metals if it is present as Tl(I); reduction of Tl(III) can be brought about with cystein [61-114] or hydrazine sulphate. Mercury can be eliminated by reduction to the metal, either with ascorbic acid [56-84] or formic acid [62-22]. Since the metal itself separates out in this process, we are dealing simultaneously with precipitation masking.

Cations in higher oxidation states generally form more stable complexes with EDTA than in the lower states. Nevertheless, masking by oxidation is often a possibility, provided the metal to be masked is transformed by oxidation into a stable oxyanion, e.g. $Cr^{3+} \longrightarrow CrO_4^{2-}$. In the same way, if Mo, W, and V occur in lower oxidation states that interfere by forming complexes with EDTA they can be masked by oxidation to molybdate, tungstate, or vanadate respectively.

7.3.4. KINETIC MASKING

There are some cations that are capable of forming stable complexonates that do not interfere in the titration of other metals simply because they react too slowly with the solution of the complexing agent. Following C. N. Reilley, they can be described as 'kinetically masked'. An example is Cr(III), in whose presence other metals can be titrated with EDTA without any interference whatsoever, even though the stability constant of CrY^- is about 10^{23}. Indeed, mixtures of Cr(III) and EDTA must be allowed to stand for a long time before an appreciable amount of complexing takes place. Use is made of this fact in dealing, *inter alia*, with mixtures of Fe–Cr–Al [57-29].

In the nature of things it is impossible to effect kinetic masking of cations by adding other substances. But chemical reactions can always be slowed down by lowering the temperature. Thus, unlike other metals, Ni is only slowly displaced by Bi^{3+} from an ice-cold solution of its EDTA-complex, NiY^{2-}, and this can form the basis of a selective determination of Ni [55-66]. Complexes of Cr(III), Co(III), and Ni(II) are more effectively kinetically masked in their complexes with DCTA than in their EDTA complexes.

A kinetic effect is also exploited in the determination of Mg, Ca, Cu, Ni, or Zn in the presence of Al masked by triethanolamine. In such a solution the masked Al reacts slowly with the indicator Erio T and blocks it; but if the temperature is reduced below 5°C there is no interference.

7.4. De-masking

It is often desirable to be able to de-mask a previously masked metal at will and thereby make it once again accessible to titration. Suppose we have a mixture of the metals M and M*; perhaps M* is first masked and M is then titrated: then M* is de-masked and determined in its turn. De-masking is thus especially important for successive titrations.

In many cases de-masking can be effected just by changing the pH. Thus Mg is masked as $Mg(OH)_2$ during the titration of Ca in strongly alkaline solution; but if the pH is then lowered to 9–10 the Mg can be titrated as well. Al that has been masked as the aluminate ion is made accessible to titration by lowering the pH to 5. In other cases de-masking is brought about by increasing the pH as, e.g., with Th masked with sulphate [60-46] or Tl(III) masked with bromide ions [61-130]. In both cases de-masking to EDTA is only effective below pH 2, and because of the concomitant increase in the effective stability constants of the EDTA complexes (due to decrease in $\alpha_{Z(H)}$) adjustment of the pH to 4–5 makes the titration of Th or Tl(III) possible despite the presence of sulphate or bromide ions respectively.

If masking has been brought about by reduction the metal concerned can again be made accessible to titration by oxidation. Thus Fe masked by ascorbic acid can be changed back to Fe(III), and Cu masked by thiourea can also be de-masked by oxidation.

If masking has been brought about by complex formation the masking agent must be removed in order to effect the de-masking. Removal may be brought about by decomposition as, e.g., in the case of thiourea by boiling with H_2O_2, or by any other suitable reaction. Masking by F^- can be overcome with Be^{2+} (whereby BeF_4^{2-} is formed), and Hg(II) masked by SCN^- can be de-masked by adding Ag^+ [59-55]. The de-masking of Zn and Cd from their cyanide complexes has been extensively exploited. This can be carried out by adding formaldehyde [52-6; 53-27], which reacts with HCN to form the nitrile of glycollic acid:

$$4H^+ + Cd(CN)_4^{2-} + 4CH_2O \longrightarrow Cd^{2+} + 4HO \cdot CH_2 \cdot CN \qquad (77)$$

Chloral hydrate reacts in the same way [53-10]. The metals Co, Ni, Cu, and Hg remain completely masked during this process, and this is also the case when Zn is de-masked from its cyano-complex by acidification to pH 6 [59-26].

Further examples of de-masking will be found in the section on consecutive titrations.

The blocking of indicators, e.g. that of Erio T by Cu, can be annulled by adding methanol, ethanol, or acetone [62-138].

7.5. Consecutive titrations

Given a solution containing several metals that have to be determined, it is especially elegant if one can be determined after the other, and if possible with the same standard solution of complexing agent.

If the metals yield complexes with effective stability constants that differ sufficiently one from the other they will become complexed in turn on running in the titrant, and it only remains to be able to detect the points at which the complexation of each individual metal is completed. The naked eye is not ideal for observing the transition colour of indicators in this case, for the colour of one indicator interferes by making it difficult to detect the transition point of the others. It is therefore of advantage to use instrumental methods for detecting the individual end-points. In this way up to three metals can be determined in one operation wherein the pH may be increased after each stage. Examples are the amperometric titration of V(III)–V(IV) [60-189]; Fe–Mn, Cu–Ca, Bi–Pb–Ca [56-78] and Cu–Ni, Ni-Zn, Cu–Zn–

Ca [62-33], where the 'complexone-wave' serves to indicate the end-point. The mixtures In–Cd and In–Pb can be determined by square-wave polarography [62-36]. The stepwise titration of Cu and Zn with tetraethylenepentamine can be carried out using two polarized electrodes, and the procedure has been used in serum analysis. An amperometric sequential titration operates with solutions of two different complexones; first Ca is titrated with EGTA and then Mg with EDTA [64-6]. Potentiometric end-point detection is used in the consecutive titrations of Zn–Cd [58-60] and Pb, Cu, Ni, and Co [58-56] in the presence of alkaline earths. The following mixtures can be determined photometrically by a consecutive titration with EDTA without adding any organic indicator (i.e. by the technique of self-indication based on changes in the optical density of the cation being titrated, or of its complex, or of the complexing agent used for the titration): Fe–Cu [53-39], Bi–Cu [54-51], Ca–Mg [62-27], Pb–Bi [55-97]. Triethylenetetramine has been used for the self-indicating consecutive titration Cu–Ni [57-87]. The consecutive titration Ca–Mg can be followed thermometrically [57-97].

If it is proposed to use a dyestuff as a colour indicator there will be obvious advantages in consecutive titrations if one and the same indicator responds to all the metals that are to be determined. If two colour indicators have to be used the first should if possible be a mono-colour indicator that changes to its colourless form at the first point of equivalence and does not interfere with the determination of the colour transition of the second indicator. If two bi-colour indicators are used, photometric methods enable the end-points to be detected more precisely. Examples are the photometric determination of Ca at pH 10 with murexide followed by that of Mg using Erio T [61-50]; Zn at pH 6·8 and then Mg at pH 10, both using Erio T [53-33]; Bi at pH 1 and Pb at pH 5, both using xylenol orange [60-62] or gallein [61-32; 64-150]. The latter indicator also permits the determination of In at pH 3 followed by Pb at pH 5 [64-111]. Other examples are Cu at pH 5·1–5·4 and Zn at pH 9 using pyrocatechol violet [60-159], wherein more Cu must be present than Zn; Fe(III) at pH 3–4 and Mn at pH 6–6·5, both with methylthymol blue [58-102]; Th at pH 2 and the lanthanides in a urotropine buffer, both using arsenazo [62-19]. The combination Th–rare earths is of practical importance, and has been investigated quite frequently, cf. [63-106; 64-78]. Zr can be determined in 0·25–1N HCl or H_2SO_4 at 80–90°C followed by Th at pH 1·5–2·5, both using xylenol orange [63-8]. In the consecutive titration Fe(III)–Al using 3-hydroxynaphthoic acid this acts first as a colour indicator and then as a fluorescence indicator [61-44]. The consecutive titration Th–rare earths with xylenol orange gives better colour transitions with tri-ethylenetetraminehexa-acetic acid than with EDTA [63-27]. In titrating Fe(III) using salicylic acid, followed by the titration of Mn or Ca, Mg using Erio T, DCTA is preferred to EDTA as the titrant, for the iron is more effectively kinetically masked and the Erio T is less easily blocked [55-89].

It is also possible to follow the direct titration of one metal by the back-titration of a second, changing the pH if needs be in between. For example, Fe(III) can be titrated at pH 2, and then to determine Al a known excess of EDTA is added; after boiling to ensure the complete complexation of the aluminium as AlY^-, and then increasing the pH to 5, the excess of EDTA can be back-titrated with a standard ferric solution: salicylic acid serves for both titrations [62-13]. In a similar way Sc can first be titrated at pH 2 (xylenol orange), and then Al can be determined by back-titration of excess EDTA with Zn [65-36]. Similarly, Th can be titrated at pH 3 (xylenol orange) and then U is determined by reduction in the presence of EDTA and back-titration of the excess complexone with a Th solution [65-51]. Wehber titrates excess EDTA with a standard Cu solution using Bindschedler's green as indicator [57-6]. The metals Fe, Al, Mn, Ca, and Mg can be determined consecutively as follows [60-169]. Fe(III) is titrated first at pH 2 using sulphosalicylic acid; the pH is then raised to 3, and Al is determined as above by back-titration of excess EDTA with Cu using PAN as indicator; the solution is then divided into two equal portions, one is adjusted to pH 10 and the total concentration of Ca + Mg obtained by a back-titration with Cu using PAN; in the other half the Ca is precipitated as oxalate and the Mg determined as before. The disadvantage of such an extended series of operations is that the errors of individual titrations can become cumulative.

It is also possible to interpose a reaction other than a change of pH between the stages of a consecutive titration. Thus to determine both Fe(II) and Fe(III) a titration is first carried out with the exclusion of air, and then the whole is oxidized and titrated again so that the Fe(II) is then included in the total Fe(III) [55-23; 61-10].

Some good possibilities are presented by masking one metal, titrating the other, then de-masking the first and continuing the titration. With a mixture of Ca and Mg, Ca is first titrated at a high pH, while Mg is masked as its precipitated hydroxide; then the pH is

F

reduced and the Mg, which then goes back into solution, is itself titrated [57-124; 60-128; 61-88]. Bi can be titrated in the presence of Hg after this has been masked with KSCN; then the Hg is de-masked by adding $AgNO_3$ and titrated in its turn [59-55]. Pb can be titrated in the presence of Zn (or Cd) if the latter is masked with KCN; then by adding formaldehyde the second metal is de-masked and then titrated [52-29; 53-27; 65-119]. The de-masking is often brought about by changing the pH, as when Fe(III) is first titrated at pH 2 in the presence of Tl(III) that has been masked by bromide, using salicylic acid as the indicator; the pH is then raised to 4–5, when the Tl(III) can also be titrated with PAN as the indicator [61-130].

One procedure which exploits the kinetics of the complexation reaction is the consecutive titration Al–Cr(III) [59-90]. Here Al is first determined by back-titration of excess EDTA with Fe(III), using Tiron as indicator, after the solution has been kept for 5 minutes at 40–50°C to complete the formation of AlY^-: then a further excess of EDTA is added, and after all the Cr(III) has been transformed by boiling into CrY^- the excess of EDTA is again back-titrated.

In another sort of combined direct and back-titration, in which admittedly the amount of one of the components is obtained by difference, the two metals are first titrated together to give their total concentration; then one of the metals is masked, whereby an equivalent amount of the complexing agent is set free, which is then back-titrated with a standard solution of an appropriate metal. Sajó [61-128] has described determinations of this nature. The sum of Pb + Hg is first measured by back-titrating excess EDTA with Zn; then by adding NaCl the mercury is displaced from its chelate complex, HgY^{2-}, and the amount of EDTA liberated is determined by a further titration with Zn. In the same way the sum of Pb + Tl(III) is first obtained, and then Tl(III) is removed from its complex by reduction with sulphite, and the EDTA thus liberated is titrated with Zn. By a combination of these procedures all three metals Pb–Hg–Tl(III) can be determined. The ternary mixture Fe–Al–Ti was analysed by Sajó [54-42] as follows. First of all a titration is carried out to determine the total concentration: then Ti is displaced from its complex with phosphate and the EDTA set free is determined complexometrically: then Al is displaced from its complex with fluoride and the EDTA liberated is determined similarly. Scribner [59-111] handled mixtures of Mg–Mn in a similar way by first determining the total concentration, then masking Mg as MgF_2 and measuring the amount of EDTA set free.

Further examples of consecutive titrations will be found in Part Two of this book in sections immediately preceding the actual working instructions for individual elements. Stepwise titrations in conjunction with measurements on aliquot portions often permit of quite astonishing feats of analysis in the determination of the components of complex mixtures. By choice of more suitable indicators, pH, and masking and de-masking agents, further advances will certainly be achieved in this field.

7.6. Indirect analyses

These are based on the fact that from the known weight of a sample containing only two titratable metals and the result of the titration for their total concentration, two simultaneous equations can be set up whose solution yields the amount of each of the two components. The method has found a practical outlet in the analysis of Zr–Hf mixtures (q.v.). With mixtures of rare earths an average atomic weight can be determined from which conclusions can be drawn as to the composition of the sample. The procedure can be extended to mixtures containing more than two components if in addition to knowing the weight of the sample and the result of the titration for total concentration further experimental data are available (e.g. a titration value with partial masking) in order to provide further simultaneous equations. Unfortunately the errors in the results of such indirect analyses are generally quite large.

Part Two

Titration Procedures and Working Instructions

Titration Procedures and Working Instructions

After discussing the preparation and manipulation of stock solutions of titrants, indicators, and buffers, the second part of this book will deal individually with the metals and groups that can be determined complexometrically. As in Tables 2 and 3, the elements (metal ions and central elements of groups) will be classified according to the Periodic System and discussed in the order shown by the arrows in Table 13, where the numbers refer to the relevant sections. Cyanides will be dealt with under carbon, and thiocyanates under sulphur.

For each element a general account will be given of the various possibilities of complexometric titrations based on the pertinent literature; this will include details of suitable indicators and possible interferences. Following this come details of titrations of other substances in which interference is caused by the metal (or group) immediately under discussion – and how such interference can be overcome by masking agents. Examples of the application of the

TABLE 13

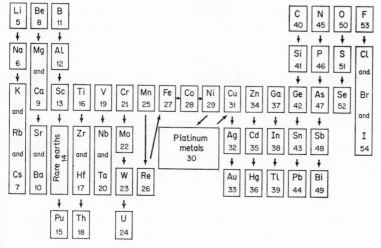

particular titration procedures to specific analytical problems then follow, with sufficient information being given to enable the desired determination to be carried out.

In the final section full working details are given for those titrations of which the authors have personal experience. In practice, these may have to be appropriately modified to take account of impurities that may be present. Detailed procedures will not be given for any of the analytical operations (e.g. precipitation) that may precede the actual complexometric titration. In many cases details of a procedure will be replaced by a cross-reference to another determination, as in the procedure for the determination of potassium, which depends on the complexometric titration of cobalt, or that for the determination of halogens, which depends on a silver titration.

1. VOLUMETRIC SOLUTIONS

1.1. Preparation

Among many other advantages of complexometry must be numbered one of very great practical importance, namely that the standard solutions can be prepared by direct weighing; provided the appropriate conditions are maintained, these solutions do not change their titre on keeping. However, a few hints and comments are worth noting.

In the vast majority of cases the standard solution of EDTA is prepared by dissolving the salt $Na_2H_2Y.2H_2O$ of formula weight 372·16. This is available commercially in a high degree of purity. According to our own observations, which confirm those of Blaedel and Knight [54-2], trade preparations have a moisture content of 0·3–0·5%. If this is taken into account direct weighing will yield a standard solution whose concentration is known with sufficient accuracy for most purposes. The moisture can be eliminated by drying at 80°C and 50% humidity (referred to 25°C), and 100% purity and constant weight is attained after four days [54-2]. Drying at a higher temperature is inadvisable, since water of crystallization is then lost. Complete dehydration is readily achieved by heating at 130–150°C. The composition of the product is strictly stoicheiometric, but unfortunately little advantage is gained, for the anhydrous salt is markedly hygroscopic. Thermogravimetric investigations [57-74; 60-85] confirm that the anhydrous sodium salt does not meet the strict requirements of a primary standard.

To prepare a 0·100M solution, 37·22 g of $Na_2H_2Y.2H_2O$ is dissolved in distilled water and made up to 1 litre. Less concentrated solutions can be prepared from this by dilution or preferably by taking smaller weights of the salt.

One extremely advantageous procedure, which, strange to say, has only rarely been used in practice, is to start with the free acid of formula weight 292·13. This acid is obtainable commercially in an analytically pure state, or it can be very readily obtained by acidifying a solution of the sodium salt with sulphuric or hydrochloric acid. The latter is preferable, as traces of iron are more easily removed. After collection on a filter, thorough washing with water, and drying at 110°C, the precipitate is non-hygroscopic, stable, and very pure indeed. Weighed portions are dissolved in aqueous alkali using two equivalents of NaOH or KOH per mole of EDTA. Ammonium hydroxide can also be used to dissolve the acid.

To prepare a 0·100M solution 29·21 g of the free acid and 9 g of NaOH are dissolved in about 400 ml distilled water. After being cooled to room temperature the volume is made up to 1 litre. Less concentrated solutions can be prepared by dilution or by weighing out proportionately smaller amounts.

1.2. Storage

Special attention must be given to the vessel in which a stock solution is stored, because in a very short time indeed quite appreciable amounts of alkaline-earth metals and other interfering cations enter the solution from the surface of the glass. Hard glass shows a lesser tendency to give up ions, especially if it has been in use for a long time or if it has been pre-treated with a hot alkaline solution of EDTA. Since flasks of artificial plastics are nowadays commonplace in laboratories, storage in such containers is unreservedly recommended, especially for very dilute solutions of titrants. Loss of titre in a plastic container is virtually zero. On the other hand, as studies by Flaschka and Sadek have shown [57-92], changes in the titre of solutions kept in glass containers can often become quite substantial, depending on their composition.

Incidentally, the use of plastic vessels is advisable for all other reagents, such as the buffers and the materials for back-titrations, especially when alkaline solutions are involved. Plastics are only unsuitable for a few indicators, such as PAN, because this indicator is soluble in organic material.

1.3. Standardization

For most purposes the molarity of a titrant can, as stated above, be calculated with sufficient accuracy from the weight of material taken, provided this comes from a reliable source. But if there is any suspicion of impurities – and these can also be introduced through the water used to dissolve it (from glass containers, defective distillation set-ups, exhausted ion-exchange resins, etc!) or if increased accuracy is essential standardization becomes unavoidable. The way in which this is carried out can, under certain circumstances, lead to remarkably discrepant results, especially when one has to deal with solutions containing impurities. For example, suppose the solution of EDTA is contaminated with a little Ca and Zn. If it is standardized against Mg at pH 10 only that amount of EDTA reacts that is not combined with Ca and Zn. But if the titration is carried out in the presence of KCN only the Ca ties up its equivalent amount of complexone. If the standardization is carried out against Cu at pH 4 only the zinc is active. Finally, had the solution been standardized against Fe at pH 2, neither Ca nor Zn interfere, and the full content of EDTA is titratable. Thus, according to the conditions used, no less than four different values for the molarity would be found. If the titre calculated from the standardization against Fe is used to calculate the result of a titration carried out at pH 10 an erroneous result would be obtained, and the standardization would have failed to achieve its aim. The golden rule in volumetric analysis is 'Standardize your solution under the same conditions as you will be using in the actual determination'. If, when preparing the standard solution, every trace of metal has really been rigorously excluded, the conditions become, of course, less critical.

1.4. Primary volumetric standards

Many substances can be used to standardize solutions of EDTA, but the following are particularly suitable. Primary standards for the determination of Ca and Mg deserve special attention, since these titrations are certainly carried out in the greatest number. Of commercially available substances, $CaCO_3$ is certainly the most suitable, for it is easily obtainable in a high degree of purity, it can be dried without having to fix the temperature within narrow limits, and it is not hygroscopic. Barcza [55-21] confirmed the carbonate as being the best in an investigation that included the oxalate, hydroxide, and sulphate.

Shead [52-45] recommended the acid calcium salt of malic acid, since this can easily be obtained pure in the form of its hexahydrate and can also be used to standardize alkali. The equivalent weight is very favourable, but the substance is not available commercially.

Among magnesium salts $MgSO_4.7H_2O$ and $Mg(IO_3)_2.4H_2O$ deserve special mention. To guarantee the exact content of water the former must be stored over a mixture of the heptahydrate and water in the proportions 5 : 1. The iodate has the larger equivalent weight, is stable on keeping, and thermogravimetry shows that it does not lose water until 110°C.

For general purposes of standardization many other substances have been recommended besides those already mentioned. A literature survey has been published by Vřeštál and co-workers [59-13], who made comparative studies and evaluated their findings statistically and called attention to a few new substances. Metals such as Cu, Ni, Zn, Fe, or Bi now come on the market in purities of 99·99% and better. However, with the exception of Bi, they all have relatively low equivalent weights. Bi has the further disadvantage that it is coated with a layer of oxide which has to be removed by washing with nitric acid. Among salts, for obvious reasons, the authors prefer those that crystallize without water of crystallization or are not hygroscopic when this has been driven off. ZnO can easily be obtained in a high degree of purity, but it has a low equivalent weight. From this point of view HgO is preferable, and it has been recommended by Barcza [59-56]. Vřeštál *et al.* regard $PbCl_2$ as a particularly valuable standard substance, for it is easily purified by recrystallization from water, contains no water of crystallization, is stable towards keeping and drying, absorbs practically no moisture, and has a high equivalent weight. $PbCl_2$ can be used for standardizations in both acid (acetate–xylenol orange) and alkaline media (tartrate–ammonia–Erio T).

Favourable equivalent weights can be achieved with salts of organic acids. Vřeštál and his co-workers [59-13] add bispyridino-cadmiumthiocyanate to the corresponding zinc salt recommended by Budĕšinský [56-83]. It was emphasized that drying under controlled conditions is necessary for the double salts of the Schönite type recommended by Flaschka [52-22].

Powell and his co-workers [60-100] recommended a primary standard of some theoretical interest. The complex CdHY (where Y = hydroxyethylethylenediaminetriacetic acid) can be obtained pure by recrystallization, is anhydrous, possesses a large equivalent weight, and can be used to standardize bases and complexones. However, for the latter purpose the tricarboxylic acid must first be destroyed by boiling with persulphate in acid solution; because of this inconvenience the procedure is likely to have little practical value.

1.5. Solutions of other complexones

Other frequently used solutions of complexones can be prepared similarly. Generally speaking, one starts out with the free acid, which is taken up in some water containing two equivalents of NaOH and then made up to a known volume. The amounts of acid needed can be calculated from the following formula weights: NTA, 191·14; DCTA, 346·34; EGTA, 380·35; DTPA, 393·18. The solutions are then standardized against a suitable metal.

1.6. Standard solutions of metals

Standard solutions of magnesium sulphate or zinc sulphate are needed for back-titrations. These are prepared by weighing out $MgSO_4.6H_2O$ or $ZnSO_4.7H_2O$ respectively and then standardizing with a solution of EDTA of known concentration. Solutions of $MnSO_4$ for back-titrations are prepared in a similar manner.

Solutions of bismuth nitrate are suitable for back-titrations in acid solution. 0·0100M $Bi(NO_3)_3$ is prepared by dissolving 2·090 g of the pure metal in about 7 ml conc. HNO_3 and diluting to 1 litre; the solution is simultaneously about 0·05M in HNO_3.

A 0·1000M solution of copper is prepared by dissolving 6·354 g of the purest metal in the least possible amount of nitric acid; the solution is then warmed to drive off nitrous gases, cooled, and made up to 1 litre with distilled water.

If the copper solution for back-titrations is to be standardized against one of EDTA it is often simpler to dissolve 25 g of $CuSO_4.5H_2O$ per litre.

With solutions of metallic salts the nature of the anion of the usual salt taken is generally unimportant. In a few procedures, however, the nature of the anion is significant, and in these cases this must be taken into account when selecting the salt that is to be used or by dissolving the metal itself in the appropriate acid.

1.7. Solutions of metal complexonates

0·1M MgY and 0·1M ZnY. The magnesium complex of EDTA is used in substitution titrations. The sodium salt $Na_2MgY.4H_2O$ is available commercially for this purpose, and provided it is sufficiently pure, this solid can be used directly. However, one must never fail to check whether the sample really does contain Mg and EDTA in the ratio 1 : 1. The pH of a solution of the salt should be between 8 and 9; after adding some $NH_3-NH_4^+$ buffer of pH 10, a drop of Erio T should yield a dirty violet coloration which is changed to blue by a single drop of 0·01M EDTA and to red by a single drop of 0·01M $MgSO_4$. The corresponding sodium zinc complexonate, $Na_2ZnY.4H_2O$, also crystallizes with four molecules of water and is treated in exactly the same way.

Solutions of MgY^{2-} or ZnY^{2-}, as well as those of CuY^{2-} and HgY^{2-} or of any of the other metal complexonates that are often needed, can also be prepared as follows. Approximately 0·1M solutions of EDTA and of the desired metal are prepared. Two or three 25-ml portions of the metal solution are then titrated with the EDTA; the average titration value is calculated, and this volume of EDTA is then added to 25·00 ml of the metal solution. In every case the correct stoicheiometry of the dissolved metal complexonate should be confirmed as described above for MgY^{2-}.

2. PREPARATION OF INDICATORS

The next section describes the preparation of solutions of the most important and most frequently used indicators. Where unstable solutions would result it is preferable to use a solid carrier [51-20].

2.1. Erio T

The dyestuff (0·2 g) is dissolved in triethanolamine (15 ml) and absolute alcohol (5 ml). Such solutions can be kept for at least a month.

To produce a dilute solid indicator a mixture of Erio T (0·25 g) and NaCl (100 g) is ground to a fine powder. Sugar can also serve as a carrier.

2.2. Murexide

About 0·5 g of the powdered red dye is suspended in a few ml of water and vigorously shaken. The undissolved portion is allowed to settle and the saturated supernatant liquor is used for titrations. Every day the old supernatant liquid is removed by decantation and replaced by pure water in order always to have a fresh indicator solution.

A finely ground mixture of 1 g murexide and 100 g NaCl (or sugar) can be used as a solid indicator.

2.3. Metalphthalein

A few drops of conc. ammonia are added to a mixture of cresol-phthaleinbismethyliminodiacetic acid (0·18 g) and naphthol-green (0·02 g; No. 5 in Schultz Dyestuff Tables) and the volume made up to 100 ml. The solution keeps for about a week.

2.4. Tiron

A 2% solution of the di-sodium salt of pyrocatecholsulphonic acid is used. The colourless solution can be kept for a long time.

2.5. Pyrocatechol violet

The commercial indicator (0·1 g) is dissolved in water (100 ml). The solution keeps for several months.

2.6. Eriochromeazurol S

The dyestuff (0·4 g; Geigy & Co.) is dissolved in water (100 ml). The solution keeps well.

2.7. Dithizone

Freshly recrystallized dithizone (25 mg) is dissolved in ethanol (100 ml). This solution must be made up afresh after several days.

2.8. Xylenol orange

The indicator (0·1 g) is dissolved in a little water by adding a few drops of 1M NaOH and the volume made up to 100 ml. The solution keeps for several weeks.

2.9. PAN

A solution of the indicator (0·05 g) in ethanol or methanol (100 ml) can be kept for months.

3. BUFFER MIXTURES

The following are recommended for the various pH ranges:

pH 1. Supplies of 1M HCl and 1M HNO_3 are kept for this purpose.

pH 2–4. Glycine and its hydrochloride. The solid aminoacid is used to buffer strongly acid solutions as required: *p*-chloroaniline can be used for the same purpose.

pH 4–6·5. Acetate–acetic acid. Stocks of 1M acetic acid and 1M sodium are kept ready to be mixed as required.

pH 6. Solid urotropine (hexamethylenetetramine) or its 20% solution.

pH 6·5–8. Triethanolamine and its hydrochloride. Solutions of 1M $N(C_2H_4OH)_3$ and of 1M HCl are prepared and mixed as required. In using this base it must be borne in mind that triethanolamine is a complexing agent and, as such, forms complexes of considerable stability with many cations. This is undesirable in most titration procedures, as it causes a decrease in the jump in pM at the end-point. The concentration of free triethanolamine must therefore be kept as low as possible. Malates and veronal have the same disadvantages in this pH region. Phosphate buffers are completely useless owing to the precipitation of metal phosphates. An ideal buffer material for the pH range from 6 to 8 would be imidazole, for its 'onium salt has a pK value almost dead on 7 and its structure is not suitable for the formation of metal chelate complexes. However, in view of its high cost, the use of imidazole as a buffering material scarcely comes into question.

In the neighbourhood of pH 7 buffering with NH_3–NH_4^+ is also very good. Naturally the buffer capacity is only of small magnitude, e.g. $[NH_4^+]$ ~0·1 and $[NH_3]$ ~10^{-3}. However, the pH can be easily adjusted by spotting on indicator paper, and a drop of NH_4OH can be added if it sinks too low. In this way almost ideal conditions are created, since there is extremely little complexing of the metal by the components of the buffer mixture.

pH 8–11. Ammonia–ammonium salts. Solutions of 1M NH_4OH and 1M NH_4Cl are kept in readiness for mixing as required. Glycine and borate buffers cannot be recommended for this pH region, since complexes of aminoacetates and borates are generally much more stable than the corresponding ammine complexes. Moreover, borates yield precipitates with some cations.

pH 10. A buffer of pH ~10 is very often required. This can be prepared by mixing ammonium chloride (70 g) with concentrated ammonia (570 ml of density 0·90) and making up to 1 litre. Only the purest chemicals should be used for this buffer. Commercial ammonia often contains small quantities of alkaline earths, and it is best to prepare the ammonia solution freshly by dissolving NH_3 from a gas-cylinder in distilled water.

4. ADDITIONAL CHEMICALS

4.1. Masking agents

KCN, Na_2S, NaF, triethanolamine, tiron, diethyldithiocarbamate, dimercaptopropanol (BAL).

4.2. De-masking agents

30% formaldehyde solution, chloral hydrate.

4.3. Reducing agents

Ascorbic acid, hydroxylamine sulphate.

4.4. Oxidizing agents

H_2O_2, $K_2S_2O_8$.

All these chemicals ought to be free from calcium and other metals. In this respect especially the KCN, Na_2S, and formalin should be tested with Erio T after buffering to pH 10.

5. LITHIUM

According to De Sousa [61-173], it is possible to determine lithium indirectly in the following way. Lithium chloride is first separated from the chlorides of sodium and potassium by well-established classical methods. It is then taken into solution and the chloride ion precipitated with silver nitrate. The silver chloride is next separated and the nickel set free by double decomposition with tetracyano-nickelate (cf. Method 32) is titrated with EDTA. The procedure has been worked out on the micro-scale and is accurate to about 0·5% with quantities of from 50 to 500 μg Li_2O with a tendency towards low results. For amounts in the range 10–50 μg the results are about 1–2% low.

6. SODIUM

Sodium can be determined indirectly by first precipitating it in the form of its triple salt, $NaM(II)(UO_2)_3$ (acetate)$_9$, and then titrating the metal M(II) when this precipitate has been redissolved. Flaschka [52-55] first described this method for a microdetermination in which the zinc triple salt was used. The indicator Erio T used for the zinc titration is blocked by uranyl ions, so that carbonate ions must be added to mask them. Masking of the uranium can be omitted if the titration is carried out in 50% aqueous ethanol using dithizone as the indicator; this is particularly advantageous in serum determination [59-36; 59-37].

Besides zinc, other metals can be used for precipitating a triple salt, e.g. magnesium [60-19], and cobalt and nickel [57-44]. Doležal *et al.* [62-122] determined 10–250 μg Na by precipitation as the cobalt triple salt followed by a determination of the uranium content of this precipitate by reduction with ascorbic acid, addition of excess EDTA, and back-titration with Th using xylenol orange as indicator.

Gagliardi and Reimers [58-94] report that a solution containing sodium ions will displace an equivalent amount of magnesium from an ion-exchanger (a strongly acidic sulphonated resin such as Lewatit S 100) in the magnesium form; the magnesium in the eluate is titrated with EDTA using Erio T as indicator. Since potassium behaves in the same way, only the total concentration of the two metals can be found in a mixture of the two.

The determination of sodium is used in the analysis of serum [59-36; 59-37], glass [62-121], and waters connected with the production of natural gas [60-19].

6.1. Procedure

REAGENTS: Zinc uranyl acetate and wash-liquors needed for dealing with the precipitate; standard solution 0·01M EDTA; Erio T; buffer of pH 10; approx. 1M HCl; approx. 1M $(NH_4)_2CO_3$ or the solid itself.

PROCEDURE: The filter with the washed precipitate of sodium zinc uranyl acetate, which should contain about 5 mg Na, is moistened with 3–5 ml HCl in a beaker, and after the addition of about 50 ml water the suspension is boiled up and passed through a filter. The filtrate and washings that amount to about 100 ml are neutralized by ammonium carbonate, and a slight excess is added to keep the uranium in solution. To the resulting slightly yellow solution is now added 2 ml of buffer and the indicator, and titration is carried out with the standard EDTA to the appropriate change of colour. Because of the presence of uranium, this will be a mixed colour, a yellowish-red before the end-point, and a greenish-blue after it.

REMARKS: The titration of the zinc can be effected anywhere in the pH range 7–10. It is therefore insensitive to the concentration of ammonium ion present, so that quite large amounts of HCl can be used to dissolve the initial precipitate. Indeed, it is not at all essential to add the buffer, and ammonia alone can be used.

The precipitation of the triple salt can be carried out according to one or other of the innumerable modifications of the classical procedure. Generally speaking, alcohol saturated with the triple salt has been recommended as a wash-liquor. However, by this treatment zinc is sometimes hydrolysed and thrown out of solution, so leading to high results; in our experience, which confirms Kathen and Lang's proposals [A-10], acetic acid saturated with triple salt is to be preferred. For determinations on the micro-scale a sintered-glass filter, or better still the use of a filter stick [59-37], is recommended. Naturally the operations can also be carried out in a centrifuge cone. In this way a few tenths of a milligram of sodium can be reliably determined, especially if the titration is carried out with 0.001M EDTA.

If the determination is only carried out occasionally it is hardly worth while making up a solution of ammonium carbonate, since this does not keep; the neutralization can be done just as effectively with the solid salt.

7. POTASSIUM (RUBIDIUM AND CAESIUM)

Potassium can only be determined indirectly. Most methods are based on the precipitation of $K_2M[Co(NO_2)_6]$ and titration of the cobalt contained in the precipitate. Flaschka *et al.* [56-122] were the first to report on such a determination for the case $M = Na$. After the precipitate obtained by a standard procedure had been dissolved in dilute HCl some urea was added to remove oxides of nitrogen and the cobalt titrated in weakly ammoniacal solution with murexide as indicator (cf. 28). A much sharper end-point is obtained by using CuY–PAN in acetic acid solution [57-86]. The cobalt titration is also possible by back-titration of excess EDTA with zinc against Erio T [57-68; 60-79], but this has to be carried out rapidly because of the danger of blocking the indicator. Sen [57-108] determines the cobalt by titration in a mixture of acetone and water containing a lot of thiocyanate.

The difficulties in this method do not derive from the titration of cobalt, for this can be carried out with sufficient accuracy. Certainly the unfavourable factor ($1Y \equiv 1Co \equiv 2K$) must be taken into account; but the real problem comes in the precipitation stage. The precipitate is relatively easily soluble, so that an appreciable excess of precipitant has to be used. This implies that thorough washing is necessary – but naturally this must not be pushed too far. An even more serious difficulty is the fact that the composition of the precipitate is intimately bound up with the experimental conditions and, moreover, the ratio K : Co = 2 : 1 expected for the stoicheiometric composition is generally not realized exactly. It is for this reason that the method has been mainly restricted to serum analysis, in which it is relatively easy to establish reproducible conditions for the precipitation.

Naturally, the use of the method is not confined to K^+, for it can be employed without modification for NH_4^+, Rb^+, and Cs^+. Sen [58-99] used M = Ag in the precipitation stage. Precipitates obtained in this way were less soluble, and their composition was closer to that of the idealized formula. The method is, however, of limited use, since chlorides interfere. In addition to ammonium and the heavier alkali-metal ions, Tl(I) can also be determined in this way.

Special care must be taken to ensure the absence of ammonia. Evaporation of the solution and fuming-off of ammonium salts is always an obvious, but a time-consuming and troublesome, possibility. If the concentration of ammonia is less than $0.02M$, or if it can be brought below this level without an unacceptable decrease in the potassium concentration, Holasek *et al.* [60–185] showed that the elimination of the ammonia is easily possible by working in the presence of formaldehyde. NH_3 combines with formaldehyde to form urotropine (hexamethylenetetramine), which does not react with the precipitant below the concentration specified. Pečar [60-1] recommends oxidation with bromine in an alkaline medium for the removal of NH_3; any excess of bromine is finally removed by reduction with formic acid under slightly acid conditions.

Gagliardi and Reimers [58-94] find that K displaces an equivalent amount of Mg from an ion-exchange resin (a strong sulphonic resin such as Lewatit S 100) in the Mg^{2+} form. The Mg in the eluate can then be titrated complexometrically using Erio T. Since Na behaves in the same way, only the total concentration of an admixture can be obtained. The process of ion-exchange must be preceded by the complete removal of all other cations, and this severely limits the practical application of the method.

De Sousa has proposed a rather roundabout method for the determination of K [60-25] and Rb [61-28]. K or Rb are precipitated as perchlorates in the usual way; the precipitates are separated, dried, and reduced to KCl (or RbCl). The chloride is then thrown down as

AgCl and determined complexometrically after the addition of tetra-cyanonickelate (cf. 32). The tetraphenylborate method devised by Flaschka and Sadek [58-37] is very accurate, and the results are substantially independent of the composition of the sample solution. The tetraphenylboron anion reacts with Hg(II) as follows:

$$B(C_6H_5)_4^- + 4HgCl_2 + 3H_2O \longrightarrow 4Cl^- + 3H^+ + 4Hg(C_6H_5)Cl + H_3BO_3$$

It is a remarkable fact that this reaction also occurs with Hg(II), even when it is combined with EDTA. This opens up the possibility of precipitating K (and other substances such as Rb, Cs, Tl(I), NH_4^+, and many organic substances that give insoluble products with the tetraphenylborate ion), removing the precipitate and dissolving it in acetone and allowing it to react with a solution of HgY. The amount of EDTA thus set free is titrated with Zn in an acetate buffer using PAN as indicator. Since $1K \equiv 1B(C_6H_5)_4^- \equiv 4Hg \equiv 4Y \equiv 4Zn$, the factor is extraordinarily favourable, so that drop errors, etc, have practically no effect on the accuracy, which is almost entirely dependent on how well the operations of precipitation and washing have been carried out.

Cigolea *et al.* [60-196] have described a determination of caesium in which it is precipitated by $KBiI_4$, and excess bismuth in the filtrate is back-titrated with EDTA.

The methods are employed in the analysis of serum [56-122; 57-86; 60-1; 60-79], urine [61-167], and erythrocytes [60-79] for potassium.

8. BERYLLIUM

By radiometric measurements of the distribution coefficient of beryl-lium on an ion-exchange resin in the presence and absence of EDTA, Merrill, Honda, and Arnold [60-107] calculate a value of log $K = 9.8$ ($\mu \sim 0.1$M) for the stability of the Be–EDTA complex. In a later research Adamovich and Napadailo [61-165] took up the same problem and investigated the ion-exchange equilibrium at pH ~ 2 and $\mu = 0.04$–0.05. According to their results, there exists a pro-tonated complex for which they calculate a stability constant $K = [BeHY^-]/[Be^{2+}][HY^{3-}] = 10^{8.8}$. They dispute the existence of a complex BeY^{2-}, since, on the basis of their measurements, this ought to have a stability constant of about 10^{17}, which is far too large in comparison with the stability of MgY^{2-}. Moreover, this is in contradiction with the fact that Be is precipitated by NH_3 even in the presence of EDTA; for the solubility product of $Be(OH)_2$ is not small enough to permit precipitation if the stability of BeY^{2-} were really as great as this. From unpublished data obtained by titrating a mixture of Be and EDTA with NaOH, H. Flaschka calculates a value of $\log K_{BeY} = 9{\cdot}2$ (room temperature; $\mu \sim 0{\cdot}1$). Only a short portion of the curve in the pH range 4–5 could be used for this calculation, since at a higher pH the form of the curve is distorted by hydrolysis phenomena, and the onset of precipitation is indicated by the solution becoming opalescent. The possible existence of hydrolysed species and protonated complexes was not taken into account in the calculation. Starostin *et al.* studied the solubility of $Be(OH)_2$ in solutions of EDTA at various pH values and at 20°C ($\mu = 0{\cdot}3$). The authors came to the conclusion that the complex BeY^{2-} exists at pH 7–7·5 and calculate $\log K_{BeY} = 10{\cdot}2$. Above pH 7·5 the complex $BeY(OH)^{3-}$ exists, and for this they give $pK_b = 5{\cdot}4$.

The relatively low stability constant together with the great tendency of Be^{2+} to be hydrolysed makes titration difficult. Be can be determined in a weakly acidic acetate medium by back-titration of excess EDTA with a standard cobalt solution in an acetone–water mixture (1 : 1) containing thiocyanate. In this case the addition of the organic solvent obviously causes a sufficient enhancement of the effective stability of the Be–EDTA complex [55-103].

An indirect method depends on the following principle [59-71]. Beryllium is precipitated as

$$[Co(NH_3)_6][(H_2O)_2Be_2(CO_3)_2(OH)_3].3H_2O$$

from a solution containing ammonium carbonate by adding a luteo-cobalt(III) salt. EDTA can be included with advantage as a masking agent to prevent the precipitation of Fe and Al. The precipitate is collected, washed, decomposed by acid, and the content of Co titrated with EDTA using murexide as indicator.

9. MAGNESIUM AND CALCIUM

Calcium and magnesium are the principle or secondary ingredients of an immense number of natural and artificial products. The classical analytical procedures for this pair are tedious and time-consuming, whereas complexometric titration provides an elegant method for both metals, a fact which explains why the introduction of these methods into current analytical practice has been so rapid. It is appropriate to discuss the two metals simultaneously, since they almost always occur together, and it is also important to know how the mixture of Ca and Mg will behave if only one of the two elements is to be determined. In view of its practical importance, the analysis of biological fluids is dealt with separately in a later section. The number of literature references actually quoted represents only a fraction of all the published papers; but this is justified by the fact that very many of these papers contribute nothing new so far as the actual complexometric titration is concerned. However, the papers to which references have been made present a good survey of the possibilities that exist, and they also point to some of the problems that still await solution.

9.1. Magnesium

The determination of Mg with EDTA was described quite early on by Schwarzenbach and his co-workers [48-5; 48-7]. The indicator Erio T then employed is still one of the most frequently used indicators, even today. Adaptation of the titration to the micro-scale [52-22] and even to microgram amounts [52-37; 57-134] presents no difficulty. The accuracy that can be achieved in the complexometric procedure [60-33] and the stoicheiometry of the titration [60-129] have been thoroughly studied.

The stability of the complexes of Mg with EDTA and with the indicator are just high enough to permit an accurate titration, and in point of fact the colour change at the end-point (wine-red to blue) is less sharp than in many other complexometric titrations. The titration must be continued until the last suggestion of a reddish hue has gone – but this is easily detectable. Near the end-point the reaction proceeds somewhat slowly, in view of which the solution should be slightly warmed.

Erio T and many analogous dyestuffs are 'blocked' by traces of heavy metals – especially by Cu. This disadvantage is easily overcome by suitable masking agents. KCN prevents interference by Cu, Ni, Co, Fe, etc. Since these interfering metals are precipitated as sulphides, Na_2S is equally effective here and also with Mn (cf. the titration of Mg in the presence of large amounts of Mn [61-81]). Al can be masked by triethanolamine, in which case it is advantageous to titrate at 5°C [55-84], since otherwise there remains the possibility of transfer of aluminium from its complex with the masking agent to the indicator.

Interference by heavy metals can often be avoided by a procedure involving back-titration. In this case the interfering metals are first complexed by the EDTA, and thus react with the indicator only slowly or not at all, so that the back-titration can be completed before any blocking begins. If, for example, one back-titrates with Zn, up to 20 mg/l of Cu can be tolerated. The comparative freedom from interference in Hahn's inverse titration (which has already been mentioned [58-76; 60-124]), in which a calculated known excess of EDTA is titrated with the sample solution, depends on the same principle.

Innumerable other indicators have been introduced besides Erio T, such as aluminon [61-2], which permits the successive titration Fe–Al–Ca–Mg, Lacquer scarlet C [57-109], dyestuffs of the acid chromium blue group [58-68], chromoxan green GG [58-89], pyrocatechol violet [56-41], and arsenazo I [62-90]. Diehl *et al.* [60-36], on the one hand, and Belcher's group [57-43], on the other hand, have examined innumerable dyestuffs for their suitability as indicators. Recently the use of calmagite [60-134] has made great strides; so far as the stability of its metal complexes and the colour changes go it behaves practically the same as Erio T, but the stock solution keeps better.

Instrumental methods deal predominantly with photometric titrations which can be carried out directly in the ultraviolet region [54-14] or in the presence of Erio T [53-45; 54-69; 55-12] or other indicators, such as chromazurol S [57-102] or calmagite [61-50]. Potentiometric titration with a mercury-drop electrode and amperometric titration [62-23] can also be used with Mg and permits the successive titrations Ni–Mg, Zn–Mg, or Bi–Mg. Conductometric [57-34] and thermometric [57-97; 63-67] procedures have also been described.

Magnesium only interferes in other titrations in alkaline media, so that it scarcely presents any problems in the determination of other metals, since titrations can be carried out in acidic solution. Masking of Mg is effected by precipitating its hydroxide in strong caustic soda solution, but precipitation of the fluoride [54-74] is also possible.

Titration of Mg in the presence of phosphate has been discussed by Collier [55-58], who recommends the removal of large amounts by extraction [54-25]. Ion-exchange resins are also a good way of removing phosphate. Often enough it is sufficient to dilute the titration solution greatly in order to retard the formation of $Mg(NH_4)PO_4$, since this compound has a pronounced tendency to supersaturation. Back-titration is another possible expedient when phosphates are present. The titration of magnesium in the presence of calcium is discussed below. At this point only one possibility will be mentioned, viz. the separation of Ca as molybdate [56-58] and determination of Mg in the filtrate; this is valid if only the amount of Mg is required.

Mg can be determined complexometrically in pharmaceutical preparations [53-12], aluminium alloys [53-47; 57-94; 57-102], electron metal [61-64], pig-iron and cast-iron [53-44; 56-7], titanium [58-69], nickel sulphate [52-3], gun-powder [56-93], soils and plant material [51-7], rocks [58-89], minerals [54-36], and slags from the manufacture of uranium [60-69]. For the titration of Mg in the presence of Ca see below.

9.2. Calcium

Calcium was one of the first metals for which a complexometric titration was described [48-7]. The titration can be carried out in very dilute solutions and also with very small quantities of Ca [52-22; 52-37; 57-134]. Murexide, the original indicator, has been thoroughly studied [49-4; 60-103] and is still widely used nowadays. However, the colour change from red to blue-violet which takes place in a strongly alkaline medium (pH 12) is not so sharp as that with many other metallochromic indicators. The murexide solution only keeps for a few hours, and it is therefore preferable to use it as a solid intimately ground up with 100 parts of NaCl. Oxidative or hydrolytic decomposition of murexide in the titration solution must also be taken into account, especially in photometric titrations, where the decomposition often makes itself unpleasantly obvious by a slow decrease in absorbancy. Mixed indicators, e.g. an intimate mixture of 0·1 g murexide and 0·5 g naphthol-green B with 100 g NaCl, have been recommended to improve the detection of the end-point [51-14]. Many other substances have been recommended as indicators for calcium, although they have never superseded murexide. A selection would include calcon [57-123], cal-red [56-26], eriochrome blue black SE (Erio SE) [58-36], acid chrome blue black [60-96], and sometimes others as well [59-75; 62-146]. All these are *o,o'*-dihydroxyazo compounds closely related to Erio T. A systematic investigation of the indicator properties of such substances has been made by Diehl *et al.* [60-36]. Very many dyestuffs have also been investigated by Belcher and his co-workers [57-43]. Laquer scarlet C [57-109], omega chrome blue green BL [59-101], phthaleincomplexone [53-1; 56-54], glyoxal-bis-(2-hydroxy anil) [55-46], chromazurol S [55-31], H-acid [59-147], acid alizarin black SN [60-155], and pyrogallol-carboxylic acid [56-1] have also been used as indicators for Ca. Aluminon [61-2] permits the successive titration of Fe–Al–Ca–Mg.

Calcichrome should also be mentioned. It was synthesized by West [60-154], but is probably identical with Hydron, recommended by Russian workers [60-108; 61-5]. Furthermore, methylthymol blue [58-34] and pyrocatechol violet [56-41] can be used for Ca.

Calcein [56-72; 57-103] can be used both on the grounds of the colour change produced by calcium and also as a fluorescent indicator in ultraviolet light [62-150]. The commercial fluorescein-complexone exhibits a residual fluorescence after the end-point which stems from impurities [62-149] and which can be masked [61-85] by adding phenolphthalein (0·25 g to 1 g of indicator). The behaviour of calcein and calcein W is similar; here acridine has been recommended to mask the residual fluorescence [62-151]. Thymolphthalexone has also been recommended as a fluorescence indicator for Ca [61-34]. To ensure against any interference in detecting the end-point Toft *et al.* commend the use of a simple piece of apparatus [63-74] which proved useful in titrations with calcein, but should be equally effective with other fluorescence indicators.

Practically all the indicators for Ca do not give a sharp colour change until high pH values are reached. There are, however, a few directly indicating systems that operate below pH = 11, such as Mg–EDTA (at least 5% of the Ca present) or ZnY in combination with Erio T, or the combinations ZnY–zincon [58-4] and CuY–PAN [56-44]. Any magnesium present is included in the titre.

Those indicators that operate at high alkalinities are often preferred, since Mg, which so often accompanies the Ca, is then precipitated as its hydroxide (see below). However, care must be taken to ensure that the alkaline hydroxide used to raise the pH contains no carbonate and that none is introduced from the atmosphere or from water or reagents, as otherwise the precipitation of $CaCO_3$ may occur. The precipitate redissolves during the titration if this is carried out slowly. However, it is advantageous and saves time if precipitates are avoided by excluding carbonate and, in view of the possibility of $Ca(OH)_2$ being thrown out, by working in sufficiently dilute solution. Back-titrations are another possibility for getting round the formation of turbidities.

Interferences in the titration of Ca have been thoroughly studied. The group Fe, Al, and Cu that occurs in innumerable products of natural or artificial origin can be excluded in a wide variety of ways. The possibility of separation by precipitation with ammonium hydroxide is always to hand, but it is often time-consuming, since a double precipitation may be necessary. For details of masking of Fe,

Al, and Mn see the sections on the individual elements. If Al alone is present no special precautions need to be taken in the determination of Ca, since at the customary high pH the Al will be present as aluminate, and in this form it does not react with the complexone. Care must nevertheless be exercised over the choice of indicator, as many dyestuffs are blocked by Al under these conditions. For the case where there is an extremely high content of Al cf. [55-46; 60-96] and for high Mn concentrations cf. [61-81].

Titanium can be masked by hydrogen peroxide (cf. Ti). The addition of KCN provides a broad basis for masking procedures, and the use of ion-exchangers opens up many possibilities. The possibility of interference due to anions must also be borne in mind. Reference has already been made to hydroxide and carbonate ions. Ferrocyanide, either present from the start or formed when masking Fe, can, because of the low solubility of its Ca salt, lead to suspensions which slowly disappear again during the course of the titration. Copious studies have been made of the interference caused by phosphate. Small amounts do not interfere with the direct titration of Ca, since owing to the phenomena of supersaturation, the titration is frequently completed before the precipitate forms. The permissible ratio P : Ca is about 4 : 1, but it depends very much on the dilution [55-58].

Large amounts of PO_4^{3-} can be tolerated if the procedure of back-titration is chosen. Cimerman [58-6] recommends a standard solution containing $0.1M$ EDTA and $0.05M$ ZnY for the determination of Ca in the presence of phosphate. Ion-exchange resins [53-5] or liquid–liquid extraction are the last resort with an extremely high PO_4^{3-} content.

Since the introduction of procedures for titration in acid media, calcium no longer interferes in the determination of other metals. For some titrations in alkaline media (but not for Mg) masking by fluoride has been recommended [54-74].

Accuracy and precision are both good in the complexometric determination of Ca, as many studies [60-33, 60-129] have shown.

Instrumental methods for Ca exist in great profusion. Particularly favoured are photometric methods, which are based on the colour change of murexide, which is not easily recognized with the naked eye [53-45; 54-10; 54-69; 55-12; 55-52; 55-69; 57-54; 58-96; 59-35]. But other indicators, such as calcon [59-112], Cu–PAN [62-147], or metalphthalein, have also been employed. Photometric titrations are naturally possible in ultraviolet light [228 mμ] [54-14] and lend themselves to automation using a variety of indicators [58-110; 59-119]. The 'slope-indication' on introducing Cu^{2+} has also been described [57-113; 57-14]. Amperometric titration at the mercury-drop electrode [62-33] enables consecutive titrations, such as Ni–Ca or Cu–Zn–Ca, to be performed; in this case the 'chelon-wave' serves as the indicator. Ca can also be determined indirectly by amperometry in a strongly ammoniacal solution by titrating the Zn^{2+} displaced by the Ca^{2+} from zinc-complexonate [52-36; 54-88; 63-73]. It is advantageous to use EGTA in the potentiometric titration of Ca with a mercury electrode, since Mg does not then interfere [57-10; 63-65]. Haslam *et al.* [60-111] have carried out automatic potentiometric titrations using a silver electrode; here again the consecutive titration Ca–Mg is possible. Radiometric [59-60] and conductometric [57-34] titrations have been described. Thermometric titrations [57-97; 63-67] are of special interest in connection with mixtures of Ca and Mg, as the heats of formation of the complexonates of these two metals not only differs in magnitude but they are also of opposite sign.

The number of practical applications of the complexometric determination of Ca is indeed legion. A limited choice from these possibilities is given below. Since the determination of Ca is often coupled with that of Mg, the reader should also consult the sub-section Ca + Mg and hardness of water. By using visual indicators the following have been analysed: stearates [53-8], sugar-beet juice [57-14], casein [61-15], water [51-14; 52-48], rain-water [55-50], pharmaceutical preparations [50-9; 51-10; 51-11; 53-12; 55-107; 56-102], tricalcium phosphate [60-179], phosphate rock [62-146], plants and vegetable matter [52-40], photographic materials [57-24], and colophony [54-98]. In addition, it is possible to determine free lime in silicates [55-25; 58-95] and Ca in caustic soda [62-145]; in the latter case an enrichment of Ca is produced with the chelating ion-exchanger Dowex-A1. A photometric titration with murexide has been recommended for determining the water-soluble constituents of gypsum [55-52] and in the analysis of natural waters [57-54]. Calcein serves as the photometric indicator for the determination of Ca in lithium salts [61-7]. Titration with EGTA and the mercury electrode is used for the analysis of feeding stuffs [63-65]. For the analysis of biological fluids refer to the relevant subsection.

9.3. Mixtures of calcium and magnesium

The resolution of Ca–Mg mixtures can be effected in several ways.

Naturally a separation [61-87] is always possible, but time-consuming; ion-exchange resins are of advantage here [54-8]. Gehrke [54-5] recommends the separation of Ca as its sulphite, but it can also be precipitated in the classical way as oxalate; after ignition the residue is taken into solution and titrated complexometrically. With very small amounts of Ca the precipitated oxalate can be dissolved in acid, and after the addition of EDTA and making alkaline, the excess of complexone can be back-titrated. However, after precipitation with oxalate the colour change of Erio T when determining Mg in the filtrate is no longer sharp, therefore the amount of oxalate used should be kept down to a bare minimum.

Procedures that avoid a separation are more elegant. The most frequently used procedure is certainly that in which Ca is titrated in strongly alkaline solution in the presence of precipitated $Mg(OH)_2$; in a second aliquot portion the sum of $Ca + Mg$ is determined (bearing in mind all the points discussed under the titration of Mg), and then the amount of Mg is determined by difference.* No difficulties arise if there is a lot of Ca and only a little Mg. If the circumstances are less favourable a number of points must be watched, and studies of these will be found in the literature [54-34; 58-1; 58-39]. The most important points, and in particular those that are of the greatest practical significance, will all be discussed in the following sections.

The precipitate of $Mg(OH)_2$ can interfere in various ways. On the one hand, there is the possibility of Ca being co-precipitated and in this way being withheld from the titration; on the other hand, the colour change of the indicator often becomes blurred, since the dyestuff becomes adsorbed in the flocculant precipitate.

Addition of sugar [54-34; 58-9] ought to prevent the co-precipitation of Ca, but this has not been confirmed in other studies [60-33]. According to Flaschka and Huditz [52-30], one certain method of keeping the extent of co-precipitation to a minimum is first to add to the initially neutral or acidic test solution a quantity of EDTA slightly in excess of that equivalent to the calcium, and only then to make alkaline. In every case the alkali should be slowly added dropwise and with good stirring. According to Lewis *et al.* [60-33], some EDTA is carried down with the precipitate, but, on standing, it is set free again owing to the recrystallization of the $Mg(OH)_2$. In order

---

* If the solution contains much ammonium salt this must be removed by fuming it off, otherwise it is impossible to achieve the high pH values that are required.

better to detect the colour change of the indicator (e.g. with murexide) it is expedient, though not absolutely necessary, to carry out the precipitation in a standard flask, to fill this up to the mark, and when the precipitate has settled out to remove a clear aliquot portion for backtitrating the slight excess of EDTA.

Baugh *et al.* [61-107] also obtained good results with some samples of high Mg content (determination of the Ca content at a level of about 0.5% in MgO) by precipitating $Mg(OH)_2$ with $0.5M$ NaOH very slowly and with good stirring (the caustic alkali contained some KCN and NH_2OH,HCl) and carrying out the titration with EDTA directly on the suspension using Cal-red indicator. Lewis and Melnick [60-33] stress the importance of slow precipitation and vigorous stirring.

As a study by Kenny and co-workers has shown [58-39] the final pH, the indicator used and its amount also have an influence on the result. In this respect the experiences of Belcher and his collaborators [58-1] are especially important. Of many indicators they tested, calcon was found to be the most suitable. The end-point is better in the presence of precipitated $Mg(OH)_2$ than in pure calcium solutions, and low values for Ca were never obtained when amounts of magnesium were present, although this very definitely caused lower results with other indicators (e.g. murexide, methylthymol blue, or calcein).

Poor colour changes due to adsorption of the indicator on the Mg precipitate can be improved by not adding the indicator until after the precipitation of the Mg and, furthermore, by waiting till the precipitate has assumed a crystalline character. According to Lott and Cheng [59-102], the end-point is improved by adding a few drops of polyvinyl alcohol. Burg *et al.* [60-140] obtained a similar effect with acetylacetone.

In summary, it can be said that there are many possible ways of improving the situation, but an absolutely universally valid procedure can hardly be expected, and if the highest accuracy is to be achieved the optimum conditions must be worked out for each individual case. It is scarcely to be wondered at that so many researches have been reported that aim at avoiding the precipitation of $Mg(OH)_2$. To this end the addition of tartaric acid has been recommended [57-105; 61-88]. According to our own experiments, and in confirmation of the findings of other authors [60-140], the effect of the tartaric acid is certainly to protect the Mg from precipitation, but the values obtained for Ca are too high when EDTA is used as the titrant. If, however, EGTA is used as the titrant [60-140] correct values for Ca are now obtained, as the Mg-complex formed by this complexone is appreciably less stable than that formed by Ca. It is interesting to refer here to the fact that even in this case the end-point with Calcon is only sharp if the ratio Mg : Ca is at least unity. If this observation is compared with the findings of Belcher *et al.* [58-1] referred to above the conclusion must be drawn that magnesium, whether in the form of a precipitate or as a metal-complex, participates in some way that as yet evades explanation in the mechanism of the end-point that occurs in the system Ca–calcon.

One of the outstanding problems in the determination of Ca in the presence of Mg is that so far no simple visual indicator for Ca has been found that can function at a pH where Mg still remains in solution. Ringbom [58-4] resolved this difficulty by securing an indirect indication of the end-point by means of the system Zn–EGTA–zincon. A pH of about 9·5–10·0 is obtained with a buffer containing 25 g borax, 3·5 g NH_4Cl, and 5·7 g NaOH per litre. Very sharp end-points and the correct calcium concentrations were obtained with pure solutions. Unfortunately, however, the ammonium-ion concentration is a critical factor when using Zn and zincon and, moreover, the ratio Ca : Zn should be about 10, which greatly enhances the difficulties of maintaining optimum conditions in actual determinations. Flaschka and Ganshoff [61-50] describe another procedure which also involves titration with EGTA at about pH 10 but using murexide as indicator. By using a photometric end-point it is then possible to determine Ca in the presence of more than 100 times as much Mg. Ca can also be determined potentiometrically at pH 10 in the presence of Mg if EGTA is used as the titrant [57-10].

Reference must also be made to Strafelda's method [60-35], wherein Mg is thrown down as phosphate at pH 9 and the calcium is determined in the presence of this precipitate by back-titrating a measured excess of EDTA potentiometrically with a standard Ca solution and a mercury electrode. The amount of phosphate added is critical. On the one hand, it must be sufficient to reduce the solubility of $Mg(NH_4)PO_4$ below the point at which it reacts with EDTA, but on the other hand, the amount of phosphate present must not be too large, otherwise $Ca_3(PO_4)$ is thrown out. No reference is made to co-precipitation of Ca.

In view of all that has been said, it is impossible to give working instructions to cover all possible eventualities, but sufficient modifications of the standard method exist for a good workable procedure to be selected for every case that turns up in practice. But it must not be forgotten that most studies have been carried out on pure solutions, whereas in practical analysis the situation is complicated by high salt concentrations, the presence of interfering elements and of the complexing agents that have been added to mask them.

Naturally the most elegant procedures are consecutive titrations: in the first place to save time, and secondly, they need less sample, a point which is often of practical importance. Researches have been carried out with this in mind, and have produced very good results, at least on synthetic mixtures. Körös [53-32] first titrates Ca at pH 13 against murexide, then acidifies the solution, whereby the murexide is cleaved hydrolytically; the pH is then brought to 10 and the Mg is titrated against Erio T. Naturally the difficulties already referred to in titrating Ca in the presence of $Mg(OH)_2$ obtain here also. Lott and Cheng [57-124] first titrate Ca at a high pH against calcon, then lower the alkalinity by adding acid and ammonium chloride and proceed with the titration of Mg using Erio T as indicator. Schmid and Reilley [57-10] avoid introducing errors through the precipitation of $Mg(OH)_2$ by first titrating Ca in a homogeneous solution at pH 9·5–10 with EGTA using Ringbom's indicator system, Zn–EGTA–zincon. Finally, KCN is added to mask the zinc and the Mg is titrated with EDTA using Erio T. Flaschka and Ganchoff [61-50] employ photometric procedures. Ca is first titrated with EGTA at pH 10 using murexide as indicator, then Erio T is added, the wavelength of the light is changed, and the Mg determined with EDTA. Submicrogram quantities of Ca and Mg can be determined from a photometric titration curve if Mg–calmagite is introduced as a self-indicating magnesium system for the 'slope-indication' of Ca [62-27].

The determination of Ca and Mg by the above-mentioned methods has been used for the analysis of a great variety of materials. For example, the haemolymph of insects [54-4], limestone [52-39; 58-110; 61-87], dolomite [58-89; 61-87], magnesite [52-30; 61-50], lime and silicate rocks [55-87; 61-180], soils [51-7; 53-61; 55-24], glass sands [61-9], glasses [54-20], ores and slags [53-52; 59-112; 61-34; 61-68], cement [57-17], steel [60-112], and various materials containing Ca and Mg [53-55] have been analysed; to these can be added rock-salt [58-85], brines [53-31; 54-64], sea-water [54-40], and other samples rich in alkali-metal salts [63-73], welding wire containing Mn [60-49], pulp [52-15], waste water from coalmines [61-10], water in general [60-111] and in particular in mineralized waters [52-8], milk [53-54; 54-21; 62-96], preserved fruit juices [58-33], pharmaceuticals [51-10; 55-107; 56-102], plant materials after ashing [50-11; 51-7; 60-86] in particular tobacco ash [57-96], animal tissues [55-9], and biological material in general [61-119].

9.4. Calcium and magnesium in biological fluids

The complexometric determination of Ca and/or Mg in blood, serum, urine, and spinal fluid is nowadays the standard method in almost all laboratories that use volumetric procedures. The number of publications concerned with this field has passed the hundred mark, but since many of the recommended procedures differ only in quite insignificant details, only a few references will be quoted here in order to illustrate the principles. Greenblatt and Hartman [51-17] were the first to determine Ca alone in serum; their procedure was based on the use of murexide in strongly alkaline solution. Other authors describe this method with minor modifications [52-2; 52-9; 52-16; 54-53] or determine the end-point photometrically [54-19; 58-45]. Other indicators have been used, e.g. calcein [57-46], especially in ultraviolet light [57-36]; this also facilitates the automatic recording of the titration curve [52-91], working with the smallest possible samples of serum (20 μl) [62-39] and the photometric determination of the end-point [59-129]. Cal-red [58-20], calcon [60-146], phthaleincomplexone [56-68], acid alizarin black SN [60-156], and fluorescence indicators [60-6; 60-81] have also been used. Extensive comparison with the classical oxalate procedure has clearly established the advantages of complexometric titration [cf. 55-53]. Ca alone in urine can be determined as in other materials by the standard EDTA method [54-19], by photometric titration [58-96], and by using fluorexon [61-182]. In view of the high phosphate content, it is often advantageous in urine analyses to dilute the test sample very considerably or to get round the precipitation of sparingly soluble materials by a back-titration.

In addition to these special methods for Ca, reference should be made to the sections for Ca + Mg, for many determinations of Ca are linked with those of Mg.

Ca and Mg in serum: Holasek and Flaschka [52-33] were the first to describe the determination of Mg in serum wherein Ca is separated out as oxalate and titrated after dissolving the precipitate, while the Mg is determined in the centrifugate. The advantage of this procedure is that both metals can be determined in a single aliquot portion. Gjessing's method [60-128], which involves consecutive titrations, offers the same advantage. In it the Ca is first titrated photometrically in caustic soda solution using murexide, whereby the small amounts of $Mg(OH)_2$, which presumably remain in colloidal solution, do not interfere. Finally, glycine is added and the solution boiled, whereupon the murexide is destroyed and the $Mg(OH)_2$ dissolves to give a solution that is titrated using Erio T. However, most methods involve two separate aliquot portions of sample solution. In the one Ca is determined at a high pH using murexide (see above) or other indicators, such as Erio SE [58-36], and the sum Ca + Mg is determined in the other. Erio T is usually used for this latter titration. The procedure also works on the ultramicro scale [51-13], and the accuracy is improved by photometric titration [59-63; 59-107; 60-175]. Automatic titration is also possible [59-118]. Ca and Mg in urine [58-93] can be determined as in serum with only minor modifications. Ca and Mg in plasma [51-9] and spinal fluid [56-4] are determined in just the same way as in serum.

9.5. Determination of the hardness of water

The determination of the hardness of water was described very early on by Schwarzenbach and his co-workers [46-2], and this was indeed the first complexometric titration procedure to be employed in practical analysis. Innumerable working recipes [50-5; 50-6; 50-7; 50-8; 52-47; 52-48; 55-47], including a microprocedure [53-62], are to be found in the literature.

It will be appropriate to distinguish between two groups of procedures, viz. the determination of total hardness and the determination of the individual hardness due to Mg and to Ca. In measuring total hardness Ca and Mg are titrated together and the sum total is calculated as Ca. This titration is usually carried out at pH 10 using Erio T as indicator. To achieve a sharp end-point it is essential for at least 5% Mg (calculated on the basis of the Ca content) to be present. Since this is not necessarily the case in all samples of water, it is essential either to add a known amount of Mg and to take this into account in the subsequent calculation or, better, to introduce it into the test sample as its Mg–EDTA complex. However, when a large number of analyses have to be made it is far simpler to use a standard solution of titrant that contains besides EDTA (H_2Y^{2-}) the necessary percentage addition of MgY^{2-}.

Interferences in these titrations have been carefully studied [53-42], and they were mainly caused by contamination by traces of heavy metals. It is readily possible to render them harmless by adding KCN and ascorbic acid, and triethanolamine as masking agent. The inter-

G

ference is not generally shown as a high titre due to the simultaneous titration of the impurities, for their amount is far too small, but it is due rather to blocking of the indicator. Na_2S [53-62] is a good masking agent for the usual interfering elements, aluminium excepted. It is often simpler to include the masking agent in the buffer. Hahn [58-76] avoided, or at least decreased, interferences by employing an inverse titration, i.e. a measured amount of EDTA is taken and titrated with the sample of water. In practice, however, this procedure is too troublesome. Fewer interferences are to be feared in titrations using chromazurol S as indicator [55-31], as this dyestuff is less blocked; however, the colour change is not so sharp as that of Erio T.

In determination of hardness due to Ca and Mg individually it is usual to adopt a procedure with two aliquot portions. In one Ca is determined alone at high pH and in the other Ca + Mg are titrated together at pH 10; Mg follows by difference. The important points are stressed in the relevant sections above. The situation in the Ca titration is generally free from complications, as in all normal waters the concentration of Ca far exceeds that of the Mg.

For analyses of waters containing polyphosphates Brooke [52-7] recommends separation using an ion-exchange resin prior to the Ca titration. Schneider *et al.* [51-18] titrates the hardness in sugar-beet juices using Erio blue-black B.

Acid–base titration of temporary hardness can precede the complexometric determination of permanent hardness. The complexometric titration is carried out after the sample has been fully titrated with alkali.

Studies have also been reported on photometric titrations [50-4], as these are of interest with samples of water with a pronounced self-colour. Photometric indication of the end-point lends itself to automation [59-119]. Lacy [63-46] has described a semi-automatic procedure in which the titration curve (using Erio T) is plotted by a recorder. Two breaks in the curve are obtained, of which the first corresponds to the Ca. In this way the hardness due both to Ca and to Mg are determinable in a single operation. According to Erdey *et al.* [62-117], two breaks occur also in the high-frequency titration curves. Conductivity titration [57-28; 57-35] proves good with turbid and coloured samples. Since the concentration of salts in natural waters is usually low, the conductometric procedure is very practicable, for here the determination is not rendered difficult by interfering background conductivity.

(a) DIRECT DETERMINATION OF MAGNESIUM USING ERIO T

REAGENTS: Standard solution 0·01M EDTA; Erio T; buffer pH = 10.

PROCEDURE: The concentration of magnesium in the test solution should not exceed 10^{-2}M. Acidic samples are first neutralized with NaOH. To every 100 ml is then added 2 ml buffer and some Erio T, and the titration is carried out to the change from red to blue. The last trace of a reddish hue should vanish with the final drops of titrant. Since complex formation does not proceed instantaneously, the titration must be carried out slowly near the end-point.

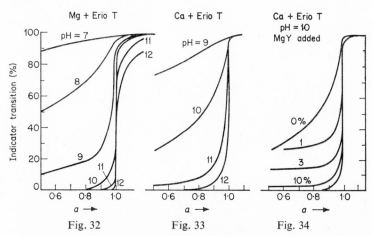

Fig. 32   Fig. 33   Fig. 34

Figure 32. Colour transition of Erio T during the titration of Mg at various pH values.

Figure 33. Colour transition of Erio T during the titration of Ca at various pH values.

Figure 34. Colour transition of Erio T during the titration of mixtures of Mg and Ca at pH 10. The Mg content of the solution is varied and shown as a percentage of the amount of Ca present.

REMARKS: Fig. 32, which is compounded from the curves of Fig. 4 and Fig. 23, shows that the pH ought to be maintained pretty accurately at pH = 10 during the titration. Both a higher and a lower pH worsen the end-point. On this account it is important to neutralize acidic samples with NaOH and not with NH_4OH before adding the buffer, to avoid introducing ammonium ions into the solution.

By the correct choice of conditions the end-point is so sharp that it is even possible to titrate with 0·001M EDTA.

(*b*) SUBSTITUTION TITRATIONS OF CALCIUM USING ERIO T

REAGENTS: Standard solution, 0·01M EDTA; Erio T; buffer at pH 10; 0·1M MgY.

PROCEDURE: The concentration of calcium in the test solution should not exceed 10^{-2}M. If this solution reacts acid it must be neutralized with NaOH. Then to every 100 ml add 2 ml of buffer, 1 ml of 0·1M MgY, and 2–4 drops of Erio T and titrate to the colour change from red to blue. The last hint of a reddish hue vanishes with the final drop of titrant. The titration must be carried out slowly near the end-point.

REMARKS: Fig. 33, which was obtained by combining Figs. 5 and 24, shows how the colour of Erio T changes when calcium is titrated without any additional magnesium. In this case there is no sharp colour change even at pH = 11, and in this strongly alkaline solution titration does not lead to a pure blue colour, for we are already in the region of transitional colour for Erio T acting as an acid–base indicator where HD^{2-}(blue) $\longrightarrow D^{3-}$(brown).

Figure 34 demonstrates the improvement that can be realized by adding Mg–EDTA. Since the calcium complex CaY^{2-} is stabler than the magnesium complex, substitution takes place, so that we are in effect dealing with a simultaneous titration of Ca + Mg which proceeds as shown graphically in Fig. 11. Fig. 34 is obtained by combining the pMg values taken from Fig. 11 with the curves showing colour changes in Fig. 23. It shows how even an addition of so little as 1% Mg improves the end-point quite appreciably. Practically the maximum effect that can be achieved is obtained with an addition of 10% Mg. Unfortunately further addition of MgY^{2-} would unnecessarily increase the ionic strength and somewhat diminish the jump in pMg.

By the correct procedure the colour change is so sharp that it is even possible to carry out microtitrations with 0·001M EDTA.

(*c*) DETERMINATION OF THE SUM OF CALCIUM AND MAGNESIUM
WITH ERIO T. TITRATION OF HARD WATER

REAGENTS: 0·01M EDTA as standardized titrant; Erio T; buffer of pH 10; 0·1M MgY and masking agents for iron and copper if required.

PROCEDURE: 2 ml of buffer is added to 100 ml of the water to be tested together with masking agents as required and some Erio T. The titration is immediately carried out to the colour change from red to blue. The last hint of red vanishes at the end-point.

If the water undergoing examination contains no magnesium, magnesium complexonate must be added and the procedure described above is followed.

Solutions obtained from minerals, soils, ashes, etc, are dealt with just the same as water after separation or masking of any heavy metals present.

REMARKS: A still sharper end-point is obtained by first acidifying the sample of water with HCl, boiling for about 1 minute to drive off CO_2, cooling and neutralizing with NaOH, and then adding the buffer and indicator.

For the influence of magnesium cf. the observations appended to procedure (*b*).

(*d*) DIRECT TITRATION OF CALCIUM USING MUREXIDE

REAGENTS: 0·01M standard EDTA; murexide; 1M NaOH; 0·01M $CaCl_2$ if required.

PROCEDURE: To each 100-ml portion of the neutral solution to be titrated (acidic samples are first neutralized with NaOH), which should not contain more than about 5 mg Ca, add 10 ml NaOH and masking agent if necessary. Next add the indicator and immediately titrate to the colour change from red to violet. The titration must be carried out immediately after the addition of NaOH because of the danger of $CaCO_3$ becoming precipitated.

REMARKS: The indicator transition at the end-point of this titration is shown in Fig. 37, which was obtained by combining values of pCa from Fig. 5 with the transition curves for murexide given in Fig. 26. From this it is abundantly clear that the titration should be conducted above pH 12. Even here the indicator still changes colour at somewhat too low pCa values determined by the very characteristic course of the transition curves (Fig. 37) with the abrupt change in direction at $a = 1$ mole EDTA per atom of Ca and 100% colour change. Although this break coincides exactly with the point of equivalence, it is difficult to recognize visually, as the solution has to be titrated until the very last trace of red vanishes. By photometric titration, i.e. by recording the optical density during the titration, best at a wavelength of 480 or 590 mμ (cf. the absorption curves of Ca–murexide, CaH_2D^-, and of murexide in alkaline media, H_2D^{2-}, shown in Fig. 19), this break can be pin-pointed, and in this way the accuracy of the method is increased significantly.

The curves for pCa in Fig. 14 determine the appropriate conditions for the titration of Ca in the presence of Mg. In the strongly alkaline solution (pH 13) magnesium hydroxide is so very sparingly soluble that the titration of the calcium left in the solution at the end-point gives as good a jump in pCa as when magnesium is absent (curve II, Fig. 14). If the solution is first made alkaline and allowed to stand for 1–2 minutes until the precipitate of hydroxide has formed and then the indicator is added and titration carried out, a very easily recognized end-point is in fact obtained. But the figures for calcium come a few per cent too low, since the precipitate of magnesium hydroxide carries down some calcium with it. Part of this calcium comes back into solution at the end-point, so that on standing the murexide turns red again, and a little more titrant can be added.

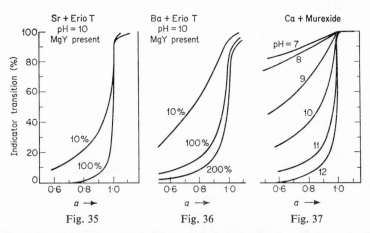

Fig. 35   Fig. 36   Fig. 37

Figure 35. Colour transition of Erio T during the titration of mixtures of Sr and Mg at pH 10. The Mg content of the solution is varied and shown as a percentage of the amount of Sr present.

Figure 36. Colour transition of Erio T during the titration of mixtures of Ba and Mg at pH 10. The Mg content of the solution is varied and shown as a percentage of the amount of Ba present.

Figure 37. Colour transition of murexide during the titration of Ca at various pH values.

However, if only very little magnesium is present the $Mg(OH)_2$ does not come out immediately, and there is time to titrate the super-saturated solution. The broken curve I of Fig. 14 corresponds to such a titration, and it can be seen that at the point where the calcium is completely complexed by EDTA ($a = 0.5$) there is a feeble jump in pCa. Nevertheless, this change is indicated by murexide in a satisfactory way, so that the correct amounts of calcium are found, although their values show more scatter than in the absence of magnesium. This is the situation in determining the calcium hardness of natural waters.

In order to obtain accurate results in the presence of considerable amounts of magnesium (e.g. analysis of magnesite) the calcium must first be complexed with a small excess of EDTA, and when this has been done the solution is made strongly alkaline, whereby pure $Mg(OH)_2$ is thrown down without any calcium being precipitated; the excess of EDTA is then back-titrated. How this is carried out is shown in the following example.

500 ml of the solution of magnesite containing not more than 20–40 mg of calcium is placed in a 1,000-ml standard flask, and an amount of EDTA equivalent to rather more than the amount of calcium present (as determined in a previous rough titration) and about 2 ml triethanolamine (to mask any iron and manganese that may be present) are added. The solution is neutralized and then made alkaline by adding 100 ml of N NaOH. After making up to the mark and mixing well the solution is allowed to stand for 30 minutes. 200 ml of the clear solution standing over the precipitate is withdrawn by a pipette, indicator is added, and the slight excess of EDTA back-titrated with $0.01M$ $CaCl_2$. The difference between the amount of EDTA added initially and the equivalent of $CaCl_2$ required is used to calculate the calcium content.

(*e*) DIRECT TITRATION OF CALCIUM USING METALPHTHALEIN

REAGENTS: Standard solution $0.01M$ EDTA; metalphthalein-napthol green; M NH_4OH; M NH_4Cl.

PROCEDURE: The concentration of the calcium in the solution to be titrated should lie between 10^{-3} and 10^{-4} gram atoms per litre. Acidic samples are first neutralized with NaOH. Then to every 100-ml portion add 0.3 ml of indicator, 5 ml NH_4OH, and a few drops of NH_4Cl if necessary and titrate until the red colour suddenly fades; the fully titrated solution is grey.

REMARKS: Fig. 38, obtained by combining Figs. 5 and 29, shows just how the percentage extinction changes during the titration. It follows that it is best to work at, or a little below, pH 11. If the pH falls to 10 the colour change is long drawn-out; if the pH is above 11 a red colour of appreciable intensity persists at the end-point. At pH $10.5–11.0$ the extinction at the end-point amounts to only some 10–13% of the initial figure, and this residual red colour is rendered invisible by the additional napthol green.

To reach pH 11 with an ammonia buffer it is essential for only small amounts of ammonium salts to be present. As a rule the protons liberated in the complexation reaction are enough to change sufficient NH_3 to NH_4^+. Should this not be the case and an intense red colour persists at the end-point, a few drops of ammonium chloride are added. The results are greatly improved by using photometry to detect the end-point.

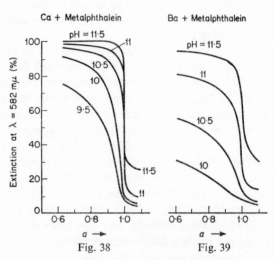

Figure 38. Colour transition of metalphthalein (XVIII) during the titration of Ca at different pH values.

Figure 39. Colour transition of metalphthalein (XVIII) during the titration of Ba at different pH values.

(*f*) DIRECT TITRATION OF CALCIUM USING CALCON [57-123]

REAGENTS: Standard solution of 0·01M EDTA; calcon; 2M KOH; diethylamine.

PROCEDURE: The test solution should be about 10^{-2}M in Ca, and if acidic it should first be neutralized by NaOH or KOH. Then to each 100 ml of the neutralized solution add 5–7 ml of diethylamine, which should be sufficient to establish a pH of 12·5. Then add calcon indicator and titrate with EDTA (immediately to avoid the precipitation of $CaCO_3$) to a permanent pure blue colour.

REMARKS: The high pH required for this titration can also be achieved by adding NaOH or KOH. Many observers find the end-point with calcon to be sharper if magnesium is present. When this is the case and Mg is not originally present in the test solution 1–2 ml of a 0·1M solution of a magnesium salt can be added. The amount of diethylamine specified above is sufficient to establish the desired pH, even in the presence of Mg. If Mg is present, the end-point often comes back on standing, and a further drop or two of titrant must be added to produce a permanent blue colour; half a minute should be allowed to pass before reading the burette.

The titration can be carried out with EGTA instead of EDTA, and this is particularly advantageous if the Ca has to be determined in the presence of much Mg, and the precipitation of magnesium is prevented by adding tartaric acid.

10. BARIUM AND STRONTIUM

The stability constant of SrY (log $K = 8 \cdot 63$) is practically the same as that of MgY (log $K = 8 \cdot 69$), whereas the corresponding barium complex is weaker (log $K = 7 \cdot 76$). Therefore the two metals can only be titrated in an alkaline solution. Consequently, under ordinary conditions of titration the precipitation of sparingly soluble carbonates is very likely to occur, so that the determination is almost always carried out by a substitution titration or a back-titration. When using Erio T as indicator the titration of Sr by a substitution reaction with MgY or ZnY gives a really good colour change [48-5]. The situation is less favourable with Ba. Indeed, reference to the relevant stability constants suggests that the titration should be barely possible. However, as Manns *et al.* pointed out [52-49], it is the stability constants of the indicator complexes that must be considered. For the reaction

$$MgY^{2-} + Ba\text{–Erio T} \rightleftharpoons BaY^{2-} + Mg\text{–Erio T}$$
$$\log K = 8 \cdot 7 \quad \sim 2 \qquad \quad 7 \cdot 8 \qquad 7 \cdot 0$$

the equilibrium constant is about 10^4, which is adequate for a titration. Moreover, there is a jump in pMg after complete complexing of the two metals, despite the fact that Mg^{2+} is bound before the Ba^{2+} (cf. Fig. 13). Sijderius [54-37] has shown that the pH of the solution and the concentration of buffer and of foreign salts greatly influences the colour change, and hence the accuracy of the titration, so that these factors ought to be kept within certain limits for optimum performance.

As barium sulphate is soluble in alkaline EDTA, barium can be separated from other cations by precipitation as its sulphate, and then after taking up in alkaline EDTA the excess EDTA can be back-titrated.

A good deal of information on the determination of Ba originates from investigations on the exploitation of this precipitation for the indirect determination of sulphate and details will be found in the relevant sections (p. 318).

Ballczo and Doppler [54-94; 56-97] have been concerned with the determination of very small amounts of Ba and report good results (accurate to $\pm 3\%$ on 50 μg) when back-titration is carried out with a solution of $MgCl_2$. A mixture of tropaeolin 00 (1 part), Erio T (2 parts), and NaCl (500 parts) was used as an indicator powder. Indicator error was allowed for by a double titration.

Theis used chromazurol S as indicator [55-31], but the colour change is not very sharp and the standard EDTA should be stronger than 0·05M. Moreover, an exactly known amount of $MgCl_2$ must be added to achieve a satisfactory colour transition. Dragusin [61-148] describes the direct titration at pH 9·5 with a mixed indicator that changes from violet to green. For this purpose 20 mg of a fresh suspension of sodium rhodizonate in 2–3 ml water is added to the test solution, and then a 1% aqueous solution of alkali blue is added dropwise until a purple coloration appears.

Very sharp colour transitions are obtained by using the metalphthalein indicator introduced by Schwarzenbach *et al.* [53-1] (cf. remarks following the working instructions below). Particularly good colour changes occur in 50% aqueous alcohol, but in view of the increased danger of insoluble carbonates being precipitated, a known excess EDTA must be added and back-titration undertaken with a standard solution of Ba.

Here again a mixed indicator (metalphthalein and naphthol green) is preferable to the metalphthalein alone. McCallum [56-64] recommended a mixed indicator of 0·1 g metalphthalein, 0·005 g methyl red, and 0·05 g dianil green (No. 576, Allied Chemical and Dye Corporation) in 100 ml solution.

Gordon and his co-workers describe the photometric titration of small amounts of Ba. 0·5–5 mg of Ba can be titrated at 650 mμ using Erio T (without substitution with MgY or ZnY) [56-10]. Still smaller amounts can be determined with metalphthalein at 570 mμ [56-71].

Polyak [57-26] determined small amounts of Ba potentiometrically by direct titration or back-titration of a known excess of EDTA with $MgCl_2$ solution using a Pt–W electrode pair. Campbell and Reilley have reported on the amperometric titration of Ba [62-33]. The phenomenon of the 'free chelon wave' (cf. the section on amperometry) is used to indicate the end-point.

The determination of barium finds practical use in the analysis of water [56-54] and curative waters [54-94; 56-97]. Kemula [61-71] determined Ba in Ba-ferrites by precipitating with sulphate, dissolving the $BaSO_4$ in EDTA, and back-titrating the excess with Mg. Low contents of Ba (down to $0·05\%$) in nickel alloys have been determined in the same way [57-26]. De Sousa [60-23] analysed minerals for Ba by precipitating the carbonate or chromate, dissolving this in EDTA, and back-titrating with a Ba solution. The mineral barite can be dissolved directly in EDTA and the excess back-titrated [61-148]. Verma *et al.* [57-120] describe the successive titration of Zn–Ba. Zinc is titrated first at pH 6·7 using Erio T; more EDTA is then added, and after the pH has been raised to 10 the Ba is determined by back-titration with Mg. The title of this paper is misleading, for it reports only results obtained on pure solutions, whereas nothing is said about the actual analysis of lithopone, where difficulties can be expected in dissolving samples for analysis.

Strontium has been titrated by Schwarzenbach and Biedermann [48-5] by a substitution reaction with MgY and using Erio T as indicator. Ogawa and Musha [60-109] determine amounts of 0·1–6 mg Sr per 50 ml to about $\pm 1\%$ by a photometric titration at pH 10–11·5 with metalphthalein using a wavelength of 580 mμ.

Volf [60-37] states that Unithiol can be used to mask Zn, Cd, Hg, Pb, and Sn during the titration of Sr and Ba. In the titration of other metals Ba, in not too large amounts, can be masked as its sulphate. In the cold the reaction between $BaSO_4$ and EDTA proceeds slowly enough for the titration to be completed before significant amounts of the precipitate goes into solution.

10.1. Substitution and back-titration of strontium and barium using Erio T

10.1.1. WITH MAGNESIUM COMPLEXONATE

REAGENTS: Standard 0·01M EDTA; Erio T; pH 10 buffer; 0·1M MgY.

PROCEDURE: The test solution is neutralized with NaOH, and to every 100 ml is added 10 ml 0·1M MgY, 2 ml buffer, and 2–4 drops of Erio T. Titration is then carried out with EDTA until the colour changes from red to blue.

REMARKS: The colour change with Sr, and especially with Ba, is appreciably less sharp than in the corresponding titration with calcium. This is a consequence of the relative stabilities of the complexes. With barium the stability constant of BaY^{2-} is only $10^{7·8}$ compared with $10^{8·7}$ for MgY^{2-}, the complex of the cation to which the indicator responds. Thus, on adding the complexing titrant, the magnesium disappears from solution before the barium, and this produces a change in the colour of the indicator, which occurs too soon, as illustrated in Figs. 35 and 36. These diagrams result from a combination of the pMg values of Figs. 12 and 13 with the transition curve for the indicator given in Fig. 23. They show that the colour already starts to change appreciably before the equivalence point and is complete when this is reached. The situation is only slightly improved by large amounts of magnesium complexonate. Barium can also be titrated with EDTA without adding magnesium if a photometric end-point is used [56-10].

10.1.2. WITH ZINC COMPLEXONATE

REAGENTS: Standard solution $0.01M$ EDTA; Erio T; conc. ammonia; $0.1M$ ZnY; some $1M$ NH$_4$Cl.

PROCEDURE: About 100 ml of the neutral test-solution (neutralization with NaOH if necessary), which contains not more than 5×10^{-4} gram atoms of barium or strontium respectively, is treated with 10 ml of zinc complexonate and 1 ml ammonium chloride. Then sufficient ammonia is added to make the mixture approx. $0.5M$ with respect to this base (3–4 ml of 25% NH$_4$OH per 100 ml of sample), and after adding indicator the titration is carried out until the colour changes from red to blue.

REMARKS: As explained in the theoretical section, sufficient ammonia must be present to increase the magnitude of the coefficient $\alpha_{Zn(NH_3)}$ to round about 10^8. At the equivalence point there is then a really well-defined jump in pZn, as illustrated in Fig. 15. This is somewhat better developed at pH 11 than at pH 10, but the higher pH value is needed in particular because the jump in pZn of Fig. 15 must be made to correspond with the transition curve of the indicator shown in Fig. 25. The Erio T changes too soon at pH 10, whereas at pH 11 the indicator transition of Fig. 25 exactly coincides with the steep rise in pZn shown in Fig. 15.

To achieve a pH of 11 with the ammonia buffer the ratio [NH$_4^+$]/[NH$_3$] can only amount to 0.02. Therefore in $0.5M$ NH$_4$OH, the most favourable ammonia concentration, the concentration of ammonium salt should not be greater than amount $0.01M$.

10.1.3. BACK-TITRATION OF BARIUM

Instead of a substitution titration it is often preferable to carry out a back-titration with $0.01M$ zinc sulphate. This is specially suited to the determination of barium in a precipitate of $BaSO_4$. The precipitate is suspended in about a two-fold excess of an accurately measured amount of 0·1M EDTA, and 4 ml of conc. ammonia (25%) is added. After this addition of base the precipitate dissolves slowly when the solution is boiled. Ammonia that evaporates off must be replaced. Finally, 1 ml of 1M NH_4Cl is added together with indicator, and after diluting to 100 ml the excess of EDTA is back-titrated with the standard solution of zinc until the colour changes from blue to red. At this point of equivalence $BaSO_4$ begins to separate out again, since the solubility product has again been exceeded. To prevent any alkaline-earth ions coming from the glass vessel during the dissolving stage it is preferable to use a Pyrex or quartz flask.

10.1.4. DIRECT TITRATION OF BARIUM (OR STRONTIUM) USING METALPHTHALEIN

REAGENTS: Standard solutions of 0·01M EDTA; and 0·01M $BaCl_2$; metalphthalein–naphthol green; conc. NH_3; ethanol; 1M NH_4Cl if required.

PROCEDURE: To every 100 ml of the neutral test-solution (acidic solutions are first neutralized with NaOH) with a concentration of metal ions between 5×10^{-4} and 5×10^{-3} gram atoms per litre add 0·3 ml of indicator and 5–10 ml ammonia. Because of the danger of barium carbonate separating out, the titration is commenced at once; the end is shown by a sudden paling of the red colour. Should the solution remain moderately red after passing the end-point, the pH must be lowered by adding some ammonium chloride. After adding a small excess of EDTA, sufficient alcohol is added to give a 50% mixture, and the whole is back-titrated with the standard 0·01M $BaCl_2$ solution. Addition of the alcohol changes the colour from grey to greenish (due to the naphthol green), and at the end of the titration there is a very sharp colour transition to red.

REMARKS: The intensity of the colour of metalphthalein during the course of the forward titration (aqueous solution) is illustrated by Fig. 39, which is derived from Figs. 6 and 30. From this it is clear that the pH of 11 during this titration must be maintained very exactly. Even at pH 10·5 the colour change is long drawn out, and at pH 11·5 some 30% of the colour intensity persists after the end-point. The situation at the end-point is substantially improved by the addition of alcohol; this addition ought not to be made before the back-titration is started, otherwise $BaCO_3$ generally precipitates at once.

11. BORON

According to Borchert [59-54], boron, in the form of boric acid, can be determined by precipitation as barium borotartrate, $Ba_5B_2C_{12}H_8O_{24}.H_2O$, and titration of its barium content. A solution containing 13 g $BaCl_2.2H_2O$, 14 g tartaric acid, and 240 g NH_4Cl per litre serves as precipitant. About 10 ml of the test solution containing $0.1-3$ mg of B is treated with ammonia in a 100-ml standard flask and made up to the mark with the precipitating solution. After 2 hours the mixture is passed through a dry filter paper and 10 ml of the filtrate is diluted to 100 ml. The excess Ba in a 25-ml aliquot portion of this dilute solution is determined either by direct titration using phthalein complexone in a strongly ammoniacal solution containing 30% alcohol or indirectly by back-titration of excess EDTA with Mg using Erio T. The blank for the precipitating solution is obtained by using an identical volume as in the determination and replacing the test solution by distilled water. The method is used to determine boron in baths used for the heat treatment of steels.

12. ALUMINIUM

Aluminium can be determined complexometrically both directly and through back-titration; but in both cases special precautions have to be taken in view of peculiarities in the properties of this element. Al shows a marked tendency towards the formation of polynuclear hydroxy-complexes which react only very sluggishly with EDTA. The situation has been treated mathematically by Wänninen and Ringbom [55-98], who show that the hydrogen-ion concentration below which the formation of hydroxy-complexes can be neglected depends on the Al concentration. The limiting concentration can be calculated approximately from the following formula:

$$[H^+] = (10^{-8} . C_{Al})^{\frac{1}{3}}$$

Thus if C_{Al} is set at about 10^{-2}M a pH of 3.3 is required for the titration if hydroxy-complexes are to be neglected. However, at this pH the value of $\alpha_{Z(H)}$ for EDTA is already so large that the effective stability constant for AlY^- is only $10^{6.7}$, so that the jump of pM in the titration is no longer a large one. Nevertheless, a direct titration of Al is still possible at pH 3, and the results are surprisingly good. At slightly lower pH values the titration of Al is no longer feasible. It is worth pointing out that an effective stability of $10^{6.7}$ makes it very difficult to explain why the titration of iron with EDTA at pH 2 gives distinctly high results in the presence of even moderate amounts of Al. This discrepancy between theory and practice needs clearing up.

The sluggishness of the complexing reactions of Al explains why initially only procedures involving back-titrations were reported. There are a considerable number of these, and the indicators and reagents used for the back-titration are varied in the extreme. Přibil [51-15] was the first to describe this determination, and he back-titrated with Fe and found the end-point potentiometrically. One of the first procedures with a visual end-point originated from the same author [53-12]. Here back-titration is carried out with Zn at pH 10 using Erio T. This must be done quickly and with a cold solution, otherwise the indicator is blocked. Following the development of procedures which function in acidic media, this method has lost much of its significance. Measurements in acidic media have the advantage of greater selectivity, and in this context it means that at least Ca and Mg and often even Mn do not interfere.

Ter Haar and Bazen [54-104] and also Chernikov [55-51] described back-titration at pH ~3·5 with a standard solution of Th using alizarin S. The results found by the authors first mentioned are about 1% too low, so that an empirical factor obtained by standardization with a solution of known Al content is suggested. When the procedure was adapted to the micro-scale by Flaschka *et al.* [53-22] no systematic errors were detected for quantities of Al below 1 mg. On the micro-scale the back-titration is carried out in boiling solution, which ensures that the Al is transformed more quickly into its EDTA complex. The reaction between Al and EDTA is not instantaneous, so that small amounts of Al can remain uncomplexed for a long time. Boiling speeds up the reaction, and so completes the complex formation. This point was first stressed explicitly by Milner and Woodhead [54-91], who recommend in their procedure boiling for about 2 minutes before starting the back-titration (with Fe). Although this period of boiling is sufficient in many cases, no general rule can be given. The length of time needed to complete complex formation with Al depends on the pH of the solution, the concentration of foreign salts [62-87], the previous history of the solution, and the excess of complexing agent. If the latter is large enough (50% or more) complex formation can be complete within a few minutes without boiling [55-64]. All these points must be taken into consideration when using the complexometric titration of Al in practice.

It is also of importance to avoid, from the very start, the formation of hydroxy-complexes which are kinetically inert. Naturally these are not present in sufficiently acidic solutions (pH <1). If, however, excess acid is neutralized with ammonia, caustic soda, or any other strong base local excess of alkali is unavoidable and hydroxy-compounds result which are subsequently degraded only very slowly. On this account Flaschka and Abdine [56-45] recommend a preliminary rough neutralization to pH 0–1. Under these conditions only moderate amounts of hydroxy-complexes are formed, and when these occur locally they are quickly broken down again. The final pH for the titration is adjusted by adding a buffer so that localized over-neutralization is eliminated. Such a procedure is especially important if the back-titration is carried out at higher pH values. EDTA is always added first before the acidic test solution is neutralized.

Fe(III) has been proposed from many quarters for back-titrations. Milner [54-91] back-titrated at pH 6·5 with sulphosalicylic acid as indicator. This has also been recommended by other authors [60-160; 62-13]. Brhaček [59-19] introduced salicylic acid as indicator. The colour transition and indicator constants are practically the same, but the sulpho-compound has the advantage of greater solubility. This is of importance in the consecutive titration Fe–Al, where a high concentration of the salicylic acid serves to complex the Al during the Fe titration (q.v.). Back-titration with Fe is also possible with 3-hydroxy-naphthoic acid [61-44], Tiron [59-9], acetyl-acetone [62-3], and chrome azurol S [60-122].

Zn is also favoured for back-titrations and as an indicator, eriochrome cyanin R [53-43; 61-64], xylenol orange [60-150; 61-131], ferri-ferrocyanide in conjunction with dimethylnaphthidine [61-42; 55-67], benzidine [54-41], or variamine blue [60-89]. The last named was used by Wehber [56-106] as a direct redox indicator with a standard Zn solution containing a little Cu. The test-solution is treated with some thiocyanate, and the redox system Cu(II)/Cu(I) produces the jump in potential to which the indicator responds. The mixed solution of Zn and Cu used for the back-titration has this advantage over a pure Cu solution that only a low concentration of the strongly coloured complex CuY^{2-} results, so that the end-point is not overshadowed. This is of particular importance when a large excess of EDTA has to be back-titrated.

Other back-titration procedures employ Pb and PAR [59-79], Cu and calcein W [59-9], and Zn with Erio T in a solution cooled below 10°C [58-5]. Banyai and Erdey [60-150] back-titrated with a standard solution of Hg(II) with diphenylcarbazide as indicator. The end-point is extremely long drawn out and appears too late. However, if some 1,10-phenanthroline is added, following a discovery made by Přibil, the colour change takes place instantaneously and at the correct point. The mechanism of this catalytic action needs much closer study.

Wänninen and Ringbom's method [55-98], used in many actual analyses, deserves special attention. The excess of EDTA is titrated with Zn in a solution containing 50% ethanol buffered with acetate to pH 4·5 using dithizone as indicator; the end-point is extraordinarily sharp. The authors give a complete mathematical treatment of the course of the titration. Many investigators have reported on their tests of the procedure and comparisons with other complexometric titrations [60-110; 62-7; 62-9]. Gottschalk [60-127] showed on statistical grounds that with 0·1–0·001M solutions of titrant the drop error hardly amounts to 0·05 ml. The addition of alcohol is necessary to keep the indicator and its Zn-complex in solution, but according to Gottschalk, it also has the welcome function of increasing the stability of AlY^- by 2–4 powers of ten.

Recently, direct titration has received special attention. The first such procedure, which gives a good end-point, originates from Flaschka and Abdine [56-45]; the titration is carried out in boiling aqueous acetic acid at pH 3 with CuY–PAN. A full examination of the procedure and its practical application has been reported [62-34]. Thikhonov [62-88] finds a slight tendency to low results, and to achieve the highest accuracy recommends standardizing the EDTA against an Al standard. Iwamoto [61-66] showed that tartaric acid causes no interference in the titration and even has a favourable effect, since clearly it prevents the formation of hydroxy-compounds. The titration is conducted in a boiling solution at pH 3·7, with PAR in place of PAN as indicator in view of its better solubility. Theis [55-26] and Paul [56-28] describe the direct titration in a medium buffered with acetate at pH 4 using chromeazurol S. The end-point appears only slowly and is not very sharp, so that Theis prefers the back-titration with an Al solution. The situation is much the same, according to Taylor [55-8], if haematoxylin is used as indicator. This author avoided difficulties over the sharpness of the end-point by conducting an inverse titration, i.e. by titrating a given amount of the complexing agent with the test-solution; the inconvenience of this procedure is obvious enough. In a brief communication Kundu [61-2] refers to the possibility of titrating Al at pH 4·4 in a boiling solution using aluminon as indicator, but he gives no further details.

Kristiansen [61-44] used a fluorimetric end-point. With 3-hydroxy-naphthoic acid in a solution buffered with glycine to pH 3 Al gives a blue fluorescence in ultraviolet light which changes to green at the end-point. Warming to 50° is necessary to accelerate the change.

Al may interfere in the determination of other cations, on the one hand, it if is co-titrated, on the other, if it should 'block' the indicator. Přibil and his co-workers have dealt with the possibilities of masking it. Triethanolamine [53-11; 54-76] masks Al in alkaline solution against EDTA, but on longer standing it does not prevent the blocking of Erio T. It is therefore quickest and best to titrate into a solution cooled below 15°. In a strongly alkaline solution, e.g. in the titration of Ca, Al is masked by transformation into aluminate, and special interest attaches to the masking of Al by fluoride [54-74] whereby AlF_6^{3-} is formed. Boiling the solution, as described in the original papers, can be omitted according to later investigations [62-87]. Complete masking against EDTA can be achieved even in the cold provided a five- to six-fold excess of fluoride is added. Masking with fluoride is not only of importance for eliminating interference by aluminium but also of especial significance in the selective determination of Al, for in a boiling solution fluoride even complexes with Al bound to EDTA. This opens up the possibility of treating the test solution with EDTA and back-titrating the excess so that all the complexable cations are bound up. If fluoride is now introduced and the solution is boiled, the EDTA bound to the Al is set free and can then be titrated. Of course, other cations that react with fluoride must be absent. This method was used by Sajó [54-41], who titrated with zinc using ferri-ferrocyanide-benzidine as indicator. Furlani [60-158] used a potentiometric end-point under otherwise identical conditions. The fluoride method has been thoroughly studied by Cimerman *et al.* [58-5], particularly from the viewpoint of practical analyses. Titration of the liberated EDTA was carried out with Zn at pH 7 in a solution cooled below 10° with Erio T as indicator.

Masking of Al by sulphosalicylic acid is possible in the titration of Th at pH 1·6 [60-70] or the rare earths at pH 7–8 [62-89] with arsenazo as indicator. Sierra [62-15] masked Al with tartrate in the titration of Zn at pH 7 using Erio T. Sajó touched on the possibility of masking Al with pyrophosphate [56-98]. An interesting example of masking stems from Halacz *et al.* [61-64], who eliminated the effect of Al in the determination of Mg in electron metal by adding sodium silicate. The Mg was determined by back-titration in alkaline solution with Zn or Mg using Eriochrome cyanine; obviously this masking reaction involved insoluble aluminium silicate.

Přibil *et al.* discuss the advantages of replacing EDTA by DCTA [55-90; 62-87]. The investigations are of less importance for the determination of Al in pure solutions than for the resolution of Fe–Al–Ti mixtures.

The combination Al–Fe, which occurs so frequently in analysis, is of special importance. As already pointed out, it would be expected on the basis of the relevant stability constants that a consecutive titration on the following lines would present no difficulties: iron is first titrated in a strongly acid solution, then the pH is increased and the Al determined. However, this is only practicable if auxiliary complexing agents are present and particular indicators are chosen. The auxiliary complexing agent may act simultaneously as the indicator. Thus Begelfer [62-13] describes the titration of Fe at pH 2–3 in the presence of much sulphosalicylic acid followed by the addition of EDTA, adjustment to pH 4·5, boiling, and after cooling, back-titration with a standard solution of Fe. Wehber [57-6] first titrates the Fe in the presence of sulphosalicylic acid, and then determines Al by adding NTA and back-titrating with Cu using Bindschedler's green as indicator. According to Wakamatsu [60-169], the titration of Fe using sulphosalicylic acid can be followed by that of Al if EDTA is added and the excess determined by back-titration with Cu using PAN as indicator. Kristiansen [61-44] first titrates Fe at pH 2 using 3-hydroxynaphthoic acid to the colour change from blue to yellow. Then the titration is continued at pH 4–5 in ultraviolet light until the colour of the fluorescence changes from blue to green.

Furlani [60-158] has described the consecutive titration Al–Cr. Al is first determined by back-titration with Zn using a potentiometric end-point with the ferri-ferrocyanide system. Then Cr is complexed with EDTA at the boiling-point, and after cooling the excess of complexone is back-titrated. Liteanu *et al.* [59-60] proceed similarly but back-titrate with Fe visually using Tiron. Patzak and Doppler [57-29] describe the resolution of mixtures of Fe–Al–Cr by using aliquot portions.

Only slight reference has been made to the possibility of masking other metals with 1,10-phenanthroline during the determination of Al. The high cost of this reagent prohibits its use in practice when more than microgram amounts of the sample have to be used.

Instrumental end-points have been used by Přibil *et al.* [51-15] in the determination of Al; they used potentiometry in back-titration of

excess EDTA with Fe. The method was adopted by Patzak and Doppler [57-29] for the analysis of Fe–Cr–Al mixtures. Sajó noted briefly
[54-41] that the back-titration of excess EDTA with zinc, using ferri-
ferrocyanide, can be followed potentiometrically, and this was studied
more throughly by Furlani [60-158].

Amperometric indication of the end-point in the back-titration
with Fe was carried out by Babenyshev [60-151] at a rotating Pt-
electrode and by Goldstein *et al.* [63-13] using a standard vanadyl
solution for the back-titration and a stationary Pt-foil electrode.

According to Ringbom [53-57], a photometric titration is possible
if excess EDTA is back-titrated with Cu at pH 5 using murexide.
Elofson [61-149] obtained excellent results by back-titrating with Zn
against chromeazurol S. Wallraff [61-42] determined Al in the presence of other metals by the fluoride procedure. The liberated EDTA
was titrated with Zn using ferri-ferrocyanide-dimethylnaphthidine as
the indicator system.

The complexometric determination of Al has found extremely
varied uses in practice. In many cases use is made of precipitations,
extraction, or ion-exchange to achieve preliminary separations. To
quote only a selection; Al can be determined complexometrically in
pharmaceuticals [53-12], Liquor alum. acet. DAB(VIII) [61-105],
slags [55-51; 55-96; 61-77], clays [63-42], cracking catalysts [60-139],
aluminate solutions [60-160], cryolites [58-75], ferrosilicon [59-19],
ores [61-45], cement [61-42], water from coalmines [61-174], zircon
sands [61-45], and silicates [55-51; 55-81; 55-96]. A host of recipes
exist for the titration of Al in alloys [54-91; 55-78; 59-9], especially
those of Mg [60-151] and Zn [61-175], also in electron metal [61-64],
high-temperature alloys [59-103], brass, bronze [55-3], and steel
[62-34].

12.1. Direct titration of aluminium using CuY–PAN

REAGENTS: 0·01M EDTA as standard solution; Cu–EDTA solution (cf. Section 1.6); PAN indicator; 0·1% bromophenol blue in
alcohol; conc. ammonia; 10% ammonium acetate; acetic acid.

PROCEDURE: The acid test solution is neutralized with ammonia
to a pH between 0 and 1 (indicator paper). A few drops of bromophenol blue are introduced and by adding ammonium acetate the
colour is brought to a blue-grey shade at which point 5 ml conc.
acetic acid is added. The pH should now be about 3 (indicator paper
or pH meter) and should be corrected, if necessary, by adding acetic
acid or ammonium acetate. After introducing a few drops of the

CuY solution and sufficient PAN to produce a strong violet coloration, the solution is heated to boiling. Titration is now carried out, keeping the solution near the boiling-point all the while. Just before the end-point the indicator colour changes to yellow, since the EDTA reacts more quickly with Cu–PAN than with Al. However, the violet colour quickly returns. As the end-point is approached the time taken for the colour to return gets longer, so that it becomes necessary to add the titrant dropwise. The end-point is reached when the yellow colour persists for more than half a minute at the boiling-point. To make sure, boil the solution for one minute more; often a slight red shade develops which can be discharged with a final drop of the EDTA solution.

REMARKS: If larger amounts of Al are present, the titration can be carried out with 0·1M EDTA. Any Fe present is quantitatively co-titrated, so that the procedure can also serve for determining the sum Fe + Al. If much iron is present it is advisable to pre-titrate this roughly with sulphosalicylic acid as indicator, since it can already hydrolyse at pH 3 on heating (brown coating on those parts of the glass vessel that have been over-heated). Alkaline earths and Mg do not interfere. Ti is hydrolysed and the precipitate does not interfere, apart from the fact that, in large amounts, it may carry down some Al. Up to about 20–30 mg Mn/150 ml does not interfere.

12.2. Determination of aluminium by back-titration with iron using salicylic acid as indicator

REAGENTS: Standard solution of 0·1M EDTA and 0·1M $FeCl_3$; solid salicylic acid; approx. 1M HCl; approx. 1M NH_4OH; approx. 1M Na acetate.

PROCEDURE: The precipitate containing aluminium (e.g. $Al(OH)_3$ produced when separating the Al from accompanying metals) with a metal content of 30–50 mg Al is dissolved in a little HCl and EDTA is added in slight excess. The mixture is adjusted to pH ~6 by drops of ammonia and spotting on methyl-red indicator paper. It is then boiled for a short time. After being cooled, it is diluted to at least 100 ml and treated with 3 ml sodium acetate. The pH of the mixture should now be about 6–6·5. Introduce 0·2 g of the solid indicator, and when this has dissolved titrate with the standard ferric solution until a red-brown colour persists for a short time. During the titration the pH must not fall below 5, and to prevent this a few more drops of ammonia should be added if necessary.

REMARKS: The ferric complex of EDTA is appreciably more

stable than the corresponding aluminium complex (cf. Table 2). Hence on reaching the equivalence point of the back-titration the following displacement occurs:

$$AlY^- + Fe^{3+} \longrightarrow FeY^- + Al^{3+}$$

This proceeds slowly, however, so that a sufficient increase in the concentration of iron takes place at the end-point to produce a reaction with the indicator. On standing, and more quickly on warming, the brown colour of the ferric salicylate fades again. Narrow limits are set to the pH values at the end-point. If it is too high the salicylic acid no longer responds to the iron(III); for the hydroxide ions (formation of $Fe(OH)^{2+}$ and $Fe_2(OH)_2^{4+}$) and acetate ions (acetato complexes) are in competition with the indicator. On the other hand, if the pH is too low the above equilibrium becomes mobile, so that as the titrant flows in, the iron merely displaces the aluminium from its complex with EDTA.

As in all complexometric titrations involving iron(III) and phenolic indicators, the precision of the method is only about 1%. The use of very dilute standard solutions is not recommended.

12.3. Determination of aluminium by back-titration with zinc using dithizone as indicator

REAGENTS: Standard solutions of 0·05M EDTA and 0·05M $ZnSO_4$; acetic acid–ammonium acetate buffer containing one mole of each per litre; dithizone in alcohol.

PROCEDURE: An excess of EDTA is added to the test-solution containing about 5–15 mg Al per 50 ml, and the pH is adjusted to 4·5 with buffer. The solution is boiled for a short time, cooled, and diluted with its own volume of ethanol. 2 ml of indicator are added for each 100 ml and the back-titration with zinc is begun. The end-point is recognizable by a very sharp colour change from a greenish violet to red.

REMARKS: Addition of alcohol is necessary because of the insolubility of dithizone in water; but it also improves the conditions for the titration, in that it increases the stability of AlY^- compared with that of ZnY^{2-}. At the end-point pZn is about 7–8. Too low results could be obtained if the buffer was added before the EDTA, especially if the stage of boiling was omitted, for aluminium forms polynuclear hydroxy-complexes which do not then react completely with EDTA.

13. SCANDIUM

Wünsch [55-91] tried to apply the method worked out by Flaschka for the rare earths (q.v.) to the direct titration of Sc in a medium of pH 10 containing tartrate and with Erio T as indicator; he found the titration to be impossible, since the indicator became 'blocked'. The titration is feasible, however, in a solution containing malic acid which has been adjusted to pH 7·5–8 by adding 1M NH_4OH and 1M NH_4Cl. Below pH 7·5 free Erio T begins to appear in the protonated red form; above pH 8 blocking of the indicator sets in. The addition of malic acid is necessary to protect the Sc from precipitation as its hydroxide. Lowering the pH also has a favourable effect on the effective constants of the indicator- and titration-complex. A similarly favourable change can be achieved at pH 10 if an organic solvent is added. Thus the reaction between Sc^{3+}, Erio T, and EDTA in 50% aqueous ethanol is reversible, and back-titration of excess EDTA with a standard solution of Mg is now possible [61-12]. Concerning the complex between Sc and Erio T cf. Wünsch [61-12].

Titrations in acidic media are more advantageous, in that they are more selective. Kinnunen [57-137] titrates Sc with EDTA in a cold solution at pH 3–4, using xylenol orange. Cheng [55-20] has reported on the formation of a coloured complex between Sc and PAN. The complex is stable over the pH range 1·5–4, but the end-point in the direct titration is not sharp. Back-titration of excess EDTA with Cu using PAN as indicator (cf. Part One, Section 3.3.3) is preferred in view of the extraordinarily sharp end-point. A very sharp end-point is obtained in the direct titration if use is made of the system Cu–EDTA–PAN (cf. Section 31.3).

According to Fritz [59-96], a really selective titration of Sc is possible if a visual indicator is relinquished in favour of a complex-forming indicator and a photometric end-point is employed with Cu as a slope-indicator (cf. Part One, Section 5.2.1.2). The titration is carried out at pH 2·7 and 745 mμ. The Sc is complexed first without any change in the optical density; after reaching the end-point the titration curve rises abruptly due to the formation of the strongly absorbing complex CuY. Cations that form EDTA complexes of high stability (e.g. Fe(III), Bi, Th, Zr, etc) interfere with the titration, but the presence of very many other metal ions can be tolerated. Thus Sc can be titrated with an average error of $\pm 0·5\%$ in the presence of Y (60), La (20), Ce(III) (40), Nd (40), Sm (50), Dy (50), Yb (10), Lu (40), Al (30), Co (40), Cd (40), Pb (15), UO_2^{2+} (70) – where the numbers in brackets denote the excess of the foreign metal. Alkaline earths have no effect, even at still higher concentrations. Fe(III) must be reduced to the bivalent state by passage through a lead reductor. A proportion higher than Sc : Fe(III) = 1 : 1 could not be accepted owing to re-oxidation of ferrous iron during the titration, but the exclusion of atmospheric oxygen by means of nitrogen or an inert gas was not investigated. The effect of adding ascorbic acid was certainly tried out. However, the quantity was found to be critical, since too large an excess reduced the copper that serves as indicator. Among the common anions, chloride, perchlorate, nitrate, and acetate do not interfere. Sulphate interferes in concentrations over 0·1M. Other authors have reported on the back-titration in alkaline media with standard solutions of Hg (potentiometric) [62-120] and Ni (using murexide) [62-6].

14. YTTRIUM AND THE RARE EARTHS (LANTHANONS)

Flaschka [55-65] was the first to report a direct titration of rare earths. This was carried out at the boiling-point in a medium containing tartrate of pH 8–9 using Erio T as indicator. The pH must not exceed 9, for otherwise the indicator-complex becomes too stable and is not broken down by EDTA. The addition of tartrate is necessary to prevent the precipitation of hydroxides, and boiling puts up the rate of reaction. The method has been tested for Y, Sm, Pr, and Gd on the micro-scale, but in view of the similarity in their chemical behaviour and the fact that the stability constants of their complexes are relatively close to one another, it can be predicted that other members of the group would behave in the same way. In point of fact, since then all tervalent lanthanons with the exception of promethium have been titrated complexometrically. Ce(IV) interferes, but can easily be reduced with ascorbic acid. There can be no doubt that Pm, too, could be determined complexometrically. The stability constant of its EDTA complex is given as $1·5 \times 10^{16}$ [25°C, $\mu = ?$] by Bruno and Barbieri [61-123]; according to these authors, the complex PmY^- occurs in the pH range 1·8–9·0. Direct titration of lanthanons is also possible in an alkaline medium (pyridine buffer) using chromeazurol S, and this has been checked with lanthanum and cerium [59-14].

Since the discovery of new indicators it is possible to titrate the lanthanons in acid medium; this increases the selectivity, for now

Mg and the alkaline earths, and in many cases Mn also, do not interfere. As indicators in such titrations bromopyrogallol red [56-88] and xylenol orange [57-137; 8-113; 59-24; 62-91] can be used in solutions buffered with acetate or urotropine. Other indicators are carminic acid at pH 3·7 [58-79], gallein in an urotropine solution [62-30], alizarin S at pH 4 [56-59] (appropriately admixed with methylene blue), and at the same acidity pyrocatechol green [60-187]. Az-oxine S [61-112] can be used as an indicator at pH 6. Excellent results are reported for the use of arsenazo [58-61]. In general, the total concentration of rare earths is obtained in these analyses for which a pH of 4·5 represents a lower limit. However, according to Gimesi *et al.* [62-90], other pH values are optimal for individual lanthanons. For Ce(III) pH 6 and for lanthanum pH 4·5 is recommended, whereas yttrium and erbium can still be titrated at pH 3. For a comparative study of direct titrations cf. Lyle and Rahman [63-49].

Back-titrations have also been reported. Flaschka [55-65] back-titrates excess EDTA with Zn either at pH 8–9 using Erio T, or at pH 5 with the system ferri-ferrocyanide-dimethylnaphthidine. Cheng [58-64] titrates with Cu at pH 3 using PAN. Takamoto [55-103] recommends back-titration with a standard Co-solution in 50% aqueous acetone containing acetate and thiocyanate. Back-titration of EDTA can also be effected with Ni in ammoniacal solution using murexide [62-6]. Mitsumi [59-87] suggests a standard solution of Mg for back-titration with Erio T as indicator and uses the method for the determination of lanthanons after precipitating them as oxalates and dissolving the latter in an alkaline EDTA solution [63-50]. Lanthanum can be determined at pH 5 in the form of its salt with dibutylphosphoric acid if back-titration is carried out with a standard La-solution using xylenol orange as indicator [63-15]. Back-titration with La is also possible at pH 8–9 with arsenazo [62-127].

Electrical methods extend to high-frequency titrations [56-59], and the amperometric titration of Eu [60-7] with DTPA at pH 5 and −0·85 volt (against a saturated calomel electrode). Extensive use has been made of photometric titration with, e.g., arsenazo as indicator. The method is very sensitive and especially suitable for microgram amounts [61-54; 62-19]. Lane and Fritz [57-25] state that it is possible to titrate satisfactorily down to $7·5 \times 10^{-7}$M Nd solutions with 10^{-5}M EDTA. Efimov [60-130] uses PAR at pH 8 for the photometric titration of Er at 504 mμ and for the total concentration of rare earths. Crouch [59-137] titrates Ce, Pr, Nd, Sm, and Gd in a solution buffered with acetate to pH 5·8 and La at pH 5·2 using xylenol orange at 580 mμ. Bril *et al.* [59-95] use photometric titration with alizarin S at 520 mμ for determining rare earths of the cerium group up to Eu, and they discuss the possibility of a back-titration procedure with a standard La-solution.

The consecutive titration Th–lanthanons deserves special attention. According to Chernikov [60-70], Th is first titrated at pH 1·5–2·5 using xylenol orange as indicator; the pH is then increased by adding urotropine, and the titration continued for the determination of the rare earths. The method works up to a ratio Th : lanthanons = 70 : 1. Small amounts of Fe can be rendered harmless by reduction with ascorbic acid. Al can be masked with sulphosalicylic acid, but Mn is co-titrated. According to Chernikov, arsenazo is unsuitable for the titration of lanthanons following that of Th, since the Th blocks the indicator at the higher pH used for the titration of the rare earths, despite the fact that the Th is complexed by the EDTA. However, Kuteinikov [62-89] states that this can be prevented by 'double-masking' if sulphosalicylic acid is also present. The combination of EDTA with sulphosalicylic acid completely inhibits the reaction between Th and arsenazo in the pH ranges 1–6 and 9–10, but less completely within the pH range 6–9. Rare earths could therefore be titrated at pH 5. Addition of sulphosalicylic acid has the further advantage of masking Al, of which up to 1·5 mg/ml can be tolerated. Up to 2 mg Fe/50 ml can be masked with ascorbic acid.

Peshkova and his co-workers [62-19] have reported on the use of photometric end-point indication in the consecutive titration of microgram quantities of Th and the rare earths. Arsenazo serves as the indicator, and for the titration of Th, which is carried out at pH 2, its concentration must be at least equivalent to that of this metal. The lanthanons are determined subsequently in a solution buffered with urotropine. It is rather extraordinary that although the arsenazo is present even in excess in this method, the blocking of the indicator as reported above for the visual titration is not referred to. Onosova [62-6] titrates Th in acid solution with pyrocatechol violet and then measures the lanthanons by back-titrating excess EDTA with Ni in ammoniacal solution using murexide. Bril *et al.* [59-95] describe the consecutive titration with photometric indication using alizarin S as indicator. Th is titrated at pH 2·8; EDTA is then added, the pH is adjusted to 3·7 with an acetate buffer, and the solution heated to 85°C; finally, excess EDTA is back-titrated with a standard La-solution.

Milner and Edwards [58-113] report on the consecutive titration

Bi–Nd (or Pr). Bi is determined at pH 2 with xylenol orange and the lanthanon at pH 6.

Many metals interfere with the titration of the rare earths. In an acidic medium Mg and the alkaline earths hardly have any effect if the process is carried out at pH 5–6. The extent to which Mn interferes depends on the pH of the solution, the concentration of Mn, and the indicator used. Masking of small amounts of iron can be effected by reduction with ascorbic acid [62-89] or hydroxylamine hydrochloride [59-95]. Al is masked by sulphosalicylic acid [58-61; 62-89]. Fritz [58-61] has shown that U(VI) present in two-fold excess over the rare earths has no effect and that Hg can be masked by iodide, and Zn, Cd, Hg, and Pb by diethylthiocarbonate and by thiourea.

The determination of rare earths and EDTA together is of importance with reference to work with ion-exchangers both on the research and production scale. Martynenko [58-114] precipitates the lanthanons as oxalates and determines them gravimetrically after ignition to oxides; in another aliquot portion the amount of EDTA not bound to metal is obtained by titration, and the total amount of complexone is then calculated from the two sets of data. Tereshin and Tananaev [62-91] solve the problem in a far more elegant fashion. The EDTA is titrated with Fe in a warm solution at pH 2 using sulphosalicylic acid as indicator. In this process the whole of the rare earths are displaced from their complexes, and the total EDTA is thus obtained. The fully titrated solution is then treated with urotropine and the lanthanons which are no longer complexed are titrated with EDTA with xylenol orange as indicator.

Sometimes the average atomic weight of a mixture of rare earths is of interest. Following Flaschka [55-65], this can be obtained simply and very accurately if a mixture of the oxides is weighed out, dissolved up, and then titrated with EDTA. The following formula holds for the calculation: average atomic weight $= (100 \ G/V) - 24$, where G is the weight of mixed oxides in mg, and V the volume in ml of 0·0100M EDTA solution required. Patrovsky has pointed out [59-24] that if larger amounts of yttrium are present, then, because of the low atomic weight of Y compared with that of the lanthanons, the yttrium content of the rare-earth mixture can be calculated directly from the average atomic weight.

Complexometric titration is used in practice for determining the rare earths in monazite sand and commercial intermediates [59-95; 62-19], in Bi–U–Nd alloys [58-113], and in fission products [62-127].

15. PLUTONIUM

The study of complex formation between Pu and EDTA is difficult because the Pu exists in various oxidation states, several of which show a pronounced tendency towards hydrolysis; furthermore, besides the usual 1 : 1 complexes, compounds of 1 : 2 composition also exist. Kabanova *et al.* [60-10] report that Pu(VI) at pH 3–5 is reduced by EDTA to Pu(V), which then forms a 1 : 1 complex with EDTA. One molecule of EDTA reduces 6 molecules of PuO_2^{2+}. In the presence of a very large excess of EDTA the reaction slowly proceeds further to give Pu(IV). From measurements of ion-exchange equilibria at pH 4–5 Gelman [59-92] estimates the stability of the EDTA complex of Pu(V) in the form of its plutonyl ion as log K = 10·2. Foreman and Smith [57-130a] used the same method, but arrived at the substantially higher value of log K = 16·4 (in 0·1M KCl). Kabanova [61-18] gives the value log K = 12·9 (in 0·1M KCl) and states that a 1 : 2 complex exists as well as the 1 : 1 complex. For Pu(III) and Pu(IV) Metz [57-64] estimates the approximate values of log K = 16 and 26 respectively. Foreman and Smith [57-130a] calculate 18·1 and 17·7 from exchange equilibria with an ion-exchange resin at 20°C and μ = 0·1M KCl. The figure for Pu(IV) refers to the hydrolysed species of the cation. In a later paper [57-130b] the authors investigated complex-formation spectrophotometrically in 1N nitric acid, where Pu(IV) occurs completely unhydrolysed. They obtained the value log $K_{Pu(IV)Y}$ = 24·2. However, in the calculation they ignored the fact that in solutions of such high acidity the EDTA also forms the protonated complexes H_5Y^+ and H_6Y^{2+}. Klygin and his co-workers [59-126] determined the stability constants for the proton complexes and included these in a fresh calculation of the stability constant for Pu(IV)Y and obtained the value log K = 26·1.

The complexometric titration of Pu(III) in its tervalent form was first carried out by Milner [56-75] and worked out purely empirically without any knowledge of the stability constants. The determination was made at pH 2·5 (checked during the titration and, if necessary, readjusted) by back-titrating excess EDTA with Th using a mixed indicator of alizarin S and methylene blue. The solution of Pu(III) was obtained by dissolving the metal in HCl. Extremely sharp endpoints were obtained with freshly prepared solutions, even at very high dilutions. On the other hand, aged solutions gave an indistinct colour change, and a completely useless colour effect was obtained in one which had been kept for three weeks. The addition of methylene blue is not absolutely necessary, but it is advisable in order always to achieve exactly the same colour change from green to purple. In its absence, and because of the blue colour of Pu(III) itself, the range of the colour transition with alizarin alone changes with variations in the amount of plutonium.

Pu(IV) was titrated by Paley [60-58] in a strongly acidic medium (0.1–0.2M in HCl or HNO_3) using arsenazo. The determination is very selective, and La, UO_2^{2+}, Pb, Cr^{3+}, and Ni do not interfere. Even Fe(III) can be tolerated provided its amount is less than 4% of that of the Pu.

Brown *et al.* [63-6] describe a direct titration of Pu using CuY–PAN or xylenol orange. Another procedure, according to these authors, is first to add excess EDTA and back-titrate with Cu using PAN as indicator; F^- is then added, and the amount of EDTA liberated in proportion to the Pu present is obtained by a further titration with Cu.

Boase and co-workers [62-65] use complexometric titration for the determination of Pu in reactor fuels; here excess EDTA is back-titrated photometrically with Zn.

16. TITANIUM

Complex formation between EDTA and titanium was first studied polarographically by Blumer and Kolthoff [52-11]. Below pH 2·5 the results are quite straightforward and the waves are practically completely reversible; the existence of Ti(IV)Y and Ti(III)Y$^-$ was established and estimates of their stability constants were reported. Pecsok and Maverick [54-96] later reported an extensive polargraphic study. Below pH 2 the half-wave potential (-0.22 volt v. the saturated calomel electrode) is independent of pH and corresponds to the reaction $TiY + e \rightleftharpoons TiY^-$. Above pH 2·5 the situation becomes complicated by the appearance of hydrolysed species; hydrolysis is so extensive that the onset of precipitation is manifested by opalescence. The stability constant of Ti(III)Y$^-$ was calculated as $\log K = 21.3$. No constant can be given for the simple complex Ti(IV)Y, since this complex is not in equilibrium with the ion Ti^{4+} but with the titanyl ion, TiO^{2+}. However, the equilibrium constant for the reaction $TiY + H_2O \rightleftharpoons TiO^{2+} + 2H^+ + Y^{4-}$ is reported to be 3×10^{-23}. The titanyl–EDTA complex, $TiOY^{2-}$, certainly exists as well, and its stability constant is given by $\log K = 17.3$.

Hydrolytic phenomena and slowness of reactions are certainly responsible for the defects in the complexometric determination of Ti if these are carried out without adding hydrogen peroxide (q.v.), and a way round through back-titration becomes a necessity. Sajó [54-42] determined titanium as follows. The acid test-solution was treated with EDTA, and after buffering with acetate the excess was back-titrated with zinc using the ferri-ferrocyanide benzidine indicator system. If other titratable metals are present the fully titrated solution is treated with sodium phosphate and boiled, whereupon all the Ti is precipitated as phosphate. After cooling, the amount of EDTA that is liberated in proportion to the Ti content is titrated with Zn. The method can be used for the determination of Ti in the presence of Fe and Al and other ions that are not masked by phosphate. Sajó [56-100] also recommends back-titration with vanadium(V) with diphenylcarbazide (or carbazone) as indicator. Precise details are lacking throughout, and it is to be expected that the accuracy would be adversely affected by co-precipitation and, furthermore, that the amount of phosphate would be critical, since if excess were added other ions could also be displaced from their EDTA-complexes.

Šir and Přibil [55-90] determine Ti in a solution buffered with pyridine by back-titrating excess DCTA with Zn using Erio T as indicator. The same authors back-titrate excess EDTA in a solution buffered with pyridine acetate (pH 4–5) with Cu using pyrocatechol violet [56-86]. Kinnunen [57-137] reports on the back-titration of excess EDTA in an acetate buffer with Tl(III) using xylenol orange as indicator, but data on the accuracy of the results, special precautions, slowness of reaction, etc, are not given. Takao and Musha [61-72] describe a direct titration with EDTA at pH 2 with amperometric indication; the precision is about 2%.

Sweetser and Bricker [54-14] found that a Ti–peroxy–EDTA complex, more stable than the normal complex, was formed by the addition of hydrogen peroxide. On back-titrating excess EDTA with Fe using salicylic acid as indicator no displacement of titanium ensued at the end-point. The back-titration with Fe was carried out photometrically at pH 1·9 and a wavelength of 540 mμ. The possibility of determining Ti photometrically in the presence of peroxides was referred to.

Direct titration was thoroughly studied by Musha and Ogawa [57-20]. They titrated at pH 0·4–4 and 365 mμ. At low pH values Zn, Al, alkaline earths, and appreciable quantities of Zr do not interfere if the concentration of the element does not exceed 200 mg per litre. Mo and W must be absent. The authors report the following constants for peroxytitanium–EDTA complexes:

$$\frac{[TiO(H_2O_2)Y^{2-}]}{[TiO(H_2O_2)^{2+}][Y^{4-}]} = 2.7 \times 10^{20}; \frac{[TiO(H_2O_2]Y^{2-}]}{[TiOY^{2-}][H_2O_2]} = 2.3 \times 10^6$$

Comparison between these values and those given above for the peroxide-free complexes clearly shows the stabilizing action. Quite apart from the increased stability of the complex, which reduces the danger of any displacement reaction at the end-point, there is the further considerable advantage that complex-formation between peroxytitanate(IV) and the complexone proceeds smoothly and hydrolytic phenomena no longer interfere.

Lieber [60-136] titrates Ti directly and visually after adding a measured amount of ferric iron to the test-solution. The Fe displays the end-point through the medium of salicylic acid as indicator; the amount of Fe added must, of course, be taken into account in the calculation. The visual end-point is not particularly sharp, so Lieber recommends photometric indication.

Generally speaking, the presence of peroxide is advisable for back-titrations as well. Bieber and Večera [61-82] back-titrate excess EDTA with Bi using xylenol orange (cf. also [63-61; 63-63]). This is done in a strongly acid medium and below 20°C; furthermore, the absence of sulphate and chloride is essential for a successful titration, otherwise the precipitation of sparingly soluble bismuth compounds occurs. However, the method permits of the consecutive titration Fe–Ti. Fe is first titrated at 70°C in the absence of peroxide with salicylic acid, which serves as indicator and simultaneously prevents the precipitation of titanium hydroxide. The titrated solution is then cooled below 20°C, treated with peroxide and EDTA, and excess of the latter back-titrated with a standard Bi-solution.

The restrictions of having to work in a cold solution, and in the absence of sulphate and chloride, disappear if, following Wilkins [59-4], the excess of EDTA in the presence of peroxide is back-titrated with Cu, using PAN; this is carried out in a solution buffered with acetate. The use of calcein W as a fluorescence indicator [59-9] has the advantage over PAN and other indicators that large amounts of Ti can be determined, for the quenching of the fluorescence at the end-point of the titration is not masked by the yellow background colour of the solution.

Lassner and Scharf [61-168] also recommend back-titration with Cu using PAN and titrate at pH 5–5·2 in a hot solution (70–80°C).

In a study of interfering elements these authors observed that Nb in the presence of peroxide forms a Nb–peroxy–EDTA complex that is sufficiently stable to make the determination of Ti impossible [60-3]. DCTA does not show this behaviour, and its use is therefore preferable, for then the titration of Ti can be carried out even in the presence of niobium and tantalum. Should any molybdenum be present (q.v.), then in the course of back-titration with Cu it is easily displaced from its complex with DCTA (but not from its EDTA complex), so that with this titration reagent the determination of Ti can also be carried out in the presence of Mo and W. In a subsequent paper Lassner and Scharf [62-50] recommend a fluorescence endpoint, since the intrinsic yellow colour of the solution has then less effect, and they propose methylcalcein and methylcalcein blue as indicators.

Giuffré and Capizzi [60-135] describe a potentiometric back-titration of EDTA with a solution of Fe(III) at pH 1·6–1·9. The adjustment of pH is critical, for it strongly influences the shape of the titration curve, the height of the potential jump, and the position of the equivalence point. The pH of the solution must be checked during the titration and corrected if necessary. The advantage of the method is that Ti can be titrated in the presence of Al. The same authors [62-80] describe a photometric titration for the determination of Ti in the presence of Al; excess EDTA is back-titrated with Fe at pH 1·5 and 510 mμ with sulphosalicylic acid as indicator.

It is worth noting that peroxytitanate is so stable in an alkaline medium that H_2O_2 can be used to mask Ti during the titration of other metals with Erio T as indicator [57-3]. Masking of Ti against EDTA by precipitating it as phosphate has already been mentioned [54-42].

The titration of Ti is used practically for the analysis of a variety of substances, although previous separation by liquid–liquid extraction or the use of ion-exchange resins is generally necessary. Extraction with cupferron precedes the determination of Ti and Fe in the analysis of slags, ferro-titanium, and ilmenite [61-82]. Magnet metal alloys are subjected to a preliminary separation by ion-exchange [59-9]. Photometric determination of titanium is used in the analysis of the raw, intermediate, and finished products of the lime and cement industries [60-136; 61-42]. Ti has been determined in cemented carbides [61-168], alloys [63-62; 63-63], and without previous separation potentiometrically [60-135] or photometrically [62-80] in the presence of Al in Al–Ti catalysts.

17. ZIRCONIUM AND HAFNIUM

The titration of Zr with EDTA was first described some time ago, but the reactions that take place during the actual titration and in preparing the test-solution for the titration are still far from being completely understood. However, this presents no obstacle to carrying out an accurate determination. According to every paper published to date, Hf behaves in exactly the same way as Zr, so that anything stated below for Zr holds equally for Hf.

The stability constant for the zirconium–EDTA complex was determined by Morgan and Justus [56-50], who find $\log K = 19 \cdot 4$ ($\mu = 0 \cdot 1 \text{M}$; 25°C). The value for hafnium was $19 \cdot 1$. The authors refer to a 1 : 1 complex, but do not establish in any way whether they are concerned with a complex of Zr^{4+}, zirconyl, or some other cationic species, and in what form the uncombined zirconium occurs in the solution at equilibrium. The value was obtained from photometric measurements of the displacement equilibrium with CuY^{2-}. Iwase [59-52] reports $\log K = 18 \cdot 0$ ($\mu = 0 \cdot 1 \text{M}$; 25°C) for the complex ZrY. These low values for the stability constant are difficult to understand, for zirconium can be titrated in a strongly acid medium. Working instructions in the literature specify the concentration of acid in the titration solution up to 2N; at pH 0·5 the value of $\alpha_{Y(H)}$ for EDTA is already $10^{19 \cdot 1}$! Either the complex is exceptionally 'robust', which, however, conflicts with the above-mentioned possibility of a direct titration, or the zirconium's well-known predilection to undergo hydrolysis plays a part in some way that is either unknown or insufficiently appreciated.

Musil and Theis [55-30] state that when using a solution freshly prepared from $ZrCl_4$ the amount of EDTA required for the titration corresponds exactly to a complex $Zr : Y = 1 : 2$. However, hydrolysis takes place if the solution is first boiled or allowed to stand for a long time, and if titration is then carried out the amount of EDTA used corresponds exactly to the ratio 1 : 1. Although this observation has not been confirmed from other quarters, it is obvious that so far as Zr is concerned there is still a lot to be cleared up. In a recent paper Kyřš and Caletka [63-51] describe experiments on equilibria between Zr and EDTA by way of absorption on silica gel in 1–5N HNO_3. They give $\log K_{ZrY} = 28 \cdot 5 \pm 0 \cdot 3$ (1N HNO_3) and $30 \cdot 6 \pm 0 \cdot 2$ (5N HNO_3). For $\mu = 0 \cdot 1$ they give a value $29 \cdot 5 \pm 0 \cdot 5$, which clearly is a calculated one. These values are more consistent with the known facts based on actual titrations.

H

Fritz and Fulda [54-45] described the first direct titration. The authors investigated carminic acid, chromeazurol S, and chloranilic acid as possible indicators, but found them to be unsuitable. Sharp end-points were obtained with alizarin cyanol RC and, better still, eriochrome cyanin RC in the pH range 1·3–1·5. The concentration of Zr should be about 3–5 × 10⁻³M. It is necessary to titrate slowly near the end-point, and on these grounds a back-titration is proposed in which excess EDTA is back-titrated with Zr in a hot solution. Fe(III) is co-titrated, and to eliminate its interference the iron should be reduced with zinc amalgam; however, in this case air must be excluded during the titration to prevent re-oxidation of the ferrous iron. As later authors have shown, reduction with ascorbic acid, and in many cases with Sn(II), is just as effective and far simpler.

Banerjee [55-92] used SPADNS as indicator and worked at pH 2. Fe can be masked with ascorbic acid. Interference is caused, *inter alia*, by Ni, Cu, Ag, molybdate, phosphate, tartrate, and sulphate. Datta [60-145] proposed the same indicator.

Musil and Theis [55-30] successfully used chromeazurol S as indicator, whereas earlier workers [54-45] had obtained no usable results with it. However, the results of direct titrations were not particularly good, for which reason it was proposed to titrate a known volume of standard EDTA solution with the test-solution of Zr containing 10% HCl. The inconvenience of this procedure ensures little hope of its being accepted in analytical practice. Goryushina *et al.* used Erio T to indicate the end-point. In a solution of Zr that is 2N in HCl, Erio T forms a blue-violet compound on heating which is decomposed at the end-point of the EDTA titration with a colour change to pink. The selectivity at this high acidity is quite considerable. Since tin causes no interference, Sn(II) can be added as a reductant to eliminate interference from small amounts of Cu, V, and Mo, as well as up to 100 mg Fe(III). Other complex-forming substances, such as phosphate, oxalate, tartrate, and fluoride, interfere. The authors also mention carminic acid as a possible indicator, a proposal that conflicts with earlier investigations [54-45]. Sun [60-13] used gallein as indicator in the pH range from 0·5 to 1·6, with the sharpest colour change at pH ~1. The indicator reaction is slow, and it was recommended to wait for about 10 minutes after reaching the end-point to check against any possible return of the transition colour. Among forty-eight ions that were investigated only the following were found to interfere: Sb(III), Ce(IV), Fe(III), Mo(III), Sn(II) and Sn(IV), Ti(III), chromate, permanganate, vanadate, molybdate, oxalate, tartrate, fluoride, and sulphate. Many of these potential sources of interference are either seldom present or easily removable or rendered harmless by treatment with oxidizing or reducing agents. Korkisch [58-91] proposes the use of Solochrome violet R for the macro- and micro-determination of Zr in 1N HCl. However, the indicator responds to very many other cations and is easily destroyed by oxidation or reduction. It is thus limited in its application to more or less pure Zr solutions such as might be obtained after separation with ion-exchange resins. The same is true for Solochrome black 6BN, which the same author recommends [60-178], for this even requires the absence of nitrate. The titration of milligram amounts of Zr is accurate to 1–2% and for microgram amounts to 5–6%.

Xylenol orange, chosen by Klygin and Kolyada [61-91] for the direct determination of zirconium in 0·25N sulphuric acid, appears to be one of the best indicators. Whereas sulphate ions interfere in the titration with many other indicators, sodium sulphate is actually added in this case, for then the end-point is much sharper. In the very strongly acid solutions even Fe(III), In, Sc, Th, Y, Ni, and Co do not interfere (in amounts up to 10 mg per 300 ml) although their EDTA complexes are very stable. Moreover, vanadate (up to 100 mg) and molybdate do not interfere, but phosphate, oxalate, and fluoride must be absent. Under the conditions of the titration Bi reacts to the indicator and therefore interferes. According to Lukyanov [62-61], considerable amounts of Cu can be tolerated if the titration is carried out in 0·5N sulphuric acid and thiourea is added as a masking agent. Xylenol orange is also recommended by Tsuchiya [62-59] for the direct titration in $0·9 \pm 0·3$N HCl. Here, too, interferences are small, and it is noteworthy that up to 1 g of U(VI) per 100 ml has no effect. According to Volodarskaya *et al.* [63-8], the consecutive titration Zr–Th is also possible if xylenol orange is used. Zr is first titrated at 80–90°C in 0·25–1N HCl or H_2SO_4, then the pH is raised to 1·5–2·5 and the Th determined. If Fe is to be masked by reduction to Fe(II) hydroxylamine hydrochloride must be used, as ascorbic acid produces low results.

According to Kononenko and Poluetkov [62-60], sulphonaphthylazoresorcinol is a more suitable indicator in exactly 2N HCl. Sulphate does not interfere at all, and Sn(II) and Sn(IV) are also without influence, so that the former can be added to reduce Fe(III) and other interfering elements.

Majumdar [60-41] uses *o*-carboxylphenylazochromotropic acid (chromotrope 2C) as an indicator for the direct titration of Zr at pH 1·4–2·8 in a solution warmed to about 50°C. Here, too, sulphate has no effect, and considerable quantities of iron can be masked with ascorbic acid.

Since the discovery of serviceable indicators for the direct titration, the procedures for back-titration that have been worked out, more especially in former years, have lost much of their significance, particularly as many of them are conducted at high pH values, where selectivity is appreciably reduced. The method recommended by Fritz and Johnson [55-108] is an exception, however, since there are relatively few interferences and, furthermore, several ions can be present, in particular tartrate, that interfere in all direct titrations. Back-titration of excess EDTA is effected with Bi at pH 2 using thiourea as indicator. Sulphate does not interfere, so that it can be added to mask Th. If tartrate is present molybdate and phosphate have no effect. Tartaric acid can be used to mask Nb and Ta, which would otherwise inevitably precipitate out as hydroxides at the pH used for the titration. Thiourea functions not only as an indicator but it also masks Cu, and especially Bi, to such an extent that titration is possible in the presence of chloride without separation of the sparingly soluble BiOCl.

A standard solution of Fe(III) is recommended by many authors for the back-titration. Milner and Pennah [54-92] back-titrate at pH 5–6 using salicylic acid, whereas Volodarskaya [60-93] used sulphosalicylic acid under the same conditions. Pande *et al.* propose 2-hydroxy-3-naphthoic acid [60-15] and *o*- or *p*-cresotic acid. In both cases an excess of EDTA is added, and after adjusting the pH to 6 and boiling for a few minutes the solution is cooled to about 30°C, buffered with acetate to pH 4, and back-titrated.

Back-titration at pH 5 with copper using PAN is proposed by Endo [60-141]. Kinnunen [57-137] back-titrates at pH 1–2 with bismuth or at pH 2–3 with thorium, using xylenol orange as indicator. In obvious ignorance of Kinnunen's work this was also proposed by Lukyanov *et al.* [60-61] some years later. The Russian authors also use arsenazo as indicator. Belcher and his co-workers [60-153] raise the possibility of determining Zr by back-titration with Cu using *o*-dianisidine- or *o*-diphenetidine-tetra-acetic acid as fluorescence indicators. Further details, sample analyses, or information about interferences were not given. According to Sajó, back-titration with Zn with the ferri-ferrocyanide benzidine system [56-99] or with vanadium using diphenylcarbazide [56-100] is also possible.

Finally, Takamoto's method [55-103] should be mentioned, in which back-titration with Co is carried out in 50% aqueous-acetone containing thiocyanate.

The following instrumental end-point indications are possible. Photometric end-points have been obtained in back-titrations with Fe(III) using salicylic acid [54-14] and benzhydroxamic acid [55-80]. Amperometric titration is feasible in back-titrating excess EDTA with Bi [56-13; 60-71] or vanadyl ion [63-13]. Amperometric titration with a rotating tantalum electrode following Khadeev and Kvashina [60-75] is interesting, for the end-point indication is due to the wave corresponding to oxidation at $+1.2$ volt (against calomel) of the EDTA set free after the end-point. Fe(III), Sb(III), Bi, Sn(IV), Ta, W, and Mn interfere with the determination. Interference from small amounts of F^-, Hg(II), and Cu can be eliminated by adding Al, Cl^-, or tartaric acid respectively or at least appreciably reduced if large amounts are present. Potentiometric titration with the mercury-drop electrode has also been described [58-12].

Practically the complexometric determination of Zr has been used, *inter alia*, for the analysis of alloys, ores, and concentrates [60-61], especially after separation with ion-exchangers [60-177], and more particularly for Cu–Zr [62-61], Th–Zr [59-110], Mg–Zr [60-93], and Nb alloys [60-71], as well as for borides and nitrides [59-85], paint driers [59-116], and Zr-sands [63-52].

In all the Zr determinations any Hf present will be co-titrated, and as far as present information goes Hf alone can be determined by any method that has been worked out for Zr. Hoshino [62-128] has described an apparently simple separation of Hf from Zr by extraction with cyclohexanone. The complexometric titration is carried out after back-extraction into an aqueous phase. Analysis of mixtures of Zr and Hf is possible by indirect analysis, taking into account their different atomic weights. First the total Zr + Hf is obtained by established methods and, after ignition, a known weight of mixed oxides is titrated with EDTA. Working with aliquot portions is also possible. The results are substituted into two simultaneous equations whose solution provides the desired information [59-86; 62-129; 62-130].

Poluetkov *et al.* [62-130] give the following formulae for the calculations:

$$\text{HfO}_2\,(\text{mg}) = \frac{0.00812A - \text{v}M}{0.00336}; \text{ZrO}_2\,(\text{mg}) = \frac{\text{v}M - 0.00476A}{0.00336}$$

where A is the total weight of mixed oxides (in mg), v the volume (in ml), and M the molarity of the EDTA solution.

17.1. Direct titration of zirconium (and hafnium) using xylenol orange

REAGENTS: Standard 0·01M EDTA; xylenol orange indicator; conc. HCl, HClO$_4$, or H$_2$SO$_4$; ammonia.

PROCEDURE: The test solution containing 30–50 mg metal per 100 ml is treated with ammonia until a definite permanent turbidity is produced. The volume is estimated and for every 100 ml, 3 ml of conc. HCl or HClO$_4$ or 1 ml of conc. H$_2$SO$_4$ is added and the solution is boiled to break down all products of hydrolysis. The solution is then cooled to about 50°C, indicator is added, and titration is carried out with EDTA (slowly near the end-point) to the colour change from red to pure yellow.

REMARKS: If the Zr (and/or the Hf) has been segregated by precipitation as hydroxide or by means of mandelic acid, etc, the precipitate can be taken up directly in approx. 6N acid and diluted with washings to about 0·3N.

17.2. Back-titration of zirconium (and hafnium) with Fe(III) using salicylic acid as indicator [54-92]

REAGENTS: 0·1M EDTA and 0·1M FeCl$_3$ standardized one against the other; solid salicylic acid; approx. 1M NH$_4$OH; approx. 1M CH$_3$·COOH.

PROCEDURE: The acidic test solution containing about 100 mg Zr per 100 ml is treated with a small excess of EDTA and heated to boiling-point. Ammonia is then added dropwise until the pH reaches about 6–7 and the boiling is continued for 2 minutes. After cooling, 3 ml of acetic acid and finally 2 ml of ammonia is added, whereupon the pH should be about 5. After adding 0·2 g of solid indicator, titration is carried out with the standard ferric solution until the first trace of a red colour.

REMARKS: As is always the case when using a phenol–ferric iron reaction to indicate the end-point, the solution must not be too dilute. However, titration can still be carried out with a 0·02M solution of titrant. The complex between zirconium and EDTA is only slowly formed, and it is therefore necessary to warm the solution to start with and not to neutralize until later, otherwise a precipitate of zirconium hydroxide would result which would subsequently be difficult to redissolve.

18. THORIUM

Cabell [52-41] showed in 1952 that thorium forms a stable complex with EDTA, and the result was exploited independently by Fritz [53-19] and ter Haar and Bazen [53-9] for its complexometric determination. Alizarin S was used for the direct titration at pH 2·8, and this has also been proposed by later workers [54-99; 60-72], and the end-point can be made sharper by adding xylene cyanol FF [57-121]. Compounds analogous to alizarin S were studied by Owens and Yoe [60-106], namely 2-quinizarin sulphonate (titration at pH 2–3·4) and 2-phenoxyquinizarin-3,4'-disulphonate; the latter was preferred, for the titration could then be carried out in the pH range from 1·4 to 3·4. Nowadays innumerable indicators are available for the titration of Th in an acidic medium, such as azodyes [A-7; 60-54; 60-55; 60-188; 61-37], chromeoxane dyestuffs [57-32], and special azo-compounds of the chromotropic acid group [55-94; 56-53; 57-51; 63-18]; other indicators are indophenolcomplexone [60-52], gallocyanine [61-73], haematoxylin [61-40], dihydroxysulphon-phthalein [60-11], hydroxyhydroquinonephthalein [61-65], stilbazo [61-147], eriochrome cyanine R [58-15], carminic acid [58-79], and thorin [62-90]. Use has also been made of Zn-dithizone at pH 3 in 60% alcohol [61-61], Fe(III)–KSCN [59-138], and the redox systems vanadium(V)–EDTA–diphenylcarbazide in an acetate buffer [63-1] and Fe(II)–cacotheline [57-11]. A very great deal has been written about xylenol orange [59-110; 60-46; 60-72; 60-152; 63-8; 63-16; 63-27], pyrocatechol violet [54-77; 56-84; 57-30; 57-132; 57-133; 58-77; 59-76; 60-72; 62-6], arsenazo [60-72; 62-89; 62-90], and chrome-azurol S [59-14; 61-74]. With the last-named indicator anthranildi-acetic acid has also been used as the titrant [59-76]. These indicators are especially valuable for the sharpness of their colour transitions and the fact that they permit of titrations at really low pH values. However, Apple [60-152] has pointed out that high concentrations of neutral salts have a very bad effect on the colour transition of xylenol orange. In such cases an improvement can be brought about by adding methylene blue [63-8].

Much has also been reported on determinations based on back-titrations of excess EDTA. Back-titration is possible with a standard Th-solution and alizarin S [53-22], with Cu and PAN [58-43; 58-108; 69-92], xylenol orange [60-152] or *o*-dianisidinetetra-acetic acid [60-153], or with Co in an acetone–water mixture containing thiocyanate [55-103]. Back-titration with Fe using 2-hydroxy-3-naphthoic acid

[60-15] or *o*- or *p*-cresotic acid [60-16] has also been described. Back-titrations are of interest when Th has been separated from interfering elements by, e.g., precipitation with oxalate [60-72] or more selectively with iodate [57-30; 60-72; 62-92]. The precipitate is simply taken up in a standard solution of EDTA and the excess back-titrated.

Generally speaking, the determination of thorium is only interfered with by the presence of metals that form very stable complexes with EDTA, viz. Bi, Zr, Hf, Fe(III), etc. What additional elements can be tolerated, and in what concentrations, depends on the indicator that is used and on the pH chosen for the titration. The lower the pH, the more selective the titration. Malat *et al.* [59-14] titrate at pH 1–2 using chromeazurol S, whereby alkalis, alkaline earths, Ag, Tl(I), Mn, Zn, and Cd have no effect. Large amounts of Co, Cr, and U interfere because of the intrinsic colour of their cations. In the presence of Al, Pb, and Cu the colour transition at the end-point is not from violet to yellow but from violet to red – which is less sharp. Fe(III) and Hg can be masked with ascorbic acid. Suk and Malat [54-77] state that the titration of Th at pH 2·5 with pyrocatechol violet as indicator is unaffected by alkalis, alkaline earths, Mn, U, Pb, Co, Ni, Al, Zn, Cd, and lanthanons. Fe(III), present in up to ten-fold excess, can be masked with ascorbic acid: however, a large excess of reductant must be avoided, otherwise the sharpness of the end-point suffers [56-84]. Lupan [58-77] used the same indicator for titrations in 60% alcohol. La, Pr, and Nd do not interfere, and Ce(III) can be tolerated up to a 500-fold excess. Banerjee [55-93] carries out the titration at pH 3·1 with SPADNS as indicator. Alkalis, alkaline earths, Cd, Co, Al, La, Ce, and V in up to ten-fold excess do not interfere: here, too, iron can be masked with ascorbic acid. Chernikov [60-72] describes the titration of Th at pH 2·4–2·6 using xylenol orange, where up to 10 mg of Zr in about 50 ml of solution can be masked with tartrate or trihydroxyglutaric acid.

The high stability of the Th–EDTA complex permits of many consecutive titrations wherein Th is first titrated in a strongly acidic solution, and then other metals are titrated after raising the pH. The consecutive titration Th–lanthanons is of special importance, and details have been given in Section 14. In adopting this general principle certain difficulties must often be taken into account with Th, since this metal reacts with some indicators at the higher pH, although it is bound to EDTA. This 'blocking' of the indicator is a special feature with arsenazo (cf. Section 14). Přibil has written about the blocking of xylenol orange [63-27] when, following on the completed titration of Th, bivalent metals of the group for which log $K_{MY} = 16$–18 are titrated at pH 5–6 with the same indicator. If complexones such as DTPA or TTHA (triethylenetetraminehexaacetic acid) are used in place of EDTA there are no interferences.

The relatively high stability of the complexes that Th forms with sulphate ions is of particular importance, for they enable Th to be masked. The effectiveness of this masking is greater the lower the pH. Přibil [60-46] finds a pH below 1 to be best of all, but it is still possible to work at somewhat higher pH values. Masking by sulphate can be exploited for a selective determination of Th. According to Milner [59-110], Th can be determined in the presence of Zr as follows. The total concentration, Zr + Th, is first determined by back-titrating an excess of EDTA with a standard solution of Bi at pH 2·5–2·7 using xylenol orange as the indicator. The fully titrated solution is then acidified to pH 1·2–1·3 and treated with about 2 g of ammonium sulphate. Finally, the EDTA set free in an amount equivalent to that of the Th is back-titrated with Bi using the same indicator: Zr can be calculated by difference. Přibil [60-46] extended Milner's procedure to Fe(III) and several other bi- and tervalent cations. Difficulties arise with metals that form less stable complexes, since during the back-titration they are displaced wholly or partially from their EDTA-complexes and are therefore either included with the titration value for the thorium or at least are responsible for high results. Rogers and Brown [63-16] employ a similar de-masking procedure. Titration at pH 3, using xylenol orange, gives the total concentration of Th and all the other metals present (in so far as they react at pH 3): a solution of fluoride buffered to pH 3 is now added and the whole is heated to boiling. The amount of EDTA equivalent to the Th is then titrated with a standard solution of Cu with PAN as the indicator. Of 27 cations studied, only Bi, Cd, Cu, Co, Ce(IV), Fe(III), Ni, Pb, Zn, Zr, Pu, and Cr(III) were included in the direct titration. Clearly this provides a highly selective titration for Th. However, their studies of the amounts of foreign cations that could be tolerated were restricted to only about 3 mg in the presence of 20–40 mg of Th, and it remains to be seen how the procedure will work in the presence of larger amounts of interfering elements. In any event, the method seems to offer high possibilities for an extremely selective determination of Th after a separation stage in which the concentration of interfering elements is greatly reduced. Hg, Ga, In, and Tl(III), for example, were not included in the investigation, but it is to be expected that these will be co-titrated without, however, reacting later with fluoride.

Instrumental methods for detecting the end-point in the determination of Th include an amperometric titration with the dropping-mercury electrode based on a displacement reaction with PbY^{2-} [57-91] and on two platinum wire electrodes polarized at 1·48 volt [60-57]. Amperometry also serves to detect the end-point in the back-titration of excess EDTA with a solution of vanadyl ion [63-13]. Hara and West [55-43] describe the high-frequency titration with very dilute Th solutions. Photometric titrations have been widely used. Banks [55-71] titrates at 585 mμ with chromeazurol S; Bril *et al.* use alizarin S for the direct titration and for the back-titration with a standard solution of Th [59-95], while Datta [62-5] prefers β-SNADNS-6 at 590 mμ and Peshkova *et al.* [62-19] recommend arsenazo I at 755 mμ. Automatic end-points were obtained by Malmstadt [54-12] directly at pH 3·1 and 290 mμ or at 290 or 320 in a back-titration with Cu, and by Menis *et al.* [58-70] using quercetin as indicator. Pang [59-84] has described a complexometric determination of Th in which CeY^- forms the titrant: the absorption produced by the cerium complexonate in the ultraviolet region is used to give the photometric end-point. The mercury-drop electrode serves for potentiometric end-point indication [58-12].

The practical uses of the complexometric titration of Th extend to ores and minerals [59-138; 61-20], in particular monazite sand [57-30; 60-72], gas mantles and glasses [56-40], alloys with Mg [60-93], Mg, Cu, and Al [63-8], U [54-99; 57-132; 57-133; 58-15], Bi [57-132; 57-133], Zr [59-110], W [58-43], and reactor fuels [60-152].

Preliminary separation by means of ion-exchange resins has often been used. For the consecutive titration Th–lanthanons, see Section 14.

18.1. Direct titration of thorium with pyrocatechol violet as indicator

REAGENTS: Standard 0·01M EDTA; pyrocatechol violet; dilute ammonia.

PROCEDURE: The acidity of the test solution with a thorium concentration not exceeding 5×10^{-3}M is adjusted to pH 2 by adding ammonia dropwise until Congo red test-paper no longer gives a pure blue colour. Pyrocatechol violet is now added when a blue colour should appear. If the colour still has a reddish tinge the pH is still too low, and this can be corrected with a few more drops of ammonia. The solution is gently warmed (to about 40°C) and titrated to the colour change to yellow. Since hydrogen ions are liberated during the titration, it is often necessary to add a few more drops of ammonia towards the end of the titration.

REMARKS: If the solution is too cold the complexing reaction is somewhat too slow, so that the end-point becomes drawn out. The dyestuff chromotrope 10B is a very good alternative indicator to pyrocatechol violet [A-7].

Sulphate causes drawn-out end-points, but the addition of $BaCl_2$ overcomes this difficulty. The end-point can be detected quite easily in the resulting milky suspension.

19. VANADIUM

Flaschka and Abdine [56-44] were the first to describe a direct complexometric determination of vanadium. Vanadate in a solution, buffered with acetate to pH $>3 \cdot 5$, is boiled with ascorbic acid and after complete reduction the VO^{2+} so formed is titrated hot with EDTA using CuY–PAN as indicator. As, PO_4^{3-}, and W do not interfere. Mo must be absent, as the resulting deep blue coloration masks the transition of the visual indicator. A further direct method is due to Kaimal and Shome [69-25]. The vanadium is reduced with sulphite in sulphuric acid solution, excess SO_2 is boiled off, and the pH adjusted to 3; after cooling and adding alcohol (about an equal volume) the solution is titrated with EDTA using *N*-benzoyl-*N*-phenylanthranilic acid as indicator. According to Hara [61-61], solutions containing more than 40% alcohol give very sharp end-points at pH $3 \cdot 5$–$5 \cdot 0$ if Zn–dithizonate is used as the indicator.

Vanadium(IV) can be determined by back-titration in the following manner. Following Kinnunen [55-4], the reduction is effected with sulphite in a weakly acidic solution, EDTA is added, and the whole is boiled to complete complex formation: the solution is then adjusted to pH 10 (ammonia–ammonium chloride buffer), some ascorbic acid is added, and back-titration is carried out with a standard solution of Mn with Erio T as indicator. In a later paper [57-137] the author describes the more selective back-titration in an acid medium (pH 2–3) in which a standard solution of Th is used in conjunction with xylenol orange. Sen [58-90] back-titrates with cobalt in an alcohol–water mixture containing thiocyanate. Cu was used for back-titrations in solutions buffered with acetate using the fluorescent indicators *o*-dianisidinetetra-acetic acid (Belcher *et al.* [60-153]) or calcein (Wilkins [59-8]). Back-titration with Pb is possible using pyrogallol red [59-20]. Fritz and his co-workers have described a very selective determination of V. The test-solution is passed through a cation-exchange resin, and after this has been washed the V is eluted selectively with acid containing about 1% H_2O_2. After boiling to decompose the peroxide the solution is reduced, treated with EDTA, and the excess determined by back-titrating with a standard solution of Zn using naphthylazoxine S. Ringbom *et al.* [57-15] and Sajó [56-100; 58-66] have studied the formation of the EDTA complex of vanadium(V). The latter propose the use of a standard solution of vanadium(V) as a reagent for back-titrations in the complexometric determination of other metals, diphenylcarbazone being used as indicator [56-100]. In a later paper the direct titration of vanadium(V) is described [62-94] and attention is drawn to various difficulties. A direct titration using diphenylcarbazone is only possible in a very narrow pH region from 6·7 to 6·9. Below 6·7 the V is present partially as decavanadate, which reacts only very slowly with EDTA. Above pH 6·9 the indicator ceases to function. It is therefore necessary to start out with an alkaline vanadate solution which is brought to the desired pH by boiling with ammonium chloride. But even the duration of the boiling must be rigidly adhered to, as otherwise the pH falls too low and the solution has to be discarded. The situation is improved if the titration is carried out in the presence of sugar, mannitol, or glycerine when the pH range 4·7–6·8 becomes available. Kakabadse and Wilson [61-16] have described researches on the direct titration of vanadium(V) in a solution about 0·03N with respect to perchloric acid using xylenol orange, but they seem to show little confidence in their own method. In contrast to the direct titration, the back-titration of vanadium(V) can be carried out in a wide pH range from 3·5 to 6. To this end it is proposed to boil the sample with excess EDTA in a solution buffered with acetate and, after cooling, to back-titrate the excess of EDTA with a standard solution of Zn using diphenylcarbazone [69-94; 62-110] or with Zn, Pb, or Cd using pyrocatechol violet [62-94]. Even now the end-points are not particularly sharp, and the addition of methylene blue ought to improve the situation for diphenylcarbazone [62-94].

According to Sajó [58-66], the effective stability constant of the vanadium(V)–EDTA complex, VO_2Y^{3-}, between pH 3·5 and 6 is practically independent of pH and $\log K^{eff\cdot}$ is about 7 or 8. Below pH 3·5 the effective stability falls off rapidly, since α increases; above pH 6 the decrease is due to the formation of metavanadate. In the specified pH range from 3 to 6 vanadium(V) is displaced from its

EDTA-complex by all the metal ions that can be titrated under these conditions; thus the system vanadium(V)–EDTA–diphenylcarbazone can be introduced quite generally to serve as an indicator [63-1]. The effective formation constant of VO_2Y^{3-} has been determined by Schwarzenbach *et. al.* over the whole pH-range [65–123].

An indirect way of determining micro-amounts of vanadium (>10 μg V_2O_5) originates from de Sousa [62-97]. Vanadate is precipitated as its silver salt and allowed to react with $Ni(CN)_4^{2-}$, whereupon the Ni set free is titrated with EDTA using murexide (cf. Section 32).

Reilley *et al.* [58-12] have described the potentiometric titration of VO^{2+} at pH 3·5 at a platinum electrode. According to Tserkovnitsa [60-189], titration can be carried out amperometrically at pH 3–4 and at +0·9 volt against a saturated calomel electrode with a vibrating platinum electrode which responds to V(III) and V(IV). If both oxidation states are present the titration curves show two breaks, the first corresponding to V(III). Goldstein *et al.* [63-13] propose the use of amperometric indication using a standard solution of VO^{2+} for the back-titration of Al, Zr, Th, and other metals in solutions buffered at pH 4.

The complexometric determination of vanadium is used in the analysis of magnet alloys [59-8] and lead vanadate [62-110]. In the latter case the sum Pb + V is obtained by back-titrating excess EDTA at pH 6 with a standard solution of Zn using diphenylcarbazone. The solution is then brought to pH 10, whereupon VO_2Y^{3-} is decomposed; the amount of EDTA liberated in amount equivalent to that of the vanadium is back-titrated with Zn using acid chrome blue as indicator.

19.1. Direct titration of vanadium using Cu–PAN

REAGENTS: Standard solution 0·01M EDTA; acetate buffer pH 4–5; Cu–EDTA solution; PAN; ascorbic acid.

PROCEDURE: The test solution containing about 20 mg of V in 50 ml is adjusted to pH 4 with acid or alkali and then treated with buffer. Ascorbic acid is added, whereupon a blue colour develops. Next heat to boiling until the blue colour changes to bluish-green. Add a few drops of CuY-solution and indicator and titrate the solution while still hot to the colour change from violet to yellow.

REMARKS: The procedure relates to solutions containing vanadium(V). Reduction is unnecessary if the V is present completely in the tetravalent state. If much V is present the addition of the large amount of ascorbic acid that will be necessary to reduce it causes a lowering of pH. The pH ought not to fall below 3·5. In such cases it is appropriate to dissolve the ascorbic acid in a little water and to neutralize with caustic soda. Such solutions do not keep!

20. NIOBIUM AND TANTALUM

Ferrett and Milner [56-123] were the first to report on the formation of a complex between Nb and EDTA on the basis of polarographic investigations. In attempts to determine Nb polarographically in an EDTA medium Kennedy [61-169] could detect no linear relationship between wave-height and concentration. Kirby and Freiser [63-12] investigated the polarographic reduction of Nb–EDTA more closely and established that a stable solution and complete complex-formation only result if the solution is boiled. Wave-height and half-wave potential are dependent on the pH of the solution and on the amount of EDTA in excess. According to the pH range either one or two protons are involved in the reduction process Nb(V) $\longrightarrow$ Nb(IV) owing to the presence of hydrolysed complexes. Nothing much can be said concerning the composition and stability of these complexes. Spauszus and Hupfer [61-170] showed that marked hydrolysis sets in from pH 4·3. In solutions destined for titration the situation is still more unfavourable, for here the concentrations are generally speaking higher than in polarography.

Lassner and Scharf [60-3] report that niobium(V) resembles titanium (q.v.) and that on adding hydrogen peroxide the complex formed with EDTA is completely stabilized. The reason for this is the replacement by $^-$O$-$OH of the $^-$OH group in the coordination site that is not occupied by the EDTA ligand: this prevents further hydrolysis. The formation of a mixed complex between Nb–H_2O_2 and EDTA is immediately obvious from the appearance of a yellow colour. The regions over which the mixed complex exists have been determined photometrically as a function of pH. At pH 3–4 the peroxyniobium(V) ion can displace copper from its EDTA-complex. Attempts to exploit this result for the determination of niobium are wrecked by the fact that the colour transition at the end-point extends over more than 1 ml of an 0·05M solution of EDTA.

However, peroxyniobium(V) ions also react with other chelating agents. With DCTA it forms a chelate at pH 3 which is so extensively dissociated by pH 5 that even in the presence of peroxyniobium(V) ions the DCTA is completely available for reaction with copper.

This result was used for the titration of titanium (q.v.) in the presence of Nb (and Ta). Semi-quantitative data on mixed complexes of Ti, Nb, and Ta have been reported by Lassner and Püschel [63-37].

The peroxy-complex of niobium also reacts with metal indicators [63-37] such as methylthymol blue at pH 5 with the formation of a deep blue compound. This has been recommended as a highly selective method for detecting Nb [62-77].

The most stable complexes of peroxyniobium(V) are those of NTA, DTPA, and N-β-hydroxyethylethylenediamine-N,N',N'-triacetic acid. This permits of a determination of Nb by back-titration with a standard solution of Cu [63-38]. The acidic solution of Nb containing hydrogen peroxide is treated with an excess of 0·05M NTA and then buffered to pH 5·0–5·5 with sodium acetate or urotropine. Finally, the excess NTA is back-titrated with 0·05M copper solution with methylcalcein as fluorescence indicator. The reacting proportion between Nb and NTA is 1 : 1. By using pure niobium solutions, a standard deviation of 0·008–0·011 ml of 0·05M titrant solution was found for 1–5 mg Nb.

The titration with N-β-hydroxyethylethylenediamine-N,N,N'-triacetic in the same way is also possible, but to ensure quantitative formation of the chelated complex the solution must be heated after the complexone has been added. The back-titration is carried out after cooling to room temperature.

An indirect complexometric titration of Nb comes from Wakamatsu [60-170]. Niobium is precipitated by zinc chloride from a 30% acetone–water mixture of pH 6–6·5 in the presence of hydrogen peroxide. The precipitate is collected, taken up in HCl, EDTA is added, and after buffering with acetate the excess EDTA is back-titrated at pH 4–5 with Cu using PAN as the indicator. The composition of the precipitate determined under identical conditions leads to the ratio Zn : Nb of about 0·4, and this is used in calculating the results of the analysis. The method has been used for determining Nb in iron and steel.

Literature data concerning tantalum are even scarcer than for niobium. Kirby and Freiser [61-171] confirm polarographically that complex formation takes place between Ta and EDTA. In the pH range 3·3–5·6 the half-wave potential is lowered from −1·23 to −1·36 volt by adding EDTA. One electron is involved in the reduction process, which is from Ta(V) to Ta(IV). No data are given on the composition or stability of the easily hydrolysable complexes.

The titration of Nb using NTA described above is also possible for

Ta (Lassner; private communication), but there are insufficient data concerning serviceability, limits of error, interferences, etc, to enable any definite conclusions to be drawn. In Wakamatsu's paper [60-170] is the note 'tantalum behaves similarly' (to Nb), but it does not follow that the indirect method described above for Nb can be used for Ta.

Interferences by Ta and Nb in the titrations of other metals are few in number. What appears to be a drawback in the titration of Nb or Ta, namely, the high tendency towards the formation of poly-nuclear complexes, is an advantage here, since both metals separate out. None the less, we must not overlook co-precipitation and the adsorption of the indicator on the hydrated oxides. Precipitation can be prevented by masking with tartaric acid or fluoride, and this may be of use if the metal to be titrated is not affected by these masking agents. Use is made of this in the titration of cobalt [59-2] and molybdenum [59-3].

21. CHROMIUM (see also under Chromate)

The velocity of complex formation between chromium(III) and EDTA (as well as other complexones of this type) is so extraordinarily small that direct titration does not come into question, and Cr can only be determined by back-titration. But here we run up against certain difficulties. The colour of the complex, CrY^-, that exists between pH 5 and 6 is an intense red-violet: above pH 6–7 hydroxy-complexes of bluish-red colour occur, so that if more than even a few milligrams of Cr are present in a titration solution of the usual volume the superposition of these colours upon the transition colours of the indicator has to be taken into consideration. It is necessary therefore either to work in very dilute solutions or to use an instrumental end-point. Fluorescence indicators have improved the situation. An interesting possibility has been described by Arbuzov *et al.* [61-75]. A narrow beam of white light is projected through the titration solution on to a white screen, and the colour change seen there is observed through a suitable filter.

Prolonged boiling is needed to transform Cr completely into its EDTA complex. According to different authors, the time required ranges from 2 to 15 minutes. How long is actually necessary depends on the nature and amount of foreign salts present and also more particularly upon the form in which the Cr itself is present, i.e. whether as aquo-, chloro-, or sulphato-complex, etc. Pavlov *et al.*

[60-105b] have concerned themselves with the influence which the ligands attached to chromium have on the amount of EDTA needed for complete complexation and find significant differences between chloro-, nitrato-, and sulphato-complexes. According to our own experience, boiling for less than 10 minutes is generally insufficient. The reaction is accelerated quite generally by reducing agents, since small amounts of chromium(II) catalyse the complex formation of chromium(III). The use of zinc dust for this purpose has been described by Irving and Tomlinson [66-35].

In this connection it is interesting to refer to an observation by Přibil [51-19], according to whom complex formation between Cr(III) and EDTA takes place if chromate is reduced in the presence of EDTA with, e.g., ascorbic acid. The explanation has been put forward that complex formation between free chromium ion and EDTA takes place before aquo-, chloro-, or other complexes have been formed by slower reactions. Aikens and Reilley [62-74] have studied the situation more closely and postulate the rapid formation of unstable intermediates of penta- and tetra-valent Cr and suggest that the final stage of reduction to Cr(III) proceeds in the complex of EDTA already formed. Small amounts of Cr(II) are also formed which act catalytically. Beck and Bardi [61-166] are, however, of the opinion that reduction proceeds by direct attack on a chromyl complex, $(CrO_2)_2Y$. Such a compound should exist by analogy with the molybdenyl–EDTA complex that has been isolated.

The complexometric determination of Cr differs from other back-titrations simply in the fact that the solution must be boiled for a longer time to achieve complete complex formation. Once the complex has been formed it is stable and kinetically inert ('robust'), so that the back-titration can also be carried out at low pH values – which is an advantage from the point of view of selectivity.

Wehber [56-111] back-titrates with Fe at pH 4–5 after only 1–2 minutes boiling: he uses Bindschedler's green as indicator and finds that aged solutions give better colour transitions. Kinnunen [55-4] recommends back-titration with Mn in ammoniacal solution with Erio T as indicator. In a later paper [57-137] back-titration with Th at pH 4–4·5 using xylenol orange is proposed: the advantage here is that alkaline earths are not included in the titration value. Both papers specify boiling for 15 minutes. Liteanu [59-90] choses 5 minutes boiling and back-titrates with Fe(III), using Tiron as indicator. Weiner [57-16] boils for 10 minutes and back-titrates with Ni at 40°C in an ammoniacal solution using murexide. Back-titration with Ni has also been employed by Pavlov [60-105a]. Wilkins [59-7] recommends boiling for 15 minutes and back-titration with Cu using calcein as a fluorescence indicator. In this way larger amounts of Cr (up to 40 or more mg per 100 ml) are determinable. Verma [60-80] also back-titrates with Cu at pH 3–5 to the point where the green fluorescence of calcein indicator is quenched, but he uses DCTA instead of EDTA. This has the advantage that Cr can be determined in the presence of bichromate (or chromate if present originally). Whereas EDTA in acid solution is oxidized by dichromate even in the cold, DCTA does not react even on boiling prolonged. Cameron [61-136] avoids difficulties in recognizing the colour changes of the indicator by using an extractive end-point. After boiling for 5 minutes and then cooling to room temperature the excess of EDTA is back-titrated at pH 8·5 with Co. Thiocyanate is used as the indicator in the presence of triphenylmethylarsonium chloride and chloroform: the mixture is shaken and the colour change can be observed in the organic phase.

Patzak and Doppler [57-29] used Fe(III) for the back-titration and determined the end-point potentiometrically: a period of 5 minutes boiling is specified. Although only check analyses for 3–4 mg of Cr were reported, this instrumental indication seems capable of dealing with the largest amounts of chromium.

Khalifa [62-75] reports a photometric method which can best be described as quasi-direct. Several equal aliquot portions are treated with increasing amounts of DCTA, boiled, and after cooling are made up to the same weight (or volume) with water. The optical density is measured and plotted against the volume of standard EDTA solution added. The usual extrapolation gives the end-point. This thoroughly inconvenient procedure affords no grounds for being considered worth while as compared with the simpler back-titration.

Interferences by chromium in other titrations have been studied and found to be easily overcome. The extraordinary inertness towards reaction is turned to advantage here, for in point of fact many other metals can be titrated completely before Cr has reacted with the titrant in sufficient amounts to interfere. Reilley has designated this frequently used procedure 'kinetic masking'. During titrations in an alkaline medium Cr can interfere by forming a hydroxide precipitate; however, the addition of tartaric acid prevents this. Přibil [61-94] has considered the chemical masking of Cr, although this is often unnecessary. By prolonged boiling (5 minutes) with triethanolamine Cr can be transformed into the deep ruby-red triethanolamine complex. Although Cr can be masked in this way against reaction with the titrating agent, it interferes by overlaying the colour changes of the indicator, unless only very small amounts of Cr are present. How much Cr can be tolerated depends on the indicator used and, moreover, on whether the succeeding titration allows the test solution to be diluted extensively. As another possibility for masking, prolonged boiling with ascorbic acid should be mentioned. The solution turns green, and on adding ammonia the precipitation of hydroxide no longer takes place: the formation of some sort of complex must be assumed. The intrinsic colour is sufficiently weak for Ca, Mn, or Ni, e.g. to be titrated in ammoniacal solution in the presence of up to 1 mg of Cr per ml. Přibil also reports that, in alkaline solution, chromate does not affect the indicators methylthymol blue and thymolphthalexone, although Erio T is destroyed by oxidation within 10 minutes. This opens up possibilities of masking Cr by changing it to chromate, though which method of oxidation should be used is not stated.

The sluggishness with which Cr reacts is used to advantage for its determination in the presence of other metals and is employed in consecutive determinations, where it is often advisable to work with aliquot portions.

Wehber [56-111] reports on the consecutive titration Fe–Cr. Iron is first titrated directly with EDTA in an acetate–monochloroacetate buffer at pH 3·5 at room temperature with Bindschedler's green as indicator. The fully titrated solution is then treated with excess EDTA and boiled for 5 minutes; after being cooled to room temperature the excess of EDTA is back-titrated with Fe. The indicator is not destroyed during the boiling.

Liteanu [59-90] described the consecutive titration Al–Cr. The test-solution is first treated with excess EDTA, brought to pH 5–6, and maintained at 50°C for 5 minutes, whereby the Al is complexed completely, but the Cr should not react. After cooling to 20°C back-titration is carried out with Fe(III) with Tiron as indicator. The fully titrated solution is again treated with excess EDTA and boiled for 5 minutes, after which it is again cooled to room temperature and back-titrated again with Fe; this gives the Cr content. The method works for about 1–6 mg of Cr_2O_3 and Al_2O_3. At the relatively high pH prescribed for the titration, interference due to the precipitation of hydroxide would seem to be unavoidable with larger amounts of metal oxides.

Patzak and Doppler [57-29] have described the resolution of

mixtures of Fe–Al–Cr. An aliquot portion is treated with EDTA at pH 1·5, warmed to 50°C, and the excess of complexone immediately back-titrated with standard ferric solution. This gives the Fe content. A second aliquot portion is treated with EDTA, boiled for 5 minutes, cooled to 40°C, and then buffered to pH 5–6. Back-titration with Fe now gives the sum of all three components. In a third aliquot portion, treated similarly, the back-titration is carried out at pH 1·5 to give the sum of Fe + Cr.

The complexometric determination of Cr is employed practically in the analysis of chrome baths [58-97]. Since EDTA is oxidized by chromic acid, precipitation with mercurous nitrate is recommended. Since Hg(I) disproportionates in the presence of EDTA, the excess of precipitant must be removed by adding chloride ions. Both precipitates are coarsely crystalline, separate out quickly, and show no tendency towards co-precipitation: they are filtered off together. Pavlov and his co-workers [60-105c] determined Cr in films of high polymers. The film is boiled with EDTA solution, whereupon rapid dissolution of the Cr takes place. Then a standard Ni solution is added and the excess titrated using murexide as indicator.

Concerning the determination of Cr in minerals and in steel after oxidation to chromate, see below.

21.1. Chromate and bichromate (see also under Chromium)

Chromate and bichromate can be determined via a chromium determination (v.s.) after reduction to Cr(III). Aikens [62-74] makes use of this possibility. The reduction can with advantage be carried out in the presence of EDTA. This avoids the prolonged boiling that is otherwise necessary in view of the extraordinarily slow reaction between Cr(III) and EDTA. For if the reduction is carried out in the presence of EDTA the chromium goes over into the complex immediately and quantitatively provided the correct conditions have been fulfilled. The reduction is effected with sodium bisulphite in a medium buffered to pH 6·6 with triethanolamine or with triethanolamine and acetate. The relative proportion of chromate to bisulphite is to some extent critical, and varies with the amount of chromate present; this implies some practical difficulties in carrying out this procedure. Back-titration of excess EDTA is carried out with Th at pH 2·5–3·5 using xylenol orange as indicator.

Like other anions, chromate and bichromate can be determined by precipitation with a cation that can be determined complexometrically. This makes possible the determination of chromium after oxidation to chromate. Isagai and Tatkeshita [55-22] exploit this possibility by precipitating chromate with a standard solution of Ba and back-titrating the excess of Ba in the filtrate with EDTA. The method has been used for the determination of Cr in steels after Fe and Cr have been separated by Dowex IR-120 in its H^+-form.

De Sousa [61-25] uses the principle to determine Cr in minerals. In view of possible interferences by other metals in the back-titration, he prefers to determine Ba in the precipitate after this has been filtered off. The precipitate is then dissolved in acid, the chromate reduced with KI, and the Cr separated as hydroxide. Ba is then the only metal present in the filtrate, and it can easily be titrated using metal-phthalein as indicator (cf. Section 10.1.4).

22. MOLYBDENUM

Molybdenum forms complexes with EDTA of the composition $Mo : Y = 2 : 1$ in both oxidation states (IV) and (V), as was first demonstrated polarographically [56-30]. The complex with Mo(V) is more stable than that of Mo(VI), and even sufficiently stable to displace Cu from its complex with EDTA [59-3]. We are dealing here with a 'robust' (kinetically inert) complex. This follows from the fact that Mo(V) cannot displace Bi^{3+} from its complex with EDTA; yet once the Mo(V)–EDTA complex has been formed, it resists decomposition by bismuth ions [59-3].

De Sousa was the first to report on an indirect complexometric determination of Mo [55-99]. The method depends on the precipitation of calcium molybdate and the determination of Ca in the precipitate. The method was later improved and simplified by Lassner and Schlesinger [57-21]. Umeda's method [60-97] also depends upon precipitation, for he throws out the molybdate with a standard solution of Pb and determines the excess of Pb in an aliquot portion of the filtrate by back-titrating with EDTA at pH 10. Methods based on precipitation have lost their practical interest, since direct titration procedures have been worked out.

When describing the determination of metals by back-titrating excess EDTA with a standard solution of Zn using the ferrocyanide–ferricyanide system to indicate the end-point, Sajó also mentions Mo [56-99]. However, there is nothing to be found in this paper concerning the working procedure, the accuracy, the limitations, and so on. Kinnunen [58-80] has reported briefly on the determination of

Mo(VI) through back-titration of excess EDTA at pH 5·5 with a standard solution of Zn using xylenol orange as indicator. He describes the end-point as so bad and lacking in sharpness as to question the practical usefulness of the method. On the basis of his titration results the formation of a 1 : 1 complex is postulated. Busev and Chan [59-62] studied the method more closely and confirmed its unsuitability. They then report on tests with many substances of the serviceability of a mixed indicator from pyrocatechol violet and indigo carmine. The titration is preferably carried out at about pH 4–5 in the warm and permits of the determination of 80–300 mg of Mo to about 2%. The formation of a 2 : 1 complex was confirmed, in agreement with the polarographic data. Interference is caused by W, Mn, Fe, Zn, Ni, Cu, and Co. Sr, Ca, and Mg have no effect in cold solutions, but lead to the formation of precipitates in hot solutions. Ba always produces a precipitate. The end-point can also be detected amperometrically, but the results are strongly dependent upon the acidity of the solution, and the optimum pH of 4·7 must be maintained very accurately.

Methods dependent upon the formation of complexes of Mo(V) are more promising. Hydrazine sulphate is used in all the recommended procedures for speedy and quantitative reduction. The composition of the intensely yellow-coloured EDTA complex is again 2 : 1. The reduction is carried out in a strongly acidic medium in the presence of EDTA, the excess of which is then measured by back-titration at some appropriate pH. Busev and Chan [59-61] back-titrate with Zn at pH 10 using Erio T as indicator. More advantageous, in that it is more selective, and especially because W does not interfere, even when present in appreciable concentrations, is back-titration in an acidic medium and Klygin and Kolida [61-90] use a solution of zirconium sulphate in N sulphuric acid. Headridge [60-83] found that back-titration with Zn at pH 5·3 using xylenol orange was unsuitable, as consistently low results were obtained. A probable reason for this is the formation of hydrogen peroxide at this relatively high pH. The back-titration should therefore be conducted at a lower pH, although, to be sure, xylenol orange can no longer serve as the indicator. However, alizarincomplexone can be used and the end-point determined photometrically. Endo and Tomori [62-125] report nothing about low values when back-titration is carried out with Pb in a solution buffered with urotropine and using xylenol orange as indicator. Back-titration with Cu at pH 4·5 using PAN has been described by Kawahata *et al.* [62-41].

Busev and Chan [59-61] mention briefly the possibility of directly titrating Mo(V) at pH 1·8, photometrically at 387 mμ.

Lassner and Scharf [59-3; 61-138; 61-139] have gone very thoroughly into the chelatometric behaviour of Mo. The working instructions given below are based on their researches.

Mo(VI) interferes with the titration of other metals if this is carried out in weakly acid solution. Partial chelation with EDTA is responsible for a long-drawn-out colour transition of the indicator, and in many cases to an over-consumption of titrant. Many indicators are completely blocked by Mo. Acetylacetone is a possible masking agent in strongly acid solutions, but it can mask very many other metals as well [60-30]. Mo(VI) does not interfere at pH 10 where it exists as an anion and, as such, does not react with EDTA. The possibility of its forming sparingly soluble metal salts must, however, always be kept in mind. Back-titration generally provides a remedy in such cases.

The titration of Mo is used in practice for the analysis of alloys [63-43], especially hard metals [59-3; 61-138], ferro-molybdenum [62-41; 62-125; 63-43], and molybdenum trioxide [62-41; 62-125].

22.1. Back-titration of molybdenum

REAGENTS: Standard solutions of 0·05M EDTA and 0·05M $CuSO_4$; 1 : 1 H_2SO_4; conc. ammonia; hydrazine sulphate; tartaric acid; alcohol; PAN indicator.

PROCEDURE: The test solution containing up to 30 mg of Mo in at most 100 ml is made approximately neutral and treated with a measured excess of EDTA solution. Then 5 g of tartaric acid and 2–5 g of hydrazine sulphate are added, and after acidification with 2 ml of 1 : 1 sulphuric acid the solution is boiled for 5 minutes. It is then adjusted to about pH 4 with ammonia and sufficient alcohol is added to bring its content up to about 30%. After adding PAN indicator the solution is titrated with the standard copper solution until it becomes violet, and then it is carefully back-titrated with EDTA until it just becomes a pure yellow.

CALCULATION: Since 1 mole of EDTA reacts with 2 gram atoms of Mo, the equivalent weight is here (and exceptionally!) equal to half the atomic weight. Thus 1·00 ml of 0·0500M EDTA corresponds to 9·594 mg of Mo.

REMARKS: Using the 'pendulum end-point' greatly decreases the drop-error here as in Section 10.1.4. The determination can be carried out in the presence of considerable amounts of W, which is completely masked by tartaric acid. In the absence of tartrate W prevents the complete reduction of Mo. Tartaric acid also masks Ti, Nb, and Ta. Fluoride can be used to mask Th, Al, Ce, and U. Interference is caused by Bi, Zn, Co, Cd, Ni, Cu, Hg, V, Cr, and Pb, for up to the present no effective masking agent is known for the particular conditions of this determination. Iron interferes because it is reduced to Fe^{2+}, which reduces Cu^{2+} in the back-titration.

The following procedure is possible for the determination of Mo in the presence of not too large amounts of interfering elements (Fe excepted!). In an aliquot portion of the solution the titration is carried out without the stage of reduction by hydrazine and with DCTA in place of EDTA. Mo(VI) is not complexed, or at least is insignificantly complexed, by DCTA, so that this titration gives the total concentration of all the metals present with the exception of Mo. In a second aliquot portion the titration with EDTA is carried out after reduction with hydrazine, and this gives the total concentration including Mo, so that the latter is obtained by difference.

The titration can also be conducted as a micro-titration with 0·005M EDTA on 0·5–2 mg of M even in the presence of a 500-fold excess of W.

With higher concentrations of Mo the intrinsic yellow colour of the Mo–EDTA complex interferes. In such cases it is preferable under otherwise identical experimental conditions (although the addition of alcohol can be omitted) to titrate in ultraviolet light with calcein as the indicator: the end-point of the back-titration is then revealed by the quenching of the intense green fluorescence. Up to 90 mg of Mo can be determined accurately in this way.

23. TUNGSTEN

Experiments to determine W by direct or indirect titrations have so far proved unsatisfactory. Kinnunen [58-80] tried out the following route. The strongly acidic solution of tungstate was treated with a measured excess of EDTA, buffered with acetate to pH 5, and the excess of EDTA back-titrated with Tl(III) using xylenol orange. However, the end-point was so bad that any practical use of the procedure could not be considered.

Indirect methods, however, have proved successful. De Sousa [53-60] precipitates calcium tungstate and titrates the calcium content of the separated and washed precipitate. Bykolvskaya [62-84] proceeds as follows. The solution containing about 1 mg of W per ml is treated with a measured amount of 0·1M lead nitrate solution, the lead tungstate is filtered off, and the excess of Pb in the filtrate is determined by adding EDTA and back-titrating the excess with Zn using Erio T as indicator.

Lassner and Scharf have studied interference caused by W in the titration of other metals. Of special interest is its effect on the titration of Mo (see above). Masking of W by tartaric acid proves successful [59–3], provided that tartrate does not prevent the titration of the element that is to be determined. Worth mentioning is the discovery [61-139] that the complex of W with DCTA is appreciably weaker than its complex with EDTA. The authors therefore recommend the use of this complexone as the titrant for the titration of other metals, since W can then be present without a masking reagent and yet cause no interference. The indicator system Cu–PAN–DCTA can be used for the direct titration, or back-titration can be carried out with Cu using PAN as indicator.

24. URANIUM

Evidence for complex formation between EDTA and the uranyl ion was first presented by Cabell [52-41]. On the basis of potentiometric titrations of solutions containing uranyl ions and excess EDTA the formation of a complex with $UO_2^{2+} : Y = 2 : 1$ was inferred. The occurrence of further buffer regions supported the idea that protonated species occur as well. The titration curves could only be evaluated up to about pH 6, after which irregularities became obvious and precipitates appeared in the solution. Davis [61-146] investigated the equilibrium between uranyl ions and EDTA polarographically, but, so far as the formation of complexes was concerned, he neither confirmed nor extended Cabell's work. By using high-frequency methods as well as potentiometric titrations on uranyl–EDTA mixtures, and also in the direct titration of uranyl with EDTA, Hara and West [55-42] obtained titration curves with sharp breaks at the ratio $UO_2^{2+} : Y = 2 : 1$. Rao and his co-workers [57-12] confirmed this composition by means of Job (continuous variation) curves. An exhaustive study was carried out by Kozlov and Krot [60-172]. On the basis of spectrophotometric measurements three different species were identified, and all of them were obtained preparatively. Below pH 2 the 'salt' UO_2H_2Y, previously described by Brintzinger and Hesse [A-9], occurs ionized in the solution. On raising the pH the compound $(UO_2)_2Y$ is formed; this is stable between pH 3 and 6 if the molar ratio $UO_2 : Y$ amounts to about $2 : 1.2$. On increasing the concentration of EDTA a second ligand anion becomes coordinated and the complex UO_2Y^{2-} results. The existence of protonated species was ruled out as improbable on theoretical grounds, but the formation of hydroxy-complexes and, above pH 7, of polymeric hydrolysis products was assumed to take place. If a sufficient excess of EDTA is present the hydrolytic precipitation of uranyl can be avoided right up to pH 8–9.5. Kozlov and Krot have calculated stability constants from their photometric results, and obtain for the species $(UO_2)_2Y$ and $(UO_2)Y^{2-}$ the values $\log K = 15.2 \pm 0.3$ and 10.4 ± 0.2 respectively for 24°C and $\mu = 0.1M$.

If we take into consideration the fact that it is necessary to work at a relatively low pH to avoid precipitation, and thus with a higher value of $\log \alpha_{Z(H)}$ into the bargain, then it is scarcely to be wondered at that with the small effective stability constants that result the titration of uranyl ions runs into difficulties.

Hara and West [55-42] report that uranyl can only be titrated oscillometrically in very dilute solutions that are essentially free from foreign salts. Lassner and Scharf [58-83] investigated the possibility of a visual titration and found that the effective stability constants could be increased sufficiently to make a titration feasible if certain organic solvents were added. Isopropanol proved to be the best, and twice its volume was added to the test solution. Titration was carried out at pH 4.4 (urotropine buffer) in a hot solution (80–90°C) using PAN: at the end-point there was a colour change from red to yellow, although it was not particularly sharp. A highly selective determination of uranyl in the presence of other cations can be carried out as follows. The titration is first carried out as described above to give the total concentration of uranyl plus all the other titratable metals. Sodium phosphate is then added, whereupon uranyl phosphate is precipitated and the equivalent amount of EDTA is set free and can then be titrated. The relatively high concentration of isopropanol limits the method to dilute solutions for the solubility of salts in a 2 : 1 isopropanol–water mixtures is quite low.

Uranyl ions scarcely interfere at all in the titration of other metals apart from the possibility of the precipitation of insoluble compounds or blocking the indicator used. Both difficulties can be avoided in alkaline solution in the cases so far known by masking uranyl with carbonate ions [52-25] or hydrogen peroxide [58-82].

The determination of uranium in its tetravalent state is far more promising. The stability of the 1 : 1 complex U(IV)Y is calculated as

$4·2 \times 10^{25}$ by Klygin and co-workers [59-126] from photometric measurements, and this order of magnitude ($6·8 \times 10^{25}$) has been confirmed from other quarters [62-81]. Kinnunen [57-137] was the first to describe the titration of U(IV). Uranyl solutions were reduced by boiling for 10 minutes with ascorbic acid in the presence of EDTA at pH 3–4; after cooling the pH is reduced to 2–3 and excess EDTA is back-titrated with a standard solution of Th or Tl(III) using xylenol orange as indicator. The method was checked by Lovasi [61-67] and used to determine the concentration of uranium solutions in the presence of impurities. Buděšinský *et al.* [62-124] reduce with $S_2O_4^{2-}$ and back-titrate excess EDTA with Th using methylthymol blue.

Klygin [61-89] reduces by boiling with amino-iminomethane sulphinic acid for 5–8 minutes and titrates directly with EDTA at pH 1·7 using arsenazo I as indicator. The pH is critical and must be maintained within $\pm 0·1$ units. Rozhdestvenskaya *et al.* [63-48] use the same reducing agent but titrate directly at pH 2 amperometrically with a graphite electrode at $+0·2$ volt with respect to saturated mercurous iodide.

Paley and Hsüe [61-79] titrate U(IV) at pH 1·0–1·8 using thoron. At this lower pH the titration becomes very selective. Alkali metals, alkaline earths, Mg, Zn, Cd, Al, La, Ce(III), Ti(IV), Mn, Cr(III), Ni, and Pb produce no interference, and Th, Ce(IV), and Fe(II) can be tolerated up to a ratio of 10 : 1. Co, Cu, Bi, and Hg(II) must be absent, and fluoride ions also interfere.

Korkisch [61-36] titrates U(IV) in solutions that are 0·01–0·2N with respect to hydrochloric acid and containing up to 1 mg of U per ml. Solochrome black 6 BN serves as the indicator and changes from blue to red at the end-point. Bi, Cu, Zr, Hf, Th, and Sn(II) interfere by forming complexes with the indicator which are also blue. The reduction to U(IV) is brought about by adding granulated zinc to the test solution or by passing it through a Zn-, Pb-, or Ag-reductor. Any U(III) formed in this process is oxidized to the tetravalent state by exposure to atmospheric oxygen by swirling the flask for 1–2 minutes. Micro-determinations can also be carried out by this procedure.

So far as can be ascertained from the literature, investigations of interferences have been carried out by adding the interfering elements to the test-solution *after* the reduction of the uranium. This naturally falsifies the position to some extent, since many elements that might exist in the original uranium solution simply do not exist in the solution after reduction, or at least they are not present in an oxidation state that interferes. For example, Ce(IV) is reduced to Ce(III), which is harmless, while Hg(II) separates as a metal in the reductor and is held back there. The method is therefore more selective than might appear at first sight.

Naturally the prior isolation of U by liquid–liquid extraction or ion-exchange is necessary in many cases: to this end there are innumerable unexceptionable and rapid methods in the literature.

25. MANGANESE

The complexometric determination of manganese was described quite early on by Biedermann and Schwarzenbach [48-5], who carried it out by back-titrating excess EDTA with Mg using Erio T as indicator. Zinc can also be used for the back-titration. Flaschka has reported on the determination of micro-amounts [52-22] after an exchange reaction with MgY^{2-}, and on direct titrations for micro- [53-24] and macro-amounts [53-25]. The direct titration using Erio T as indicator presents no difficulties at all if it is carried out at pH 10 and provided a reducing agent is added to prevent the oxidation of Mn(II). Ascorbic acid and hydroxylamine have been used for this purpose.

With large quantities of manganese $Mn(OH)_2$ is formed if the usual pH 10 buffer is used. Admittedly the turbidity disappears during the course of the titration, but this becomes very slow, and for this reason Flaschka recommended [53-24; 53-25] adding tartrate as an auxiliary complexing agent to prevent this precipitation. However, since the tartrato-complex reacts slowly, it is necessary to work at a higher temperature. Too high a concentration of tartrate must be avoided, otherwise the sharpness of the colour transition is adversely affected. On this account Wehber [57-2] titrates at pH 8, where no precipitation of hydroxide takes place. However, the buffer capacity of the ammonium chloride–ammonia mixture that is used is not very large at this lower pH value. Přibil [54-79] keeps Mn in solution with triethanolamine. Addition of ascorbic acid or hydroxylamine is especially important in this case, as otherwise there is oxidation to a green Mn(III)–triethanolamine complex, which is inactive towards complexometric titration. The simplest procedure would seem to be to dilute the test solutions extensively; this has no effect on the accuracy of the results, since both the indicator complex and the EDTA complex are sufficiently stable to ensure an excellent colour transition even at very high dilutions.

So far as the complexometric titration is concerned, Mn behaves very much like Mg, and practically all indicators for Mg can be used equally for titrations of Mn under the same conditions, provided only that a reducing agent is added. Thus o,o'-dihydroxyazodyestuffs [53-42], pyrocatechol violet [56-41], and many other indicators can be used, and because of the higher stability constants of the Mn-complexes the end-points are generally sharper. For this reason Kinnunen [55-4] recommends replacing the standard solution of Mg by one of Mn in back-titration procedures. Mn can also be titrated with other indicators, e.g. phthaleincomplexone [58-2], methyl-thymol blue [58-102], eriochrome red B [57-2], vanadium–diphenyl-carbazide [63-1], fluoresceincomplexone [59-22], methylcalcein, and methylcalcein blue [60-65], which, if used with Mg, would give either no or only bad results.

The determination of Mn is also possible by back-titration of excess EDTA with Co in a 50% acetone–water mixture in the presence of KCNS [60-169]. Instrumental end-point detection has also been used in, e.g., conductimetric [54-32], photometric [53-45; 54-69; 54-70], potentiometric (mercury-drop electrode) [57-8; 58-11; 58-12; 62-29], and high-frequency titrations [54-47]. The stability constant of the complex MnY^{2-} (log $K_{MnY} = 13 \cdot 8$) is too low to be able to titrate Mn directly in the presence of alkaline earths. On the other hand, the value is high enough for Mn to interfere in the titration of heavy metals with the exception of those lying in the very stable group with log $K > 20$. This is of importance in practical analyses, when Mn occurs either as an interfering element or when its determination in the presence of other elements is rendered difficult or impossible if the accompanying metals are not easily masked. The combinations Ca, Mg, Mn and Al, Fe, Mn are of special importance, since they often turn up in the analysis of silicates, where, according to the way in which the separations have been carried out, the Mn can be driven into the first or the second group. There have been innumerable studies of different ways of resolving each of these groups, and there are papers reporting more or less successful attempts to carry out consecutive titrations.

Separations, of course, are always possible: such as that of Mn from alkaline earths and Mg with thioacetamide [55-62] or from iron by precipitation of MnO_2 by using $KClO_3$ [55-68], or the removal of Fe by liquid–liquid extraction [54-39]. Naturally, masking or a consecutive titration is more elegant and less time-consuming. The first attempts to mask Mn originated with Přibil [53-11], who introduced triethanolamine. In alkaline solution air-oxidation yields a deeply coloured emerald green complex of Mn(III) which is stable to EDTA. However, on account of the intense colour of this complex, the amount of Mn that can be masked in this way is limited, and if more than 3–4 mg of Mn per 400 ml is present in the final volume the colour change of most indicators is swamped to such an extent that they are unusable. Přibil titrated Ca in the solution containing the masked Mn by using murexide as indicator.

Masking of Ca and Mg with fluoride, first described by Přibil [54-74], was chosen by Wehber [57-2] for the determination of Mn. Up to 10 mg of Mg can be rendered ineffective; in the presence of increasing amounts of Ca low results for Mn soon appear, and with 15–20 mg of Ca the errors are unacceptable. As Povondra and Přibil found out [61-84], this is due to the absorption of Mn by the CaF_2. This undesirable situation can be overcome by adding an excess of EDTA to the solution containing Mn, Ca, and Mg, making the solution alkaline, and then introducing the fluoride. Since Ca and Mg are now complexed by the EDTA, no precipitate separates to start with. If the mixture is now back-titrated with a standard solution of Mn (containing some hydroxylamine) first of all the excess of EDTA is bound up by the Mn; then Ca and/or Mg are displaced from their complexonates and precipitated as fluorides. In this process no Mn is absorbed, since it is all present as the anionic complex MnY^{2-}. The completion of the titration is revealed by a sharp transition in the colour of Erio T from blue to red.

Radko [60-77] resolved a mixture of Mn, Ca, and Mg by titrations of three separate aliquot portions. The total concentration of the three cations was obtained by titrating at pH 10 in the presence of hydroxylamine using acid chrome dark blue as indicator. Ca was then determined at pH 12 after Mn had been precipitated by Na_2S and Mg as hydroxide. Finally, Mn alone was titrated after Ca and Mg had been masked as fluorides. In rock analyses Matsui [61-47] determined Mn by the analysis of two aliquot portions. In the first the sum of Ca + Mg + Mn is determined after adding triethanolamine and hydroxylamine. In the second triethanolamine only is added so that Mn is masked and the sum Ca + Mg is determined. Methylthymol blue is used for both titrations. According to Radko [60-76], the consecutive titration Fe–Mn can be carried out as follows: Fe is titrated first at pH 2–3 using sulphosalicylic acid; then tartrate and hydroxylamine are added and the pH increased to 10 and titration carried out with acid chrome blue (Erio SE). Similarly,

Al–Mn or (Al + Fe)–Mn can be determined if the Al (or the sum Al + Fe) is first determined by back-titration of excess EDTA with Fe(III) using sulphosalicylic acid: then Mn is titrated as described above. See also Wakamatsu [60-169].

Although the stability of the Mn–EDTA complex is more than enough for a sharp titration, some attention has nevertheless been devoted to the use of DCTA, which forms a more stable complexo-nate. This interest stems less from the increased absolute stability of the complexes than from the greater difference between the value of the constant for the Mn complex and that for Ca and Mg. By using DCTA Mn can be titrated, according to Körös [58-102], even at pH 6·0–6·5 using methylthymol blue, and, moreover, even in the presence of moderate amounts of Ca and Mg (several mg) without having to mask them. Přibil [55-89] showed that the use of DCTA was advantageous for the consecutive titration Fe–Mn. Fe is first titrated at pH 2–3 using salicylic acid and then the Mn at pH 10 using Erio T. In the same titration with EDTA as the titrant the Fe leaves the complex and blocks the indicator.

Mn forms a cyano-complex which is far too weak to interfere with its reactions with EDTA or the indicator, at least under ordinary conditions of titration. Thus KCN can be introduced as a masking reagent for the other metals that form cyano-complexes [53-25], and then one proceeds in the same way as with Mg. However, according to Přibil [61-93], masking of Mn by cyanide can be effected under rather special conditions. The solution which contains triethanol-amine is first made strongly alkaline, and all the Mn is allowed to undergo atmospheric oxidation to the Mn(III)–triethanolamine complex. Then KCN is added and the pH reduced to 10 by introducing acetic acid. The solution becomes practically colourless, since the hexacyano-complex of Mn(III) is formed. If Fe and Al are absent Ca can now be titrated. Up to 80 mg of Mn can be masked in this fashion. The same paper also contains a very interesting way of masking Mn should Fe be present. The acidified test-solution is treated with tri-ethanolamine and then made alkaline with 20–30 ml of conc. ammonia; KCN is then added. The ferricyanide thus formed oxidizes Mn(II) by the following reaction

$$Fe(CN)_6^{3-} + Mn(CN)_6^{4-} \rightleftharpoons Fe(CN)_6^{4-} + Mn(CN)_6^{3-}$$

Of course, the amount of iron must be more than equivalent to that of the Mn. The cyano-complexes of Fe(II), Mn(III), and of the excess Fe(III) are complexometrically inert. The titration of alkaline earths runs into difficulties, however, when, e.g., the sparingly soluble calcium hexacyanomanganate separates, for this only reacts slowly with EDTA.

Pb and Bi can be masked by BAL [54-79], whereafter Mn can be titrated in the usual way with Erio T as indicator.

Practical applications of the complexometric titration of Mn include its determination in metallurgical slags, in metals and inorganic raw materials [61-84], silicate rocks [61-47], adhesives [61-143], alloys [54-39], spiegeleisen [55-68], and ferromanganese [59-30; 60-92].

Direct titration of manganese

REAGENTS: 0·01–0·1M EDTA; 20% triethanolamine hydrochloride; buffer pH 10; Erio T indicator; ascorbic acid.

PROCEDURE: The acid test solution is treated with 10–15 ml triethanolamine solution and a spatula full of ascorbic acid; it is then neutralized and adjusted to pH 10 with a few ml of buffer. The indicator is then introduced and the titration carried out to a sharp colour change from red to blue.

REMARKS: The triethanolamine masks small amounts of iron and aluminium also: for larger amounts KCN must be added as well. In the latter case it is advisable to add more ascorbic acid to reduce all the ferricyanide that is formed, a process that can with advantage be speeded up by warming for a short while. However, the actual titration of manganese must be carried out on a cold solution, as otherwise aluminium blocks the indicator. Magnesium and alkaline earths are titrated with the manganese.

26. RHENIUM

According to Hamaguchi and Sugisita [61-172], the direct titration of rhenium in the form of its perrhenate can be carried out as follows. Perrhenate is precipitated as $TlReO_4$ from a slightly acid or neutral solution, the precipitate is separated by centrifugation, washed, and dissolved in a mixture of bromine and hydrochloric acid. Thallium is hereby oxidized to the tervalent state, and after adjusting the pH to 5 it can be titrated with EDTA using xylenol orange. Molybdate, permanganate, thiocyanate, bromate, iodate, and halogens must be absent. Chlorate and perchlorate do not interfere.

27. IRON

Schwarzenbach [48-7] described the complexometric determination of Fe(III) a long time ago. The titration is carried out at pH 2–3 with Tiron as indicator. The rate of complex formation is somewhat slow with Fe, so that gentle warming of the solution is recommended, although not absolutely necessary. This also holds for titrations with other indicators. The end-point with Tiron is characterized by the disappearance of the last trace of a green shade and a transition to yellow (the colour of FeY^-), and in macro-titrations it is sufficiently sharp. This indicator is still used nowadays for many analyses. However, if the iron titration is to be followed by the determination of other metals at higher pH values it is necessary to bear in mind that under these circumstances Tiron also forms 1 : 2 and 1 : 3 complexes of red and violet colour respectively [51-5]. The 1 : 3 complex is more stable than the Fe–EDTA complex, so that in alkaline solution the iron is withdrawn from its complexonate and EDTA is set free.

In his adaptation of the iron titration to the micro-scale, Flaschka [52-22] recommends sulphosalicylic acid as the indicator. Other authors also use this indicator [57-6; 59-97; 60-162; 62-13]. Salicylic acid can also be used. The end-point is practically identical, but the sulphonic acid derivative has the advantage of greater solubility, which is not only more convenient but of actual importance in the presence of Al, for a masking action can be exerted on it for which large amounts of salicylate are often needed.

Very many other compounds with phenolic OH-groups have been proposed as indicators for iron, e.g. *o*-carboxyphenylazochromotropic acid [60-41], 3-hydroxy-2-naphthoic acid [60-14; 61-44], or hydroxamic acid [61-103]. The last named permits of titration at a very low pH (1–1·5), which is of importance when consideration is given to interfering elements, especially Al. A similar low pH (1·7) is possible with α,5,5′-trimethylformaurine-3,3′-dicarboxylic acid (Structure IX of Part One, Section 3.3.2 with CH_3 in place of ring C, methyl groups in positions 5 and 5′, and carboxyl groups in positions 3 and 3′). Other indicators are kojic acid, 5-hydroxy-2-(hydroxyethyl)-1,4-pyrone [56-82], salarcide (salicylaldehyde semicarbazone) [60-118], CuY–PAN [56-45; 61-66], CuY–PAR [61-66], acetylacetone [61-1; 61-116; 62-2], chromeazurol S [55-29], and thiocyanate [47-4; 53-18; 57-4; 60-161; 61-63]. Quite early on the advantages and

I

disadvantages of the last named were evaluated critically on theoretical grounds [53-18], and the suitability of an extractive end-point has also been discussed. Various triphenylmethane dyestuffs have been examined as possible indicators [57-59].

It must be admitted that the colour transitions in the complexometric titration of Fe(III) are generally less sharp than those of, e.g., the familiar redox titration. This is particularly true of the use of indicators of phenolic character, of which amounts must often be added that are quite substantial when compared with the usual concentrations of indicators (cf. Part One, Section 3.2). The absolute stability constant of the Fe(III)–EDTA complex is very high (log $K_{FeY} = 25.1$), so that Fe(III) can be titrated at low pH values with comparative freedom from interference. It must be noted, however, that the effective stability constant under the actual conditions of a titration is not always so large. The lowest pH for an EDTA titration with an indicator is generally about 1, for owing to the high α-value the effective stability constant is reduced to log $K^{eff.} = 7.9$. Although log α decreases rapidly with increasing pH, the situation is complicated by the formation of hydroxy-complexes. Just above pH 10 Fe(III) comes out of its EDTA complex and is precipitated as hydroxide. This has to be watched in consecutive titrations, for the displacement of Fe(III) by another metal can take place at quite low pH values.

Redox indicators can also be used for the titration of Fe, e.g. variamine blue [54-60; 60-89], Bindschedler's green [56-107], and the combinations Fe(II)–cacotheline [56-27] and KCNS–*p*-anisidine [60-99].

Descriptions of back-titration procedures are also to be found in the literature, e.g. in acid solution with Th using xylenol orange [57-137], with Cu using *o*-dianisidinetetra-acetic acid (fluorescence indicator) [60-153] and in ammoniacal solution with Zn using Erio T [55-90] and using DCTA. These and other back-titration methods are primarily of importance for titrations of the total concentration of Fe(III) plus other metals.

The stability of the Fe(III)–EDTA complex is appreciably lower (log $K = 14.3$) than that of the ferric complex. Nevertheless, the titration of Fe(II) is feasible, for example at pH 4 with DCTA using methylthymol blue [58-102] or with EDTA at a somewhat higher pH using pyridine-2-aldoxime as indicator [63-57]. This is only of interest in special circumstances, for under normal working conditions the Fe always occurs in the tervalent state. However, there remains the possibility of titrating Fe(III) in the presence of Fe(II) or of determining both in the same solution [55-23]. Fe(III) is determined first using salicylic acid as indicator: then Fe(II) is oxidized with persulphate and the titration continued. When Fe(II) is present it is essential to work with the exclusion of air, since in the presence of complexones Fe(II) becomes a really powerful reducing agent. In the absence of EDTA the normal potential of the couple Fe(III)/Fe(II) is $+0.77$ volt, whereas it falls to 0.1 volt in the presence of excess of the complexone. The solution is so strongly reducing that some substances that may also be present, e.g. Ag^+, are reduced [52-14]. The enormous difference between the stabilities of the complexonates of Fe(II) and Fe(III) forms the basis for the reductive masking of iron with ascorbic acid or hydroxylamine.

As already mentioned, the high stability of the Fe(III)–EDTA complex permits of a relatively selective determination of Fe in a strongly acidic medium. The lower the pH, the more selective the titration. The determination of Fe in the presence of Al is of special practical importance. On the basis of the difference between their stability constants it could be anticipated that Al would have no influence upon the titration of Fe. However, this is not the case unless it is carried out at a very low pH – below 2. In no sense can the situation be regarded as completely cleared up, and there are contradictory statements to be found in the literature concerning interference by Al. The overall measure of this interference (a high result for Fe due to co-titration of Al) is not a function of acidity only, for it also appears to depend on the salt concentration of the test-solution, on the temperature (Al is only slowly complexed), and on the indicator used. The lower the stability of the Fe-indicator system, the better. Very often the situation can be improved by replacing EDTA by another complexone, best of all by DCTA. A selective masking of Al (cf. Section 12) offers another route which cannot, however, be followed in a consecutive titration.

Consecutive titrations of Fe–Al are possible by a variety of methods. For example, Wehber [57-6] titrated Fe first using sulphosalicylic acid and then determined Al at pH 5 by back-titrating an excess of NTA with Cu using Bindschedler's green as indicator. Kristiansen [61-44] titrates Fe first at pH 2 using 3-hydroxy-2-naphthoic acid in a solution that has been cooled down with a few pieces of ice. After reaching the Fe end-point (blue to yellow) the solution is heated to 50°C, adjusted to pH 3, and the Al titrated using the same indicator in ultraviolet light until the colour of the fluorescence changes from blue to green. Check analyses with mixtures of Fe and Al in the proportion 1 : 2 to 2 : 1 gave very good agreement with the correct values. Fe can also be titrated at pH 2 using sulphosalicylic acid, and then the Al can be determined by adding a known excess of EDTA and back-titrating the excess at pH 4–5 with a standard solution of Fe [62-13]. The sulphosalicylic acid has a masking effect on the Al. Larger amounts of Al can be tolerated if the Fe is first titrated at pH 1–2 with aluminon as indicator: the pH is then increased to 4·4 and the Al is titrated using the same indicator. The method worked out by Kundu [61-2] even permits of the subsequent titration of Ca and/or Mg at pH 10. Instrumental methods for detecting the end-point are specially valuable here, such as a photometric titration at pH 1 or even slightly below using sulphosalicylic acid as indicator [60-144].

Xylenol orange (XO) imparts high selectivity in reactions where some theoretically interesting reactions take place [59-27]. In a solution containing practically only Fe(III) the reaction taking place at the end-point, viz. Fe(III)–XO (blue) + EDTA $\rightleftharpoons$ Fe(III)Y + XO (yellow), proceeds so sluggishly that even at the boiling-point the indicator must be regarded as being 'blocked'. If, however, some Fe(II) is present, then after the complexation of the Fe(III) (i.e. all that which is not bound to XO), some Fe(II)Y is formed, and at the end-point the following redox reaction takes place by a fast one-electron transfer: Fe(III)–XO + Fe(II)Y $\rightleftharpoons$ Fe(II)–XO + Fe(III)Y. The indicator complex, Fe(II)–XO, thus formed is unstable at the low pH used in the titration and dissociates at once to form free XO, whose yellow colour reveals the end-point. The lower the pH, the more Fe(II) must be present. Thus about 2 g of ferrous sulphate are needed for the titration of 20 mg of Fe(III) at pH 1·5. The temperature also has an effect on the amount of Fe(II) needed. Titrations of about 50 mg of Fe(II) at room temperature can be carried out successfully in the presence of 0·5 g Al, provided an amount of F^- equivalent to the Al is added. Zr can also be masked with F^-. Th does not interfere, since the amount of sulphate ion introduced through the ferrous salt is sufficient for masking purposes. The amount of Cr that can be tolerated is limited by the intensity of the interfering colour of the Cr itself. Among the usual tervalent ions that might be present only Bi and Tl interfere. Bivalent cations have no great effect. Naturally the titrations must be carried out with the exclusion of air if errors due to the co-titration of oxidized Fe(II) are to be avoided.

Instrumental methods in connection with the determination of Fe were described at an early stage. Přibil [51-15] recommended potentiometric titrations in an acetate buffer at pH ~5, whereby steady potentials are achieved instantaneously: with a Pt-electrode the potential jump at the end-point amounts to some 350 mV per 0·04 ml of EDTA. Potentiometric back-titration of excess EDTA with Fe(III) provides a method for determining many metals, and by using aliquot portions the resolution of mixtures of Fe, Al, and Cr can be carried out [57-29]. The redox behaviour of the iron couple in the presence of EDTA was studied by Belcher *et al.* [57-33] from the standpoint of its analytical importance. By amperometric titration [57-33] at a rotating Pt-electrode [58-41; 59-44] and by the square-wave procedure [61-35] Fe(III) can be determined with EDTA and also subsequently Fe(II) with ferrocyanide in the same solution.

In view of the intrinsic colour of the complex FeY^-, photometric titration can be carried out very easily with self-indication [53-37], e.g. in the consecutive titration Fe–Cu [53-39]. Alternatively, it can be carried out after adding an indicator such as sulphosalicylic acid [60-144] or ferron [59-123]. Photometric end-point indication also enables the titration to be automated [60-78].

Taking into consideration the fact that Fe(III), which is the principle or secondary component of innumerable materials, forms very stable complexonates, we can easily see why its masking is of the greatest possible practical importance. The problem was tackled early on and with success. Many methods are now available for masking Fe(III) in alkaline media. Probably the most frequently used method is its transformation into a complex with triethanolamine, a device introduced by Přibil [53-10; 54-76; 55-89; 59-16]; this inhibits the reaction between Fe and complexones and almost all indicators as well. Certain points must be kept in mind when putting this into practice. The addition of triethanolamine basifies the solution and often leads to the precipitation of iron hydroxide, which then reacts only slowly with the masking agent even if the solution is warmed. For this reason it is advisable to start out from a strongly acidic test-solution and to adjust this slowly to the desired pH after adding the triethanolamine or, alternatively, to add the triethanolamine in the form of its hydrochloride.

Triethanolamine also masks several other cations, e.g. Al and Mn, so that masking by cyanide according to Flaschka and Püschel [54-62] is important in certain cases. This is carried out as follows. The acidic test-solution is treated with some tartrate to prevent any precipitation of $Fe(OH)_3$ when it is subsequently made alkaline. The KCN and some ascorbic acid are added and the solution is heated to boiling, whereby Fe is transformed into ferrocyanide. At the end of the reaction the solution is practically colourless. Subsequent demasking of Zn and Cd is possible by adding formalin [56-47]. In the presence of larger amounts of iron this process can lead to the formation of suspensions, since the ferrocyanides of many cations, including those of Ca and Mg, have low solubilities. In such cases one can either work slowly and in hot solutions, whereupon the turbidity gradually disappears during the course of the titration, or one can employ the principle of back-titrating excess of EDTA.

Přibil has reported on an interesting method for the simultaneous masking of Fe and Mn in the presence of KCN [61-93]. Fe can be masked in an almost neutral solution (pH 7–8) by sulphosalicylic acid if rare earths are to be titrated [62-89]. Ascorbic acid is an outstanding reagent for masking Fe in acid solution: the method [56-84] depends on the fact that Fe(III) is reduced to Fe(II), which, in view of the low stability of its complexonate, causes no interference when other metals that form more stable complexes are titrated. Other reducing agents, such as hydroxylamine hydrochloride, could also be used. The exclusion of atmospheric oxygen is necessary. Fe(III) can also be titrated complexometrically with ascorbic acid, whereby it is simultaneously determined and masked, following which other complexometric titrations can be carried out.

Fe(III) is also reduced by thiourea, which is used for masking Cu. To prevent this reduction F^- can be added, for this has the effect of masking Fe(III) against thiourea.

Practical use of the Fe titration extends to the analysis of ferrosilicon and silicates [59-19], ferrochrome and chromites [61-181], blood [54-26], haemoglobin [55-70], ashes from animal tissues [55-9], boiler scale [55-61], limestone [52-46], soils [53-61], and cement [61-42]. Other substances analysed include electroplating solutions [62-52], chrome baths [57-49], paper-pulp [57-27], adhesives [60-131], slags [61-77], Cu [54-67], and W alloys [59-1]. An interesting procedure is the determination of metallic iron in oxide slags [61-46]. The sample is treated with $HgCl_2$ in ethanol in an atmosphere of CO_2, whereupon only elementary Fe reacts. After filtration the solution of Fe(II) is oxidized to Fe(III), and this is titrated at pH 2 in a solution containing Cl^- with Tiron as indicator.

27.1. Direct titration of iron(III) using Tiron, salicylic acid, sulpho-salicylic acid, or chromeazurol S

REAGENTS: Standard solution of 0·1M EDTA; Tiron (salisalicylic acid, sulphosalicylic acid, or chromeazurol S); persulphate; solid glycine.

PROCEDURE: The test solution should contain about 20–70 mg of Fe per 100 ml and not too much excess HCl. If there is any fear that part of the Fe is in the bivalent condition, a few crystals of persulphate are added and the solution is boiled for a short time. Excess glycine is now added to achieve a pH of 2–3 (0·2–0·5 g per 100 ml should suffice), and about 2 ml of Tiron and the warm solution (40°C) is titrated until the blue colour is bleached. The end-point is reached when the last trace of a greenish hue disappears and the solution becomes a pure yellow.

REMARKS: The colour change becomes drawn out if very dilute EDTA or cold solutions are used. When an acetic acid buffer is used the concentration of acetate must be kept low, as otherwise the Fe is transformed into acetato-complexes and the clarity of the end-point becomes even poorer.

Thiocyanate can also be used as the indicator [47-4; 53-18]. If salicylic acid is used about 0·2 g of solid acid are used per 100 ml of test solution, and the titration is carried out at pH ∼4.

With chromeazurol S as indicator the titration is carried out at pH ∼3 in the warm: the colour change is from greenish-blue to orange.

27.2. Direct titration of iron(III) using variamine blue

REAGENTS: Standard solution of 0·01M EDTA; variamine blue; dilute ammonia; glycine.

PROCEDURE: The test solution should contain about 5 mg iron per 100 ml. If excess acid is present it must be neutralized to the first slight change of Congo red paper (pH 2–3). The addition of 0·2–0·5 g of glycine is also worth while. A few drops of indicator are then added and the titration started. Just before the end-point the initially blue-violet solution becomes grey, and with the last few drops it changes to yellow (the colour of the iron–EDTA complex).

REMARKS: The change in redox potential of this titration can be established with even greater accuracy by potentiometry with a bright platinum electrode [51-15].

28. COBALT

The possibility of titrating cobalt complexometrically with thio-cyanate as indicator was first noted in a patent specification [A-8] and in a paper by Biedermann and Schwarzenbach [48-5]. Various authors have studied this method in greater detail. Takamoto [55-100; 55-101] worked with a 1 : 1 acetone–water mixture and extended the back-titration of excess EDTA with a standard solution of Co to the determination of almost all the heavy metals [55-102; 55-103]. According to his results, the adjustment of pH is critical, and 3 ml of 6M ammonium acetate solution should be added for every 35 ml of test-solution. On the other hand, Sen [58-90] finds that the pH is without any significant influence, and correct results are obtainable in the pH range from 6 to 12. Sen Sarma [55-115] stresses the importance of maintaining precise conditions for the titration of small amounts of Co, since otherwise the intensity of the colour of the Co–SCN complex, and even its composition, are subject to undesirable variation. A concentration of 30–35% acetone (v/v) and 5% ammonium thiocyanate were recommended as optimal. Under these conditions it was possible to determine 25 μg of Co per 100 ml visually with satisfactory accuracy [60-171]. For smaller amounts, and to achieve the highest accuracy, a photometric end-point is recommended. Amyl alcohol can be used in place of acetone with the advantage that it can extract the Co–SCN complex whereby a selective determination becomes possible. For in this way Co can be determined even in the presence of not too large amounts of Ni. The direct titration of Co using murexide has been written up by Flaschka [52-22] as a micro-method, but it can be used equally well for macro-quantities. Numerous other indicators exist for the Co determination. To name a few: pyrogallol red [56-85] and its bromo-derivative [56-88], naphthol violet [57-81], pyrocatechol violet [54-78], xylenol orange [57-79], and methylthymol blue [58-34]. Sharp end-points are also given with indicator systems such as CuY–PAN [56-43], CuY–naphthylazoxine [57-122], and vanadium–EDTA–diphenylcarbazone [63-1].

Much has been written about back-titrations, which are often worth-while ways of getting round some problems. Harris and Sweet [54-102] back-titrate with Zn in ammoniacal solution with Erio T as indicator. Kinnunen commends Mn as a reagent for back-titrations [55-4]. However, Erio T is easily blocked by Co, so that the titration must be carried out quickly: any volumetric solutions in which the indicator has been blocked must be discarded. Back-titration in an acidic medium is preferable, since it is more selective. Here back-titration can be carried out with Cu using PAN [56-43; 56-46] or with vanadium(V) using diphenylcarbazide or diphenylcarbazone as the indicator. Lukazsewsky *et al.* [57-84] use dithizone as the indicator and back-titrate with Zn in a 50% alcohol–water mixture. Cameron and Gibson [61-135] use an extractive end-point. The titration is carried out in aqueous solution which is shaken up with thiocyanate and triphenylmethylarsonium chloride in a layer of chloroform. Excess EDTA is back-titrated with a standard solution of Co, and the end-point is revealed by the appearance of a blue colour in the organic solvent. This back-titration procedure is available not only for Co but for a whole series of other metals [61-136].

In an alkaline medium and in the presence of complexones, Co(II) is easily oxidized by hydrogen peroxide to Co(III). With EDTA a deep blue complex $Co(III)(OH)Y^{2-}$ is formed which takes up a proton on acidification and goes over into the violet-red complex $Co(III)(H_2O)Y^-$. On adding a base the reaction is reversed. On standing the aquo-complex loses water and there results the anion $Co(III)Y^-$, which also has a violet-red colour, but which can now only be transformed back into the blue hydroxy-complex in a strongly alkaline medium and only slowly. All complexes of Co(III) are kinetically extremely inert ('robust') and are stable even in strongly acidic media; this can be exploited for determinations with enhanced selectivity. After Co has been oxidized in an alkaline medium Sajó back-titrates with Zn using the ferri-ferrocyanide benzidine indicator system [56-99]. However, the amount of hydrogen peroxide is critical. Bad end-points are obtained if exactly the right amount of H_2O_2 needed to oxidize the Co is not taken. Kinnunen [57-137] used Th or Tl(III) with xylenol orange for the back-titration. In both procedures the amount of Co is strictly limited by the fact that the intense red colour of the Co(III)Y easily masks the colour change of the indicator. Flaschka therefore recommends a photometric end-point (q.v.). The use of fluorescence indicators is also advantageous. Of special interest is the determination of Co in the presence of Ni and the resolution of Ni–Co mixtures, a topic which will be dealt with under Ni in Section 29.

Instrumental methods for determining Co include amperometric [54-29], conductrimetric [54-32; 57-34], thermometric [57-97], and high-frequency titrations [55-41; 55-117]. Potentiometry with the

Hg-drop electrode is possible on both the macro- [58-12] and ultramicro-scale [57-8]. Automation of the Co determination has been worked out by Malmstadt [60-78]. The back-titration of EDTA with a Co solution in an acetone–water mixture containing thiocyanate gives very good results for Co and also for other metals, especially on the micro-scale [60-171]. Photometric indication is specially advantageous when dealing with Co in the tervalent state. Flaschka and Ganchoff [61-51] oxidize Co in an alkaline medium and back-titrate excess of EDTA with Bi at pH 1 and below using pyrocatechol violet: the only metals that interfere are those that form very stable complexes, such as Bi, Fe(III), Zr, or Th.

Co(II) is co-titrated during the determination of other metals with stability constants ranging from 10^{16} to 10^{20}. However, it does not interfere with the determination of metals that form more stable complexes if the titration is carried out at a low pH. Masking of Co with cyanide is possible, and the yellowish cyano-complex of Co(III) results from atmospheric oxidation. Should large amounts be masked in this way, interference can be caused by the yellow colour if visual indicators are used. The cyano-complex of Co(III) is not broken down by formaldehyde.

Practical applications of the complexometric titration of Co extend to its determination in cemented carbides [60-5], colophony [55-35], metal naphthenates [55-116], and other paint driers [58-25; 61-143], and in magnet alloys [57-39]. Lacourt [55-5] describes an interesting possibility of titrating Co in a single spot after it has been separated from Cu and Zn by paper chromatography.

28.1. Direct titration of cobalt using murexide

REAGENTS: Standard solution of 0·01M EDTA; murexide; approx. 1M ammonia.

PROCEDURE: The acid test solution which should not contain more than about 25 mg Co per 100 ml is neutralized with ammonia to pH 6. The indicator is added next to the still acidic solution, and this assumes an orange yellow colour. The Co–murexide complex (probably CoH_4D^+) now serves directly as an acid–base indicator for further neutralization in which ammonia is added until the colour changes to yellow (to CoH_2D^-). Titration with EDTA is then carried out to a sharp colour change from yellow to violet. Since the solution is only slightly buffered, the protons set free during the formation of the Co complex can cause the pH to fall so low that the orange indicator complex is formed again. In such cases a drop or so of ammonia is added, whereupon the clear yellow colour is restored and the titration can proceed.

REMARKS: The relative stabilities of the ammine and murexide complexes of Co are such that the indicator complex is broken down again by an excess of ammonia. By keeping to the above working instructions the conditions for a good end-point are at their best, since with the very low concentration of ammonia that prevails the Co that is not bound to EDTA exists almost exclusively as the (hydrated) Co^{2+} ion. In this way the danger of the metal becoming oxidized by the air is prevented. Should this happen, the very inert cobalt (III) ammine complexes would be formed and this is unable to react with EDTA.

29. NICKEL

The direct titration of nickel in a strongly ammoniacal solution using murexide as indicator was described by Schwarzenbach as long ago as 1948 [48-7]. The method was adapted by Flaschka and co-workers to the micro- [52-22] and ultramicro-scale [57-125], and it gives an extraordinarily sharp end-point. In this case, as with practically all Ni titrations, it is necessary to titrate slowly near the end-point, as the rate of formation of the Ni-complex is not very high. Warming the solution often helps to reduce this trailing of the reaction. Pyrocatechol violet [56-41] and chromeazurol S [59-14] have been recommended as additional indicators for titrations in an alkaline medium. The colour transition with the latter indicator can be sharpened considerably by keeping the ammonia concentration low and adding pyridine. One disadvantage of titrations in an ammoniacal medium is that Mg and alkaline earths are co-titrated.

This interference does not occur if the titration is carried out in an acid medium. According to Flaschka *et al.* [56-44], this is possible in a direct titration in a boiling solution using the indicator system CuY–PAN. Conaghan [56-57] titrates at pH 5 with sodium 1-nitroso-2-naphthol-3,6-disulphonate. Fritz [61-112] used naphthylazoxine S in a medium buffered with acetate or at pH 8. In the latter case citrate does not interfere, a result which is important for its use as a masking agent for Sn or Th; for example, SNAZOXS [60-67] is another indicator for the titration of Ni in acetate solution, and the system VY–diphenylcarbazone can also be used [63-1].

Ni can also be determined by way of back-titrations, and this has the advantage that the lower rate of reaction is now of no consequence. Back-titration is possible with copper and PAN [56-43; 56-46], calcein W [59-9], or *o*-dianisidinetetra-acetic acid [60-153]. The last two are both fluorescence indicators. It is also possible to titrate with Zn using an indirect redox indication of the end-point with ferri-ferrocyanide and dimethylnaphthidine [55-67] or benzidine [56-99]. Kinnunen recommended back-titration with Th or Tl(III) using xylenol orange [57-137], whereas ter Haar and Bazen [56-5] titrate with Th at a relatively low pH (2·8) using alizarin S. Takamoto [55-102] recommended back-titration with Co in an acetone–water mixture containing thiocyanate. Cameron and his colleagues [61-136] use the same metal for the back-titration and add thiocyanate, triphenylmethylarsonium chloride, and chloroform; the colour change is observed in the organic phase.

The above is only a short list of existing possibilities. Practically any indicator for Cu can be used and Cu introduced for the back-titration. The same holds for Zn as a reagent for the back-titration.

Back-titration in an alkaline medium is less favourable, since alkaline earths and Mg are co-titrated. However, the method is important, for nickel can be precipitated very specifically as its dimethylglyoxime complex; this can be taken up in HCl and the metal allowed to react with EDTA, after which the excess is back-titrated with Zn using Erio T as indicator [53-26]. In the absence of Pd this combination of analytical procedures offers a specific and a rapid method of determining Ni. In comparison with the gravimetric procedure, it has the enormous advantage that co-precipitation of reagent glyoxime is unimportant.

The following instrumental methods have been introduced in the determination of Ni. Sweetser and Bricker [54-14] use a photometric titration which is self-indicating at pH 4 and 1,000 mμ. The method was used by Brake *et al.* [57-63] for the titration of Ni in the presence of Co that had been masked with nitroso-R-salt. Ringbom titrated with murexide as the indicator [53-45], and the same indicator was used by Dewald [58-38] to determine Ni in the presence of Fe that had been masked with pyrophosphate. Silverstone [62-16] has described a potentiometric titration after separating Ni as its complex with dimethylglyoxime. Titrations are possible with the mercury-drop electrode, and they have been carried out on the ultramicro-scale [57-8] and permit of consecutive titrations. The latter can also be carried out with an amperometric end-point at the potential of the 'complexone-wave' [62-33].

In view of the position of Ni in the table of stabilities for EDTA complexes, it can readily be predicted that the determination of Ni will be interfered with by a large number of other metals and that Ni, for its part, will interfere in many other complexometric titrations. Masking of Ni by cyanide is simple and effective. The resulting tetracyano-nickelate complex is stable and 'robust' (kinetically inert), so that formaldehyde can be used to de-mask cyanide complexes of Zn and Cu without decomposing the $Ni(CN)_4^{2-}$. The problem of analysing mixtures of Ni and Co has attracted a great deal of thought. Brake *et al.* [57-63] treated the test solution with nitroso-R-salt, whereupon the corresponding complexes of both metals are formed. On boiling with nitric acid the excess of reagents and the Ni-complex are broken down, although the Co-complex remains untouched: the Ni is then determined photometrically. Flaschka and Püschel [55-66] exploit the slowness of its reactions for a selective determination of Ni. The solution containing Ni and Co is titrated with EDTA using pyrocatechol violet in a solution lightly buffered to pH 10. Then by introducing nitric acid the pH is reduced to 2 and the solution is cooled by adding ice. If titration is now carried out with a standard solution of Bi the Co is immediately displaced from its complex, whereas the Ni-complex remains unchanged. This method can also be used to determine Ni in the presence of other metals, e.g. Cu, Cd, Zn, Pb, etc, and also permits of their determination by difference. Přibil *et al.* [54-80] proceed on the following lines. Ni and Co are transformed into their EDTA complexes, the solution is made alkaline, and $Co(II)Y^{2-}$ is oxidized to $Co(III)Y^{-}$ with hydrogen peroxide. On adding KCN, only the Ni reacts and is removed from its complexonate. The amount of EDTA thus set free in amount equivalent to the Ni present is then titrated with a standard solution of Mg using Erio T. However, in view of the intense colour of $Co(III)Y^{-}$, this method is limited to small amounts of Co. Replacing EDTA by DCTA is of advantage [55-114].

The resolution of another pair of elements, Ni and Cu, which frequently occur together, is possible if the Cu is masked by thiosulphate. Cheng [58-63] introduced this method of masking. PAN serves for the Ni titration, but the end-point is long drawn out, even in a solution warmed to 50°C. Kalinichenko [58-16] therefore titrates at pH ∼8 with murexide as indicator. Kitagawa [62-45] exploits the more selective titration in an acidic medium and avoids the trailing end-point by back-titrating with In using PAN as indicator. Kinnunen [57-137] uses thiourea to mask Cu and back-titrates with Th with xylenol orange as indicator.

The masking of iron with pyrophosphate is possible [56-98; 58-38]. The titration of Ni is not interfered with if other metals are masked by fluorides [57-74; 61-112].

The complexometric titration of Ni has been employed practically for its determination in electroplating baths [54-11; 57-62; 61-102], in accumulators [62-10], ferrites [61-70; 62-45], manganese catalysts [57-47], and alloys [62-86], especially those of Cu [58-16], Fe [59-31], Alnico [58-42], and Al–Ti–Ni [59-9], by procedures in which ion-exchange resins are often used. After a previous separation Ni has been determined in Co and its salts [55-11; 56-87], and after precipitation with DMG in steel [52-42] and Nb alloys [62-85].

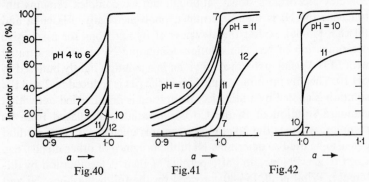

Figure 40. Colour transition of murexide (CIII) during the titration of Ni at different pH values.

Figure 41. Colour transition of murexide (CIII) during the titration of copper at various pH values.

Figure 42. Colour transitions of Erio T (LIX) during the titration of Zn at various pH values.

(*Attention is drawn to the fact that the scale of the abscissae in these graphs is greatly enlarged as compared with those of Figures 32–8.*)

29.1. Direct titration of nickel using murexide

REAGENTS: Standard 0·01M EDTA solution; murexide; approx. 1M NH_4Cl; conc. NH_4OH; masking agents if required.

PROCEDURE: The acidic test solution, which should not contain more than about 15 mg of Ni per 100 ml, is neutralized and indicator added, together with a masking agent if this is necessary. 10 ml of NH_4Cl is then added to the approximately neutral solution. If the pH is still below 7 the indicator will have an orange yellow colour (due to NiH_4D^+) and ammonia must be added dropwise until the colour changes to yellow (due to NiH_2D^-). The titration is started and continued until the end-point is near. Should the colour revert to orange because of a fall in pH, a few extra drops of ammonia are added and the titration is continued. Just before the end-point the solution is made strongly ammoniacal (by the addition of 10 ml of conc. ammonia) and the titration is continued to the brilliant colour change from yellow to bluish-violet.

REMARKS: Sufficient data are available to present a quantitative picture of what is going on. Figure 40 results from combining the curves of Figs. 7 and 27. It shows that a really good colour change is obtainable throughout the entire pH range from 7 to 12, though it will be sharpest and most brilliant in strongly alkaline solution. However, it is not feasible to bring the solution to pH 10 or 12 before the start of the titration, for this would induce a copious turbidity, since nickel hydroxide or a basic salt would separate out. These considerations determine the details of the forgoing recipe, in which the titration is started in a neutral solution which is only made strongly alkaline just before the end-point. It is very convenient that murexide can be used in the presence of nickel as a pH indicator to adjust the solution to the point of neutrality. However, the murexide must not be introduced directly into the initially strongly acidic solution, for it would then be quickly broken down by hydrolysis.

29.2. Determination of nickel by back-titration (after precipitation with dimethylglyoxime)

REAGENTS: Standard $0.01M$ EDTA solution; $0.01M$ $MgSO_4$ or $ZnSO_4$; Erio T; approx. 2M HCl; approx. 2M NaOH; methyl red; pH 10 buffer.

PROCEDURE: The precipitate of nickel *bis*-dimethylglyoximate is taken up in the minimum amount of hot HCl. A small excess of EDTA is added and the solution neutralized with NaOH using methyl red as the indicator. After dilution to give a nickel concentration less than 5×10^{-3}, 2 ml of buffer per 100 ml of test solution is added and back-titration is carried out with Mg or Zn solution to the colour change from blue to red. Over-titrated solutions have to be discarded.

REMARKS: This method works solely because the reaction between NiY^{2-} and the dyestuff indicator only proceeds very slowly provided EDTA is present in excess. As soon as the end-point is reached pNi decreases abruptly and the complex between Ni and Erio T forms: this is a complex that cannot subsequently be broken down.

Sharper end-points than those obtained in back-titrations with $MgSO_4$ or $ZnSO_4$ can be achieved with a standard solution of $MnSO_4$ [55-4].

30. PLATINUM METALS

There are not many data concerning EDTA complexes of the platinum metals. The definite existence of an EDTA complex of Ir has been established [56-52], and this has been used for its spectroscopic determination. The complexing action between Pd and EDTA forms the basis for its titration [55-1; 55-59; 58-81; 60-186; 62-78]. The stability of the complex [55-60], which has the composition PdY^{2-} within the pH range 4–9, was determined by potential measurements as $\log K = 18 \cdot 5 \pm 0 \cdot 6$ (25°C; $\mu = 0 \cdot 2$). Spectrophotometric evidence for a complex between Rh and EDTA has been published [58-78] and the compound has been isolated, but nothing further has been reported for a long time about its use in analysis. The existence of complexes of Ru and Pt may be inferred from the way in which they interfere in the titration of Pd.

Flaschka [53-21] uses the exchange reaction

$$Ni(CN)_4^{2-} + Pd^{2+} \rightleftharpoons Ni^{2+} + Pd(CN)_4^{2-}$$

for determining micro-amounts of Pd. The amount of Ni set free is titrated in ammoniacal solution using murexide. An analogous reaction is used for the indirect determination of silver (cf. Section 32), where the exchange proceeds almost instantaneously. With Pd, however, the reaction velocity is low in an ammoniacal medium. The mixture is therefore allowed to stand for some time or the reaction is speeded up by gentle warming. Alternatively, the solution is acidified (HCN!) and then immediately basified by adding ammonia. In an acidic medium the exchange is virtually instantaneous.

Kinnunen and Merikanto [55-1] have used the method for the analysis of alloys, but did not find the colour transition of murexide particularly sharp. They therefore recommend either a photometric titration or the determination of the liberated Ni by adding EDTA and determining the excess by back-titrating with a standard solution of Mn. In this case it is necessary to work fast, as otherwise there is the danger of the indicator (Erio T) becoming 'blocked' by the Ni. To avoid this a recommendation is made in a later paper to cool the solution down to 5°C before adding the indicator and then to titrate immediately with the Mn. The colour transition is then appreciably sharper.

Whatever modification of the tetracyanonickelate procedure is adopted, Ru, Os, Rh, and Ir do not interfere. Pt does interfere, since it also reacts with tetracyanonickel(II). So does gold; whether completely or partially is not known; in any case, no procedure for determining Pt in this way has been reported.

The first method for a complexometric determination of Pd based on the formation of a Pd–EDTA complex was published by Mac-Nevin and Kriege [55-59]: excess EDTA is back-titrated with Zn using Erio T. Since Pd forms an extraordinarily stable ammine complex, the usual ammonia–ammonium chloride buffer cannot be used in the back-titration and the pH must be adjusted to 10 with caustic soda. If the titration is carried out in the presence of chloride ions, Pt does not interfere, since the chloro-complexes of Pt(II) and Pt(IV) are sufficiently strong to prevent the formation of an EDTA complex. Rh interferes and can produce values for Pd up to 50% low! – but nothing is known about the cause of this remarkable interference. If the titration is completed at room temperature and without delay the high values obtained in the presence of Ru and Ir lie within acceptable limits. With amounts of Ru and Ir amounting to about 30% of the Pd present the results are 2–3% high.

A procedure involving back-titration in a strongly alkaline solution (pH 10·5–14) with a standard solution of Ca originates from Tserkasevitch [60-186]. To mark the end-point the substance described as 'H-indicator' is used; this appears to be identical with T. S. West's 'calcichrome'. According to *Chem. Abstracts*, the accuracy is given as $\pm 15\%$ for 1–10 mg of Pd, but this must be an error. The back-titration in an alkaline medium can also be followed potentiometrically with Hg [62-120].

Titration of Pd is also possible in an acidic medium. Once again this is a back-titration procedure, for as yet no metallochromic indicator for Pd is known. Kinnunen and Merikanto describe two methods. The first [55-1] depends on the back-titration of excess EDTA with Bi at pH 1–2 using pyrocatechol violet. However, this procedure is not described so precisely as the tetracyanonickelate method, and it suffers, moreover, from the disadvantage that chloride ions must be absent; on the other hand, it has the advantage that Pt does not interfere. The second method [58-81] employs back-titration with a standard solution of Th at pH 3, or with a Tl(III) solution at pH 4–5 using xylenol orange as indicator.

Yurist and Tyukova [62-78] describe back-titration with Zn at pH 6 with xylenol orange as indicator, and they discuss the resolution of mixtures of Pd and Pb. An aliquot portion of the test-solution is titrated at pH 6, as described above, thus giving the sum Pb + Pd. The Pb is determined in a second aliquot portion of the ammoniacal solution containing tartrate: here acid chrome blue or methylthymol blue serve as the indicator. Pd, which is masked as its ammine complex in the second titration, is obtained by difference.

Practical applications of these various procedures extend to the determination of Pd in Pt–Pd alloys [58-81], mixtures of Pt and Pd salts [55-1], pharmaceuticals [60-186], and Pd–Pb pastes [62-78].

30.1. Indirect determination of palladium

REAGENTS: 0·01M EDTA; conc. ammonia; murexide–NaCl indicator powder; solid $K_2Ni(CN)_4$.

PROCEDURE: The test-solution containing up to 20 mg of palladium in its bivalent form is made approximately neutral and then treated with 10 ml conc. ammonia and 0·2–0·25 g of dipotassium tetracyanonickel(II), warmed for a short time and diluted to 150–200 ml. After adding murexide indicator it is titrated with EDTA to the colour change from yellow to violet.

CALCULATION: Since one gram atom of Pd liberates one gram atom of Ni we have: Pd (mg) = 106·4vM, where v (ml) is the volume and M the molarity of the EDTA solution.

REMARKS: It is essential for the palladium to be present as Pd(II), and if necessary this can be achieved by boiling with hydrochloric acid. Solutions of alloys in aqua regia must be fumed down repeatedly with conc. HCl. Of the other noble metals Ir, Os, Rh, and Ru do not interfere: but Ag, Au, and Pt must be absent.

31. COPPER

The complexometric determination of copper presents scarcely any difficulties. It can be effected equally well by direct or back-titration and, depending on the conditions and the indicator, in the entire pH range from about 5 to 10. The titration in ammoniacal solution with murexide as indicator, originally described by Schwarzenbach [48-7] and adapted to the micro-scale by Flaschka [52-22], is still favoured nowadays in many practical applications in view of its simplicity and the sharpness of the colour change. Even the minutest amounts of Cu can be titrated in the spot obtained after paper chromatography [56-8]. The only point to which attention must be given is to avoid adding too much ammonia, otherwise the end-point becomes unsharp or missing altogether. This is true not only for murexide but when many other indicators are used as well, and it is a consequence of the fact that the stability of the ammine complexes of Cu is considerable when compared with those of the relevant indicator complexes. To avoid a damaging excess it is generally simplest to add ammonia dropwise until the initial precipitate of $Cu(OH)_2$ has just redissolved to give the deep blue tetrammine: when this has taken place the solution is diluted strongly with water. This dilution is necessary not only to reduce the concentration of ammonia but it also has the objective of decreasing the concentration of Cu as well so as to prevent the colour change due to the indicator being swamped by the relatively intense blue colour of the Cu–EDTA complex.

Besides murexide, other indicators can be used for the titration of Cu in ammoniacal solution, such as pyrocatechol violet [54-81], eriochrome cyanine [57-72], nitrosochromotropic acid [61-19] best admixed with xylene cyanole FF [62-137], chromeazurol S [55-27], fast sulphon black F [57-42; 58-3], etc.

In view of the greater selectivity, titration in an acidic medium is often preferable. Here the indicator PAN introduced by Cheng [55-18] certainly has the sharpest colour transition and is the most often used. PAN can also be used in ammoniacal solution, but the colour change there is not so sharp by far. A few points should be watched when using PAN in an acidic medium. Both the indicator and its complex with Cu (and with other metals too) is very sparingly soluble in water. The end-point is rather protracted, since the titrant only reacts slowly with the Cu–PAN complex, which is probably present as a colloid. It may also be that the complex reacts slowly *per se*. Whatever the cause, some speeding up is desirable from the practical point of view, and this can be achieved either by titrating in a hot solution or by working with a solution containing about 50% of ethanol, methanol, or acetone. Other organic solvents would certainly have a similar effect.

Because of the increased sharpness of the colour transition, Cu is the favoured metal for back-titrations in conjunction with PAN [56-43], although it is often advantageous to use a standard solution containing Zn as well as Cu for the back-titration in order to limit the blue colour produced by the Cu itself.

Since the Cu-complex of PAN is more stable than the complexes with other metals, the system Cu–PAN can be exploited to indicate the end-point in many other titrations [56-44]. To give a selection of other indicators for the titration of Cu in acid solution the following may be noted: xylenol orange [56-39], pyrocatechol violet [54-81], certain *o*-nitrophenols [56-57], violuric acid [63-56], chromeazurol S [55-27], PAR [57-5], SNAZOXS [60-67], azoxine [61-112], glycine-thymol blue [62-23], and in 75% alcoholic solution *N,N'*-bis(2-hydroxyethyl)-dithio-oxamide [62-4]. In addition, 7-(2-pyridylazo)-8-hydroxyquinoline [63-70] at the optimum pH of 2·9.

In addition to the end-point indication with metallochromic indi-cators there are also fluorescence indicators such as fluorescein-complexone [59-22] or calcein W [59-11]: in the latter case the use of trien as the titrant produces a significant increase in selectivity. Redox indicators have also been reported. Variamine blue [55-17] can be used, and here the addition of some KCNS increases the sharpness of the end-point [56-109], especially with large amounts of Cu. Another combination is KCNS and *p*-anisidine [61-115].

Back-titrations, like that with Cu using PAN [56-43], often have the advantage that warming or the addition of organic solvents is unnecessary. Kinnunen [57-137] recommends Th or Tl(III) for back-titrations with xylenol orange, whereas Takamoto [55-102] back-titrates with Co in an acetone–water mixture in the presence of thio-cyanate. Cameron *et al.* [61-136] use the combination KCNS + tri-phenylmethylarsonium chloride + chloroform for the back-titration with Co using an extractive end-point.

A study of the electrometric properties of the system Cu(I)/Cu(II) in the presence and absence of EDTA has been carried out by Belcher *et al.* [55-104], according to whom a direct potentiometric titration is possible with a Pt-electrode. Reilley and co-workers titrate using the mercury-drop electrode [58-11; 59-113; 62-29], with which consecutive titrations present no difficulties. Titrations with two polarized mercury electrodes can be used for the determination of Cu in serum and permit the consecutive titration Cu–Zn for which penten is used as the titrant, as this does not react with alkaline earths or Mg [62-99]. Amperometric titration is also feasible [62-33] and makes possible the consecutive titrations Cu–Ni or Cu–Zn–Ca. Papers have also been published on radiometric [60-133], high-frequency [55-41], conductimetric [57-34], coulometric [56-77; 62-25], and thermometric titrations [57-97; 63-67].

Photometric titrations are specially favourable for Cu, since it can easily be used in a self-indicating system [53-37; 62-1]: titrations with indicators can also be used, e.g. with pyrocatechol violet [59-21], which permits of the consecutive titration Bi–Cu, or murexide [53-45; 54-43; 54-69; 54-70]. As a self-indicating system Cu has often been used as a slope-indicator (Part One, Section 5.2.1.2) in the titration of other metals [57-13; 57-14; 61-113; 62-26]. Successive titrations of, e.g., Fe–Cu [53-39] or Cu–Bi [54-51] are based on this. The use of other complexing agents, especially trien [57-87; 57-88], in the photometric determination of Cu leads to increased selectivity. With trien the determination of Cu becomes practically specific. Photometric indication of the end-point enables the determination to be automated [59-120; 60-78].

From the position of Cu in the list of stability constants of metal–EDTA complexes it is not difficult to infer that the titration of Cu will be interfered with by many other metals and that many other metals will be co-titrated with Cu. Interferences due to copper itself are not due solely to co-titration, for besides this the presence of even traces of Cu can cause 'blocking' of a number of indicators. Apart from the use of masking agents, such blocking can be avoided, at least with Erio T, by working in solutions containing methanol, ethanol, or propanol [62-138].

Cu can be masked easily and certainly with KCN, and full information about this has been available for a long time. Unfortunately many other metals commonly associated with Cu are masked simultaneously. De-masking with formaldehyde improves the situation so far as Zn and Cd are concerned (q.v.). If Erio T is to be used for the titration of a metal accompanying Cu it must not be added until the masking of Cu by cyanide is complete, otherwise oxidative decomposition of the indicator takes place to a yellow compound [59-146]; alternatively, a reducing agent (such as ascorbic acid) must be present to prevent the interference. Other indicators also would probably be attacked in a similar way. Cheng [58-63] and subsequently Karoly [59-49] describe the use of thiosulphate as a relatively selective masking agent for Cu in weakly acid solution when Zn is to be titrated. However, too large an excess of masking agent must be avoided, otherwise Zn becomes masked as well. It is not difficult to measure the correct amount to be added, as the progress of the masking reaction can be followed by the bleaching of the blue Cu colour. Lystsova [61-109] showed that Cu can also be masked by thiosulphate in the titration of Ni. Körös [57-39] described the masking of Cu by adding KI or KSCN and ascorbic acid in the determination of Zn. Cu is reduced to Cu(I) in this process and precipitated as the corresponding sparingly soluble compound.

The determination of Cu and other metals in one and the same solution has been described by many authors. The procedures use instrumental end-point indication, masking, and de-masking, or both in turn. Reference has been made above to the systems Cu–Zn, Ni, or Co. Přibil *et al.* [55-89] describe the determination of Cu in the presence of Fe, Ni, Co, and Mn using DCTA. Mixtures of Hg and Cu can be analysed [57-65] if the sum of the two metals is first determined by back-titration with Cu using murexide as indicator; then Hg is masked with KI, and the EDTA set free is determined by a further titration with Cu. Numerous binary and a few ternary mixtures of copper with other metals have been analysed by Khalifa *et al.* using potentiometric back-titration with Hg and calling on the devices of masking and working with aliquot portions [58-57; 59-124; 59-125; 60-163; 61-14]. An interesting procedure is that of Aoki [60-159] who determined Cu–Zn in a simple consecutive titration. The Cu is first titrated at pH 5 using pyrocatechol violet, and then Zn is titrated at pH 9 with the same indicator. However, the procedure only works if less Zn than Cu is present.

The practical applications of the Cu determination cover a multiplicity of materials such as coins [55-76], alloys in general [54-67], especially those of Al [55-78] or W [59-1]. Ores have been analysed for Cu by precipitating with xanthate and titrating after taking up the precipitate [62-66]. Cu has also been determined in linoleates [59-89], in Co and its salts [56-87], copper–ethylenediamine solutions [58-106], and electrometrically with two polarized mercury electrodes in serum [62-99]. There are a particularly large number of papers dealing with the determination of Cu in electroplating baths for copper and brass [56-51; 57-62; 59-49; 61-145]. Any cyanide that may be present is removed by fuming down with acid and H_2O_2. Zn is determined after de-masking with formaldehyde, and after the total concentration has been obtained Cu is determined by difference.

31.1. Direct titration of copper with murexide as indicator

REAGENTS: Standard solution of 0·01M EDTA; murexide; approx. 1M NH_4OH; approx. 1M NH_4Cl.

PROCEDURE: The test-solution should not contain more than about 20 mg of Cu per 100 ml. Acid solutions are neutralized with ammonia and only a slight excess is used to give a pH of about 8. With only weakly acidic solutions about 10 ml of NH_4Cl is added as well for every 100 ml of solution to ensure that the ammonia does not produce too high a pH. Indicator is then added and the titration with EDTA carried out to the colour change from yellow to violet.

REMARKS: The most favourable conditions for the titration can be deduced from Fig. 41, which results from combining Figs. 8 and 28. In complete contrast to the corresponding titration of Ni it is better with Cu to work at low pH values. This arises from the fact that ammine complexes of copper are appreciably more stable than those of nickel, so that the copper–murexide complex can be broken down again by too much ammonia (cf. Fig. 3). High concentrations of ammonia must therefore be avoided. Some ammonia must, of course, be present to keep the metal in solution. The danger of precipitating basic salts, e.g. $3Cu(OH)_2.CuCl_2$, is further diminished by the presence of much NH_4^+, which keeps the pH down. Naturally ammonia is a poor buffer for the region pH 7–8; but triethanolamine cannot be used, since its complexes with Cu are appreciably more stable than those with ammonia.

31.2. Titration of copper using PAN as indicator

REAGENTS: Standard 0·01M EDTA solution; 1% PAN in methanol or ethanol; acetate buffer pH 5; methanol or ethanol.

PROCEDURE: The test-solution containing up to about 20–30 mg of Cu per 100 ml is, if necessary, pre-neutralized and treated with 5 ml of buffer and 3–5 drops of indicator solution. The solution is then heated to boiling and titrated with EDTA to a very sharp colour change from deep violet to canary yellow. Instead of heating, the solution can be made up to 30–50% in an alcohol and titrated at room temperature.

REMARKS: If the titration is not carried out in an alcoholic solution a fine film of PAN often forms on the surface or adheres to the glass surface of the titration flask. Simply to rinse the vessel does not suffice to remove the dyestuff and to avoid an unwanted colour effect when the flask is used next time dissolution in acid or alcohol is necessary.

31.3. Titration of various metals using the system CuY–PAN

REAGENTS: Standard solution of 0·01M EDTA; solution of Cu–EDTA complex (Part Two, Section 1.7); PAN; acetate buffer pH 5.

PROCEDURE: The test solution is neutralized if necessary and then treated with 5 ml of acetate buffer per 100 ml, 3–5 drops of the CuY solution, and a few drops of PAN solution. The solution is heated to boiling and titrated with EDTA until the colour changes from deep violet to canary yellow.

REMARKS: This general procedure is applicable to a large number of metals, e.g. Pb, Hg, Zn, Cd, Ni, and Co; vanadium requires the addition of ascorbic acid. Al can be titrated in strong acetic acid; likewise Fe or the sum of Al + Fe. The indicator system (CuY–PAN) can also be used for titrations in ammoniacal solution; however, the colour transition is not so sharp. Even Mg and the alkaline earths can be titrated. The advantage over many other indicators lies in the fact that no 'blocking' of the indicator takes place. It is also possible to titrate in the cold provided alcohol has been added, not, however, in the titration of Al.

32. SILVER

The stability of the Ag–EDTA complex (log $K = 7 \cdot 2$) [53-46] is too small to permit of a direct titration with a complex-forming indicator. However, silver can be titrated potentiometrically with a silver wire as an electrode [62-17]. The titration is carried out in a medium buffered with borate to pH 9. Although the potential jump is quite well developed, the method has little practical significance, since all metals that form more stable complexes are co-titrated.

Ag can be determined indirectly through a reaction with an amalgam [58-13]. Aseni made use of this possibility [61-154] to determine Ag in the mixtures Ag–Bi and Ag–Cd using a Bi or Cd amalgam respectively. An amount of Bi (or Cd) equivalent to the Ag leaves the amalgam and enters the solution and is then titrated with EDTA (cf. Part One, Section 6.7).

Another route to an indirect titration is through an exchange reaction with tetracyanonickelate, $Ni(CN)_4^{2-}$, whereby an equivalent amount of Ni is set free which can be titrated with EDTA using murexide [52-28] or thymolphthalexone and dimethyl yellow [62-123] as indicators. The exchange also takes place quantitatively with sparingly soluble silver salts, so that precipitates obtained during the course of a separation can be submitted to titration without more ado [52-28]. The titration of silver has thus achieved considerable importance in that it makes all those anions that can be precipitated by Ag^+ accessible to a complexometric titration (cf. Section 54 on the halogens). The method has been extended to the micro-scale [52-26] and yields excellent results for a few mg of Ag, especially when the end-point is obtained photometrically.

Practical use is made of the exchange method in the analysis of $AgNO_3$ [62-123], coins containing silver [55-76], and photographic materials [57-24].

32.1. Indirect determination of silver

REAGENTS: Solid potassium nickel cyanide, $K_2Ni(CN)_4$ (for the preparation of which cf. Flaschka [52-26]); standard solution of 0·01M EDTA; murexide; conc. NH_4OH; approx. 1M NH_4Cl.

PROCEDURE: The solution to be titrated or the solid silver compound containing about 100 mg of Ag is treated with 10 ml conc. ammonia, 10 ml of NH_4Cl, and 0·2 g of tetracyanonickelate, whereupon everything should go into solution. If necessary (for instance, with AgI) this can be assisted by gentle warming. Now dilute to 100–200 ml, add indicator, and titrate to the colour change from yellow to violet.

CALCULATION: Since according to the exchange reaction (whose equation is given below), two gram-atoms of Ag are required to liberate one gram-atom of titratable Ni, the equivalent weight is equal to twice the atomic weight.

REMARKS: The overall stability constant of the complex ion $Ni(CN)_4^{2-}$ is 10^{27} and that of $Ag(CN)_2^-$ is 10^{21} [A-5]. This gives a value of 10^{15} for the equilibrium constant of the reaction

$$2Ag^+ + Ni(CN)_4^{2-} \rightleftharpoons 2Ag(CN)_2^- + Ni^{2+}$$

which implies that the reaction would go completely to the right. If the nickel complex is present in excess the concentration of free silver ions drops to about 10^{-10}, which is sufficiently low to ensure the solution of sparingly soluble salts such as AgCl, AgBr, and AgSCN. Moreover, silver cyanide cannot separate out. There is, however, the danger of forming sparingly soluble nickel cyanide by the reaction

$$Ni^{2+} + Ni(CN)_4^{2-} \longrightarrow Ni[Ni(CN)_4]$$

To avoid this precipitation sufficient ammonia is added to transform the cyanide-free nickel into its ammine complexes. Furthermore, any silver that is not bound to CN^- will also occur as the ammine complex $Ag(NH_3)_2^+$. If then the ammonia concentration is about 1M, which is roughly the situation if the above working instructions have been followed, we obtain the following value for the actual exchange equilibria

$$[Ni]' \cdot [Ag(CN)_2]'^2 / [Ag]'^2 \cdot [Ni(CN)_4] = 2·6 \times 10^9$$

Here $[Ni]'$ and $[Ag]'$ signify the total concentrations of nickel and silver respectively that are not bound to cyanide. Thus:

$$[Ni]' = \sum_{j=0}^{6} [Ni(NH_3)_j] = 2·6 \times 10^8 [Ni]$$

$$[Ag]' = \sum_{j=0}^{2} [Ag(NH_3)_j] = 10^7 [Ag]$$

Hence, in the presence of excess tetracyanonickelate the value of [Ag]' drops to about 10^{-8} and the concentration of free silver ions, i.e. [Ag$^+$], to about 10^{-15}, and this is sufficiently low to bring even the very sparingly soluble AgI into solution.

33. GOLD

Kinnunen and Merikanto [55-1] have reported on the indirect complexometric determination of gold. Alloys are taken up in aqua regia, and after filtering off any AgCl that may be present the gold is extracted repeatedly with ether and the combined ether extracts evaporated down. The aqueous solution that remains behind is neutralized and the gold determined as described under Pd or Ag by reaction with Ni(CN)$_4^{2-}$ and titration by EDTA of the Ni thereby set free (cf. Section 32.1 above).

34. ZINC

The complexometric determination of zinc presents no difficulties, and was described by Schwarzenbach and Biedermann [48-5] quite early on. This procedure involves titration in an ammoniacal solution buffered to pH 10 with Erio T as indicator, and an extremely sharp end-point results. Micro- [52-22] and ultramicro-determinations [52-37; 57-134] are easily possible. Titrations can even be carried out on the spot obtained after separation by paper chromatography [55-5; 56-8; 59-40]. Almost all the dyestuffs related to Erio T that have been reported as indicators for Mg can be used just as well for the titration of Zn. 8-Hydroxyquinoline-5-sulphonic acid [58-19] can serve as a fluorescence indicator. Additional indicators for the titration in alkaline solution include pyrocatechol violet [54-78; 56-41; 60-159], zincon [55-2; 58-4], naphthol violet [57-81], methylthymol blue [58-34], and murexide [53-8; 54-46; 60-132]. One advantage of the last named indicators as compared with Erio T and most of its analogues is that traces of heavy metals do not cause any 'blocking' of the indicator. When using zincon the concentration of ammonia is critical, and it must be kept small. With murexide ammonia is best kept away altogether; the titration is carried out at pH 8·0–8·5 or, alternatively, in a solution that has been adjusted to pH 13 by adding ethanolamine [63-20]. If the titration is carried out at a lower pH, such as about 6·8 for Erio T [53-33], appropriately in a malic acid buffer, or pH 8 for murexide, the interferences are fewer and, above all, quite considerable amounts of Mg can be tolerated.

From the standpoint of selectivity titration in an acidic medium is to be preferred. A range of suitable indicators are available, such as PAN [55-18] or Cu–PAN [56-44], xylenol orange [57-79; 59-26; 61-108; 63-14; 63-21], and methylthymol blue [58-34; 61-108], which is used in a solution buffered with acetate or urotropine. Dithizone in a 50% alcoholic medium at pH 4–5 gives a particularly sharp end-point [55-98]. The end-point in a solution buffered with acetate to pH 4–5 can also be used for a direct visual redox indication. In such a solution containing ferri- and ferro-cyanide, benzidine [56-99], 3,3'-dimethylnaphthidine [53-34; 53-35; 55-67], or variamine blue [57-117] serve as redox indicators. In view of the slowness of the reaction at the end-point in the direct titration, back-titration with a solution of Zn is advisable with this type of end-point indication. Back-titration is also possible with a standard solution of Co in an acetone–water mixture containing thiocyanate [55-102].

Instrumental indication of the end-point has been used extensively in the titration of Zn. Amperometric determination is possible at the potential of zinc itself [51-16] or with the 'complexone wave' [57-101; 62-33], and has been described for the determination of Zn in oils [53-48] and alloys [56-3]. The consecutive titrations Zn–Mg, Zn–Ca, and Cu–Zn–Ca are possible [62-33]. Electrometric indication with two polarized electrodes with tetraethylenepentamine as titrant has also been reported [62-99].

Potentiometric titrations have been carried out via a back-titration with Fe(III) at pH 5–6 [51-16] or in a direct titration with a static mercury-drop electrode [58-12]. In the latter case the determination can include microgram quantities of zinc [57-8]. High-frequency [54-2; 55-41], radiometric [60-133], and thermometric titrations [57-97; 62-44] present further possibilities. Photometric titrations deserve special mention, whether they are self-indicating in the ultraviolet region [54-14] or involved indicators (generally Erio T) [53-45; 54-69; 56-25; 56-73]; they can also be automated [59-120]. For the photometric determination of Zn in the presence of much Cd see under cadmium (Section 35).

A glance at the table of stability constants of EDTA complexes shows quickly that the titration of Zn will be interfered with by very many other metal ions. However, there are fortunately several possible ways in which the determination can be made selective, especially in the presence of those metals, including iron, which frequently accompany zinc in natural and artificial materials [56-47]. The demasking of zinc cyanide complexes is of particular importance in this connection. Formaldehyde [52-6; 53-27] and chloral hydrate [53-10] are available for this. In ammoniacal solution formaldehyde reacts with free cyanide and with cyanide bound to Zn (and Cd!) with the formation of the nitrile of glycollic acid so that the zinc that was originally masked against indicator and EDTA is set free and can be titrated. Other metals that form complex cyanides, such as Fe(II), Fe(III), Hg, Cu, Ni, and Co, which would otherwise be titrated complexometrically, are only slowly or not at all broken down, and so remain masked. However, traces of them, especially Cu, can be set free and then cause difficulties, not because of their being co-titrated, for the amounts concerned are too small, but by blocking the indicator action of the Erio T which is customarily used. There are various precautions that must therefore be taken (cf. the working instructions) to avoid this. Replacing Erio T by an indicator that is not susceptible to blocking, e.g. by pyrocatechol violet or murexide [60-132], is advantageous, but the colour transition is, generally speaking, not so sharp. When working with formaldehyde as a de-masking agent it is also necessary to bear in mind that the aldehyde reacts with ammonia to form urotropine, whereby a decrease in pH can be produced.

Janousek [59-26] reported an interesting de-masking of zinc cyanide by acid. At pH 6 the cyanide complex is so strongly dissociated that Zn reacts with xylenol orange and with EDTA, while Cu, Ni, and Co remain masked. The evolution of hydrogen cyanide certainly proceeds slowly at the pH specified and is moderate in amount; however, it is necessary to work in a fume-chamber.

The combination Cu–Zn is of especial importance in practice, for apart from the cyanide method already mentioned, there are other possibilities. Aoki [60-159] describes a consecutive titration in which the Cu is first titrated at pH 5·1–5·4 using pyrocatechol violet; then the pH is increased to 9 and Zn is determined using the same indicator. The method fails when more Zn is present than Cu, for then Zn is partially co-titrated with Cu. According to Cheng [58-63], another possibility is to mask Cu with thiosulphate. However, the amount of masking agent must be measured out exactly (which can be recognized by the disappearance of the blue colour of the aquo-cupric complex), since otherwise the Zn is complexed as well, and the colour transition of the PAN indicator starts to be drawn out. The procedure also works for micro-determinations [63-21], for which xylenol orange serves as the indicator. The sum Cu + Zn can be obtained by titration with murexide as indicator and the amount of

Cu calculated by difference. Copper can also be masked by reduction with ascorbic acid or KI (in the presence of thiocyanate) [57-39].

An increase in the selectivity of the Zn determination can also be achieved by replacing EDTA by a more selective titrant. For this purpose polyamines, which will only react with a limited number of cations, are pre-eminently suitable. The use of such titrants is possible now that a selective indicator has been found in the form of zincon. Reilley *et al.* [57-9] have reported on the use of triethylene-tetramine. Johnson [62-99] has carried out titrations with tetra-ethylenepentammine electrometrically with two polarized electrodes.

EGTA has acquired a certain importance in connection with the combination Cd–Zn, since the complex of this complexone with Cd is appreciably more stable than that with Zn. This permits of the determination of microgram amounts of both metals to be based on a photometric titration curve using the principle of 'slope indication' (Part One, Section 5.2.1.2) with murexide as indicator [63-11].

Fabregas *et al.* [62-98] exploit the difference in stability constants in the following way. In the presence of much sulphate ion Pb can be displaced from its complex with EGTA by Cd but not by Zn. The lead sulphate so formed is filtered off, and the Zn in the filtrate is titrated with EDTA using Erio T as indicator. The method works for Cd : Zn ratios up to about 20 : 1. However, Zn can be titrated visually with xylenol orange as indicator in the presence of up to a 300-fold excess of Cd, provided the latter is masked (following Flaschka and Butscher [63-81]) by a high concentration of iodide. According to un-published researches by the same authors, a several thousand-fold excess of Cd can be tolerated if the titration is carried out photo-metrically with DTPA as titrant.

In addition to cyanide, which has already been dealt with fully, zinc can also be masked by BAL [54-79] and with unithiol [60-59; 60-60].

As masking agents for other metals during the titration of Zn, mention should be made of acetylacetone [63-14] and triethanol-amine [54-76] for Al, as well as fluoride for Al, Ca, and Mg [54-74], since these are important in analytical practice.

Since Zn is easy to titrate and gives a sharp end-point, it is often preferred as a primary standard for the standardization of solutions of complexones. The starting-point is pure zinc metal or a zinc salt; here the zinc–pyridine–thiocyanate complex [56-83] is particularly suitable in view of its high equivalent weight. Zinc is also favoured as a reagent for the back-titration in the determination of other metals.

Furthermore, because of the sharpness of its colour transitions with certain indicators, it has become well established for displacement indication in the system ZnY–zincon [58-4] in alkaline, and the system ZnY–dithizone in acid and neutral media [61-61].

The complexometric determination of zinc has been used for the analysis of the most varied sorts of materials. It has been determined in metallurgical products and alloys [52-6; 53-6; 53-40; 56-3], and especially in those of aluminium [53-6; 54-89; 56-87; 61-122; 61-175] as well as in In–Zn–Sn [63-19], brass [55-3; 57-39; 59-49], bronze [55-3; 57-39], copper alloys in general [54-67], U–Zn [54-46], and ferrites [61-70; 61-104]. Zinc has been determined in ores and concentrates [55-77; 58-55], water [58-88], steam condensate [54-71], zinc salts in general [61-108], spinning baths [52-44], oils [53-48], zinc stearate [53-8], and after ashing in animal [55-9] and plant [58-19] tissues. Zinc has been titrated in cyanide plating baths after de-masking with formaldehyde [54-24; 54-52; 56-56; 58-21; 59-120]. In the paint and lacquer industries zinc titrations have been used in the analyses of pigments [55-44], especially ZnO [52-35; 56-67] and lithopone [57-120], as well as in paint-driers [61-143] and colophony [54-98]. There are innumerable procedures for the analysis of pharmaceutical preparations [53-12; 57-113; 60-98; 60-173; 61-6], in particular insulin [53-53; 54-82; 58-73]. The determination of metallic Zn in the presence of ZnO in zinc dust is interesting theoretically too. The sample is boiled with a solution of $HgCl_2$ at pH 6, whereupon only metallic Zn reacts and goes into solution. After filtration the mercury is removed by reduction in ammoniacal solution with hydroxylamine hydrochloride and the Zn is titrated using Erio T as indicator.

34.1. Titration of zinc and cadmium in the absence of Cu, Co, and Ni with Erio T as indicator

REAGENTS: Standard solution of 0·01M EDTA; Erio T; pH 10 buffer.

PROCEDURE: The test solution should not contain more than about 25 mg of Zn (or 50 mg of Cd) per 100 ml. Acidic solutions are first neutralized with NaOH, if necessary after a preliminary separation of tervalent metals with acetate. Now to every 100 ml is added 2 ml of buffer and then indicator: titration is carried on until the colour changes from red to blue. The last suggestion of a reddish hue disappears with the final drops of titrant.

REMARKS: Fig. 42 for the zinc titration is compounded of Figs. 9 and 25. From this it follows that the zinc titration can be carried out in the entire pH range from 7 to 10. It gives extraordinarily sharp colour changes, so that even a 0·001M EDTA solution can be used successfully. The solution should not be too cold (30–40°C). It is not feasible to work much below pH 7, for then the metal-free Erio T begins to occur as the red species H_2D^-. As shown in Fig. 42, the jump in pZn rapidly becomes small above pH 10. Actually the situation is not as bad as Fig. 42 leads one to expect, since really good colour transitions occur even in the presence of much NH_3. The reason for this is the formation of the mixed dyestuff complex $ZnD(NH_3)_x$.

Since the stability of the cadmium–Erio T complex has never been determined, no diagram corresponding to Fig. 42 can be drawn for the cadmium titration. It is known, however, that Cd can be titrated in the entire pH range from 7 to 10.

34.2. Titration of zinc and cadmium in the presence of Cu, Ni, and Co using Erio T as indicator

REAGENTS: Standard solution of 0·01M EDTA and 0·01M $MgSO_4$; Erio T; conc. ammonia; 1M KCN; 1M chloral hydrate (or 1M formaldehyde); 1M NH_4Cl.

PROCEDURE: The total content of heavy metals and alkaline earths is first determined in a preliminary titration as follows. The acidic test-solution is treated with excess EDTA (a ml); then for every 100 ml of solution 2 ml of NH_4Cl is added and the pH is brought to 10 (thymolphthalein should be blue), and after adding Erio T the mixture is titrated to a red colour (rapidly to avoid 'blocking') with the standard solution of $MgSO_4$, of which suppose b ml are required. The difference, $(a - b)$, gives the total concentration of polyvalent metals. To the titrated liquid is now added 2 ml of KCN solution, whereby Cu, Co, Ni, Zn, and Cd are transformed into their cyano-complexes and the equivalent amount of EDTA is set free; after adding more indicator if necessary this is finally titrated with $MgSO_4$. The titration value, c ml, is a measure of the content of heavy metals.

For the main titration sufficient of the test-solution is taken to have a total concentration of heavy metals between 1 and 5×10^{-4} gram-atoms: this is calculated from the value of c obtained in the pre-titration. This solution is neutralized with NaOH, which is added until the solution is on the point of precipitation: then 2 ml of KCN is added and the whole diluted to 100 ml. Had the pre-titration indicated the presence of alkaline earths ($a - b - c > 0$), Erio T is added and titration carried out with EDTA till the colour changes from red to blue. The zinc is next de-masked by adding 6 ml of chloral. One minute after adding the chloral the pH is brought down by adding 8 ml of NH_4Cl; a few drops more of ammonia are then added, and the red solution is titrated with EDTA until it turns sharply to blue, adding more Erio T if needs be. The concentration of the Zn (or Cd) is calculated from the titration value.

If formaldehyde is used for the de-masking only twice the theoretical amount is used (4 ml of 1M formaldehyde), and only a few seconds need elapse before starting the back-titration.

REMARKS: The cyano-complexes of Zn and Cd have overall stability constants of about 10^{19} [A-2; A-4; 54-103]. If these metals are not to react with Erio T their pM values must be at least 14 at pH 10 (cf. Fig. 25). This can only be achieved if the concentration of free cyanide ion in solution is at least 10^{-2}M. Under these conditions the concentrations of the various ions are $[Zn^{2+}] \sim 10^{-14}$, $[CN^-] \sim 10^{-2}$, and $[Zn(CN)^{2-}] \sim 10^{-3}$. Alkaline earths that may be present cannot be titrated free from any interference from Zn and Cd unless there is an excess concentration of cyanide of this order.

To effect the final de-masking of the two metals the cyanide concentration must be reduced to about 10^{-4}, whereupon the concentration of zinc rises to 10^{-6}, when, as shown in Fig. 25 for the value pZn = 6, the Erio T exists in the form of its zinc-complex, i.e. in the red form. However, at the same time the pH rises, for according to the equation

$$HCN + RCH=O \longrightarrow RCH(OH)CN$$

HCN disappears from the system, and this can only occur in consequence of the equilibrium

$$CN^- + H_2O \rightleftharpoons HCN + OH^-$$

being displaced to the right. The rise in pH can even bring about the precipitation of zinc or cadmium hydroxide. This is the reason for adding the ammonium chloride, which restores the solution to the pH region favourable for the titration.

The stability constant of $Ni(CN)_4^{2-}$ is about 10^{27} and that of $Cu(CN)_4^{3-}$ about 10^{28} [50-140]. With a concentration of free chloride ions of 10^{-2}M, pNi is about 22, and after the addition of aldehyde if we calculate with $[CN^-] = 10^{-4}$, pNi will have decreased to 14. Experience shows that with these low concentrations of Ni^{2+} the metal is unable to react with Erio T. However, pNi must not be allowed to fall too low, for even when pNi ~ 9 the nickel–dyestuff complex is irreversibly formed. We have therefore little room for manoeuvre, and in fact interference does occur if too much Ni or too much Cu is present. Furthermore, the addition of aldehyde must be carefully handled. A large excess of this de-masking agent leads to blocking of the indicator by Ni and Cu that have also been de-masked. This is almost always found to occur if the fully titrated solution is allowed to stand, even when the titration itself has been carried out without interferences. Chloral de-masks more slowly than does formaldehyde, and thus introduces less danger from blocking.

If only small amounts of Zn or Cd are to be determined in the presence of large amounts of other metals it is better to employ an extraction procedure [53-40; 54-38]. The sample solution is treated with thiourea (to mask Cu and Ni) and fluoride (to mask Fe) and large amounts of NH_4CNS; the thiocyanates of Zn and Cd are then extracted by hexone (methyl isobutyl ketone). After separation the organic phase is diluted with water using acetone to prevent the separation of two phases, and titration is carried out directly according to procedure 34.1.

Zinc can also be precipitated as ZnS after masking Cu and Ni with KCN. After being washed the precipitate is redissolved and the zinc determined by procedure 34.1 [54-89].

34.3. Titration of zinc using dimethylnaphthidine [53-34]

REAGENTS: Standard solution of 0·1M EDTA; 1% $K_4Fe(CN)_6$; 1% 3,3'-dimethylnaphthidine in acetic acid; approx. 1M Na-acetate; approx. 1M HCl.

PROCEDURE: The test solution should be about 0·01M with respect to Zn. Acidic solutions are neutralized with NaOH (methyl red test-paper). To every 100 ml are added a drop of ferrocyanide and 2–3 drops of naphthidine, followed by 3 ml of sodium acetate and about 0·5 ml HCl. The pH of the mixture should now be between 5 and 6, and the colour should be a violet red. Titration with EDTA is now carried on until the colour begins to become decidedly weaker. Towards the end the titrant must only be added very slowly, waiting some 15 seconds after the addition of each fresh drop. With the last drop the final trace of a rose colour vanishes within this specified period of waiting and the solution becomes colourless.

REMARKS: The end-point is reached slowly because the reaction between the solid zinc ferrocyanide that has been formed and the EDTA is only a sluggish one. It is therefore preferable to back-titrate excess EDTA with a standard solution of Zn. The end-point

K

is quicker in this case, since the formation of the zinc ferrocyanide takes place quickly with the first drop of Zn solution in excess. Cu, Ni, Al, Cd, Pb, etc, can also be determined by this variant of the back-titration. Although this method, which is so interesting theoretically because of its reaction mechanism, has lost much of its importance since the introduction of metallochromic indicators that can be used in acid solution, it has nevertheless been used till very recently in a number of practical procedures. In place of the naphthidine compound, benzidine or variamine blue can also serve as indirect redox indicators.

35. CADMIUM

Cadmium behaves towards EDTA and to almost all the indicators so far studied in a manner completely analogous to that of zinc, and its complexometric determination presents no difficulties. Papers dealing with the determination of Cd on the macro- [48-5] and on the microscale [52-22], both using Erio T, were published quite early on. However, the cadmium complex of Erio T is less stable than that of Zn, so that the amount of ammoniacal buffer added must be kept low, otherwise the end-point becomes drawn out. This is equally true for many other indicators. The following represents a selection from among the many indicators that have been proposed for the determination of Cd. Pyrocatechol violet [54-78], naphthol violet [57-81], methylthymol blue [58-34], gallein [62-30], glycinethymol blue [59-17], all of which work in a pH 10 buffer. According to Belcher *et al.* [58-2], phthalein purple can also be used as an indicator in ammoniacal solution, but it is blocked by many cations, including Zn. Vrchlabský and Okáč [62-46] state that glyoxal-bis(2-hydroxyanil) in an alkaline medium forms a violet complex with Cd but not with Zn. The non-appearance of a red colour with Zn is attributed to the higher stability of the zinc ammine complex. According to our own investigations (H. Flaschka) of attempts to exploit this indicator for the consecutive titration Zn–Cd, it does appear to form a complex with Zn; but this is so extensively dissociated that the indicator appears to undergo hydrolytic or oxidative decomposition. In its stronger Cd complex the indicator is stabilized.

The following indicators can be used in a weakly acidic medium, pH 5–6: PAN [56-43] or CuY–PAN [56-44], azoxine [57-122], SNAZOXS [60-67], xylenol orange (urotropine buffer) [57-79], and VY-diphenylcarbazide [63-1]. Bovalini [53-16] uses a citrate solution at pH 5·3 and dithizone as indicator without adding any alcohol, whereas O'Hara [61-61] prescribes a pH of 4·5 and a medium of 60% ethanol.

Back-titration procedures have also been described, e.g. with Cu using PAN [56-43], with Co in 50% acetone–water mixtures containing thiocyanate [55-102], with Zn using zincon [55-2], and with zinc using the indirect redox indication involving ferri-ferrocyanide and dimethylnaphthidine [54-57; 55-67], benzidine [56-99], or variamine blue [57-117].

Descriptions of instrumental end-points appear in many papers. Amperometric titration has been described for the classical mercury-drop electrode [51-15; 56-94] and can be employed with high accuracy [56-22] even in strongly diluted solutions (10^{-7}M!). Reilley *et al.* [62-33] use the appearance of the 'complexone wave' as an end-point indication (cf. Part One, Section 5.6 on amperometry). Amperometric indication is also employed in the titration of EDTA that has been generated coulometrically [56-77]. Square-wave polarography [62-36] and methods which superimpose small constant currents on the polarographic current [54-22] are modifications of the amperometric method.

Potentiometric indication of the end-point with the mercury-drop electrode [58-12; 58-62] and for back-titrations with a standard solution of Hg [58-57; 58-60] are of special interest in connection with increased selectivity and the possibility of consecutive titrations. There have also been papers on high-frequency [55-41; 55-112], conductimetric [54-32; 57-34], and thermometric [57-97] titrations. Indirect determination of Cd is also possible by amalgam reduction [58-13], and this is a procedure which makes the resolution of binary mixtures possible.

Although the titration of Cd itself is not particularly selective, increase in selectivity is possible up to a point by using instrumental methods. Preliminary separations, as for example, by the liquid–liquid extraction of the thiocyanate complex [54-38], have been used. The use of masking procedures and manipulation with separate aliquot portions also permits the resolution of mixtures of several components. In this context particular use is made of the possibility of masking Cd with cyanide and later de-masking with formaldehyde [52-6; 52-29]. This makes it possible to determine Cd in the presence of Pb, alkaline earths, and other metals, and, most important of all, in the presence of considerable amounts of iron [56-47]. High concentrations of KI will mask Cd against EDTA and thus permit the determination of Zn, which is not so affected. Other masking agents for Cd are BAL [54-79] and unithiol [60-37; 60-38; 60-59; 60-60].

The determination of Cd in the presence of Zn is of special interest. Sweetser and Bricker [54-14] titrate photometrically as a self-indicating system at 236 mμ in a solution made strongly alkaline with NaOH. Here zinc occurs as zincate and the Cd is prevented from precipitating as its hydroxide by adding a small but critical (!) amount of KCN. However, the problem can be solved more simply and in a way that affords practical advantages. Since the ammine complexes of Zn are much more stable than those of Cd, a differentiation between the effective stability constants in ammoniacal solution can be achieved because of the difference in the two $\alpha_{M(NH_3)}$ factors. Reilley *et al.* [61-113] titrate with DTPA using Cu as a photometric 'slope-indicator'. The conditions are much more favourable if, following Flaschka [62-26], EGTA is used as the titrant, for here the stability constants of Zn and Cd differ by some 2 log units right from the start. Here, too, Cu^{2+} is used as the indicator in a strongly ammoniacal solution. The Cu is titrated after the Cd and before or simultaneously with the Zn, and this produces the change in slope in the titration curve.

The determination of micro-amounts of Cd and Zn is possible using a single photometric titration curve if murexide [63-11] or zincon are used as indicators. The amount of dyestuff added is in excess of that based on the Zn. The Zn–dyestuff complex then serves as a slope indicator for Cd. The titration values for the Cd and the Zn can be read off from the two breaks in the titration curve.

The principle described above in which Cu ions are used to indicate the end-point can also be used in an amperometric titration [63-9] if the titration is carried out at the plateau potential for the Cu wave (-0.30 volt versus the saturated calomel electrode). Correct results for Cd can still be obtained when the ratio Zn : Cd = 500 : 1.

The complexometric determination of Cd has been used practically for the analysis of plating baths, formaldehyde being used for de-masking with cyanide baths [54-24; 54-52; 56-56; 58-21]. Cd can be determined in amalgams after Hg has been reduced to the metal with formic acid and filtered off [62-22]. Among other analyses those of alloys [54-38; 57-110], accumulator compositions [62-10], and special alloys Sn–Pb–Cd–Ni [62-62] have been reported. Cheng [62-51] analysed Hg–Te–Cd compounds by obtaining the sum Hg + Cd by titration; after adding bismuthiol (when Hg is precipitated) the EDTA set free is titrated with Zn using Erio T as indicator.

The working instructions for the titration of cadmium have already been given in the section on zinc.

36. MERCURY

Mercury can only be determined complexometrically in the bivalent state, since Hg_2^{2+} disproportionates in the presence of EDTA to Hg(II) and metal. The first methods to be reported were indirect and involved either a back-titration [48-5] or a quantitative displacement reaction with MgY^{2-} [52-22]. Hg can be titrated directly in acetic acid solution by using the system CuY–PAN as indicator. A variety of indicators respond directly to Hg and can be employed for its complexometric titration. Xylenol orange [57-135; 57-136; 61-21] and methylthymol blue [56-44] give sharp colour transitions in solutions buffered with urotropine. Moderate amounts of chloride ions can be tolerated. The quality of the colour transition falls off with increasing quantities of chloride until with high concentrations the formation of an indicator complex is prevented completely. Elimination of interference due to chloride has been reported for titrations using xylenol orange by the device of adding silver nitrate. The precipitate that results need not be filtered off, but after adding the indicator the titration must be carried out immediately and rapidly, since the dyestuff sensitizes the silver chloride, and blackening of the precipitate by a photochemical reaction makes the colour transition indistinct.

Erdey *et al.* [60-150] propose diphenylcarbazide as another direct indicator. However, the reaction between EDTA and the indicator complex is extraordinarily slow, although it can be speeded up and made virtually instantaneous by adding a small amount of 1,10-phenanthroline. The mechanism of this catalytic action has not been cleared up.

The potentiometric titration of Hg is excellent. For this reason Khalifa *et al.* have proposed Hg for back-titrations in the determination of other metals. Reilley, Schmid, and their co-workers have used HgY^{2-} in conjunction with a mercury-drop electrode for the complexometric determination of other cations. The theory and uses of this procedure have been described in the section on potentiometry (Part One, Section 5.3). Further details will be found in the section on amperometric titrations, where its importance is discussed. We are concerned here only with the direct amperometric titration of Hg at a conventional mercury [58-86] and at a tantalum [60-74] electrode.

Interference by Hg in the titration of other metals is overcome relatively easily, for Hg can be complexed fairly selectively. The possibility of masking Hg by iodide has been known for a long time [48-5]. Ueno [57-65] exploited this for the analysis of solutions containing Cu and Hg. First of all an excess of EDTA is back-titrated with Cu using murexide as indicator, and in this way the sum Hg + Cu is found. After adding KI the EDTA equivalent to the Hg is set free and measured by a further titration with Cu.

Of more recent date is the possibility of masking Hg by reduction to the metallic state: this can be effected by boiling with ascorbic acid in acid solution [56-84; 57-83]. Masking by unithiol [60-59; 60-60] has also been reported. Cheng has described masking by precipitation with bismuthiol [62-52], and uses this for the analysis of samples containing Cd and Hg. The sum Hg + Cd is obtained first by titration, and then the Hg is precipitated with bismuthiol and the amount of EDTA thus set free is back-titrated with Zn using Erio T as indicator.

Thiocyanate has also been proposed as a masking agent, and according to Barcza [59-55] this makes the consecutive titration Bi–Hg possible. The Bi is determined first with Hg masked by KSCN: then the Hg is de-masked with silver nitrate and titrated in its turn.

The complexometric titration of Hg has been used very extensively in pharmaceutical analyses. It has been determined in salves [53-12; 57-113; 59-55] and official Hg compounds [56-18; 57-100; 57-135; 57-136; 61-21]. After combustion by the oxygen flask method the Hg in organomercury compounds has been titrated visually [61-110] or amperometrically [58-86]. The analysis of Fahlore has been reported [63-24]. Michal [56-69] determined small amounts of Hg in ores and concentrates by distilling out the metal, dissolving it up, and titrating complexometrically.

36.1. Substitution titration of mercury using Erio T as indicator

REAGENTS: 0·01M EDTA; 0·1M MgY; Erio T; buffer of pH 10.

PROCEDURE: The test solution should not contain more than about 50 mg of Hg per 100 ml. 5 ml of MgY solution is added and the mixture neutralized with NaOH to methyl red indicator paper. Then 2 ml of buffer and some indicator is added and the titration carried out with EDTA to the colour change from red to pure blue. Gentle warming speeds up the reaction at the end-point.

REMARKS: If metals are present that would be co-titrated with Hg by the above procedure the fully titrated solution is treated with 1–2 g of solid KI, whereupon the HgY^{2-} formed during the titration is transformed into HgI_4^{2-} and an amount of EDTA equivalent to the Hg is set free. This can now be back-titrated to a red colour with a standard solution of Mg or Zn.

Hg can also be determined very satisfactorily by back-titration. After an excess of EDTA has been added to the acidic test-solution this is neutralized by NaOH, pH 10 buffer and indicator is added, and the excess of EDTA is back-titrated with a standard solution of Zn or Mg.

Excellent results are obtained by titration in a weakly acidic medium if CuY–PAN is used as the indicator. The procedure here is that given in the general working instructions of Section 31.3 above.

37. GALLIUM

Flaschka and his co-workers [54-61] have reported on the visual titration of Ga on the micro-scale; an excess of EDTA is back-titrated with Zn, Mg, Pb, or Mn with Erio T as the indicator. If the usual NH_3–NH_4^+ system is used to buffer to pH 10 the colour transition in the back-titration is not sharp, and it often comes back. However, if the titration is carried out at pH 9·0–9·5 or below excellent results are obtainable. Mn can be recommended as the best reagent for the back-titration for it gives a sharper end-point than Pb and particularly than Mg, and has the advantage over Zn that masking by cyanide can be permitted. In contrast to Al, Erio T is not blocked by Ga. The adjustment of pH is carried out by adding the usual pH buffer and finally adding several ml of 10% ammonium chloride solution. The same authors have also back-titrated excess EDTA in a weakly acidic solution with a standard solution of Zn with the ferri-ferrocyanide–dimethylnaphthidine redox system. Flaschka and Abdine [56-44] have described the direct titration in a hot solution buffered with acetate to pH 3–5 using CuY–PAN as indicator. The method is suitable for macro- or micro-titrations. Mee and Corbett [61-78] later studied this method more closely and show that the nature of the buffering material has a large effect on the course of the titration. Trailing end-points and low results were obtained with a phthalate buffer of identical pH. Acetate obviously works as an auxiliary complexing agent and masks Ga against the formation of hydroxy-complexes which react only slowly or not at all. In the absence of acetate, Ga begins to precipitate as its hydroxide by pH ~3, whereas in the presence of acetate precipitation does not start until pH 6. Mee and Corbett lowered the titration pH to 1·6–2·0 (CuY–PAN) as indicator, and thus increased the selectivity. Fluoride does not interfere in the titration, and can therefore be used to advantage for masking Al should this be present.

Innumerable other procedures have been reported. Among these should be mentioned the back-titration of excess EDTA with Co in a water–acetone mixture containing much thiocyanate [55-103], direct titration at pH 2 with gallocyanine as indicator for an end-point from blue to red [55-79] and a determination carried out at the same pH with azoxine as indicator for a colour change from yellow to pink [62-48]. Kinnunen [57-137] described the direct titration with xylenol orange as indicator, but he obtained a very long-drawn-out end-point, and therefore advocates back-titration of excess EDTA with a standard solution of Th or Tl(III). Pyrocatechol violet has also been used as an indicator [55-36] in solutions buffered with acetate; in a direct titration the end-point is revealed by a colour change from blue to yellow. According to Flaschka and Sadek [56-42], this same indicator can be employed to great advantage for the back-titration of excess EDTA with Bi in a strongly acid medium. The titration is still possible at pH values just under 2, and therefore shows a very high degree of selectivity. Only Bi, In, Fe(III), Hg, Th, and Zr interfere, since they form especially stable complexes with EDTA. Fe and Hg can be masked by reduction with ascorbic acid.

Quite early on Patrovsky [53-38] described the titration of Ga in ultraviolet light using morin as a fluorescence indicator. The titration is carried out either in a solution buffered with acetate or in a medium containing acetate and tetrafluoroborate [55-36] at pH 3·5: in the latter case up to 40-fold excess of aluminium can be masked. The end-point is revealed by an abrupt quenching of the fluorescence. Gregory and Jefferey [62-20] find that the intensity of the fluorescence of the Ga–morin complex is increased by the addition of many organic substances (e.g. methanol or acetone) and also by lowering the pH (from 4 to 1·7). They also find that the addition of some methyl-red indicator is advantageous, for it masks the slight residual fluorescence that persists after the end-point has been reached. In this way the end-point is considerably sharpened, so that the titration can be carried out without an ultraviolet lamp and merely in bright sunlight.

Very small amounts of Ga (50–1,000 μg) have been titrated fluorimetrically by Crawley [58-40] in a tartrate medium of pH 2·5–3·5 using oxine in ultraviolet light. Further investigations on the fluorescence reactions of Ga are due to Korenman *et al.* [60-195].

Chudina [59-130] has reported on an amperometric titration of Ga carried out at pH 1·5–2·5 and about −0·9 volt (against the saturated calomel electrode). At this low pH the titration is very selective, and in particular the presence of phosphate caused no interference.

The fluorescence titration of Ga with morin has been used practically in the analysis of bauxites [53-38] and technical solutions of Al salts [62-20]. CuY–PAN serves as the indicator in the determination of Ga in electrolytic baths [60-165] and As–Sb–Ga alloys [61-121]. A back-titration with Zn using Erio T has been used in the analysis of ores [56-49]. In most cases a preliminary concentration of Ga is necessary as well as its separation from accompanying metals by the aid of liquid–liquid extraction and/or ion-exchange.

38. INDIUM

The direct visual determination of In on the macro- [53-28] and micro-scale [53-23] was first reported by Flaschka and his co-workers. Titration was carried out at pH 10 with Erio T as indicator and tartaric acid present as an auxiliary complexing agent to prevent the precipitation of $In(OH)_3$. The way of carrying out this titration is identical with that given for Pb (Section 44.1), save that it is conducted in a boiling solution. Boiling is necessary because the reaction between EDTA and the indium–tartrate complex proceeds slowly at room temperature.

According to Doležal *et al.* [56-89], the titration can be carried out at room temperature if ethylenediamine is used as an auxiliary complexing agent; simultaneously this serves to adjust the pH. The advantage gained by this procedure is appreciably diminished by the fact that the amount of diamine is critical. Too low a concentration produces a long-drawn-out end-point that keeps coming back; whereas with an increasing excess of diamine the end-point worsens and finally fails completely. The same authors also propose a back-titration in an alkaline medium with a standard solution of Mg or Zn using Erio T as indicator. The addition of tartrate is unnecessary here. The possibility of back-titrating with Cu in a medium buffered with pyridine and with pyrocatechol violet as indicator was also reported.

The advantage of titration in an alkaline medium is that cyanide can be used for masking a whole group of cations (in particular, Zn and Cd) which so often accompany In. Unfortunately Ca and Mg are co-titrated, and in view of this titration in an acidic medium is preferable.

Cheng [55-19] titrated directly at pH 2·3–2·5 using PAN. The same indicator can be used for titrations in a medium buffered at pH 7–8 with ammonium acetate and with tartrate as an auxiliary complexing agent. Masking with cyanide can be used under these circumstances, and the effect of Ca and Mg is also considerably reduced. The colour transition is from red to yellow. The use of PAN–CuY at pH 2·5–3 [56-44] gives an appreciably sharper end-point with a colour change from deep violet to yellow.

Kinnunen [57-137] titrates In directly at pH 3–4·5 using xylenol orange, or alternatively, measures excess EDTA by back-titration with a standard solution of Th or Tl(III). Phosphates do not interfere in the latter case, but fluoride must be absent. Kopanica and Přibil [60-48] recommend the same indicator at pH 3–3·5, and by working in warm solution $(50 - 60°C)$ a sharp end-point transition is obtained. Interferences by many metals were eliminated with 1,10-phenanthroline; however, the high cost of this reagent prohibits its routine use as a masking agent save for analyses on the micro-scale.

The high-stability constant of the InY^- complex $(\log K = 24·9)$ should enable a titration to be carried out at pH 2 or below. Unfortunately no indicator is at present known for a direct titration under these conditions. However, the determination of In in a strongly acid solution is possible by carrying out a back-titration with Bi using pyrocatechol violet as the indicator. Flaschka and Sadek [56-42] have used this procedure and find, in accordance with expectation, that there are far fewer interferences. For example, In can be determined in the presence of a 350-fold excess of Cd and Zn. Pb in 50-fold excess and Ni, Co, and even Cu in 10-fold excess can be tolerated; Mg and the alkaline earths have no effect at all. The titration can even be carried out in the presence of moderate amounts of Al provided the operations are carried out slowly and at 60°C. According to unpublished observations by Flaschka, this titration can be carried out at pH values somewhat below 2 if a photometric method is used to detect the end-point. Under these conditions still larger amounts of interfering elements can be tolerated.

Busev and Talipova [62-49] used 7-naphthylazo- or 7-(4-sulpho-1-naphthylazo)-8-hydroxyquinoline as indicators. The end of the titration at pH 2·5–3·0 is shown by a change of colour from yellow to red: Zn, Cd, Al, Mn, and Mg do not interfere.

Patrovsky [53-38] was the first to introduce fluorescence indication by titrating in ultraviolet light with morin as the indicator. The end-point is revealed by the quenching of the intense green fluorescence. Belcher *et al.* [60-153] used *o*-anisidinetetra-acetic acid at pH 5·5–6·0 and back-titrated in ultraviolet light with a standard solution of Cu^{2+} until the very intense fluorescence was quenched. Since very many other metals can be determined in this way, the selectivity is low so far as In is concerned.

According to Hamm and Furse [62-36], amperometric titration and changes in pH enable the consecutive titrations In–Cd and In–Pb to be carried out. Treindl [56-13] titrates amperometrically at 5°C with KI as the base electrolyte. Under these conditions the procedure works with a 10,000-fold excess of Cd.

The complexometric determination of Cd has been used in practice for the analysis of electrolytic baths [60-165; 60-166] with CuY–PAN as indicator at pH 2·5. Alloys of Ag–In [56-89], In–As–Se [61-41], and In–Zn–Sn [63-19] have been analysed. Separation by extraction and an amperometric titration were both used in the determination of In in sphalerite [58-100]. For the analysis of concentrates [62-114] In was first extracted into ether from 1·5 H_2SO_4 containing 1M KI prior to complexometric titration with PAR as the indicator.

39. THALLIUM

The first complexometric determination of thallium was due to Flaschka [52-27]. The solution of Tl(III) was treated with the Mg–EDTA complex and neutralized: after the addition of a little pH 10 buffer the Mg set free was titrated with EDTA using Erio T as indicator. Sharp end-points and good results were obtained with amounts of Tl from 30 μg to 2 mg. Following a proposal from Flaschka, a really selective determination of Tl is possible if Tl(I) is precipitated as its iodide and the precipitate dissolved by fuming off with conc. HNO_3; Tl(III) results by oxidation, and this can then be titrated by the procedure given above. In a later investigation, however, Přibil [61-92] showed that the oxidation by nitric acid was incomplete, and he employed aqua regia. Should unoxidized Tl(I) be present, very long-drawn-out end-points are obtained, which is scarcely surprising in view of the lower stability constant for the Tl(I)–EDTA complex. Bouton and his associates [57-33] report log $K_{Tl(I)Y} = 5\cdot81$ (20°C; $\mu = 1$).

Nevertheless, the stability of the thallous complex is high enough for a photometric titration to be successful. Foley and Pottie [56-118] exploit this possibility and titrate a solution buffered to pH 10 with ammonium chloride–ammonia and measure the optical density at 222 mμ. The break at the end-point is sharp, and in its neighbourhood the titration curve shows hardly any curvature worth mentioning. An indirect method for determining Tl(I) originates from Sen [58-99], who measures the amount of Co obtained in a precipitate of $Tl_2Ag[Co(NO_2)_6]$ by dissolving this up, adding EDTA, and determining the excess by back-titration with a standard solution of Co in a 50% acetone–water mixture containing thiocyanate as indicator. However, the equivalent weight for the thallium (408·74) is unattractively high.

All other titrations concern Tl in its tervalent state. The stability of TlY^- is quite considerable (log $K = 21·5$ at 20°C; $\mu = 1$) [60-180], and this permits of titrations in a strongly acidic medium with the concomitant advantages from the point of view of selectivity. Tl often occurs in the monovalent state and, as mentioned above, oxidation to Tl(III) is easily effected with aqua regia [61-92], a procedure that is also commended by Kinnunen [57-147]. If the titration is carried out on precipitated TlI it is a point of practical importance to note that any AgI precipitated at the same time does not interfere [52-27]. The oxidation can also be effected with bromine, or what comes to the same thing, a mixture of KBr and $KBrO_3$ [60-181]. However, when this oxidant has been used, it must be remembered that the titration can no longer be conducted at pH 2–3 or below, since Tl is masked by bromide ions. Masking is least towards xylenol orange, PAN, and PAR [61-130]. The masking effect of bromide ions can be compensated by raising the pH to 4–5. Ammonium peroxydisulphate should be mentioned as an alternative oxidizing agent [58-107].

Busev [60-181] has described the determination of Tl(III) by back-titrating excess Zn in a solution buffered to pH 8 using Erio T as indicator. This modification is specially suitable when, after oxidation with a bromide–bromate mixture and addition of bis-antipyrinylmethane, the precipitate of composition $C_{23}H_{24}O_2N_4.HTlBr_4$ is taken up in EDTA. A further possibility is the back-titration of excess EDTA at pH 3·5 with a standard solution of Th [56-118] using alizarin S as indicator.

Kinnunen [57-137] has described the direct titration of Tl(III) at pH 4–5 with xylenol orange as the indicator. The end-point is so sharp that Tl(III) has been proposed as a reagent for back-titration in the determination of other metals. Tl has the substantial advantage over Th, which has also been recommended for this purpose, that phosphate, sulphate, and fluoride do not interfere, so that these ions can be introduced as masking agents for other metals. Přibil [61-92] recommended the same titration and noted that Tl(III) could be determined in the presence of Tl(I) under the conditions specified.

Busev and his co-workers have discussed the direct titration of Tl in a series of papers. From an investigation of complexes of Tl(III) with xylenol orange, PAN, and PAR [60-167] they report stability constants of $8·0 \times 10^4$, $1·9 \times 10^2$, and a $1·7 \times 10^4$ respectively for the 1 : 1 complexes formed in each case at pH 2 with absorption maxima (at the same pH) of 580, 560, and 520 mμ respectively. The lowest pH values at which a successful titration can be carried out are 2·0, 1·8, and 1·7 respectively. In view of the low stability of the indicator complex, titrations of Tl(III) with PAN showed the highest selectivity.

Busev and his co-workers [61-157] also investigated the applicability of 8-hydroxy-7-(2-pyridylazo)quinoline as an indicator. The colour change from violet to yellow is very sharp and takes place over the range from pH 1·8 to 3·5; Zn, Cd, Mn, Mg, and Al do not interfere. The determination of Tl and Bi in one and the same solution is possible using these indicators. The total concentration of the two metals is first determined at pH 1·8–2·0 in a direct titration. Then Tl(III) is reduced by sodium sulphite and the amount of EDTA equivalent to the Tl thus set free is titrated with Cu using the same indicator. Tl can also be determined in the presence of Zr if the latter is masked by fluoride.

The determination of Bi and Tl using the principle stated above is equally possible with PAN as the indicator [58-107]. It is worth noting that in this procedure the titration for total concentration is carried out at pH 4–5, which is established with solid sodium acetate: no hydrolysis of Bi should take place. This method is far less selective, of course, than that described above, since Zn, Cd, and Al are co-titrated.

The authors also examined naphthylazoxine as an indicator [62-68], but the colour change (from violet to yellow) is only sharp enough in the presence of larger amounts of thallium.

Busev [61-130] also described the determination of Tl by the back-titration of excess EDTA using a standard solution of Fe(III) at 40–60°C with sulphosalicylic acid as indicator. This indicator also permits of the consecutive titration Fe–Tl. The Fe is first titrated at pH 1·8–2·0 in a warm solution containing KBr, for Tl is masked under these conditions. Then more EDTA is added, the pH raised to 4–5 with sodium acetate, and excess EDTA is back-titrated with Fe. In this same paper a very interesting theoretical possibility for the determination of Tl is described which depends upon the reversibility of the reaction $Tl^{3+} + 2I^- \rightleftharpoons Tl^+ + I_2$, which occurs to some extent. The test-solution containing tartaric acid is first neutralized to the alkaline colour of tropaeolin OO, and then the indicator is adjusted just to its change point (pH $\sim$7·5) with 1M tartaric acid. Then KI solution is added, when the above reaction proceeds some way to the right so that iodine is liberated. If titration with EDTA is now begun the Tl(III) is complexed and the above reaction is reversed. As soon as the bleaching of the iodine colour becomes detectable, when almost all the Tl(III) has reacted with EDTA, starch solution is added and the titration continued until the blue colour disappears.

Belcher and his co-workers [60-153] have discussed the possibility of determining Tl by back-titrating excess of EDTA with Cu using *o*-dianisidinetetra-acetic acid as a fluorescence indicator. Sajó determines Tl by back-titration of excess EDTA with Zn using ferri-ferrocyanide and benzidine [56-99] or with vanadium(V) and diphenylcarbazide or diphenylcarbazone [56-100].

Instrumental end-point indication was introduced by Busev [58-107], with amperometric titration at a rotating Pt electrode. Reilley and Schmid [58-12] determine Tl(III) by potentiometric back-titration of excess EDTA with Zn using a mercury-drop electrode. The working instructions in this paper specify an acetate buffer of pH 4 as the titration medium, whereas in a comprehensive table Tl occurs within the group of metals which can be back-titrated at pH 8–10.

The complexometric determination of thallium has been used in practice for the analysis of Tl–Zr alloys [61-157]. The alloy is dissolved in sulphuric acid, the Tl oxidized with persulphate, and the solution diluted until the molarity of the acid is about 0·5; KF is then added to mask the Zr, and the Tl is titrated using 7-(2-pyridylazo)-8-hydroxyquinoline as indicator. Alloys containing Sb, As, and P [60-181] are dissolved up, and after the Tl has been oxidized with bromide–bromate the Sb is masked with tartaric acid and the Tl precipitated with bis-antipyrinylmethane. The precipitate is then taken up and submitted to a complexometric titration as described above. The determination of Tl in briquette-Cd and slags has also been described [58-107]; after oxidation with persulphate the Tl is titrated amperometrically at pH 2 with high selectivity.

39.1. Substitution titration of thallium with Erio T as indicator

REAGENTS: Standard solution of $0.01M$ EDTA; Erio T; pH 10 buffer; aqua regia; approx. $1M$ NaOH; approx. $0.1M$ MgY.

PROCEDURE: The precipitate of thallium iodide containing about 10–30 mg of Tl is filtered off, washed with KI solution, and taken up in a few drops of aqua regia; the iodine formed and the nitrous gases are driven off by evaporation almost to dryness. The residue is taken up in water and diluted to 100 ml, and after adding 5 ml of MgY, neutralizing with NaOH, and adding 2 ml of buffer and some indicator, titration is carried out with EDTA until the colour changes from red to blue. The last trace of a reddish hue vanishes with the last drop of titrant.

REMARKS: The procedure has been worked out as a titration which follows the precipitation of thallous iodide. If Ag had been present in the solution it would be precipitated at the same time, but it does not interfere with the titration so long as the amounts are not so large that difficulties are caused by the turbidity that remains. Silver can thus be used as a carrier to scavenge very small amounts of Tl. If Pb is present in the original solution the precipitation of Tl $(+Ag)$ iodide can be carried out in the presence of EDTA in a solution buffered with acetate [52-21]; under these conditions PbI_2 is not precipitated.

40. CARBON (CARBONATES, CYANIDES, ORGANIC COMPOUNDS)

40.1. Carbonates (carbonic acid)

The determination of C in steels according to Kawahata *et al.* [62-42] is carried out by combustion to CO_2, which is trapped in a measured amount of $Ba(OH)_2$. The excess of Ba is measured complexometrically with EDTA and Erio T without removing the $BaCO_3$ that is formed. Amounts of C below 0.05% are accurately determinable to $\pm0.0028\%$ (absolute).

Berbenni [60-142] determines CO_2 in natural waters by driving off the CO_2 in an aliquot portion and then determining the Ca content by titration in strongly alkaline solution using murexide. A measured amount of $Ca(OH)_2$ solution is then added to a second aliquot portion, $CaCO_3$ is precipitated, and the Ca that remains in the solution is titrated. The content of CO_2 is calculated from the difference between the two titration values, taking into account the amount of Ca added. According to Zenin [55-82], aggressive carbonic acid can be determined likewise by an EDTA titration.

Laux [61-144] determines carbonates by heating the sample solution with an alkaline solution of Ca on a water-bath; precipitated $CaCO_3$ is filtered off and its Ca content determined complexometrically. The procedure is used for the analysis of exhaust gases from ethanolamine gas-treatment plants. Szekeres *et al.* [62-136] describe the determination of alkali carbonate in the presence of hydroxide wherein the former is precipitated with a measured amount of $Sr(OH)_2$ and the excess of Sr is back-titrated with bromopyrogallol red after adding a little of a Mg salt.

40.2. Cyanides

Huditz and Flaschka [52-32] have shown that murexide is an outstandingly good indicator for the direct determination of cyanide with a standard Ni solution. The sharpness of the end-point compensates for an unfavourably large equivalent weight. In view of the high volatility of HCN, it is, however, advisable to use an indirect method and to back-titrate an excess of Ni with EDTA using murexide as indicator. The advantage over Liebig's method is that halogen and thiocyanate ions do not interfere, and even small amounts of ferri- and ferro-cyanides can be tolerated. In the presence of the two ions last mentioned, and also if larger amounts of cyanide are involved, a slight turbidity is produced; this disappears during the course of the titration, which, in this case, should be carried out slowly.

Šaršunová [56-121] has used this method to determine the cyanide content of oil of bitter almonds.

For the analysis of cyanide–halide mixtures, refer to the section on halogens.

40.3. Indirect titration of cyanide

REAGENTS: Stock solutions of 0·01M EDTA and 0·01M $NiSO_4$ standardized one against the other; murexide; conc. NH_4OH; approx. 1M NH_4Cl.

PROCEDURE: 50 ml of the 0·01M $NiSO_4$ are treated with 10 ml ammonium chloride and 10 ml conc. ammonia and an aliquot portion of the sample solution which contains about 40 mg cyanide. Drops of the indicator are next added, whereupon a yellow colour is produced, and without delay titration is carried out with EDTA till the colour changes to violet.

REMARKS: If the cyanide is introduced into a neutral solution of nickel sulphate there is a danger of forming the sparingly soluble nickel cyanide $Ni[Ni(CN)_4]$. Therefore, as described above, the solution must be made strongly ammoniacal right from the start. To be sure, the ammoniacal nickel solution itself is not indefinitely stable, for in time hydroxide can separate out. For this reason the titration is started as soon as the mixture has been prepared.

A rough calculation shows that the stability of the tetracyano-nickelate has just the right value to make this titration feasible. To bring murexide to its transition point at pH 10 we need a value of at least 11 for pNi (cf. Fig. 27). We will assume that the concentration of the cyanide complex of nickel is about $10^{-3}M$ at the end-point. Then from the overall stability constant $\beta_4 = 10^{27}$ for this complex we obtain the value 2×10^{-5} for the concentration of free cyanide ion, which is about 0.5% of the total concentration of cyanide (which is $4 \times 10^{-3}M$). Thus at the end-point (pNi = 11) the EDTA has already displaced 0.5% of the total cyanide from its complex with nickel. But we must recognize that all these equilibria are probably reached slowly, so that there is time to complete the titration before the amount of excess nickel is supplemented by nickel from the cyano-complex.

40.4. Organic compounds (e.g. pharmaceutical preparations)

The following sections will deal with the determination of organic compounds and functional groups, ignoring, however, any inorganic components that may be present, details of whose determination are to be found under the headings of the particular elements. However, there are a few organometallic compounds that fall into an intermediate category, and they are mentioned here because on the basis of their precise stoicheiometry the determination of their metal content enables the amount of organic substance present to be inferred directly.

Of course, such determinations of organic substances can only be carried out indirectly through the titration of a metal that reacts stoicheiometrically with the substance that is to be analysed. To this end they are caused to react with a cation or metallic element, whereupon precipitation, complex-formation, addition, reduction, oxidation, or more than one of these processes can take place simultaneously. Precipitation reactions are the most widely used.

The question often arises whether it is preferable to determine the amount of metal in the precipitate after it has been isolated or to back-titrate the excess of metal remaining in the filtrate. There is no single answer to this problem, and it all depends on the particular circumstances which procedure should be preferred. The determination of the metal content of the precipitate has the advantage that only one standard solution is required, provided, that is, that the metal can be titrated directly. If the precipitate has to be taken up in EDTA, or if a direct titration is impossible for other reasons, then a further standard solution is needed to back-titrate the excess of EDTA, and this will not necessarily contain the same metal as that used to bring about the precipitation. The successful application of this method demands a very thorough washing of the precipitate; this may introduce a number of difficulties, as, for example, if filtration is slow or if washing leads to losses due to the appreciable solubility of the precipitate, losses which have to be minimized by the use of special wash-liquors.

These latter disadvantages are overcome if one adopts the procedure of back-titrating the excess of precipitating agent, and there is often a great saving of time as well. With compounds that are not very sparingly soluble quite a considerable excess of precipitating agent may be needed, with the result that the back-titration will require a lot of EDTA. The use of only small aliquot portions for the back-titration serves to multiply the titration error and reduces the overall accuracy of the titration. Furthermore, in this aliquot procedure the amount of sample material is quite critical: at the lower end of the scale it is limited by the solubility of the precipitate in the final volume of solution, and at the other end by the amount of precipitating solution that can be used. Finally, we must also pay attention to the possibility of the sample containing any cations that are determinable complexometrically, for these would be included in the titration of the filtrate. The organic substance to be determined can often be made to react with a metal complexonate, and its determination is completed by titration of the complexone set free thereby.

Although the number of organic materials that can be determined complexometrically is already very large, the possibilities are far from being exhausted, and we can confidently expect a significant increase in the number of papers in this field. The procedures are, of course, in no sense specific: on the contrary, they often exhibit very low selectivity. Nevertheless, a general application of one and the same recipe is scarcely possible; for small modifications in the working instructions (e.g. pH of the solution, amount of excess metal, re-action time, washing procedure, etc) are necessary from substance to substance, often even when they come from the same class of samples. No general procedures will therefore be set down, and atten-tion must be drawn to the original literature. Among the more recent literature surveys, that by Hennart [61-162] should be mentioned; this includes information on the author's unpublished methods and also deals with the determination of the inorganic components of organic compounds. Borchert [61-124] has published a survey of the analysis of pharmaceuticals which records determinations of organic substances as well as procedures for carrying out the determination of metals.

40.4.1. PRECIPITATION WITH CALCIUM

After saponification of the starting material, pectin is precipitated as its calcium salt [54-54] and the Ca titrated in a solution of the pre-cipitate. Fatty acids, obtained if needs be by saponification, are precipitated as calcium soaps and the excess of calcium back-titrated in the filtrate. In this way the total fatty acids in soaps [55-111] and the amount of free and combined oleic acid in other pro-ducts [60-157] have been determined. The determination of esters and acid chlorides of oxalic acid [57-127] depends on the precipita-tion of calcium oxalate and back-titration of excess Ca: preliminary hydrolysis is necessary.

40.4.2. PRECIPITATION WITH COPPER

Kydd [52-38] used precipitation with copper(I) to determine acetyl-ene. The cuprous acetylide was isolated, taken up into solution, and the metal titrated using murexide. This was the first example of the use of a complexometric titration for the determination of an organic substance. Anti-Lewisite gives a copper compound and can be determined through a high-frequency titration of Cu [55-112]. According to Budĕšinský [56-35], quinine is precipitated by a solution of cuprous chloride in a 1 : 1 mixture of acetone and benzene, where-after the copper in the precipitate is determined complexometrically. Leptazol [56-29], theophyllin [59-133], and derivatives of mandelic acid used pharmaceutically can also be precipitated by Cu and determined by complexometric titration of the metal in the precipi-tate after it has been separated and redissolved. Captax (2-mercapto-benzthiazole) can be precipitated by adding Cu and titrating the excess in the filtrate [63-58].

40.4.3 PRECIPITATION WITH MERCURY

Helmstadter [59-133] has reported on the determination of amino-pyrin, nicotinamide, nicotinic esters, and 4-isopropylphenazone, which were precipitated with $HgCl_2$. Mercuric acetate is used to precipitate theobromine. The excess of mercury in the filtrate is determined by adding a known excess of EDTA and back-titrating with Zn using Erio T. Barbiturates were determined in the same way [61-163]. The method gives good results for barbital, phenobarbital, and cyclobarbital, but fails with narkosan and gives bad results for mesobarbital. For the back-titration in the determination of barbiturates Roushdi *et al.* [61-57] use a standard solution of Mn instead of Zn or titrate the mercury directly at pH 5 using CuY–PAN indicator. Propyl gallate [59-57] and ethyl gallate [61-134], which are used as antioxidants, are isolated, if necessary by extraction, and boiled with a standard solution of Hg, after which the excess Hg is determined by adding a known excess of EDTA and back-titrating with Zn using Erio T.

40.4.4. PRECIPITATION WITH SILVER

Hennart precipitates sulphonamide [62-70] and 6-mercaptopurine [62-71] with silver nitrate and determines the excess metal in the filtrate complexometrically by the nickel tetracyanide method (cf. Section 32.1). Šaršunová *et al.* [55-113] determine bromoform by hydrolysis with a silver salt and complexometric determination of the AgBr so formed by the nickel cyanide procedure. $CHCl_3$ can be determined in the same way [61-156].

40.4.5. PRECIPITATION WITH NICKEL

Arikawa [54-23] has worked out a method for determining alcohols which depends upon transforming them into their xanthates and precipitating these as their nickel salts. The latter are isolated, taken up in ammonia, and the Ni-content determined complexometrically with murexide as indicator. Hennart uses the complexometric titration of Ni for the determination of dioximes [61-159]. Other compounds can be determined in the same way as Ni precipitates after they have been fully oximated. The determination of diacetyl [59-132], monoximes [59-158], *o*-acylphenols [61-160], and vicinal diketones [61-161] s possible in this way. Precipitation with nickel diethyldithiocarbamate was exploited by Nebbia for the determination of secondary amines [53-50] and for the determination of hexamethyleneimine in the presence of hexamethylenediamine [57-31]. Long-chain alkoxylsulphonates can be precipitated by tetrammine nickel ions [57-128], and this has been worked up for the analysis of detergents [59-145]. The precipitation of pyramidone with Ni and its titration using murexide is said to provide a specific determination.

40.4.6. PRECIPITATION WITH BISMUTH

Propylgallate can be determined by precipitating it with bismuth and back-titrating excess metal in the filtrate using xylenol orange or pyrocatechol violet. Special note should be taken of the possibility of precipitation with BiI_4^-, which serves for a large number of organic compounds. Budĕšinský [56-32] has made extensive use of this procedure for the determination of caffein, 8-methylcaffein, urotropine, amidopyrine, and various quarternary ammonium-, phosphonium-, and arsonium bases. The precipitation is carried out under the appropriate conditions in a standard volumetric flask; this is then filled up to the mark, and aliquot portions of the filtrate are taken to determine excess of Bi by direct titration with EDTA, which gives a sharp end-point by bleaching the intense yellow colour of the tetraiodobismuthate ion. Before the titration is started it is necessary to add a small amount of thiosulphate to reduce any iodine present. In another paper Budĕšinský and Vaničková [56-34] describe the use of this method for the determination of theophylline, quinine, Analergen, and Divascol. Aneurine can also be determined by this procedure [57-85]. Other authors have adopted the principle to the determination of chelidonine and the total alkaloids present in the roots and in the plants and tincture of *Chelidonium Majus*. Antihistamines (Antazolin–HCl, Diphenhydramine–HCl, and Chloropyramine) can also be determined in this way [62-73]. Euphylline has been precipitated with $KBiI_4$ and the Bi-content of the separated precipitate measured complexometrically [59-135]. In the determination of piperazine salts the excess of Bi in the filtrate was titrated amperometrically [56-60].

Budĕšinský [56-36] simplified the procedure by the following modification. The organic base (papaverine, narcotine, codeine, strychnine, and brucine) is precipitated in acid solution as the tetraiodobismuth compound by adding bismuth–EDTA complex and KI solution. After making up to a known volume the mixture is filtered and the amount of EDTA set free is determined in an aliquot portion of the filtrate by titration with Zn at pH 9·1 (borate buffer) using Erio T as indicator. The end-point is affected adversely by the occurrence of the displacement reaction $Zn^{2+} + BiY^- \rightleftharpoons$

$ZnY^{2-} + Bi^{3+}$. The method has been improved in a later paper by Körbl [59-29] by titrating with Th using methylthymol blue as indicator. The determination of caffein, thiamine (aneurine) succinylcholine, thiospasmine, brucine, papaverine, and morphine was reported.

40.4.7. PRECIPITATION WITH CADMIUM

Komaritskaya [61-17] precipitates Larusan with cadmium acetate and titrates the metal in the precipitate. The method can be used for the determination of drugs in powders and tablets. Castiglioni [57-48] precipitates pyridine with a concentrated alcoholic solution of $CdCl_2$ and titrates the Cd-content of the precipitate: the method serves for the rapid determination of pyridine in denatured spirits. Leptazol can also be determined by precipitation with $CdCl_2$ [60-197]. Buděšinský has recommended precipitation with cadmium iodide for the determination of various quaternary ammonium bases [57-56]. Saitzev [57-112] determined quinine hydrochloride by this method, which was finished by back-titrating the excess of Cd in the filtrate. A simplification in this kind of determination is brought about by adding a solution of Cd–EDTA in the presence of KI and, after filtration, determining in an aliquot portion the amount of EDTA set free by titration with Ca using methylthymol blue as indicator. Among other organic bases, Flexadil, Chloropromazine, and quinine can be determined in this way.

Precipitation with cadmium thiocyanate has been very widely used: Groebel and Schneider [55-48] determined pyramidone in this way by separating the precipitate, taking it up in ammonia, and titrating the Cd, using Erio T as indicator. In this case the precipitation is more selective than that with BiI_4^- and permits the determination of pyramidone in pharmaceutical preparations which contain caffein, barbiturates, codeine, aspirin, phenacetin, and chloral as well. Buděšinský determined dimethylaminoantipyrine by backtitration of excess Cd in an aliquot portion of the filtrate. This procedure can be used for the determination of aminopyrine [55-40; 58-111], isonicotinicacidhydrazide [55-38; 55-40], and urotropine [55-38].

40.4.8. PRECIPITATION WITH ZINC

Buděšinský determined 8-hydroxyquinoline and some of its derivatives that give insoluble compounds with metal ions by precipitating the Zn salt and back-titrating excess Zn in an aliquot portion of the filtrate using Erio T as indicator: cf. also [62-132]. The determination of some barbiturates worked out by Roushdi *et al.* [61-57] depends on the same principle.

40.4.9. PRECIPITATION WITH LEAD

According to Lada [61-3], pyrocatechol is precipitated by lead acetate, and the excess Pb in an aliquot portion of the filtrate is obtained by adding a measured excess of EDTA and back-titrating with Zn using Erio T as indicator.

40.4.10. DETERMINATIONS INVOLVING OXIDATION–REDUCTION REACTIONS

Anti-oxidants in fats have been determined by Sedláček by the reduction of silver ions. In the direct method the metallic silver so formed is collected by filtration, redissolved, and titrated; in the indirect method the unreacted excess of Ag^+ in the filtrate is titrated. In both cases the Ag-determination is carried out by the tetracyanonickelate procedure (Section 32.1). Nordihydroguajaretic acid [60-39], butylhydroxyanisole [60-40], ascorbin palmitate, and several gallates [61-133] have been determined by this method. Buděšinský [61-83] dealt with aldehydes, hydroxylamine, hydrazine, and some of its derivatives by oxidation with a boiling solution of HgY^{2-}. A quantity of EDTA is set free equivalent to that of the Hg reduced, and this can be titrated with Pb using methylthymol blue. For the determination of organic derivatives of hydrazine [62-72] by way of the reduction of Tl(III) see Section 45. The determination of sugars also depends on a redox process. Bultasova [54-7] has described one such method for the quantitative determination of glucose in biological fluids. An alkaline solution of Cu(II) is reduced to hydrated cuprous oxide in the well-known manner, the precipitate filtered off, washed, and dissolved in nitric acid. The copper in this solution is then titrated complexometrically with murexide as indicator. As in other procedures based on this reduction, an empirical factor has to be used in the complexometric modification. A similar method has been worked up by Potterat and Eschmann [54-93] and used to determine the sugar content of chocolates. This Potterat–Eschmann method was chosen by Rentschler [57-95] for determining the sugar content of sweetened and fermented drinks. Street [58-98] has described a simplification in the determination of sugar (in biological materials) wherein the excess of unreduced Cu^{2+} is titrated without prior separation of the copper(I) precipitate. For the application of

Cu-reduction to the determination of D-glucosamine and its derivatives cf. Tamura *et al.* [63-59]. Ascorbic acid reduces Hg^{2+} to Hg_2^{2+}, which is precipitated as Hg_2Cl_2. After isolation and oxidative resolution of this precipitate its Hg content is titrated [62-133]. Several polyalcohols and α-hydroxyacids can be determined by oxidizing them with PbO_2 and measuring complexometrically the amount of Pb^{2+} formed [63-22]. Concerning the indirect determination of organic compounds by reduction with amalgams [58-13] see Chapter Six, Section 6.7.

40.4.11. DETERMINATION OF FUNCTIONAL GROUPS

According to Döring [56-120] the determination of carboxyl groups in cellulose can be carried out as follows. After being boiled with HCl the thoroughly washed cellulose is stirred up with a standard solution of zinc, filtered and washed, and the excess of zinc contained in the filtrate is titrated. An empirical factor is needed to calculate the number of carboxyl groups. Buděšinský [57-57] studied the conditions under which nitro- and nitroso-groups can be reduced stoicheiometrically by a metal. Cadmium proved the most suitable. The amount of cadmium ion formed (which is equivalent to the amount of the particular group reduced) is determined complexometrically. The procedure is carried out on the semi-micro-scale. The same author [57-58] also reports a method for the determination of –C=C– bonds. The sample is treated with Hg acetate in methanolic solution in the presence of BF_3-etherate as a catalyst, whereupon Hg adds on to the double bond. Excess of Hg remaining in the solution is then determined by adding EDTA and back-titrating the excess with Zn using Erio T as indicator.

40.4.12. MISCELLANEOUS METHODS

According to Buděšinský, the determination of aminoacids and peptides is possible in the following manner [57-55]. The solution of the sample is shaken with freshly precipitated copper phosphate, whereupon Cu^{2+} is bound as a complex by the aminoacid. After removing excess copper phosphate by centrifugation the supernatant liquid is decanted and the copper bound to the aminoacid is determined complexometrically using PAN as the indicator. According to Holasek and co-workers [60-182; 61-164], the method can be modified if a precipitate of copper hydroxide, produced *in situ*, replaces the copper phosphate. The procedure has been applied to the determination of albumin in serum and urine, and an empirical factor is used in the calculation. Sjöström and Ritter [56-119] determine alkaloids by passing the test solution down a column of a cation-exchange resin in the Mg^{2+} form and titrating the amount of Mg (which has been exchanged quantitatively) in the eluate using Erio T as indicator. Pilz [58-112] determined the plant preservative Meta-systox-i by hydrolysis with alkaline plumbite, extraction of the lead mercaptide that is formed, decomposing this with HCl, and finally determining the amount of lead by back-titration of a measured excess of EDTA with Zn. Buděšinský [60-44] showed that sulphides, and thiourea and its derivatives, can be determined by reaction with CdY^{2-}. After precipitation of CdS, an amount of EDTA equivalent to the sulphide sulphur is set free, and this can be titrated with Ca using methylthymol blue as indicator. According to Bürger [61-69], organometallic dialkyl- and diaryl-tin compounds can be titrated directly using pyrocatechol violet. Hennart [62-69] assayed organo-magnesium halides (Grignard reagents) by determining the Mg content after they had been decomposed with HCl. Adamová and Zýka state that codeine phosphate can be assayed by a complexo-metric determination of the phosphate through precipitation with Mg and titration of this metal in the precipitate. The separation of antipyrine as $[Fe(ant)_3(SCN)_3]$ followed by titration of the Fe^{3+} contained in the precipitate is said to provide a specific determination [62-134].

41. SILICON (SILICIC ACID)

Jenik [61-22] has reported an indirect micro-method for determining 0.5–5 mg silicic acid. The silicic acid is precipitated as $CoSi_4O_9$ from an alkaline solution containing acetone by adding $Co(NO_3)_2$. The precipitate is centrifuged off, washed repeatedly with aqueous methanol, and taken up in ammoniacal EDTA. After adding a buffer of pH 10 and some Erio T (and if necessary tropaeolin OO to improve the end-point) the excess of EDTA is back-titrated with Mg. This gives the quantity of cobalt present in the precipitate. The procedure takes 20 minutes, and should have a standard deviation of only 2.7%.

Datsenko [62-76] worked out a procedure for determining silicic acid in slags by making use of results from two complexometric titrations. One 0.1-g sample is extracted with water and KOH using a shaking machine. After dilution with water the Ca contained in the extract is titrated with EDTA, using murexide. This titration and reference to a calibration curve gives the 'basicity' as it is called. A second 0·1-g sample is decomposed with HCl, the solution neutralized, and the Ca contained in the extract is again titrated, this time using murexide in a strongly alkaline solution in the presence of triethanolamine. The content of silica is calculated from the formula: $\%SiO_2 = \%CaO/\text{basicity}$, and a relative precision of $\pm1\cdot4\%$ is claimed.

42. GERMANIUM

Kim and Rim [62-154] have given an indirect method for the determination of Ge. The test solution containing about 0·2–0·5 mg Ge(IV) per ml is treated with 10 ml each of 5% tartaric acid and 5% $BaCl_2$ solution, and then the pH is adjusted to 9–10 with NH_4OH–NH_4Cl. 1 ml of a 0·25% solution of polyvinylalcohol is added to stabilize the precipitate and the whole allowed to stand for about 5 minutes. The precipitate of barium germanotartrate is filtered off, washed three to four times with the precipitating solution (a mixture of equal volumes of tartaric acid and $BaCl_2$ solution), and sucked dry. The precipitate is then taken up in 1 : 1 HCl, transferred to a volumetric flask, and made up to the mark (50–100 ml). An aliquot portion of appropriate volume is removed, adjusted to pH 10, and the Ba contained in the solution is titrated using Erio T. The analytical procedure resembles that for boron (q.v.), and has been used for the analysis of ores. Since the precipitation stage is highly selective, no interference is caused by the usual metals that accompany the Ge. However, the washing described in the procedure does not appear to be completely satisfactory.

43. TIN

Tin forms complexes with EDTA in both its oxidation states. The stability constant of the Sn(II) complex has been calculated by Smith [61-39] as $\log K = 22\cdot11$ through displacement equilibrium between Sn(II) and ThY: for a bivalent element this is uncommonly high. Besides the 1 : 1 complex a 2 : 1 complex exists as well, and there is evidence for the existence of hydroxy complexes also. Because of the ease with which Sn(II) is hydrolysed and oxidized, the equilibria are complicated and need further clarification. Reference may be made here to polemics between Langer [62-140] and Smith [62-141]. Even less is known about complexes of Sn(IV).

The first complexometric determination of tin was described by Takamoto [55-102]: the concentration of tin was obtained by adding a known excess of EDTA and back-titrating at pH 4–5 in an acetone–water mixture containing much thiocyanate. Dubský [59-25] describes a direct titration at pH 5–6 using methylthymol blue as indicator, in which Sn(IV) can be masked by a mixture of tartrate and fluoride, and Sb by tartrate alone.

Kinnunen [57-137] was the first to describe the determination of Sn(IV). The strongly acid solution was treated with EDTA and brought to pH 2–2·5 by adding ammonium acetate: finally, the excess EDTA was back-titrated with Th using xylenol orange as indicator. Jankovský [57-80] adopted a similar procedure (cf. working instructions 43.1). Both authors refer to the principle difficulty of the method, which is to neutralize the test solution without causing the precipitation of hydroxide. It would seem that the speed with which the tin becomes bound in complexes is slightly less than the rate of hydrolysis. On this account special care has to be taken over the neutralization stage.

During the Sn-titration Cu can be masked with thiourea. Interference by Sn(IV) during the titration of other metals can be masked by fluoride in conjunction with tartrate [59-25], triethanolamine [57-80], and unithiol [60-59; 60-60], provided, of course, that these metals are not masked themselves by these reagents.

Complexometric determination of tin is employed in finding the Sn content of Cu alloys [57-107]. Avoiding heating as much as possible, the alloys are taken up in aqua regia, and after the solution has been diluted and EDTA added it is neutralized and the excess complexone is back-titrated. Sn–Pb solder [62-28] is decomposed by fuming down with sulphuric acid, whereby $PbSO_4$ separates. After adding EDTA the pH is adjusted to 2·0–2·2, the $PbSO_4$ is filtered off, and excess EDTA back-titrated with Bi using xylenol orange. It is worth noting that in this method (and probably in many others) it is immaterial whether the tin is in one or other of its oxidation states or as a mixture of the two: both oxidation states afford 1 : 1 complexes of high stability. Sb does not interfere in the determination.

Wakamatsu [62-144] used Kinnunen's method for the determination of Sn in steel after removing the iron, and Furuya [63-72] used it for the determination of Sn in phosphor bronze. Konishi [62-143] modified Kinnunen's method for the determination of Sn(II) and Sn(IV) in electroplating baths. Yurist *et al.* [62-62] analysed Sn–Pb–Cd–Ni alloys by determining the total concentration of metals in an aliquot portion by back-titrating excess EDTA with Fe using sulphosalicylic acid as indicator. In a second aliquot portion Sn was masked by tartrate, and the total concentration of all the other metals was measured at pH 10: Sn followed by difference.

43.1. Determination of tin by back-titration with zinc using pyrocatechol violet [57-80]

REAGENTS: 0·05M EDTA; 0·05M zinc acetate; acetic acid; 1 : 1 ammonia; 3M sodium acetate: pyrocatechol violet; thymol blue.

PROCEDURE: The strongly acid test solution containing up to 150 mg tin is treated with 30·00 ml EDTA and 2 ml acetic acid and neutralized slowly and with very vigorous stirring with 1 : 1 ammonia to the colour change of thymol-blue indicator. After the addition of 10 ml sodium acetate the pH of the solution should be about 5. It is then diluted to 150–200 ml, heated to 70–80°C, and after the addition of pyrocatechol violet titrated with zinc acetate till the colour changes to blue.

REMARKS: The neutralization must be carried out very slowly. The addition of acetic acid is intended to prevent over-neutralization by its buffering effect: even so, local excess of alkali must be avoided by vigorous stirring and working slowly. If a turbidity does appear it is best, in our experience, to re-acidify the solution strongly and to start again with the neutralization and to take much greater care over it.

The end-point is shown by the reaction between Sn and pyrocatechol violet. This reaction is practically instantaneous, but the displacement reaction $SnY + Zn^{2+} \rightleftharpoons ZnY^{2-} + Sn^{4+}$ proceeds slowly in the cold, hence the titration is carried out at an elevated temperature.

Sb(V), alkaline earths, and magnesium do not interfere, but Bi, As(III), As(V), Sb(III), W, Mo, Fe, Ti, Al, Mn, Cr, Zr, and complexing anions like tartrate, citrate, and oxalate, do interfere. Ag^+ interferes by oxidizing the indicator, but the titration is possible in the presence of silver if the indicator is added just before the end-point, or renewed, and the determination is carried out rapidly to its end. Pb, Cu, Ni, Zn, and Cd are included in the titration value. However, by a difference titration it is possible to determine Sn in the presence of moderate amounts of these metals. The sum total of these metals, including the tin, is determined in an aliquot portion by the above procedure. In a second aliquot portion the tin is masked with triethanolamine and the excess of EDTA back-titrated with zinc at pH 10 using Erio T as indicator: Sn follows from the difference in titration values.

The difficulties that occur in the neutralization stage can be largely avoided if the neutralization is carried out in homogeneous solution. This is described in the following procedure (based on unpublished investigations by H. Flaschka and W. Wolfram), which has been worked out for the determination of the content of analytically pure tin salts.

43.2. Back-titration with copper using PAN as indicator

REAGENTS: 0·1M EDTA; 0·1M copper nitrate; acetate buffer pH 5; 20% urotropine in water; PAN; thymol blue.

PROCEDURE: The solution to be analysed is made strongly acid with HCl, treated with 30·00 ml EDTA and some thymol-blue indicator, and heated to boiling. When boiling steadily 20% urotropine solution is added dropwise quite slowly and with stirring until the colour of the solution turns to yellow. Next 20 ml of a buffer of pH 5 are added, and after dilution to 200 ml the solution is cooled down to room temperature. After adding some PAN indicator titration is carried out with a copper solution to a violet colour, and finally with EDTA to greenish-yellow.

44. LEAD

The complexometric titration of lead can be carried out directly through a displacement reaction or by a back-titration in acid as well as in alkaline solution, and it presents no difficulties. The first direct complexometric titration of Pb was described by Flaschka, who titrated the Zn (or Mg) set free by exchange with ZnY [52-22] (or MgY [52-24]) using Erio T as indicator. Micro- and macro-amounts of lead can be determined accurately in this way. The replacement reaction can be avoided if lead is titrated at pH 10 in the presence of tartrate [52-24; 52-29], which acts as an auxiliary complexing agent and prevents the precipitation of $Pb(OH)_2$. Admittedly, the complex of Pb with Erio T does not possess the brilliant red colour of the corresponding Mg or Zn complex, but is a dull blue-violet. Nevertheless, the disappearance of the last trace of a reddish hue at the end-point can be detected far more distinctly than in the Mg titration due to the higher stability of the Pb complex that is involved. Care must be taken, however, to avoid too great an excess of tartrate, as otherwise the effective stability constants of the indicator and the EDTA complex would be reduced too much and the end-point would lose its sharpness. On this account Goetz [55-74] recommends the use of DCTA, since, because of the higher stability of its lead complex, the concentration of tartrate and the adjustment of pH are less critical. Besides Erio T, compounds of the acid chrome blue group have been recommended by Russian authors [60-87]. Like Erio T, methylthymol blue can be used in alkaline media [58-34; 61-127], with the advantage that traces of Cu and other heavy metals do not block the indicator transition.

Very many indicators can be used for the direct titration of Pb in acid solution. Methylthymol blue [58-34] gives a sharp transition from blue to yellow in a medium buffered with urotropine. Under similar conditions with xylenol orange there is a pregnant colour change from red to yellow [56-39; 57-79]. Buffering by acetate is also possible, provided its concentration is kept low. With increasing amounts of acetate the end-point becomes noticeably more drawn out. The transition in a urotropine buffer is so excellent that the back-titration of excess EDTA by Pb has been proposed for the determination of many other metals. The direct titration of Pb in a weakly acid medium is also possible using gallein [61-31], pyrocatechol violet [57-78], pyrogallol red and its dibromo-derivative [56-90], and brilliant Congo blue [57-77]. CuY–PAN [56-44] and VY-diphenyl-carbazide [63-1] have also been proposed. In an acetate buffer of pH 7–8 or in a urotropine buffer 6,13-dihydro-6,13-dihydroxy-1,4,8,11-pentacenequinone-2,9-disulphonic acid [62-47] gives a blue Pb compound which is decomposed on titration with a sharp colour change to light red at the end-point. However, the amount of urotropine and the length of the period of boiling preceding the titration are critical.

Dithizone has been proposed by several authors for the direct determination of Pb, but the end-point is not stoicheiometric unless an alcohol–water mixture is used. Bovalini [53-16] reports low values in aqueous solution, and therefore recommends the inverse titration (a solution of complexone is titrated with the sample solution containing the lead); the titration values show the usual scatter round the correct result. Minami and Sato [55-33] describe a microtitration with an extractive dithizone end-point: for correct results the titre of the EDTA solution must be determined against a lead standard. Kotrlý [57-82] employs a 50% alcohol–water mixture and obtains correct results for 0·3–6 mg Pb. The same experience is reported by Hara [61-62] for titrations at pH 4·5 in aqueous ethanol.

A large number of back-titrations are also reported in the literature. Biedermann and Schwarzenbach [48-5] back-titrate with Mg or Zn in a solution buffered to pH 10 using Erio T as indicator. Kinnunen [55-4] recommends Mn for titrations under practically the same conditions on the grounds that it gives a sharper end-point than Mg and has an advantage over Zn, in that masking by KCN can be used. In an acidic medium both micro- [56-46] and even ultramicro-titrations [57-125] give excellent results if the back-titration is carried out with Cu at pH 5–6 using PAN. Blood *et al.* [63-40] rate this indicator system very highly, since it permits the determination of lead in the presence of phosphate and alkaline earths. Takamoto [55-102] back-titrates with Co in an acetone–water mixture (1 : 1) in the presence of thiocyanate.

Back-titration with an indirect redox end-point has been proposed using ferri-ferrocyanide and dimethylnaphthidine [55-67], benzidine [56-99], or variamine blue [60-89]. A fluorescence end-point is available in the back-titration with Cu using *o*-dianisidine- or *o*-diphenetidine-tetra-acetate as indicator [60-153].

Ciogolea *et al.* [62-109] has described an indirect method in which metallic zinc is allowed to react with the lead solution basified with sodium hydroxide and, after filtering, the Zn^{2+} which goes into solution is titrated.

Instrumental indications include amperometric square-wave titrations [62-36], photometric titrations self-indicating in the ultraviolet region [55-97] or against xylenol orange [60-47], thermometric [57-97], and potentiometric methods. In the latter case procedures with the mercury-drop electrode or with a bimetallic electrode [55-73] have ranged from the macro- [58-12] right down to the ultramicro-scale [57-8].

The successive titration Bi–Pb has provoked an exceptional amount of study. This is understandable, for this combination of elements occurs in many natural products and in alloys, and the analysis by classical methods is both difficult and time-consuming. The stability constants of the corresponding EDTA complexes are sufficiently far apart to allow titration with an indicator and the situation is particularly favourable since both metals can be titrated against one and the same indicator. The same principle applies to all the titrations. Firstly, Bi is determined in a strongly acidic solution (pH ~1); then the pH is adjusted to about 5 by adding acetate or urotropine and the Pb is titrated. Underwood used a photometric titration with self-indication in the ultraviolet region [55-97]. The titration can also be carried out potentiometrically without an indicator by using the mercury-drop electrode [58-12]. Using an indicator, e.g. xylenol orange, titration can be effected visually [60-62] or photometrically on the micro-scale [60-47]. Consecutive titration with gallein [61-32] is also possible.

Kusakina and Yakimets [62-110] have described the determination of Pb and V in the same solution. The total concentration is first obtained by back-titration of excess EDTA with Zn at pH 6 using diphenylcarbazide; then the pH is raised to 10 when vanadium leaves its EDTA complex, and the complexone thus liberated is measured by a further titration with Zn using acid chrome blue as indicator.

In many cases the precipitation of Pb as its sulphate permits a smooth and simple separation from mixtures containing many other substances. The precipitate can be dissolved directly in an acetate buffer or in a solution of MgY, ZnY, or EDTA itself, and then a titration is carried out. Hisada [60-91] used this method with CuY–PAN as indicator in a direct titration, while Karolev [59-70] preferred xylenol orange or methylthymol blue. Very many metals can be masked with KCN during this Pb titration [52-29] and do not interfere with the determination of lead even when present in substantial amounts. Since masking by cyanide is only possible in alkaline solution, alkaline earths are included in the titration value.

The alkaline earths are not titratable, however, in acid solution, so that the combination of masking with titrations of aliquot portions in both acid and in alkaline solution affords excellent possibilities for resolving the components of a complex mixture. In titrating lead, Fe, Al, and Mn can be masked with triethanolamine. For the use of admixtures of different masking agents for the analysis, e.g. of a Pb–Mn–Zn mixture, see Přibil's papers [54-73; 54-79]. Přibil also reports on the use of BAL in the analysis of mixtures [54-79] and of 1,10-phenanthroline for masking [59-23]. According to Hara [61-62], Pb can be masked by boiling with 3-mercaptopropionic acid when titrating Zn. Morachewskii [60-59; 60-60] discusses the use of unithiol as a masking agent.

The determination of lead finds practical use in the analysis of motor fuels and lead tetraethyl [53-49; 54-31; 57-23; 59-78], ores and concentrates [54-100; 55-74; 58-55; 59-70; 60-147], paint driers [58-25; 61-143], zinc oxide pigments [52-35], mineral oil additives [53-7], and fluoroborate electrolyte solutions [57-45; 60-176]. It has found particular use in the analysis of alloys [54-100; 62-62], and particular examples are: Bi–Pb [60-62], type metal [60-116], solder

[62-28], bronze and brass [55-3], alloys with a Cu- [54-67] or Pb-basis [55-34], special steels [61-155], Pb–Ag [62-62], low-melting Pb–Bi–Cd–Zn alloys [59-32], Pb–Sn–Sb–Cu alloys [62-111], and ferrites [61-98]. Lead can also be determined in its chromate [61-60] and vanadate [62-110], in pharmaceutical preparations [54-82; 57-113; 61-6] and by a high precision method in Pb tellurite [61-127].

44.1. Direct titration of lead using Erio T [52-24; 52-29]

REAGENTS: Standard 0·01M EDTA; Erio T; pH 10 buffer; 1M Na–K tartrate; if necessary KCN as a masking agent.

PROCEDURE: The sample solution, which should not contain more than about 30 mg of Pb per 100 ml, is treated with 5 ml of tartrate or 1 : 4 triethanolamine (or the $PbSO_4$ that is to be titrated is taken up in tartrate or triethanolamine) and made approximately neutral with NaOH. If necessary a masking agent such as KCN is added at this stage. Then 2 ml of buffer and indicator are added and the titration carried on to the colour change from red to blue. The solution must not be too cold (approx. 40°C), and the last trace of a red hue vanishes with the final drops of titrant.

44.2. Substitution titration for lead [52-35]

REAGENTS: Standard 0·01M EDTA solution; Erio T; pH 10 buffer; approx. 0·1M MgY^{2-}; approx. 1M NaOH.

PROCEDURE: The solution to be analysed, which should not contain more than about 30 mg Pb per 100 ml, is treated with 5 ml MgY (in the case of $PbSO_4$ this is dissolved directly in MgY) and made roughly neutral with NaOH. 2 ml of buffer and some indicator are then added (and masking agent if required) and titration is carried out to the colour change from red to blue.

REMARKS: Back-titration is also a good method for lead. In this case a small excess of EDTA is added, and after neutralization and addition of indicator titration is carried out with $ZnSO_4$ or $MgSO_4$.

44.3. Direct titration of lead using xylenol orange [56-39; 57-79]

REAGENTS: 0·01M EDTA; acetate buffer pH 5; 20% urotropine; 0·1% xylenol orange.

PROCEDURE: The sample solution which should not contain more than about 30–50 mg Pb per 100 ml is first adjusted to pH 2–3 with NaOH, should this be necessary, before adding 3 ml of acetate buffer or better 10 ml of urotropine buffer. At this stage the pH should be about 5. After adding a few drops of indicator solution

L

titration with EDTA is carried out to the colour change from reddish-violet to pure yellow.

REMARKS: If the lead content is high the pH can drop substantially during the titration. On this account it is advisable to check the acidity in the neighbourhood of the end-point and to correct it, if necessary, by adding more urotropine. High concentrations of acetate buffer must be avoided, otherwise the end-point becomes long drawn out. The procedure can be modified for a back-titration in which $PbSO_4$ is taken up in excess EDTA, and after neutralization and the addition of urotropine buffer back-titrated with a standard volumetric solution of Pb till there is just a slight red colour. Since $PbSO_4$ dissolves relatively quickly in EDTA, the excess of EDTA in the amount of complexone solution added can be kept quite small; in this way any large change in pH in the back-titration can be avoided.

45. NITROGEN DERIVATIVES

Concerning the determination of ammonia by the precipitation of ammonium ion refer to Section 7 on potassium.

According to Budĕšinský, the determination of hydrazine, hydroxylamine, and many of their derivatives is possible by an indirect redox method [61-83]. The sample is boiled with HgY^{2-} in acid solution, whereby mercury obtained by reduction separates as the metal, and the equivalent amount of EDTA simultaneously liberated is titrated with Pb using methylthymol blue as indicator.

Barka and Busev [62-72] used the reaction between hydrazine and Tl(III)

$$(2Tl^{3+} + N_2H_4 \longrightarrow 2Tl^+ + N_2 + 4H^+)$$

as a basis for the complexometric determination of hydrazine. A two-fold excess of a standard solution of Tl(III) sulphate is added to the acidified sample solution and then ammonia is added dropwise until there is a permanent turbidity. This is then cleared by adding acetic acid and the excess of Tl(III) is titrated with EDTA using PAN as indicator. Berka [63-35] used this procedure for derivatives of hydrazine, such as the hydrazide of isonicotinic acid, semicarbazide, and benzoylhydrazine. PAR served as the indicator in these titrations.

46. PHOSPHORUS

46.1. Phosphates

The indirect determination of ortho-phosphates can be carried out by precipitation as magnesium ammonium phosphate hexahydrate followed by a complexometric determination of the Mg. This method has been written up by Flaschka and his co-workers for determinations on both the micro- [52-23] and the macro-scale [52-31]. For micro-amounts the precipitate is taken up in acid and the solution diluted strongly and neutralized; after adding buffer the Mg is titrated directly with EDTA, using Erio T as indicator. However, with large amounts of phosphate the precipitate comes out again when the solution is basified: in this case, therefore, excess standard EDTA solution is added to the still acid solution and the excess is determined by back-titration with Mg after neutralizing and adding the buffer. Almost at the same time and quite independently De Lorenzi and Aldrovani [52-10] worked out a phosphate method for determining the content of phosphoric acid and for the analysis of parmaceutical phosphorus compounds. They used a standard Mg solution to precipitate the phosphate and determined the excess of Mg in the filtrate.

Huditz, Flaschka, and Petzold [52-31] have looked into possible interferences, and include EDTA as a masking agent during the precipitation of the magnesium ammonium phosphate. If EDTA is added before the precipitation only polyvalent cations present are complexed, and when the magnesia mixture ($MgCl_2 + NH_4Cl + NH_3$) is added the magnesium first binds up the excess of masking agent and then what is left over is used to precipitate the phosphate. This is possible because the Mg–EDTA complex possesses a low stability constant, and therefore Mg can only displace a very few metals from their EDTA complexes. This kind of masking is specially effective when phosphate has to be determined in the presence of Ca. However, the method fails for Fe, Al and Tl, and citrate must be added to mask these metals. Be can be kept in solution with sulphosalicylic acid. Under these circumstances the determination of phosphate is possible in the presence of 10–100 times the amount of one or all of the interfering elements already mentioned and also in the presence of Cu, Ni, Co, Zn, Cd, Pb, Bi, and other metals. The determination stands or falls by the exact stoicheiometric composition of the precipitate, and on avoiding the co-precipitation of other metal ions that can be determined complexometrically: consequently, all the precautions that have to be taken in the gravimetric determination of phosphate must be implicitly observed here as well. Reprecipitation has already been recommended by the original authors as a route to achieving the highest accuracy, especially if small amounts of phosphate are to be determined in the presence of large quantities of interfering elements and foreign substances. Zemlyanskii [61-111] has recently stressed this point. The possibility of removing interfering cations by ion-exchange was mooted in the first papers on the subject. Sporek [58-104] describes a practical analytical procedure involving ion-exchangers and also a preliminary separation of phosphate with molybdate in strong nitric acid solution. Precipitation with Mg is then carried out with an alkaline solution of the phosphomolybdate. Mo does not interfere here nor with the EDTA titration. Sporek also describes the elimination of As by evaporating the test solution in the presence of HBr until it becomes syrupy. Excess HBr is finally removed by boiling with nitric acid.

Eschmann and Brochon [56-55] modified the Mg-method: they accomplished the precipitation at room temperature by neutralizing the acidic test-solution containing Mg with ethanolamine. Under these conditions the long period of standing previously needed to complete the precipitation becomes unnecessary. However, in this procedure any Al and Fe present must be removed by a preliminary extraction with cupferron. According to these authors, any delays in the precipitation of small amounts of phosphate can be avoided by deliberately adding an exactly known amount, which naturally has to be taken into account in the calculation.

Kato *et al.* [55-11] use the Mg method to determine orthophosphate remaining in the filtrate after precipitating pyrophosphate with Zn. Bennewitz *et al.* [59-109] and Hozumi [61-153] have described a modification of the Mg-procedure which avoids filtration. The method has been worked out for the determination of P in organic material after wet ashing, and precipitation is carried out with a standard $0.005M$ solution of Mg. After this has been added, the solution is made 50% in ethanol, which has the effect of speeding up the precipitation and making it quantitative: furthermore, the solubility of the precipitate is lowered to such an extent that no reaction takes place with it and the complexing agent during the back-titration. Buss *et al.* [63-53] have reported on their detailed studies of the determination of phosphate through $Zn(NH_4)PO_4$. The procedure has been worked out in three modifications for micro-, semimicro-, and macro-scales, and consists essentially of a precipitation at pH

6·5–7·0, an acidity which is established by adding monoethanolamine. The advantage of precipitating with zinc is that during the subsequent titration of Zn at pH 10 the precipitate does not come out of solution again and, moreover, Zn can be titrated to give a much sharper end-point with Erio T than Mg does. Besides Mg and Zn, Th and Bi have been considered for the precipitation and indirect determination of phosphate. In their case the precipitation and titration can both be carried out in acidic solution, which has the advantage of improving the selectivity. Kinnunen [57-147] boils the weakly acid test solution after adding a measured quantity of thorium nitrate solution, cools, makes up to the mark, and determines Th in an aliquot portion of the filtrate by titration using xylenol orange as indicator. Check analyses or detailed data on pH, amount of phosphate, limitation of the method, sources of error, etc, are not reported. Compare also [63-55].

Precipitation with Bi was first proposed by Genge and Salmon [57-73], the excess Bi in an aliquot portion of the filtrate being back-titrated using pyrocatechol violet. Veda *et al.* [61-24] proceed in the same way, but use xylenol orange as indicator. The acidity of the solution is rather critical if the precipitation is to be successful. The Bi phosphate is soluble in too strongly acidic solutions; but if the acidity is insufficient flocculent precipitates result which are difficult to filter and which include Bi^{3+} and its hydrolysis products. Socolovschi and co-workers [62-55] prescribe a pH of 0·7–1·0 for the precipitation and report on the possibility of masking Fe^{3+} by reduction with ascorbic acid and Hg by adding tannic acid. They use pyrocatechol violet as indicator when back-titrating the excess of Bi.

Riedel [59-74] elegantly solves the problem of adjusting to the optimum acidity of 0·2N in perchloric acid by evaporating the test solution to dryness with nitric acid and taking up the residue in 13 ml of N $HClO_4$; this is also concentrated until fumes of perchloric acid begin to come off. Then the solution is cooled, a standard solution of Bi is added from a pipette, and the whole heated to boiling. Then, with constant stirring, 60 ml of boiling water is slowly introduced, whereby a coarsely crystalline powdery precipitate separates out by what really amounts to the process of precipitation from homogeneous solution. This precipitate need not be filtered off, because it does not interfere with the titration of excess bismuth. The back-titration is carried out with pyrocatechol violet as indicator after diluting with water to almost 300 ml. The concentration of the perchloric acid must be carried out very carefully and broken off as soon as acid fumes are evolved, otherwise a loss of acid takes place and any exact adjustment of the acidity cannot be relied on. This procedure has the further advantage that if the foregoing instructions are followed, interfering chloride ions are automatically eliminated. Moreover, any sulphate present (which would also interfere) can be removed by precipitation as $BaSO_4$ before the stage of evaporation with nitric acid. Excess of Ba does not interfere. Interference by Fe is rendered harmless by reduction with hydrazine. Since hydrazine is only marketed as its salt with hydrochloric or sulphuric acid, both of which anions interfere, hydrazine hydrate must be used, and a dilute solution of this can be neutralized by $HClO_4$ without danger. The reduction could probably also be carried out with a solution of sodium ascorbate. In any case the iron must be reduced before the Bi is added. According to Riedel, interference from Hg and Sb is easily overcome by removing them as sulphides. Among the interfering elements that remain it is only necessary to name In, Ga, Zn, and Th, which are only rarely present. Check analyses show deviations of only a few tenths per cent for amounts of ~ 5 mg PO_4^{3-} or above per 100 ml of solution. Even smaller amounts can be determined if the quantities specified above are appropriately reduced. Riedel gives a useful working approximation for calculating the amount of Bi needed, viz. ml 0·01 Bi solution (0·5–0·7 in HNO_3) required $=$ mg PO_4^{3-} expected $+$ 10.

The complexometric determination of phosphate is used in practice for the analysis of serum [52-34], pharmaceuticals [55-88], cast iron [63-54], ferrophosphor [57-106], P–Cu alloys [63-55], organic substances after combustion in a Schöniger flask [58-24; 59-109], perphosphates [54-87], wines after ashing [54-86], foodstuffs [56-55], uranium concentrates [58-105], ores, slags [63-54], and fertilizers [63-54].

46.2. Metaphosphate
Eschmann and Brochon [56-55] determine metaphosphates by boiling with sulphuric acid to hydrolyse them to orthophosphates and then determine these by the Mg method. According to Vereph and Fülöp [62-131], hexametaphosphate can be determined if inositol phosphate is precipitated by Cu in weakly acid solution and the excess metal in the filtrate is back-titrated with EDTA using PAN as indicator.

46.3. Pyrophosphates and tripolyphosphates
Pyrophosphates are hydrolysed to orthophosphates on boiling with sulphuric acid and can then be determined by the Mg method [56-55].

The time-consuming hydrolysis can, however, be avoided and pyrophosphate itself precipitated, for which purpose zinc has been proposed [55-11]. The precipitation is quantitative at pH 3·8–4·1 [55-110]. The advantage of this procedure is that after separation and washing the $Zn_2P_2O_7$ dissolves up directly in an ammonia buffer, and the zinc can be titrated using Erio T as indicator. Furthermore, any orthophosphate that may be present in the filtrate can then be determined by the Mg method. Tripolyphosphates behave in much the same way, and they can also be thrown down at pH ∼4 and determined by a Zn titration [55-110]. Raemaekers [56-117] has given detailed instructions for the determination of tri- and pyrophosphates in solutions of tripolyphosphates. Nielsch and Giefer [55-109] prefer Mn for the precipitation of pyrophosphate; the precipitate is obtained at pH 4·1 and filtered off after 16 hours. Excess Mn in the filtrate is titrated with EDTA at pH 10 in the presence of hydroxylamine using Erio T as indicator. The precipitation is quantitative if the concentration is at least 0·2 mg of $P_2O_7^{4-}$ per ml and not less than a 5-fold excess of Mn is used. Orthophosphates must be absent.

46.4. Hypophosphites

According to Ciogolea *et al.* [62-56], hypophosphites are first oxidized by boiling with nitric acid; then precipitation is effected with a standard solution of Bi and the excess Bi in the filtrate is back-titrated using pyrocatechol violet. There is an elegant procedure by Fülöp *et al.* [62-57] which exploits the reducing power of hypophosphites. The test-solution is boiled with mercury complexonate, HgY^{2-}, whereby complexed Hg(II) is reduced to the metal and an amount of EDTA equivalent to the hypophosphite is set free and is later titrated with Cu using PAN as the indicator.

46.5. Indirect determination of phosphate

REAGENTS: Approx. 2M EDTA (for masking); approx. 1M $MgSO_4$ (as precipitant); conc. ammonia; solutions of 0·1M EDTA and 0·1M $MgSO_4$ standardized one against the other; Erio T; approx. 1M NaOH; buffer of pH 10; 1M tartaric acid; KCN; ascorbic acid; sulphosalicylic acid.

PROCEDURE: The acidic test solution should not contain more than about 350 mg of PO_4^{3-}. It is treated with sufficient 2M EDTA to transform its entire content of polyvalent cations into their complexonates. If Fe, Al, or Bi are present, add 1–2 ml of tartaric acid solution. Be is masked with sulphosalicylic acid. Now add an excess of the

1M $MgSO_4$ and slowly introduce ammonia dropwise until the amount of precipitate no longer increases: then add a further quantity of ammonia and leave to digest for a few hours. Collect the precipitate and wash it thoroughly with 1M ammonia. For very exact results the precipitate should be taken up in acid and then reprecipitated.

The filter-paper containing the precipitate is next transferred to a precipitation beaker, overlaid with about 10 ml of 1M HCl, and after adding about 50 ml water the suspension is heated to boiling and passed through a second filter. After washing the remains of the first filter-paper the combined filtrate should amount to about 200 ml. A small excess of EDTA solution is now added, followed by some KCN and ascorbic acid. The solution is neutralized with 1M NaOH, 2–4 ml of buffer is introduced, and after adding the indicator the back-titration is carried out with the standard solution of $MgSO_4$ until the colour changes from blue to red.

REMARKS: If the amount of phosphate to be determined is very small a direct determination is possible. In this case the solution of the precipitate in HCl can be neutralized and buffer added without any immediate precipitation taking place, and the resulting supersaturated solution of $Mg(NH_4)PO_4$ can be titrated directly with EDTA. For such small amounts it is better to use 0·01M EDTA and to decrease the final volume and the amounts of reagents to match. The stage of neutralization and the addition of buffer can be saved by dissolving the precipitate in exactly 10 ml of HCl and adding 20 ml of 3M ammonia before the titration. This gives a solution in which the relative proportions of NH_3 and NH_4^+ correspond to a buffer of pH 10.

Pyrophosphate can be determined in just the same way. Precipitation with Zn or Mn is carried out in acetic acid solution and the precipitate collected and washed; the $Zn_2P_2O_7$ is taken up directly in the ammonia buffer, and $Mn_2P_2O_7$ is dealt with as described above for the Mg method.

47. ARSENIC

Arsenic can only be determined indirectly by complexometry. Methods based on the most varied principles are to be found in the literature. Malinek and Rehak [56-31] adopt the procedure worked out for phosphoric acid (see above for details), for arsenic acid behaves in the same way. Arsenate is precipitated with Mg from a strongly ammoniacal solution in the usual way and the well-washed precipitate is taken up in acid, and the solution is then treated with excess of a standard solution of EDTA. After neutralizing and buffering to pH 10 the excess of EDTA is back-titrated with Mg using Erio T as indicator. A direct titration of the magnesium contained in the precipitate is also possible if the As(V) in the acidic solution of the precipitate is reduced to As(III) by boiling for a short time with 2 g of hydrazine sulphate and 0·5 g of KBr. After basification the arsenic (now present as arsenite) does not interfere with the Mg titration. As in the determination of phosphate, EDTA, tiron, and tartaric acid can also be included as masking agents at the precipitation stage. The method has been used for the analysis of alloys with a high As content [56-21]. Zn can be used in place of Mg to precipitate the arsenate [60-12]. The advantage of this procedure is that the precipitate, after collecting and washing, can be dissolved directly in a pH 10 buffer and the Zn titrated in this. The method has been worked out in detail for the analysis of medicines, but it could obviously be used more widely. However, the masking agents mentioned above for eliminating interferences at the precipitation stage could not be used without further modification.

The procedure described by Vasiliev and Anastasescu [60-94] is conceived in terms of pharmaceutical analyses, and it can be applied without modification to a wider range of materials. An excess of a standard solution of Bi is added to the acidified sample, and the precipitated BiAsO₄ is collected and the excess of Bi in the filtrate is back-titrated with EDTA. Ueda *et al.* [62-108] proceed in a similar way but dissolve up the precipitate of BiAsO₄ and titrate its Bi content using xylenol orange as indicator; Ca, Mg, Zn, Co, Ni, and Cu do not interfere. Our own experience shows that the Bi method only gives satisfactory results within a limited range of concentrations; for the solubility of the precipitate should the solution be too acid, or the co-precipitation of bismuth oxide or basic bismuth salts if it is insufficiently acid, impose rather rigorous demands for the precise maintenance of the optimum pH range.

Procedures mentioned so far can only be used for the determination of As(V). If As(III) is present it must first be oxidized. However, As(III) can be determined without oxidation by Černy's method [58-10], in which arsenic is precipitated as the trisulphide. The precipitate is collected and washed, dissolved in an ammoniacal buffer, and treated with a standard solution of Cd; without removing the precipitate of CdS the excess of Cd is back-titrated with EDTA using Erio T as the indicator. The procedure has been used for the analysis of pig- and cast-iron and for ores.

Stefanac [62-58] has described an interesting possibility for the determination of As(III) and As(V) or a mixture of the two. The procedure was worked out for the determination of As in organic substances after their combustion in a Schöniger flask, but it is capable of wider exploitation. The underlying principle is the precipitation of arsenite and arsenate as silver salts, reaction of the precipitate after isolation with $Ni(CN)_4^{2-}$, and titration of the Ni set free according to the procedure in Section 32. Since both As(III) and As(V) need 3 gram-equivalents of silver for their precipitation, it is immaterial in which form the As is present, and mixtures of the element in both oxidation states give correct results for the total content of As. The disadvantage of the method is that halogens and other anions which form precipitates with Ag^+ must be absent.

Arsenic does not interfere with the complexometric determination of other metals, apart from the possible precipitation of insoluble compounds. Masking of As with unithiol is effective [60-59; 60-60].

Procedures described for phosphorus are used for arsenic too.

48. ANTIMONY

There have been reports on the preparation of complexes of tervalent antimony with EDTA [59-68; 59-131] and other complexones [59-131]. Attempts to determine antimony complexometrically have been reported, but they are not completely satisfactory. A determination by back-titration of excess EDTA with Co in a 50% water–acetone mixture containing thiocyanate gives a bad end-point [55-100; 55-103; 58-80]. However, the procedure has been used in the analysis of Sb–Bi–tellurides [63-39]. If the back-titration is carried out with Zn at pH 5–5·5 the end-point with xylenol orange as indicator is almost useless [58-80]. Dubský [59-25] has reported a coloration which appears with methylthymol blue in the presence of Sb(III), but this is useless for a titration indication, since the complex is obviously too weak and the Sb(III) is wholly or partially displaced by the metal used for the back-titration. Jardin [59-69] has described a direct titration in which the initially strongly acidic test solution is treated with KCl and HCl and neutralized with ammonia to pH 2–3. In this way a turbidity is produced due to SbOCl3, which can be made to disappear by the slow addition of titrant. It is necessary to wait for 5 minutes after each addition of a small portion of EDTA solution and to carry on in this way until the solution is clear. It is easy to see why such a procedure is rarely adopted in practice!

Apparently As(V) gives no complex with EDTA.

Not much is known of the extent to which antimony interferes in other titrations. Masking of Sb(III) by tartaric acid is possible in the titration of tin [59-25] or bismuth [63-39], and antimony is one of the metals that can be masked with unithiol [60-59; 60-60].

49. BISMUTH

Miklos *et al.* [61-53] determined the stability constant of the Bi–EDTA complex by polarographic measurements of the exchange equilibrium with FeY$^-$ and report log $K_{BiY} = 27\cdot94$ (20°C; $\mu = 0\cdot1M$ KNO$_3$). By a compensation of errors this does not differ substantially from the more recent value of $28\cdot2 \pm 0\cdot2$ found by Beck and Gergely [66-74], who also report log $K = 31\cdot2 \pm 0\cdot2$ for the Bi(III)–DCTA complex. The titration of Bi can therefore be carried out at really low pH values and it is relatively selective. Fairly acidic solutions are needed in the titration, for otherwise the determination is ruined by reason of hydrolysis. Quite a long time ago Schwarzenbach [48-7] discussed the possibility of a direct titration where the end-point was detected by the bleaching of the very intensely yellow-coloured BiI$_4^-$ complex. This method was studied more thoroughly by Cheng [54-6], who found that a pH of $1\cdot5$–$2\cdot0$ and about 10 ml of a $0\cdot5\%$ KI solution in 50 ml of total solution volume gave the best results. About 1–2 ml of the KI solution is added at the beginning of the titration and the remainder when fading of the original colour indicates the approach of the end-point. Grönquist [53-20] uses the bleaching of the Bi–thiourea complex, which also has a yellow colour and adds some gentian violet to sharpen the end-point. The titration is carried out at pH $2\cdot5$–$4\cdot0$ in the presence of high concentrations of thiourea ($2\cdot5$–$3\cdot0$ g per 50 ml) and potassium hydrogen phthalate (1 g per 50 ml). Fritz improved on this method, firstly, by omitting the buffer, and secondly, by reducing the concentration of thiourea to $0\cdot5$–$1\cdot0$ g per 50 ml. Under these conditions Bi can be titrated at pH 2 or below, and with far fewer interferences. Even chloride, sulphate, and what is rather remarkable, tartrate, have no effect. Indeed, the latter can be added to mask Sb(III). Th can be masked by sulphate, and Be, Nb, and Ta by fluoride. Interference by Fe(III) is eliminated by reduction with ascorbic acid. Pb and the other common cations do not interfere, but Sn(IV) and Zr must be absent. The visual indication given by the bleaching of the yellow iodide- or thiourea-complex is not so sharp as the transition of a colour indicator. Dyestuff indicators have therefore been preferred, although the possibilities of interference are greater than in the thiourea method.

Innumerable metallochromic indicators have been proposed for the determination of Bi. Pyrogallol red [56-85], Dalzin (i.e. bis-(allylthiocarbonyl)hydrazine) [60-121], thoron [56-74], PAN [57-131], haematoxylin [59-136; 60-117], stilbazo [61-147], gallein [61-30], methylthymol blue [59-55], pyrocatechol violet [59-21], and xylenol orange [60-47] are only a selection. According to Sajó [63-1], the end-point can also be detected with the system vanadium(V)–EDTA–diphenylcarbazide.

Gattow and Schott [62-82] have made the complexometric determination of Bi the subject of a thorough and critical investigation. Titration using pyrocatechol violet was chosen as the standard method, and determinations with xylenol orange, PAN, PAR, thoron, thiourea, haematoxylin, methylthymol blue, and brompyrogallol red were all compared to it. PAR was found to be the best indicator, for it gave correct results with the lowest standard deviation for concentrations from 0.5 μg to 500 mg of Bi per ml. Methylthymol blue and haematoxylin gave equally accurate results, but the scatter was somewhat greater. Xylenol orange, thoron, and thiourea gave smaller standard deviations, but slightly high results compared to the standard method.

In a second paper [62-83] the authors examined the influence of hydrolysis and the formation of polynuclear cations on the results of the titration. According to their findings, the neutralization of the acidic test solution to the pH required for the titration should not be carried out with NaOH, ammonia, or pyridine. Local excess of alkali conduces to the formation of polynuclear complexes which react with EDTA either very slowly or not at all, and this leads to low results or long-drawn-out colour transitions at the end-point. Moreover, when once formed, polynuclear cations redissolve only very slowly when the solution is re-acidified. There are no difficulties if the neutralization is effected with an approximately molar solution of sodium bicarbonate or acetate.

To date, pyrocatechol violet is certainly the indicator most to be recommended. In the pH range from 1.5 to 2.0 the colour of the free indicator is a pure yellow, but with Bi it forms a complex of a deep pure blue colour with the composition Bi : indicator $= 2 : 1$. With increasing acidity the colour of this complex becomes violet. This makes it possible to carry out the neutralization of the test-solution before the titration by using the colour of the indicator-complex in its character as an acid–base indicator. In the actual titration no excess of free bismuth exists in the solution immediately before the end-point, and the 2 : 1 complex is decomposed to a red 1 : 1 complex. Thus warning is given of the approach to the end-point by the appearance of a red colour in the solution or at least in the neighbourhood of each drop of EDTA as it falls into the solution. The transition to pure yellow is exceptionally sharp.

Several back-titrations have been proposed. Landgren [52-1] back-titrated excess EDTA with Mg in a solution buffered with borax using Erio T as indicator. Belcher *et al.* [60-153] used *o*-anisidine-tetra-acetic acid as a fluorescence indicator and back-titrated with Cu. Ter Haar and Baazen [54-17] used Th and alizarin S. These procedures have no advantages over the direct titration unless interfering anions, especially Cl^-, are present.

Instrumental indication was introduced by Přibil [51-15] quite early on with an amperometric titration. Usatenko [60-95] and Zhdanov *et al.* [60-42] confirm that good results are obtainable by this method. Reilley *et al.* [62-33] also used an amperometric titration, but did not use the potential of the Bi plateau but that of the 'complexone wave' (cf. Part One, Section 5.6). Potentiometric titration is also possible with the mercury-drop electrode. Photometric titration was studied by Underwood. The titration is carried out either to the point of the disappearance of the absorption due to the thiourea complex or with Cu as a 'slope indicator', a procedure which is also used for the consecutive titration Bi–Cu. Self-indication can be used in the ultraviolet region [55-97], and this is also possible in the consecutive titration Bi–Pb. An investigation of the colour reaction between Bi and xylenol orange led Kotrlý and Vřeštál [60-47] to work out a photometric titration for microgram quantities of Bi and the possibility of a consecutive titration for Bi–Pb on the micro-scale.

Solutions of Bi were recommended by Flaschka and his co-workers for back-titrations in strongly acid solutions [56-42]. The determination of Ni in the presence of Co [55-66] and, *inter alia*, of Co (after oxidation to Co(III)) in the presence of Ni can be effected photometrically. On account of its high equivalent weight and the extremely sharp end-point with pyrocatechol violet, Přibil recommended analytically pure bismuth metal as a titrimetric standard for the standardization of EDTA solutions in an acid medium.

The complexometric determination of Bi can be carried out in the presence of considerable amounts of Pb (up to 1 : 5,000), and thereby provides a truly elegant solution to an old analytical problem. By using the appropriate indicators even the consecutive titration Bi–Pb can be carried out. Bi is titrated first in strongly acid solution (pH 1–2) and lead after increasing the pH to 5–6. This is possible with, e.g., xylenol orange either visually [60-62] or photometrically [60-47]. The increase in pH is best effected by adding urotropine. Gallein [61-32] and methylthymol blue [59-55] can also be used for such determinations. Of course, instrumental methods are of especial value here. Reilley *et al.* (cf. Part One, Section 5.2.3.) were able to determine all three elements in a group, e.g. Bi, Pb, Ca.

The Bi titration is very selective. Interferences can only occur with elements that form very stable complexes, e.g. Fe, Hg, In, Zr, Hf, etc. Which element can be tolerated depends on the indicator used, on the degree of acidity, and on the presence of masking agents. The advantages that Fritz's method presents from this point of view have been discussed above. Reduction with ascorbic acid eliminates not only iron but mercuric ions too (by reduction to metal, for which warming is necessary); however, Hg(II) can also be masked with thiocyanate [59-55].

The complexometric determination of Bi is used in practice for the analysis of alloys. Examples and full procedural details are to be found in many articles [54-101; 57-132; 57-133; 58-113; 59-32; 60-62]. Usatenko titrated Bi in lead alloys amperometrically in the presence of Pb that had been precipitated as $PbSO_4$ [60-95]. Complexometric titration of Bi is the method employed for analyses of many pharmaceutical preparations [52-1; 53-20; 54-85; 55-72; 57-119; 58-51; 59-55; 61-125]. Bi serves indirectly for the determination of the organic ingredients by way of precipitation with BiI_4^- (cf. Section 40.4).

49.1. Direct titration of bismuth

REAGENTS: Standard solution of 0·01M EDTA; pyrocatechol violet; solid $NaHCO_3$.

PROCEDURE: The sample for analysis is dissolved in a little conc. HNO_3 and the excess of acid is fumed off. Then water is added until the solution contains not more than 20–30 mg of Bi per 100 ml. Indicator is then added, when a pure blue colour should result. If the solution is violet or tinged with red the acidity is too high and $NaHCO_3$ must be added until the colour becomes a pure blue; this is the situation at about pH 2·5. Titration with EDTA is then carried out to a sharp change in colour to yellow. Since hydrogen ions are liberated during the complex formation, the pH may decrease too much during the titration, and this must be corrected by adding a few more grains of $NaHCO_3$.

Above pH 4 the change in colour is directly from blue to yellow. If the acidity is somewhat greater (pH $\sim$3) an intermediate colour is encountered just before the end-point, to wit, a violet colour which only persists while a few drops of titrant are being added and then changes to yellow with quite exceptional sharpness with the final drop.

REMARKS: If chloride is present (Br^- and I^- are removed by fuming with HNO_3) there is trouble in achieving the optimum pH range as BiOCl always separates. In such a case it is preferable to use back-titration. A small known excess of standard EDTA is added to the acid test solution, and after adjusting to pH 2·5–3·0 the excess of complexone is back-titrated with 0·01M $Bi(NO_3)_3$.

50. OXYGEN

Roskam and de Langen [63-71] have described a method for the determination of dissolved oxygen in water. 100 ml of the sample is placed in a flask with a ground-in stopper and the following reagents added in the order: 2 ml of 15% sodium salicylate, 150–200 mg of ferrous ethylenediamine sulphate, and 0·5 ml of tris-(hydroxymethyl)-aminomethane. The flask is immediately closed in such a way that no bubbles of air are included over the liquid and the contents are well mixed. The flask is then opened, and immediately 5 ml of a 15% ethanolic solution of maleic acid is introduced. Finally, the solution is titrated with 0·02M EDTA until the red-violet colour of the Fe(III)–salicylate complex vanishes.

It is specially important that the Fe(II) ethylenediamine sulphate should be completely free from Fe(III). The procedure gives the same results on samples of pure water as Winkler's method, but with somewhat lower precision. With samples of polluted water the complexone method appears to be superior, as there are fewer interferences.

51. SULPHUR

According to Gimesi, Rady, and Erdey [62-79], elementary sulphur can be determined as follows. The sample containing 30–80 mg of S is treated with about 25 ml of isopropanol in the warm. An accurately measured amount of 0·2M KCN solution (10–20 ml) is then added and the whole is allowed to cool: S and CN^- react to form CNS^-. The excess cyanide is then determined by pipetting in a 0·05M nickel solution, buffering to pH 10, and back-titrating excess Ni with 0·05M EDTA using murexide as indicator. The strength of the KCN solution is standardized against that of the Ni solution in a blank experiment. 1 ml of 0·2M cyanide corresponds to 6·4128 mg of S. In the cyanide titration 1Ni reacts with 4CN (cf. Section 40.2).

The determination of S in organic compounds, alloys, and other materials is generally carried out after combustion or wet-ashing, and concludes with a complexometric titration of sulphate (q.v.).

51.1. Sulphides

Kivalo [55-13] was the first to report on the indirect complexometric determination of sulphides. The test portion of alkali sulphide is pipetted into a neutral or weakly acidic solution of copper per-chlorate, CuS is filtered off, and the excess of Cu in the filtrate is back-titrated with EDTA using murexide. Solutions of Zn or Cd cannot be used, as the precipitated sulphides are non-stoicheio-metric. The same procedure has been described by Maurice [56-80]. Bartels *et al.* [59-128; 60-101] found from more detailed studies of this method that with small amounts of sulphides it is unnecessary to filter off the precipitate. The complexometric determination of sulphides often gives lower values than the iodometric method: the above-named authors attribute this to the fact that solutions of sulphides can contain thiosulphate and sulphite, which certainly react with iodine but not with Cu. The complexometric method ought therefore to give the true content of sulphide. Buděšinský [60-44] has described another method. The test solution is boiled with a solution of CdY^{2-} in caustic soda, whereby CdS is thrown down quanti-tatively and is then filtered off. The amount of EDTA that has been set free in an amount equivalent to the sulphide is titrated in the filtrate with Co using methylthymol blue as indicator. Sulphite, thiosulphate, or thiocyanate do not interfere in this procedure, which can be extended to the determination of sulphur in thiourea and some of its derivatives.

De Sousa has described the determination of sulphide in admixture with sulphate [61-179] and thiosulphate [61-26].

Sulphide can also be oxidized to sulphate and determined com-plexometrically in this form.

51.2. Sulphates

Munger, Nippler, and Ingols [50-10] were the first to describe a complexometric determination of sulphate. The method depends on the following principle. A measured amount of a standard solution of barium chloride is added to the test solution and the excess of barium is back-titrated with EDTA. The method has been studied and modified by innumerable authors, for the procedure is associated with some fundamental difficulties. In the first place, the colour transition in the Ba titration is not particularly sharp, and all the remarks made in this connection under Ba itself are valid here, and indeed to an even greater extent, as the titration has to be carried out in the presence of the precipitated $BaSO_4$. Secondly, there is a fundamental source of error in the precipitation of sulphate itself, since it is very difficult to prepare $BaSO_4$ entirely pure and with a stoicheiometric composition. Certainly there are plenty of working instructions for gravimetric determinations which give exact results which are, however, due to the compensation of positive and negative errors. The compensations found in the gravimetric procedures are not, however, effective in the complexometric determination, since the calculation is based on the actual amount of Ba titrated.

An improvement in the colour transition at the end-point has been achieved by replacing the Erio T originally used by other indicators. Particularly good end-point transitions are obtained with phthalein-complexone [53-1; 58-117; 60-101], with which it is best to work in a solution containing alcohol. The alcohol also depresses the solubility of the $BaSO_4$, and so reduces the possibility of a reaction between the precipitate and the excess of EDTA. Phthaleincomplexone is best employed in the form of a mixed indicator. Admixture with naphthol green (see the working instructions) or a solution of 0·1% phthalein-complexone, 0·005% methyl red, and 0·05% dianil green have been recommended [56-54]. The following mixtures have been devised for rhodizonic acid. Mustafin [59-83] uses a mixture of 1 part by volume of 0·05% sodium rhodizonate and 5 parts of 0·5% 1-methylamino-4-(*p*-tolylamino)-anthraquinone in 50% alcohol. However, the rhodizonic acid solution does not keep and must be made up freshly every day. Dragusin [60-190] recommends a freshly prepared suspension of about 20 mg of sodium rhodizonate in 2–3 ml water and a few drops of a 1% solution of alkali blue.

For very small amounts of sulphate, Ito and Abe [59-42] recommend a photometric titration at 640 mμ with Erio T as indicator. The BaSO$_4$ is not filtered off but kept in suspension by constant stirring, so that variations in the background absorption are reduced to a minimum. Boos [59-108] back-titrates the excess of Ba potentiometrically with a mercury-drop electrode.

During the back-titration of excess Ba all the metals present in solution that can be titrated complexometrically are naturally included in the titration value. In order to arrive at the concentration of sulphate present, it is necessary in a second aliquot portion of the test-solution to determine the amount of EDTA equivalent to all these metals so as to be able to include this in the calculation [50-10]. This is not conducive to accuracy, especially with small amounts of sulphate. Alternatively, separations must be carried out. All cations can be held back by an ion-exchange resin [53-1; 59-108], and the sulphate can then be determined in the eluate, but this procedure often leads to an undesirably high dilution of the test solution. In such cases the procedure advocated by Belcher and his co-workers [54-1] and worked up by Jackson [54-56] may be of use. The authors make use of an observation made by Přibil *et al.* [52-13] that BaSO$_4$ is soluble in alkaline EDTA. Following Belcher, the precipitate of BaSO$_4$ is filtered off, washed, and taken up in a measured amount of ammoniacal standard EDTA. Unfortunately the rate of solution is very slow, so that prolonged boiling is necessary and the ammonia that evaporates off has to be replaced from time to time. Other authors overcome this adverse feature by using 5% triethanolamine [59-51] or, still better, caustic soda [59-51; 59-72] to make the solution alkaline. Under these circumstances the precipitate redissolves completely in 5–10 minutes in even only a modest excess of EDTA. However, the method conceals an intrinsic hazard that during this process titratable cations may be dissolved from the walls of the vessel. Thus vessels of hard glass that have been pre-treated with hot alkaline EDTA, or better still vessels of quartz or platinum, must be employed. These sources of error could also be got round by using vessels of heat-resistant plastics.

Another, and in many ways a more advantageous, method for determining sulphate depends on the precipitation of PbSO$_4$, a procedure first described by Schneyder [56-101]. Lead sulphate is coarsely crystalline, shows a lower tendency to co-precipitation, and is easier to obtain stoicheiometrically pure than BaSO$_4$. It dissolves without difficulty in ammoniacal EDTA, and Pb can be titrated to a very sharp end-point. On all these grounds the method is simple and substantially free from interferences; but it suffers from the relatively high solubility of lead sulphate. Addition of alcohol reduces this and improves the situation. Schneyder uses 20% ethanol. At pH ∼0·6 sulphate can even be determined in the presence of phosphate. Sporek [58-104] used isopropanol. Ashbrook [61-106] recommended 50% alcohol and standing for about 3 hours to complete the precipitation. Odler [61-52] used a 20% methanol solution: the precipitate of $PbSO_4$ was filtered off, washed, and dissolved in ammoniacal EDTA, after which excess complexone was back-titrated with Zn using Erio T as indicator [56-101; 58-104; 61-106]. The back-titration can be avoided in Odler's procedure [61-52], where the precipitate of $PbSO_4$ is taken up in a solution of sodium acetate and the Pb titrated using xylenol orange as indicator. Calleja *et al.* [61-150] bring the lead into solution with tartrate and back-titrate the lead directly at pH 10 using Erio T.

The complexometric determination of sulphate has found practical applications in an extraordinarily large number of cases, but this is hardly to be wondered at, since a quick volumetric determination had been found as a substitute for a tedious gravimetric procedure. To name only a few examples: sulphate or sulphur content has been determined in natural waters [50-10; 57-116; 64-139], in rock salt [58-85], pickling brines [54-64], petroleum products [56-12], cement [61-150], viscose coagulating fluids [52-44], baryta [62-104], gypsum [63-2], potash [62-107], iron and steel [55-106; 63-33], coal [57-93], reagent-grade chemicals [64-135], chromate baths [65-102], rayon fibre [56-79], and manganese leach liquors [61-11]. The method has also been used for the determination of SO_3 in the air [59-51] and for the analytical control of sulphonation reactions [61-140]. The analysis of binary mixtures of sulphate and persulphate [61-27], sulphide [61-179], or sulphite [62-106] has been reported. Papers on the determination of sulphur in organic compounds have been particularly numerous [54-15; 59-128; 60-44; 62-103]. In this connection the use of complexometry as a finish to combustion by the oxygen flask (Schöniger) method has been particularly successful [56-20; 59-108; 63-34]. The determination of sulphur as a finish to Pregl's method [59-66] and after dissolving the wad of silver wool from the combustion tube [61-178] has also been studied. Körbl *et al.* [56-33] report on a method for determining S in organic compounds that does not depend upon precipitation with Ba or Pb. Here the combustion tube is packed with silver permanganate and the amount of manganous sulphate formed (which is equivalent to the sulphur content of the sample) is leached out and the Mn(II) determined complexometrically.

An analytical scheme for the determination of S^{2-}, SO_3^{2-}, and SO_4^{2-} in waters has been worked out by Burriel-Marti *et al.* [64-138].

51.3. Indirect determination of sulphate

REAGENTS: Volumetric solutions of 0·01M EDTA and 0·01M $BaCl_2$ standardized one against the other; metalphthalein–naphthol green indicator; conc. ammonia; approx. 1M NaOH; ethanol.

PROCEDURE: A portion of the test solution containing 20–50 mg of sulphate is made slightly acid and heated while a slight excess of 0·01M $BaCl_2$ is added dropwise. Every ml of the Ba solution precipitates about 1 mg of sulphate. The solution is allowed to digest on the water-bath for some time to render the precipitate coarsely granular and insoluble; it is then cooled down and the suspension is neutralized with NaOH. The volume should now be about 100 ml. 0·3 ml of indicator and 5–10 ml of ammonia are introduced and the red suspension titrated with 0·01M EDTA until the red colour suddenly pales. After the end-point a small excess of EDTA is added, since the equivalence point can be established more precisely by back-titration in alcoholic solution. An estimate is made of the volume of the fully titrated solution, and an equal volume of alcohol is added. This gives a greenish-coloured solution. Finally, the second back-titration is carried out with 0·01M $BaCl_2$; the end-point is easily detected by the sudden reappearance of the red colour.

CALCULATION: The sulphate content is calculated from the difference between the equivalents of the total volumes of $BaCl_2$ and EDTA used.

REMARKS: Refer to observations given under the titration of Ba, Section 10.1.4.

51.4. Sulphites

The sulphite ion can be determined by reaction with Tl(III) and titration of the excess of this ion with EDTA with xylenol orange as indicator [66-55]. For its determination in water in the presence of other sulphur compounds see [64-138].

51.5. Peroxydisulphates

Persulphate ions can be determined by an amalgam reduction followed by titration of the metal ion so generated [64-67].

51.6. Thiocyanates

Thiocyanates can be determined by the method described under the halogens (Section 54), in which the silver salt is precipitated, collected, and allowed to react with $Ni(CN)_4^{2-}$. For the analysis of mixtures of halogens and thiocyanates see also under halogens. According to Ralea and Simininc [60-104], 1–6 mg of thiocyanate can be determined by adding the test solution to a measured quantity of $0.01M$ copper solution and leading in SO_2 for a few minutes. After removing the precipitate of CuSCN the filtrate is adjusted to pH 10, and excess Cu is back-titrated according to procedure 31.1 using murexide. Iodide and cyanide interfere with this determination, but it can be carried out without any trouble in the presence of chloride and bromide.

52. SELENIUM

According to Liu and Cheng [58-31], the determination of selenium in the form of selenate can be carried out as follows. The selenate is precipitated with lead at pH 2–3 in 30% alcoholic solution, filtered off, and washed. The precipitate is then redissolved in a solution containing ascorbic acid, ammoniacal buffer of pH 10, and a measured amount of a standard solution of EDTA. The excess of EDTA is then back-titrated with a standard solution of Mn using Erio T as indicator. Selenite can be determined in the same way if it is first oxidized to selenate with potassium permanganate, excess of this being destroyed with hydrogen peroxide. Small amounts of Cu, Fe, Ni, Co, and TeO_4^{2-} do not interfere, but Bi, Hg, SO_4^{2-}, and AsO_4^{3-} must be absent. According to Erdey and his co-workers [62-79], elementary selenium can be determined by allowing it to react with a measured amount of cyanide ($CN^- + Se \longrightarrow SeCN^-$). When this reaction is complete a Ni solution is added to determine the excess of cyanide and the excess of Ni is back-titrated with EDTA (cf. the determination of cyanides in Section 40.2).

53. FLUORINE

According to Belcher and Clark [53-14], indirect determination of fluorine is possible through a calcium titration. Fluoride is treated with a standard solution of Ca at pH 4.5, the CaF_2 is filtered off, and the excess Ca in the filtrate is back-titrated with EDTA using Erio T. As has been proved often enough, the end-point in the titration of a pure Co solution using Erio T as indicator is a bad one, but it can be made sharp by adding some MgY^{2-}. However, the authors make no use of this possibility. The difficulties and the possibilities of error in this determination lie less in the determination of Ca than in its precipitation. Since calcium fluoride is not very sparingly soluble, an appreciable excess of Ca must be added to ensure complete precipitation. This means that a large volume of titrant will be needed for the back-titration, and this has an adverse effect on the reproducibility. Calcium fluoride separates out only very slowly, so that standing for a few hours – preferably overnight – is needed for complete precipitation. In comparison with the more usual direct titration with a solution of thorium that was formerly used, the advantage of the complexometric procedure is that sulphate and phosphate do not interfere, even when present in 10-fold excess. Check analyses by the authors gave good results for 5–65 mg of F. For amounts below 5 mg low results of rapidly increasing inaccuracy were obtained.

Belcher and Clark found that the addition of alcohol to the test-solution to improve the speed and completeness of the precipitation did not improve the situation. This result contradicts that of Okada and Sugiyama [57-118], who worked with 75% alcoholic solutions and were able to determine micro-amounts of F with an accuracy of about 1%. The precipitation was conducted with a 2·5 or $5 \times 10^{-3}M$ Ca solution, but apart from this their method was analogous to that of Belcher and Clark.

Laszlovsky worked out another indirect method for the determination of F through the complexometric titration of lead contained in PbClF. The lead titration gives much sharper end-points, but once more the difficulty lies in the precipitation stage. Admittedly PbClF crystallizes well, but it is not completely insoluble, and once again a suitable excess of precipitant is needed. The conditions for the precipitation must be adhered to very strictly, especially with micro-amounts, otherwise the composition of the precipitate will not correspond to the stoicheiometric ratio Pb : F = 1 : 1. The method has been worked out for 0·2–2 mg quantities of F which can be determined with an accuracy of 2–3%. Standing for 24 hours is necessary to ensure complete precipitation.

Vřeštál and his co-workers [58-46] modified Laszolovsky's procedure to a back-titration of excess Pb in the filtrate. This procedure has been worked out for 4–40 mg of F and needs a standing period of 1 hour only.

Belcher and Clark's procedure has been used for the determination of F in organic compounds after decomposition in a bomb or sealed tube or by any of the other customary procedures [57-129].

53.1. Determination of fluorine with lead chloride

REAGENTS: A 0·75% (approx. 0·03M) solution of lead chloride. This is prepared by allowing a solution saturated with $PbCl_2$ at the boiling-point to cool down to room temperature, filtering, and diluting the filtrate with one-tenth of its volume of water: the exact concentration is obtained by standardization against 0·05M EDTA; 0·1% methyl orange; dilute HNO_3; dilute NaOH; sodium potassium tartrate; ammonia–ammonium chloride buffer of pH 10; 1% urotropine; Erio T–NaCl indicator powder.

PROCEDURE: The sample containing 5–40 mg of fluorine (as alkali salt) in not more than 50–70 ml of solution is placed in a 250-ml standard flask with 2–3 drops of methyl orange solution and brought to the transition colour by adding dilute NaOH or HNO_3. 150 ml of lead chloride solution (accurately measured) is allowed to flow in slowly with good stirring: the pH falls and the solution is made neutral again, this time with urotropine. After an hour the solution is made up to the mark, well shaken, and filtered through a dry quantitative filter-paper. The first portion of the filtrate is discarded, and from the remainder 100 ml (or with small amounts of fluoride, 150 ml) is pipetted into a titration flask and treated with about 1 g of sodium potassium tartrate. After adding 10 ml of buffer and some Erio T indicator the titration with EDTA is carried out to a colour change from a reddish violet to a blue from which the last trace of a reddish hue has vanished.

CALCULATION: (Ml . Pb × Molarity of Pb − ml . EDTA × Molarity of EDTA × 250/A) × 18·998 = mg fluorine
where A is the volume of the filtrate (ml) taken for the titration.

REMARKS: The above procedure is a modification of that of Vřeštál based on unpublished work by Flaschka and Wolfram. Vřeštál's original method prescribed titration in a solution buffered with urotropine and using pyrocatechol violet or xylenol orange as indicators. Because of the low buffer capacity of urotropine an ammoniacal solution of ethylenediaminetetra-acetic acid of precisely adjusted acidity is needed for the standard solution of the titrant: the trouble in preparing this is only repaid if a large number of analyses have to be made. In place of the solution of $PbCl_2$ prescribed above one of lead acetate or lead nitrate can be used, but in such cases a spatulaful of NaCl must be added to the test-solution and the precipitation completed in the warm. The results, however, are liable to larger variations. An accuracy of 1–2% is attainable by following the working instructions given above. For a series of analyses of materials of similar composition it is advisable to calculate the titre of the standard solution from the analysis of a standard sample. Quantities of sulphates in excess of about 3–4 mg (calculated as ammonium sulphate) produce definitely high results.

54. HALOGENS (EXCLUDING FLUORINE)

The indirect determination of the halogens (with the exception of fluorine) is possible through the silver titration already described in Section 32 [52-28]. The silver halide is precipitated in the usual way, filtered, and washed. The filter-paper or the sintered-glass filter with the precipitate on it is transferred directly into a strongly ammoniacal solution of $Ni(CN)_4^{2-}$ and does not interfere with the titration. Micro-amounts also give excellent results [52-26].

The procedure is used for the determination of chloride in biological fluids [56-62] and for halogens in organic substances after decomposition in a Carius tube [59-47] or combustion by the oxygen flask (Schöniger) method [56-20].

According to Aleksandrovich-Melnikova [62-112], chloride can be determined as follows. 20 ml of a 0·01M standard solution of bismuth nitrate is added to 10 ml of the test solution containing 2–4 mg of Cl, dropwise at first until a turbidity results. The mixture is then diluted to 100 ml, and the excess Bi remaining in the solution is titrated with EDTA. Pyrocatechol violet serves as the indicator or 1–2 drops of a saturated alcoholic solution of haematoxylin (0·5 g is dissolved in hot alcohol and allowed to stand for a week). Al, Ni, Zn, Cu, Mn, Ca, Mg, Ba, Sr, and Fe(II) do not interfere, and Fe(III) can be masked with ascorbic acid. The method serves for the analysis of table salt.

Chloride [62-113] and bromide [63-41] can also be determined via a lead determination according to Gusev *et al.* In a solution containing Pb^{2+}, chloride and bromide give precipitates with 2-(2-hydroxy-1-naphthyl-methyleneamino)pyridine ($C_{16}H_{14}ON_2$) of the composition $RPbX_2$, where R is the organic ligand and X = Cl or Br. The precipitate is isolated, dissolved in nitric acid, and after buffering with urotropine the lead present is titrated with EDTA using xylenol orange as indicator.

In a series of papers de Sousa has reported on the use of the silver method in conjunction with other procedures for the analysis of mixtures of halide ions and/or oxyacids of the halogen group. Thiocyanates can also be included. To analyse, for example, a mixture of bromide and bromate, Br^- is precipitated as AgBr in an aliquot portion and titrated after reaction with $Ni(CN)_4^{2-}$. In a second aliquot portion bromate is reduced to bromide: subsequent precipitation with silver now gives the sum of bromide plus bromate, from which the latter can be obtained by difference. For mixtures of chloride and iodide the total halogen is determined in one aliquot portion by precipitation with silver and subsequent titration. A second aliquot portion is treated with bichromate and heated, whereupon chloride is oxidized to chlorine and boiled off. The iodate formed from the iodide is then reduced, precipitated as AgI, and titrated. Chloride is calculated from the difference in the volumes of EDTA used [60-22]. Chlorate and perchlorate are reduced by fusion with NH_4CL in one aliquot portion and determined: in a second portion only chlorate is reduced by treatment with Fe(II) and determined complexometrically [60-21]. Ternary mixtures of chloride, cyanide, and thiocyanate can be analysed as follows. In one aliquot portion of the test solution all three components are precipitated as their silver salts. This precipitate is boiled with HNO_3, whereby AgCN is decomposed, thiocyanate is oxidized to sulphate, and AgCl remains unchanged. The AgCl is filtered off and determined complexometrically. In the filtrate the amount of sulphate equivalent to the thiocyanate is determined complexometrically by method 51.3. In another aliquot portion cyanide is determined by back-titration of excess Ni with EDTA according to procedure 40.3 [61-29].

Bibliography

A (before 1945)

A-1   RUGGLI, P., ZIMMERMANN, A., and KAPP, F., Zur Konstitution von Eriochromschwarz T. *Helv. Chim. Acta*, **13**, 784 (1930).

A-2   BRITTON, H. T. S., and DODD, E. N., Physico-chemical studies of complex formation involving weak acids. V. Solutions of complex cyanides of Ag, Zn, Cd, Hg and Ni. *J. Chem. Soc.*, 1932 (1940).

A-3   BJERRUM, J., Metal ammine formation in aqueous solution. Dissertation, Copenhagen (1941).

A-4   LEDEN, I., Potentiometrisk Undersökning av några Kadmiumsalters Komplexitet. Doctoral Thesis, Lund (1943).

A-5   GAUGIN, R., Étude potentiométrique des propriétés du cyanure d'argent. *J. chim. Phys.*, **42**, 28 (1945).

A-6   SCHWARZENBACH, G., and WILLI, A., Die Stabilität der Tironkomplexe von Mg, Ca, Mn, Co, Ni, Cu, Zn, Cd. Unpublished work.

A-7   SCHWARZENBACH, G., and ANDEREGG, G., Die Komplexe der Chromotropfarbstoffe. Unpublished work.

A-8   SCHWARZENBACH, G., and SCHNORF, P., Swiss Patents No. 660,408, U.S.A. Patents Nos. 2,583,890 and 2,583,891. Property of Chemische Fabrik Uetikon, Uetikon (Switzerland).

A-9   BRINTZINGER, H., and HESS, G., Cu, Ni and Uranylverbindungen der AeDTE. *Zeit. anorg. Chem.*, **249**, 113 (1942).

A-10   KATHEN, H., and LANG, K., Na-Bestimmung im Serum. *Biochem. Z.*, **318**, 308 (1927).

A-11   GUTZEIT, G., Sur une méthode d'analyse qualitative rapide. I. De quelques réactions spécifiques et spéciales des cations et anions les plus usuels. *Helv. Chim. Acta*, **12**, 713 (1929).

1945

45-1   SCHWARZENBACH, G., KAMPITSCH, E., and STEINER, R., Komplexone. I. Über die Salzbildung der Nitrilotriessigsäure. *Helv. Chim. Acta*, **28**, 828 (1945).

45-2   SCHWARZENBACH, G., KAMPITSCH, E., and STEINER, R., Komplexone. II. Das Komplexbildungsvermögen von Iminodiessigsäure, Methyliminodiessigsäure, Aminomalonsäure und Aminomalonsäurediessigsäure. *Helv. Chim. Acta*, **28**, 1133 (1945).

45-3   SCHWARZENBACH, G., Säuren, Basen und Komplexbildner. *Schweiz. Chem. Ztg.*, No. 9 (1945).

1946

46-1   SCHWARZENBACH, G., KAMPITSCH, E., and STEINER, R.,

Komplexone. III. Uramildiessigsäure und ihr Komplexbildungs-vermögen. *Helv. Chim. Acta*, **29**, 364 (1946).

46-2 SCHWARZENBACH, G., BIEDERMANN, W., and BANGERTER, F., Komplexone. VI. Neue einfache Titriermethoden zur Bestimmung der Wasserhärte. *Helv. Chim. Acta*, **29**, 811 (1946).

1947

47-1 SCHWARZENBACH, G., WILLI, A., and BACH, R. O., Komplex-one. IV. Die Acidität und die Erdalkalikomplexe der Anilindi-essigsäure und ihrer Substitutionsprodukte. *Helv. Chim. Acta*, **30**, 1303 (1947).

47-2 SCHWARZENBACH, G., and ACKERMANN, H., Komplexone. V. Die Aethylenediamintetraessigsaure. *Helv. Chim. Acta*, **30**, 1798 (1947).

47-3 KOLTHOFF, I. M., and STENGER, V. A., *Volumetric Analysis.* Volume II, pp. 204, 284, 331. Interscience Publishers Inc., New York (1947).

47-4 BIEDERMANN, W., Dissertation Zürich (1947).

47-5 KOLTHOFF, I. M., and STENGER, V. A., *Volumetric Analysis.* Volume II. Interscience Publishers Inc., New York (1947).

1948

48-1 SCHWARZENBACH, G., and BIEDERMANN, W., Komplexone. VII. Titration von Metallen mit Nitrilotriessigsäure. Endpunktsindika-tion durch pH-Effekte. *Helv. Chim. Acta*, **31**, 331 (1948).

48-2 SCHWARZENBACH, G., and BIEDERMANN, W., Komplexone. VIII. Titration von Metallen mit Uramildiessigsäure. Endpunkts-indikation durch pH-Effekte. *Helv. Chim. Acta*, **31**, 456 (1948).

48-3 SCHWARZENBACH, G., and BIEDERMANN, W., Komplexone. IX. Titration von Metallen mit Aethylenediamintetraessigsäure. Endpunktsindikation durch pH-Effekte. *Helv. Chim. Acta*, **31**, 459 (1948).

48-4 SCHWARZENBACH, G., and BIEDERMANN, W., Komplexone. X. Erdalkalikomplexe von o,o'-Dioxyazofarbstoffen. *Helv. Chim. Acta*, **31**, 678 (1948).

48-5 BIEDERMANN, W., and SCHWARZENBACH, G., Komplexone. XI. Die Komplexometrische Titration der Erdalkalien und einiger anderer Metalle mit Eriochromschwarz T. *Chimia (Switz.)*, **2**, 1 (1948).

48-6 SCHWARZENBACH, G., and ACKERMANN, H., Komplexone. XII. Die Homologen des Aethylenediamintetraessigsäure und ihre Erdalkalikomplexe. *Helv. Chim. Acta*, **31**, 1029 (1948).

48-7 SCHWARZENBACH, G., *Komplexon-Methoden.* Siegfried and Co., Zofingen (1948).

1949

49-1 SCHWARZENBACH, G., Komplexone. XIII. Chelatkomplexe des Kobalts mit und ohne Fremdliganden. *Helv. Chim. Acta*, **32**, 839 (1949).

49-2   ACKERMANN, H., and SCHWARZENBACH, G., Komplexone. XVI. Die Bestimmung besonders stabiler Komplexe der Iminodiessig-säure-Derivate. *Helv. Chim. Acta*, **32**, 1543 (1949).

49-3   SCHWARZENBACH, G., and ACKERMANN, H., Komplexone. XVII. Die Diaminocyclohexan-tetraessigsäure als Komplexbildner für Erdalkalien. *Helv. Chim. Acta*, **32**, 1682 (1949).

49-4   SCHWARZENBACH, G., and GYSLING, H., Metallindikatoren. I. Murexid als Indikator auf Ca und andere Metallionen. Komplex-bildung und Lichtabsorption. *Helv. Chim. Acta*, **32**, 1314 (1949).

49-5   GYSLING, H., and SCHWARZENBACH, G., Metallindikatoren. II. Beziehungen zwischen Struktur und Komplexbildungsvermögen bei Verwandten des Murexids. *Helv. Chim. Acta*, **32**, 1484 (1949).

49-6   SCHWARZENBACH, G., Die Sonderstellung des Wasserstoffions. *Chimia (Switz.)*, **3**, 1 (1949).

49-7   SCHWARZENBACH, G., ACKERMANN, H., and RUCKSTUHL, P., Komplexone. XV. Neue Derivate der Iminodiessigsäure und ihre Erdalkalikomplexe. *Helv. Chim. Acta*, **32**, 1175 (1949).

1950

50-1   BJERRUM, J., On the tendency of the metal ions towards complex formation. *Chem. Rev.*, **46**, 381 (1950).

50-2   SCHMIDLIN, H. U., Dissertation (Studienabschlussarbeit), Zürich (1950).

50-3   PRUE, J. E., and SCHWARZENBACH, G., Metallkomplexe mit Polyaminen II: mit Triamino-triäthylamin = 'tren'. *Helv. Chim. Acta*, **33**, 963 (1950).

50-4   BETZ, J. D., and NOLL, C. A., Total hardness determination by direct colorimetric titration. *J. Amer. Water Works Assoc.*, **42**, 49 (1950).

50-5   MARCY, V. M., New water hardness test is faster and gives more accurate results. *Power*, **94**, 105 (1950).

50-6   MARCY, V. M., Rapid test for calcium hardness. *Power*, **94**, 92 (1950).

50-7   DIEHL, H., GOETZ, C. A., and HACH, C. C., The versenate titration of total hardness. *J. Amer. Water Works Assoc.*, **42**, 40 (1950).

50-8   GOETZ, C. A., LOOMIS, T. C., and DIEHL, H., Total hardness in water. *Analyt. Chem.*, **22**, 798 (1950).

50-9   MATTOCKS, A. M., and HERNANDEZ, H. R., Assay of officinal calcium preparations with the Schwarzenbach method. *J. Amer. Pharmaceut. Assoc.*, **39**, 519 (1950).

50-10   MUNGER, J. R., NIPPLER, R. W., and INGOLS, R. S., Titration of sulphate with barium and versene in waters. *Analyt. Chem.*, **22**, 1455 (1950).

50-11   WILSEN, A. E., Volumetric determination of Ca and Mg in leaf tissues. *Analyt. Chem.*, **22**, 1571 (1950).

1951

51-1   SCHWARZENBACH, G., and HELLER, J., Komplexone. XVIII. Die Eisen(II) und Eisen(III)-komplexe der AeDTE und ihr Redox-gleichgewicht. *Helv. Chim. Acta*, **34**, 576 (1951).

51-2 SCHWARZENBACH, G., and FREITAG ELSI, Komplexone. XIX.
Die Bildungskonstanten von Schwermetallkomplexen der NTE.
Helv. Chim. Acta, **34**, 1492 (1951).

51-3 SCHWARZENBACH, G., and FREITAG ELSI, Komplexone. XX.
Stabilitätskonstanten von Schwermetallkomplexen der AeDTE.
Helv. Chim. Acta, **34**, 1503 (1951).

51-4 SCHWARZENBACH, G., and HELLER, J., Komplexone. XXI. Die
Eisenkomplexe der NTE. *Helv. Chim. Acta*, **34**, 1889 (1951).

51-5 SCHWARZENBACH, G., and WILLI, A., Metallindikatoren. III.
Die Komplexbildung der Brenzcatechindisulfosäure (= Tiron)
mit dem Eisen(III)-ion. *Helv. Chim. Acta*, **34**, 528 (1951).

51-6 HELLER, J., and SCHWARZENBACH, G., Metallindikatoren. IV.
Die Aciditätskonstanten und die Eisenkomplexe der Chromo-
tropsäure. *Helv. Chim. Acta*, **34**, 1876 (1951).

51-7 CHENG, K. L., and BRAY, R. H., Determination of Ca and Mg in
soil and plant material. *Soil Science*, **72**, 449 (1951).

51-8 MARTELL, A., and CALVIN, M., *Chemistry of the Metal Chelate
Compounds*. Prentice Hall Inc., New York (1951).

51-9 BUCKLEY, E. S. JR, GIBSON, J. G., and BARTOLOTTI, T. R.,
Simplified titrimetric technique for the assay of Ca and Mg in
plasma. *J. Lab. Clin. Med.*, **38**, 751 (1951).

51-10 LANDGREN, O., Bestimmung von Calcium und Magnesium in
medizinischen Präparaten. *Svensk. Farm. Tidskr.*, **55**, 578 (1951).

51-11 GRIFFENHAGEN, G. B., PFISTERER, J. L., and SLOTH, S. K., Assay
of pharmaceutical tablets containing Ca by the Schwarzenbach
method. *J. Amer. Pharmaceut. Assoc.*, **40**, 359 (1951).

51-12 DAVIES, C. W., The electrolytic dissociation of metal hydroxides.
J. Chem. Soc., 1256 (1951).

51-13 SOBEL, A. E., and HANOK, A., A rapid method for determination of
ultramicro quantities of Ca and Mg. *Proc. Soc. Exp. Biol. Med.*,
77, 737 (1951).

51-14 KNIGHT, A. G., The estimation of Ca in water. *Chem. and Ind.*,
1141 (1951).

51-15 PŘIBIL, R., and MATYSKA, B., Use of complexones in chemical
analysis. XVa. The amperometric determination of Bi and certain
other metals. *Coll. Czech. Chem. Commun.*, **16**, 139 (1951).

51-16 PŘIBIL, R., KONDELA, Z., and MATYSKA, B., Use of complexones in chemical analysis. XIII. Potentiometric determination
of certain cations by means of Complexone III solution. *Coll.
Czech. Chem. Commun.*, **16**, 80 (1951).

51-17 GREENBLATT, I. J., and HARTMAN, S., Determination of Ca in
biological fluids. *Analyt. Chem.*, **23**, 1708 (1951).

51-18 SCHNEIDER, F., and EMMERICH, A., Härtebestimmung von Ca
und Mg in Zuckerfabriksäften. *Zucker Beihefte*, **1**, 53 (1951).

51-19 PŘIBIL, R., SIMON, V., and DOLEŽAL, J., Use of complexones in
chemical analysis. XXIII. Iodometric determination of Ce(IV),
MnO_4^- and CrO_4^{2-}. *Coll. Czech. Chem. Commun.*, **16**, 573 (1951).

51-20 FLASCHKA, H., and SCHÖNINGER, W., Über das Arbeiten mit

Indikatoren geringer Haltbarkeit. *Z. analyt. Chem.*, **133**, 321 (1951).

51-21 PŘIBIL, R., Use of complexones in chemical analysis. XIV. Review of some methods of separation and estimation. *Coll. Czech. Chem. Commun.*, **16**, 86 (1951).

1952

52-1 LANDGREN, O., Bestimmung von Wismuth mit Versene. *Svensk. Farm. Tidskr.*, **56**, 241 (1952).

52-2 HOLTZ, A. H., and SEEKLES, L., Direct titration of Ca in blood serum. *Nature*, **169**, 870 (1952).

52-3 LONGFORD, K. L., Determination of Mg in nickel sulphate. *Electroplating*, **5**, 41 (1952).

52-4 HOL, P. J., and LEENDERTSE, G. C., Bestimmung von Natrium- und Kalium-ferrocyanid mit einer Zink-Komplexonmethode. *Chem. Weekbl.*, **48**, 181 (1952).

52-5 PORTER, J. D., A stable form of Eriochrome Black T indicator. *Chem. Analyst*, **41**, 33 (1952).

52-6 KINNUNEN, J., and MERIKANTO, B., Volumetric determination of Zn in metallurgical products by use of EDTA. *Chem. Analyst*, **41**, 76 (1952).

52-7 BROOKE, M., and HOLBROOK, M., Determination of hardness in waters containing polyphosphate. *Chem. Analyst*, **41**, 80 (1952).

52-8 BOTHA, G. R., and WEBB, M. M., The versenate method for the determination of Ca and Mg in mineralized waters containing large concentrations of interfering ions. *J. Inst. Water Engineers*, **6**, 459 (1952).

52-9 ELLIOT, W. E., Volumetric determination of Ca in blood serum. *J. Biol. Chem.*, **197**, 641 (1952).

52-10 DE LORENZI, F., and ALDROVANDI, R., Phosphatbestimmung in Pharmazeutika. *Farm Sci. e Tec. (Pavia)*, **7**, 309 (1952).

52-11 BLUMER, M., and KOLTHOFF, I. M., Das polarographische Verhalten von Ti(III) und Ti(IV) in Aethylenediamintetraacetat. *Experientia*, **8**, 138 (1952).

52-12 BRUNISHOLZ, G., GENTON, M., and PLATTNER, W., Sur le dosage complexométrique du Ca en présence de Mg et de phosphate. *Helv. Chim. Acta*, **36**, 782 (1952).

52-13 PŘIBIL, R., and MARINCOVÁ, D., Komplexone in der chemischen Analyse. XXXV. Die gravimetrische Bestimmung von Ba und Sulfat. *Chem. Listy*, **46**, 542 (1952).

52-14 PŘIBIL, R., SIMON, V., and DOLEŽAL, J., Reductiometric detection and determination of Ag. *Sbornik Celost. Pracovni Konf. Anal. Chem. Prague*, 90 (1952).

52-15 SMITH, L. E., Rapid determination of Ca in pulp. *Pulp and Paper*, **26**, 86 (1952).

52-16 NIELSEN, H., Determination of serum-Ca. *Nord. Med.*, **48**, 1059 (1952).

52-17 SCHWARZENBACH, G., Chelate complex formation as a basis for titration processes. *Anal. Chim. Acta*, **7**, 141 (1952).

52-19 SCHWARZENBACH, G., ANDEREGG, G., and SALLMANN, R., Komplexone. XXIII. Der Phenolsauerstoff als Koordinationspartner. *Helv. Chim. Acta*, 35, 1785 (1952).

52-20 HELLER, J., and SCHWARZENBACH, G., Metallindikatoren. V. Die Eisenkomplexe der 2,3-Dioxynaphthalinsulfosäure. *Helv. Chim. Acta*, 35, 812 (1952).

52-21 SCHWARZENBACH, G., ANDEREGG, G., and SALLMANN, R., Metallindikatoren. VI. Ein auf Metallkationen ansprechendes Nitrophenol. *Helv. Chim. Acta*, 35, 1794 (1952).

52-22 FLASCHKA, H., Mikrochemische Titrationen mit AeDTE. I. *Mikrochem. ver. Mikrochim. Acta*, 39, 38 (1952).

52-23 FLASCHKA, H., and HOLASEK, A., Mikrotitrationen mit AeDTE. II. Bestimmung kleiner Mengen von Phosphat. *Mikrochem. ver. Mikrochim. Acta*, 39, 101 (1952).

52-24 FLASCHKA, H., Mikrotitrationen mit AeDTE. III. Die direkte Bestimmung von Blei. *Mikrochem. ver. Mikrochim. Acta*, 39, 315 (1952).

52-25 FLASCHKA, H., Mikrotitrationen mit AeDTE. IV. Bestimmung von Na. *Mikrochem. ver. Mikrochim. Acta*, 39, 391 (1952).

52-26 FLASCHKA, H., Mikrotitrationen mit AeDTE. V. Indirekte Bestimmung von Silber und der Halogene. *Mikrochem. ver. Mikrochim. Acta*, 40, 21 (1952).

52-27 FLASCHKA, H., Mikrotitrationen mit AeDTE. VI. Bestimmung von Tl. *Mikro. ver. Mikrochim. Acta*, 40, 42 (1952).

52-28 FLASCHKA, H., and HUDITZ, F., Indirekte titrimetrische Bestimmung von Silber und den Halogenen mit AeDTE. *Z. analyt. Chem.*, 137, 104 (1952).

52-29 FLASCHKA, H., and HUDITZ, F., Titration von Pb mit Komplexon in Gegenwart anderer Metalle und deren Gesamtbestimmung. *Z. analyt. Chem.*, 137, 172 (1952).

52-30 FLASCHKA, H., and HUDITZ, F., Schnellbestimmung von Ca in Magnesit. *Radex Rundschau*, 181 (1952), 21 (1954).

52-31 HUDITZ, F., FLASCHKA, H., and PETZOLD, INGRID, Fällung des Phosphations mit Magnesium bei Gegenwart von Calcium und anderer Kationen und die analytische Bestimmung mit AeDTE. *Z. analyt. Chem.*, 135, 333 (1952).

52-32 HUDITZ, H., and FLASCHKA, H., Die Titration von Cyanid mit Nickellösungen. *Z. analyt. Chem.*, 136, 185 (1952).

52-33 HOLASEK, A., and FLASCHKA, H., Neue Methode zur Bestimmung von Mg und Ca in Blutserum. *Z. physiol. Chem.*, 290, 57 (1952).

52-34 FLASCHKA, H., and HOLASEK, A., Neue Methode zur Bestimmung des anorganischen Phosphors im Blutserum. *Z. physiol. Chem.*, 289, 279 (1952).

52-35 FLASCHKA, H., Mikromethoden im Farben- und Lacklaboratorium. VIII. Analyse von ZnO. *Fette und Seifen*, 54, 267 (1952).

52-36 PŘIBIL, R., and VIČENOVA, E., Use of complexones in chemical analysis. XXXIII. Polarographic and polarimetric determination of calcium. *Chem. Listy*, 46, 535 (1952).

52-37 DEBNEY, E. W., Micro determination of magnesium, calcium and zinc with EDTA. *Nature*, **169**, 1104 (1952).

52-38 KYDD, P. H., Investigation of the formation of acetylene in methane–oxygen flames. *J. Amer. Chem. Soc.*, **74**, 5536 (1952).

52-39 BANKS, J., Volumetric determination of Ca and Mg by EDTA method. *Analyst*, **77**, 484 (1952).

52-40 MASON, A. S., Determination of small amounts of Ca in plant material. *Analyst*, **77**, 529 (1952).

52-41 CABELL, M. J., Complex ions formed by Th and U with complexones. *Analyst*, **77**, 859 (1952).

52-42 HARRIS, W. F., and SWEET, T. R., Volumetric determination of Ni in steel. *Analyt. Chem.*, **24**, 1062 (1952).

52-43 BANEWITZ, J. J., and KENNER, C. T., Determination of Ca and Mg in limestones and dolomite. *Analyt. Chem.*, **24**, 1186 (1952).

52-44 UENO, K., Rapid analysis of viscose coagulating liquor using Schwarzenbach method. *Analyt. Chem.*, **24**, 1363 (1952).

52-45 SHEAD, A. C., Calcium acid malate hexahydrate as a primary standard. *Analyt. Chem.*, **24**, 1451 (1952).

52-46 CHENG, K. L., KURTZ, T., and BRAY, R. H., Determination of Ca, Mg, and Fe in limestone. *Analyt. Chem.*, **24**, 1640 (1952).

52-47 DISKANT, E. M., Stable indicator solution for complexometric determination of total hardness in water. *Analyt. Chem.*, **24**, 1856 (1952).

52-48 KNIE, K., Bestimmung der Gesamthärte und des Calciumgehaltes in einem Wasser mit Komplexon III. *Österr. Wasserwirtschaft*, **4**, 13 (1952).

52-49 MANNS, TH. J., RESCHOWSKY, M. U., and CERTA, A. J., Volumetric determination of barium with versene. *Analyt. Chem.*, **24**, 908 (1952).

52-50 SCHWARZENBACH, G., Der Chelateffekt. *Helv. Chim. Acta*, **35**, 2344 (1952).

1953

53-1 ANDEREGG, G., FLASCHKA, H., SALLMANN, R., and SCHWARZENBACH, G., Metallindikatoren. VII. Ein auf Erdalkaliionen ansprechendes Phthalein und seine analytische Verwendung. *Helv. Chim. Acta*, **37**, 113 (1953).

53-2 SCHWARZENBACH, G., and MOSER, P., Metallkomplexe mit Polyaminen. X: mit Tetrakis-(β-aminoäthyl)-äthylenediamin = 'penten'. *Helv. Chim. Acta*, **36**, 581 (1953).

53-3 SCHWARZENBACH, G., and SANDERA, JIRI, Die Vanadiumkomplexe der AeDTE. *Helv. Chim. Acta*, **36**, 1089 (1953).

53-4 WHEELWRIGHT, E. J., SPEDDING, F. H., and SCHWARZENBACH, G., The stability of the rare earth complexes with EDTA. *J. Amer. Chem. Soc.*, **75**, 4196 (1953).

53-5 BRUNISHOLZ, G., GENTON, M., and PLATTNER, E., Sur le dosage complexométrique du Ca en présence de Mg et de phosphate. *Helv. Chim. Acta*, **36**, 782 (1953).

53-6 FALLER, E. F., Titrimetrische Schnellbestimmung von Zn in

Aluminium und Aluminiumlegierungen. Z. analyt. Chem., **139**, 15 (1953).

53-7 BIONDA, G., Determinazione del plombo negli additivi e negli oli minerali mediante EDTA. Atti R. Accad. Sci. Torino, **88**, 195 (1953).

53-8 BIONDA, G., Sulla determinazione del contenuto in metallo (Ca e Zn) di alcuni stearati mediante il metodo del complessone. Atti R. Accad. Sci. Torino, **88**, 384 (1953).

53-9 TER HAAR, K., and BAZEN, J., The titration of Complexone III with thorium nitrate at pH 2·8–4·3. Anal. Chim. Acta, **9**, 235 (1953).

53-10 PŘIBIL, R., Komplexometrische Titrationen (Chelatometrie). I. Einführung und Übersicht. Coll. Czech. Chem. Commun., **18**, 783 (1953).

53-11 PŘIBIL, R., Komplexometrische Titrationen (Chelatometrie). II. Maskierung von, Al, Fe, Mn. Coll. Czech. Chem. Commun., **19**, 58 (1954); Chem. Listy, **47**, 1333 (1953).

53-12 PŘIBIL, R., ČIHALIK, J., DOLEŽAL, J., SIMON, V., and ZÝKA, J., Komplexometrie in der pharmazeutischen Analyse. I. Die Bestimmung von Hg. Czech Farmazie, **2**, 38 (1953). II. Die Bestimmung von Hg in Salben. ibid., **2**, 75 (1953). III. Die Bestimmung von Zink. ibid., **2**, 113 (1953). IV. Die Bestimmung von Calcium. ibid., **2**, 147 (1953). V. Die Bestimmung von Magnesium. ibid., **2**, 184 (1953). VI. Die Bestimmung von Aluminium. ibid., **2**, 223 (1953).

53-13 PŘIBIL, R., ČIHALIK, J., DOLEŽAL, J., SIMON, V., and ZÝKA, J., Komplexometrische Titrationen in der pharmazeutischen Analyse. Pharmazie, **8**, 561 (1953).

53-14 BELCHER, R., and CLARK, S. J., Determination of fluoride by titration with calcium chloride. Anal. Chim. Acta, **8**, 222 (1953).

53-15 BELL, R. P., and GEORGE, J. H. B., Incomplete dissociation of some Tl(I) and Ca salts at different temperatures. Trans. Faraday Soc., **49**, 619 (1953).

53-16 BOVALINI, E., and CASINI, A., Il ditizone come indicatore nelle titolazioni per precipitazione e per complessazione. Determinazione del piombo e del cadmio. Ann. Chim. (Roma), **43**, 287 (1953).

53-17 BOULANGER, F., Détermination des complexes formés avec les trilons et les terres rares. Chim. analyt., **35**, 253 (1953).

53-18 LYDERSEN, D., and GJEMS, O., Titration von Eisen(III) mit Versenate. Z. analyt. Chem., **138**, 249 (1953).

53-19 FRITZ, J. J., and FORD, J. J., Titrimetric determination of Th. Analyt. Chem., **25**, 1640 (1953).

53-20 GRÖNQUIST, K. E., Direct titration of bismuth by means of complexone. Farm. Revy, **52**, 305 (1953).

53-21 FLASCHKA, H., Mikrotitrationen mit AeDTE. VII. Bestimmung von Pd. Mikrochim. Acta, 226 (1953).

53-22 FLASCHKA, H., TER HAAR, K., and BAZEN, J., Mikrotitrationen mit AeDTE. VIII. Die Bestimmung von Thorium und Aluminium. Mikrochim. Acta, 345 (1953).

53-23 FLASCHKA, H., Mikrotitrationen mit AeDTE. IX. Bestimmung von In. *Mikrochim. Acta*, 410 (1953).

53-24 FLASCHKA, H., and AMIN, A. M., Mikrotitrationen mit AeDTE. X. Direkte Titration von Mn in reiner Lösung und bei Gegenwart von Fremdmetallen. *Mikrochim. Acta*, 414 (1953).

53-25 FLASCHKA, H., Direct volumetric determination of bivalent Mn with EDTA in presence of other metals. *Chem. Analyst*, **42**, 56 (1953).

53-26 FLASCHKA, H., EDTA in metallurgical analysis, especially determination of Ni and Zn. *Chem. Analyst*, **42**, 84 (1953).

53-27 FLASCHKA, H., Die Spezifische Titration von Zn und Cd mit Komplexon in Gegenwart anderer Kationen. *Z. analyt. Chem.*, **138**, 332 (1953).

53-28 FLASCHKA, H., and AMIN, A. M., Die maßanalytische Bestimmung des In mit AeDTE. *Z. analyt. Chem.*, **140**, 6 (1953).

53-29 STRAFFORD, N., Determination of Zn by titrating with EDTA. *Analyst*, **78**, 733 (1953).

53-30 WILLIAMS, R. J. P., A systematic approach to the choice of organic reagents for metal ions. *Analyst*, **78**, 586 (1953).

53-31 DE SOUSA, A., Le dosage par voie complexométrique du Calcium et du Magnesium dans les milieux très riches en chlorures (Sel marin brut et saumures). *Anal. Chim. Acta*, **9**, 305 (1953).

53-32 KÖRÖS, E., Bestimmung von Calcium und Magnesium nebeneinander. *Magyar Kém. Folyóirat*, **59**, 317 (1953).

53-33 BROWN, E. G., and HAYES, T. J., The simultaneous volumetric determination of zinc and magnesium with EDTA. I. The use of pH control. *Anal. Chim. Acta*, **9**, 1 (1953).

53-34 BROWN, E. G., and HAYES, T. J., The simultaneous volumetric determination of zinc and magnesium using EDTA. II. The use of the zinc–ferrocyanide–ferricyanide redox system. *Anal. Chim. Acta*, **9**, 6 (1953).

53-35 BROWN, E. G., and HAYES, T. J., Complexometric determination of Zn in presence of large quantities of Mg. *Anal. Chim. Acta*, **9**, 408 (1953).

53-36 ERDEY, L., and BODOR, A., Ein neuer Redoxindikator: das 4-Amino-4′-methoxydiphenylamin (Variaminblau). *Z. analyt. Chem.*, **137**, 410 (1953).

53-37 SWEETSER, P. B., and BRICKER, C. E., Spectrophotometric titrations with EDTA. I. Determination of Fe, Cu, Ni. *Analyt. Chem.*, **25**, 253 (1953).

53-38 PATROVSKÝ, V., Use of morin in chemical analysis. II. Volumetric determination of gallium and indium with Complexone III. *Chem. Listy*, **47**, 1338 (1953).

53-39 UNDERWOOD, A. L., Simultaneous determination of Fe and Cu with EDTA – spectrophotometric end-points. *Analyt. Chem.*, **25**, 1910 (1953).

53-40 KINNUNEN, J., and WENNERSTRAND, B., Rapid EDTA titration of Zn following thiocyanate extraction. *Chem. Analyst*, **42**, 80 (1953).

53-41 KINNUNEN, J., and WENNERSTRAND, B., Analysis of nickel sulphate. *Chem. Analyst*, **42**, 30 (1953).

53-42 STYUNKEL, T. B., YAKIMETS, E. M., and SAVINOVSKII, D. A., Kationenstörungen bei der Wasserhärtebestimmung. *Zhur. analit. Khim.*, **8**, 163 (1953).

53-43 SAJÓ, I., Rapid determination of Al by a volumetric method. *Magyar Kém. Folyóirat*, **59**, 319 (1953).

53-44 SAJÓ, I., and RÉPÁS, P., Bestimmung von Magnesium in Gußeisen. *Kohászati Lapok*, **8**, 225 (1953).

53-45 RINGBOM, A., and SANDÅS, P. E., Fotometriska titreringar med EDTA. *Finska Kemistamfundets Medd.*, 14 (1953).

53-46 RINGBOM, A., and LINKO, E., A colorimetric method for the determination of silver ions. *Anal. Chim. Acta*, **9**, 80 (1953).

53-47 SERGEANT, J. C., A rapid complexometric method for the determination of magnesium in aluminium alloys. *Metallurgia*, **48**, 261 (1953).

53-48 PICKLES, D., and WASHBROOK, C. C., The determination of Zn in oils by amperometric titration. *Analyst*, **78**, 304 (1953).

53-49 GRÜNWALD, A., Schnellbestimmung von Bleitetraäthyl mit Komplexon. *Erdöl u. Kohle*, **6**, 550 (1953).

53-50 NEBBIA, L., and GUERRINI, F., Die Bestimmung sekundärer Amine durch komplexometrische Nickeltitrationen. *Chimie e Industrie*, **35**, 896 (1953).

53-51 GOFFART, J., MICHEL, G., and DUYCKAERTS, G., Etude polarographique du mercure en présence d'acide éthylène-diamine-tetraacétique. I. *Anal. Chim. Acta*, **9**, 184 (1953).

53-52 GRAUE, G., and ZÖHLER, A., Schnellbestimmung von Ca und Mg in Erzen und Schlacken. *Angew. Chem.*, **65**, 532 (1953).

53-53 WEITZEL, G., FRETZDORFF, A. M., STRECKER, F. J., and ROESTER, U., Zinkgehalt und Glukagoneffekt kristallisierter Insulinpräparate. *Z. physiol. Chem.*, **293**, 190 (1953).

53-54 JENNES, R., Titration of Ca and Mg in milk and milk fractions with EDTA. *Analyt. Chem.*, **25**, 966 (1953).

53-55 JONCKERS, M. D. E., Dosage complexométrique du Ca et du Mg dans les matériaux calcaires et magnésiens. *Chim. analyt.*, **5**, 101 (1953).

53-56 HILLEBRAND, G. E. F., BRIGHT, H. A., and HOFFMAN, J. J., *Applied Inorganic Analysis*, 2nd edition, John Wiley and Sons, Inc., New York (1953).

53-57 RINGBOM, A., Komplexbildningreaktioner som bas för kemisk analys. *Nord. Kjemikermote, 8th Meeting, Oslo*, 96 (1953).

53-58 SCHWARZENBACH, G., and FLASCHKA, H., *Komplexonmethoden.* Siegfried and Co., Zofingen (1953).

53-59 BELL, R. P., and GEORGE, J. H. B., Incomplete dissociation of some Tl(I) and Ca salts at different temperatures. *Trans. Faraday Soc.*, **49**, 619 (1953).

53-60 DE SOUSA, A., La détermination indirecte du W dans les minerals par voie complexométrique. *Anal. Chim. Acta*, **9**, 309 (1953).

53-61 CHENG, K. L., BRAY, R. H., and KURTZ, T., Determination of Ca,

Mg and Fe in soils by EDTA titration. *Analyt. Chem.*, **25**, 347 (1953).

53-62   FLASCHKA, H., Mikrobestimmung von Wasserhärte. *Melliand Textilber.*, **34**, 1059 (1953).

1954

54-1   BLECHER, R., GIBBONS, D., and WEST, T. S., The evaluation of barium sulphate precipitates by a titrimetric method. *Chem. and Ind.*, 127, 850 (1954); cf. *Analyst*, **80**, 751 (1955).

54-2   BLAEDEL, W. J., and KNIGHT, H. T., Stoicheiometry of titration of metals with EDTA, using high frequency technique. *Analyt. Chem.*, **26**, 743 (1954).

54-3   BLAEDEL, W. J., and KNIGHT, H. T., Purification and properties of the disodium salt of EDTA as a primary standard. *Analyt. Chem.*, **26**, 741 (1954).

54-4   VAN ASPEREN, K., and VAN ESCH, J., A simple microtitration method for the determination of Ca and Mg in the haemolymph of insects. *Nature*, **174**, 927 (1954).

54-5   GEHRKE, CH. W., AFFSPRUNG, H. E., and LEE, Y. C., Direct EDTA titration methods for Mg and Ca. *Analyt. Chem.*, **26**, 1944 (1954).

54-6   CHENG, K. L., Complexometric titration of bismuth. *Analyt. Chem.*, **26**, 1977 (1954).

54-7   BULTASOVA, H., and HORAKOVA, E., Komplexometrische Bestimmung von Glucose in biologischem Material. *Chem. Listy*, **48**, 1698 (1954).

54-8   CAMPBELL, D. N., and KENNER, C. T., Separation of Mg from Ca by ion exchange chromatography. *Analyt. Chem.*, **26**, 560 (1954).

54-9   CARINI, F. F., and MARTELL, A. E., Thermodynamic quantities associated with the interaction between EDTA and alkaline earth ions. *J. Amer. Chem. Soc.*, **76**, 2153 (1954).

54-10   CHALMERS, R. A., Spectrophotometric micro titration of Ca. *Analyst*, **49**, 519 (1954).

54-11   LONGFORD, K. L., Determination of Ni in nickel plating solutions with EDTA. *Electroplating*, **7**, 46 (1954).

54-12   MALMSTADT, H. V., and GOHRBRANDT, E. C., Automatic spectrophotometric titration of milligram quantities of thorium. *Analyt. Chem.*, **26**, 442 (1954).

54-13   MARTELL, A. E., and CHABEREK, S., Use of chelating agents as reagents in titrimetric analysis. *Analyt. Chem.*, **26**, 1692 (1954).

54-14   SWEETSER, P. B., and BRICKER, C. E., Spectrophotometric titrations with EDTA. II. Determination of Mg, Ca, Zn, Cd, Ti and Zr. *Analyt. Chem.*, **26**, 195 (1954).

54-15   TETTWEILER, T., and PILZ, W., Die maßanalytische Bestimmung von Schwefel in organischen Substanzen. *Naturwiss.*, **41**, 332 (1954).

54-16   VODÁK, Z., and LEMINGER, O., Sulfonphthaleine. I. Über das Brenzcatechinsulfonphthalein (Brenzcatechinviolett). *Coll. Czech. Chem. Commun.*, **19**, 925 (1954).

54-17   TER HAAR, K., and BAZEN, J., The titration of Al with Complexone III at pH = 3·5. *Anal. Chim. Acta*, **10**, 23 (1954).

54-18   SUZUKI, S., The formation of complex ions used in analytical chemistry. IX. Studies on the complexes of copper, zinc and cadmium cyanides. *J. Chem. Soc. Japan, Pure Chem. Sect.*, **75**, 962 (1954).

54-19   POULIE, N. J., Direct determination of Ca in serum and urine with the aid of the photoelectric colorimeter. *Chem. Weekblad.*, **50**, 698 (1954).

54-20   CLULEY, H. J., The rapid determination of lime and magnesia in sodalime glasses. *Analyst*, **79**, 567 (1954).

54-21   CHRISTIANSON, G., JENNESS, R., and COULTER, S. T., Determination of ionized calcium and magnesium in milk. *Analyt. Chem.*, **26**, 1923 (1954).

54-22   ADAMS, R. N., Potentiometric titrations with controlled current input. *Analyt. Chem.*, **26**, 1933 (1954).

54-23   ARIKAWA, Y., and KATO, T., Indirekt komplexometrische Bestimmung von alkohol. *Technol. Reports Tôhoku Univ.*, **19**, 104 (1954).

54-24   MAHR, C., and OTTERBEIN, H., Direkte Titration von Zink und Cadmium in Cyanidbädern. *Metalloberfläche*, **8B**, 117 (1954).

54-25   COLLIER, R. E., Extraction of phosphates prior to EDTA titration of Ca and Mg. *Chem. Analyst*, **43**, 41 (1954).

54-26   HÄBERLI, E., Eine Methode zur titrimetrischen Bestimmung des Fe in Blut mit Hilfe von Komplexon. *Experientia*, **10**, 34 (1954).

54-27   SCHWARZENBACH, G., GUT, R., and ANDEREGG, G., Komplexone. XXV. Die polarographische Untersuchung von Austauschgleichgewichten. Neue Daten der Bildungskonstanten von Metallkomplexen der AeDTE und der Diaminocyclohexantetraessigsäure. *Helv. Chim. Acta*, **37**, 937 (1954).

54-28   SCHWARZENBACH, G., and ANDEREGG, G., Die Verwendung der Quecksilberelektrode zur Bestimmung von Stabilitätskonstanten von Metallkomplexen. *Helv. Chim. Acta*, **40**, 1773 (1954).

54-29   MICHEL, G., Etude polarographique du mercure en présence d'acide éthylènediamine-tetraacétique. II. Titrage ampérométrique des cations par le complexon basé sur l'onde de dépolarisation anodique. *Anal. Chim. Acta*, **10**, 87 (1954).

54-30   FABER, J. S., De Mogelijkheid ener complexometrische Sulfaatbepaling in Geneesmiddelen. *Pharm. Weekblad.*, **89**, 705 (1954).

54-31   MILNER, O. J., and SHIPMAN, G. F., Determination of tetraethyllead in gasoline by titration with EDTA. *Analyt. Chem.*, **26**, 1222 (1954).

54-32   HALL, J. L., GIBSON, JR, J. A., WILKINSON, P. R., and PHILLIPS, H. O., Conductometric standardization of solutions of common divalent metallic ions. *Analyt. Chem.*, **26**, 1484 (1954).

54-33   BRUNISHOLZ, G., Indicateur de murexide stabilisé pour la dosage cérométrique du fluor. *Helv. Chim. Acta*, **37**, 1546 (1954).

54-34   BOND, R. D., and TUCKER, B. M., The titration of calcium with EDTA in presence of Mg. *Chem. and Ind.*, 1236 (1954).

54-35 MOSER, J. H., and WILLIAMS, M. B., Gravimetric determination of murexide. *Analyt. Chem.*, **26**, 1167 (1954).

54-36 MURACA, R. F., and REITZ, M. T., Direct volumetric determination of Mg in calcareous minerals. *Chem. Analyst*, **43**, 73 (1954).

54-37 SIJDERIUS, R., A direct titrimetric determination of Ba with EDTA. *Anal. Chim. Acta*, **10**, 517 (1954).

54-38 KINNUNEN, J., and WENNERSTRAND, B., Rapid EDTA titration of Cd following thiocyanate extraction. *Chem. Analyst*, **43**, 34 (1954).

54-39 KINNUNEN, J., and MERIKANTO, B., Determination of manganese in copper and ferrous alloys. *Chem. Analyst*, **43**, 93 (1954).

54-40 DE SOUSA, A., La détermination rapide du Calcium et du Magnesium dans l'eau de mer. *Anal. Chim. Acta*, **11**, 221 (1954).

54-41 SAJÓ, I., Volumetric determination of Al with Complexone III. *Magyar Kém. Folyóirat*, **60**, 268 (1954).

54-42 SAJÓ, I., Über eine titrimetrische Bestimmung des Titangehaltes. *Magyar Kém. Folyóirat*, **60**, 331 (1954).

54-43 FORTUIN, J. M. H., KARSTEN, P., and KIES, H. L., Theoretical treatment of the spectrochemical titration of bivalent cations with Complexone III and metal-specific indicators. *Anal. Chim. Acta*, **10**, 356 (1954).

54-44 FRITZ, J. S., Titration of Bi with EDTA. *Analyt. Chem.*, **26**, 1978 (1954).

54-45 FRITZ, J. S., and FULDA, M. O., Titrimetric determination of Zr. *Analyt. Chem.*, **26**, 1206 (1954).

54-46 FRITZ, J. S., FULDA, M. O., and MARGERUM, S. L., Analysis of U–Zn alloys. *Anal. Chim. Acta*, **10**, 513 (1954).

54-47 HARA, R., and WEST, P. W., High frequency titrations involving chelation with EDTA. I. Chelation studies. *Anal. Chim. Acta*, **11**, 264 (1954).

54-48 QUENTIN, E., Die Bestimmung von Barium und Strontium in der Wasseranalyse. *Z. Lebensm.-Untersuch.*, **99**, 85 (1954).

54-49 HARRIS, W. F., and SWEET, T. R., Volumetric determination of Co in cobalt-nickel solutions. *Analyt. Chem.*, **26**, 1648 (1954).

HARRIS, W. F., and SWEET, T. R., Volumetric determination of cobalt. Complexometric titration with EDTA. *Analyt. Chem.*, **26**, 1649.

54-50 UNDERWOOD, A. L., Photometric titrations. *J. Chem. Educ.*, **31**, 394 (1954).

54-51 UNDERWOOD, A. L., Titration of Bi with EDTA – Spectrophotometric end-point determination. *Analyt. Chem.*, **26**, 1322 (1954).

54-52 LEFTIN, J. P., Direct determination of Zn and Cd in cyanide solutions. *Metal Finishing*, **52**, 74 (1954).

54-53 KENNY, A. D., and TOVERUD, S. U., Non-interference of phosphate in an EDTA method for serum Ca. *Analyt. Chem.*, **26**, 1059 (1954).

54-54 HOLT, R., Volumetric determination of pectin as calcium pectate. *Analyst*, **79**, 623 (1954).

54-55 IRVING, H., WILLIAMS, R. J. P., FERRETT, D. J., and WILLIAMS,

A. E., The influence of ring size upon the stability of metal chelates. *J. Chem. Soc.*, 3494 (1954).

54-56 JACKSON, P. J., Complexometric determination of barium sulphate. *Chem. and Ind.*, 435 (1954).

54-57 BROWN, E. G., and HAYES, T. J., Simultaneous determination of Cd and Mg with EDTA. *Analyst*, **79**, 220 (1954).

54-58 BROWN, E. G., Application of complexones in metallurgical analysis. – Ferrous field. *Metallurgia*, **49**, 101 (1954). Nonferrous field. *Metallurgia*, **49**, 151 (1954).

54-59 FLASCHKA, H., AMIN, A. M., and ZAKI, M. R., Microvolumetric determination of manganese. *Chem. Analyst*, **43**, 67 (1954).

54-60 FLASCHKA, H., Mikrotitrationen mit AeDTE. XI. Bestimmung des Eisens mit visueller Redoxindikation. *Mikrochim. Acta*, 361 (1954).

54-61 FLASCHKA, H., and ABDINE, H., Mikrotitration mit AeDTE. XII. Die Bestimmung von Gallium. *Mikrochim. Acta*, 657 (1954).

54-62 FLASCHKA, H., and PÜSCHEL, R., Über die Ausschaltung größerer Eisenmengen bei komplexometrischen Titrationen. *Z. analyt. Chem.*, **143**, 330 (1954).

54-63 CHARLES, R. G., Heats and entropies of reaction of metal ions with EDTA. *J. Amer. Chem. Soc.*, **76**, 5854 (1954).

54-64 HOL, P. J., Complexometric determination of SO_4 and of Ca ($+Mg$) in NaCl solutions. *Chem. Weekblad.*, **50**, 21 (1954).

54-65 VICKERY, R. C., Lanthanon complexes with EDTA. IV. *J. Chem. Soc.*, 1181 (1954).

54-66 KINNUNEN, J., and MERIKANTO, B., Analysis of nickel sulphate. *Chem. Analyst*, **43**, 13 (1954).

54-67 KINNUNEN, J., and WENNERSTRAND, B., Simultaneous determination of Cu, Fe, Pb and Zn in copper-base alloys. *Metallurgia*, **50**, 149 (1954).

54-68 LASZLOVSZKY, J., Bestimmung kleiner Fluoridmengen mit Komplexon III. *Magyar Kém. Folyóirat*, **60**, 209 (1954).

54-69 RINGBOM, A., Fotoelektriska komplexbildningstitreringar. *Svensk kem. Tidskr.*, **66**, 159 (1954).

54-70 RINGBOM, A., and WÄNNINEN, E., Theory of photoelectric complex formation titrations using metal indicators. *Anal. Chim. Acta*, **11**, 153 (1954).

54-71 SAVINOVSKII, D. A., STYUNKEL, T. B., and YAKIMETS, E. M., Determination of zinc in steam and condensates. *Elektr. Stautsii*, **25**, 49 (1954).

54-72 ÅGREN, A., Complex formation between iron(III)-ion and sulfosalicylic acid. *Acta Chem. Scand.*, **8**, 266 (1954).

54-73 PŘIBIL, R., *Komplexometrie*. Brochure from Chemapol, Prague 1954.

54-74 PŘIBIL, R., Komplexometrische Titrationen (Chelatometrie). III. Maskierung von Al, Ca, Mg mit Ammoniumfluorid. *Coll. Czech. Chem. Commun.*, **19**, 64 (1954).

54-75 MALÁT, M., SUK, V., and RYBA, O., Komplexometrische Titration (Chelatometrie). IV. Brenzcatechin als neuer, spezifischer Indikator. Die Bestimmung von Bi. *Chem. Listy*, **48**, 203 (1954); *Coll. Czech. Chem. Commun.*, **19**, 258 (1954).

54-76 PŘIBIL, R., Komplexometrische Titrationen (Chelatometrie). V. Die Maskierung von Al und Fe bei Titrationen gegen Eriochromschwarz T als Indikator. *Coll. Czech. Chem. Commun.*, **19**, 465 (1954); *Chem. Listy*, **48**, 382 (1954).

54-77 SUK, V., MALÁT, M., and RYBA, O., Komplexometrische Titrationen (Chelatometrie). VI. Brenzcatechinviolett, ein neuer spezifischer Indikator. II. Die Bestimmung von Thorium. *Coll. Czech. Chem. Commun.*, **19**, 679 (1954); *Chem. Listy*, **48**, 533 (1954).

54-78 MALÁT, M., SUK, V., and JENIČKOVÁ, A., Komplexometrische Titrationen (Chelatometrie). VII. Brenzcatechinviolett, ein neuer spezifischer Indikator. Die Bestimmung von Ni, Co, Mn, Zn, Mg and Cd. *Coll. Czech. Chem. Commun.*, **19**, 1156 (1954); *Chem. Listy*, **48**, 663 (1954).

54-79 PŘIBIL, R., and ROUBAL, Z., Komplexometrische Titrationen (Chelatometrie). VIII. Maskierung von Kationen mit 2,3-Dimercaptopropanol. *Coll. Czech. Chem. Commun.*, **19**, 1162 (1954); *Chem. Listy*, **48**, 818 (1954).

54-80 PŘIBIL, R., Komplexometrische Titrationen (Chelatometrie). IX. Beitrag zur Bestimmung von Ni neben Co. *Coll. Czech. Chem. Commun.*, **19**, 1171 (1954); *Chem. Listy*, **48**, 825 (1954).

54-81 SUK, V., MALÁT, M., and JENIČKOVÁ, A., Komplexometrische Titrationen (Chelatometrie). X. Brenzcatechinviolett, ein neuer spezifischer Indikator; die Bestimmung von Kupfer. *Coll. Czech. Chem. Commun.*, **20**, 158 (1954); *Chem. Listy*, **48**, 1511 (1954).

54-82 PŘIBIL, R., ČIHALIK, J., DOLEŽAL, J., SIMON, V., and ZÝKA, J., Komplexometrie in der pharmazeutischen Analyse. VII. Die Bestimmung von Blei. *Czech. Farmazie*, **3**, 84 (1954). VIII. Die Bestimmung des Zink im Insulin. *Czech. Farmazie*, **3**, 242 (1954).

54-83 SCHMITZ, B., Neue Titrationsmethoden im pharmazeutischen Labor. *Deut. Apoth. Ztg.*, **94**, 532 (1954).

54-84 SÉRIS, G., Dosage de l'EDTA dans le vin. *Ann. Fals. Fraudes*, **47**, 29 (1954).

54-85 ZIMMER, P., and HERZOG, K., Über die Brauchbarkeit komplexometrischer Titrationen für Artikel des DAB 6. *Pharmazie*, **9**, 628 (1954).

54-86 SCAVO, V., Volumetrische Bestimmung von Phosphat in Wein nach Veraschung. *Boll. Lab. Chim. Provinziali (Bologna)*, **5**, 65 (1954).

54-87 SCAVO, V., Maßanalytische Bestimmung von Phosphat in Gegenwart von Al, Ca und Fe. Amwendung auf die Analyse von Perphosphaten. *Boll. Lab. Chim. Provinziali (Bologna)*, **5**, 63 (1954).

54-88 LAITINEN, H. A., and SYMPSON, R. F., Amperometric titration of Ca. *Analyt. Chem.*, **26**, 556 (1954).

54-89 SERGEANT, J. C., A complexometric method for the determination of Zn in Al-alloys. *Metallurgia*, **50**, 252 (1954).

54-90   RUSH, R. M., and YOE, J. H., Colorimetric determination of zinc and copper with 2-carboxy-2'-hydroxy-5'-sulfoformazylbenzene. *Analyt. Chem.*, **26**, 1345 (1954).

54-91   MILNER, G. W. C., and WOODHEAD, J. L., The volumetric determination of Al in non-ferrous alloys. *Analyst*, **79**, 363 (1954).

54-94   MILNER, G. W. C., and PHENNAH, P. J., A volumetric procedure for the determination of Zr in its binary alloys with uranium. *Analyst*, **79**, 475 (1954).

54-93   POTTERAT, M., and ESCHMANN, H., Application des complexons au dosage des sucres. *Mitt. Lebensm. Hyg. (Bern)*, **45**, 312 (1954).

54-94   BALLCZO, H., and DOPPLER, G., Natriumphosphat als Aufschluß-mittel. Die maßanalytische Bestimmung kleiner Bariumsulfat-mengen. *Mikrochim. Acta*, 403 (1954).

54-95   MACNEVIN, W. M., and KRIEGE, O. H., Chelation of Pt-group metals. Spectrophotometric determination of Pd with EDTA. *Analyt. Chem.*, **26**, 1768 (1954).

54-96   PECZOK, R. L., and MAVERICK, E. F., A polarographic study of Ti–EDTA complexes. *J. Amer. Chem. Soc.*, **76**, 358 (1954).

54-97   BURTNER, D. C., Some UV-spectrophotometric titrations with disodium versenate. Thesis, University of Washington. *Diss. Abs.*, **14**, 755 (1954).

54-98   POKORNÝ, J., and PRIBYL, J., Bestimmung von Ca und Zn in gehärtetem Kolophonium. *Chem. Zvesti*, **8**, 329 (1954).

54-99   FURBY, E., Determination of Th in U-alloys. *A.E.R.E. Report* C/R 1435 (1954).

54-100   KINNUNEN, J., and WENNERSTRAND, B., Determination of Pb in metallurgical products. *Chem. Analyst*, **43**, 65 (1954).

54-101   KINNUNEN, J., and WENNERSTRAND, B., Determination of Bi in metallurgical products. *Chem. Analyst*, **43**, 88 (1954).

54-102   HARRIS, W. F., and SWEET, T. R., Volumetric determination of Co. Complexometric titration with EDTA. *Analyt. Chem.*, **26**, 1649 (1954).

54-103   SUZUKI, S., The formation of complex ions used in analytical chemistry. IX. Studies of the complexes of Cu, Zn and Cd cyanide. *J. Chem. Soc. Japan, Pure Chem. Sect.*, **75**, 96 (1954).

54-104   TER HAAR, K., and BAZEN, J., The titration of Bi with Complexone III at pH 2·0–2·8. *Anal. Chim. Acta*, **10**, 108 (1954).

1955

55-1   KINNUNEN, J., and MERIKANTO, B., EDTA titration of Pd and Au in presence of Pt. *Chem. Analyst*, **44**, 11 (1955).

55-2   KINNUNEN, J., and MERIKANTO, B., EDTA titrations using zincon as indicator. *Chem. Analyst*, **44**, 50 (1955).

55-3   KINNUNEN, J., and MERIKANTO, B., EDTA determination of Al, Pb and Zn in bronze and brass. *Chem. Analyst*, **44**, 75 (1955).

55-4   KINNUNEN, J., and WENNERSTRAND, B., Improvement of end point in EDTA titrations through use of manganous salts. *Chem. Analyst*, **44**, 33 (1955).

55-5   LACOURT, A., and HEYNDRICKX, P., Séparation chromatographique sur papier du Co, Cu et Zn. II. Mise au point du micro-dosage quantitatif du Co chromatographié. *Mikrochim. Acta*, **61** (1955).

55-6 ŠARŠÚNÓVÁ, M., and ČIČMANCOVÁ, Die Bestimmung von Chloroform in Syrup. *Cesk. Farm.*, **4**, 187 (1955).

55-7 BERTSCHINGER, J. P., Elektro-komplexometrische Titration einiger Calciumsalze. *Schweiz. Apoth. Ztg.*, **93**, 410 (1955).

55-8 TAYLOR, M. P., Use of haematoxyline as an indicator for the volumetric determination of Al with EDTA. *Analyst*, **80**, 153 (1955).

55-9 HAMM, R., Über die Bestimmung von Ca, Mg, Zn und Fe in tierischem Gewebe. *Biochem. Z.*, **327**, 149 (1955); *Mikrochim. Acta*, 268 (1956).

55-10 SIGGIA, S., EICHLIN, D. W., and RHEINHART, R. C., Potentio-metric titrations involving chelating agents, metal ions and metal chelates. *Analyt. Chem.*, **27**, 1745 (1955).

55-11 KATO, T., HAGIWARA, Z., SHINOZAWA, S., and TSUKADA, S., Analytical chemistry of phosphates. I. New titrimetric determina-tions of pyro- and orthophosphates. *Technol. Reports Tôhoku Univ.*, **19**, 93 (1954); *Japan Analyst*, **4**, 84 (1955).

55-12 KARSTEN, P., KIES, H. L., VAN ENGELEN, H. TH. J., and DE HOOG, P., Spectrophotometric titration of Ca and Mg with Complexone III and metal-specific indicators. *Analyt. Chim. Acta*, **12**, 64 (1955).

55-13 KIVALO, P., Complexometric determination of sulphide. *Analyt. Chem.*, **27**, 1809 (1955).

55-14 POTTERAT, M., Die Zuckerbestimmung nach Potterat-Eschmann in Schokolade. *Int. Fachschr. Schokolade Ind.*, **10**, 1 (1955).

55-15 WEHBER, P., Amalgame als Hilfsmittel indirekter Mikrobestim-mungen mit AeDTE. *Mikrochim. Acta*, 911 (1955).

55-16 WEHBER, P., Chelatometrie. I. Maßanalytische Mikrobestim-mung der AeDTE mit visueller Redoxindication. *Mikrochim. Acta*, 812 (1955).

55-17 WEHBER, P., Chelatometrie. II. AeDTE und das Cu(I)/Cu(II) Redoxsystem. Titration von Kupfer mit Variaminblau. *Mikro-chim. Acta*, 927 (1955).

55-18 CHENG, K. L., and BRAY, R. H., 1-(2-pyridylazo)-2-naphtol as a possible analytical reagent. *Analyt. Chem.*, **27**, 782 (1955).

55-19 CHENG, K. L., Complexometric titration of indium. *Analyt. Chem.*, **27**, 1582 (1955).

55-20 CHENG, K. L., and WILLIAMS, JR, T. R., Complexometric titration of scandium with PAN as indicator. *Chem. Analyst*, **44**, 96 (1955).

55-21 BARCZA, L., Bestimmung des Faktors von Komplexonlösungen mit Hilfe verschiedener Calciumverbindungen. *Acta. Pharm. Hung.*, **25**, 102 (1955).

55-22 ISAGAI, K., and TATESHITA, N., Determination of chromate ion by the use of an ion exchange resin and EDTA. *Japan Analyst*, **4**, 222 (1955).

55-23 HOSHIKAWA, G., Selective determination of Fe(III) and Fe(II) mixtures by EDTA method. *Japan Analyst*, **4**, 582 (1955).

55-24 TUCKER, B. M., Phthaleincomplexone as an indicator for calcium and magnesium in the titration of soil extracts. *J. Austral. Inst. Agr. Sci.*, **21**, 100 (1955).

55-25 VERMA, M. R., BHUCHAR, V. M., THERATTIL, K. J., and SHARMA, S. S., A note on the determination of calcium oxide or hydroxide in lime and silicate products. *J. Sci. Ind. Res. B. India*, **14**, 192 (1955)

55-26 THEIS, M., Die direkte maßanalytische Bestimmung des Al mit AeDTA (Komplexon III). *Z. analyt. Chem.*, **144**, 106 (1955).

55-27 THEIS, M., Die komplexometrische Bestimmung des Cu in saurem und ammoniakalischem Medium unter Verwendung von Chromazurol S als Indicator. *Z. analyt. Chem.*, **144**, 275 (1955).

55-28 MACEK, K., and PŘIBIL, R., Use of complexones in chemical analysis. XLV. Contribution to the electrophoresis of some metals. *Coll. Czech. Chem. Commun.*, **20**, 715 (1955).

55-29 THEIS, M., and MUSIL, A., Die komplexometrische Bestimmung des Fe mit Chromazurol S als Indikator. *Z. analyt. Chem.*, **144**, 351 (1955).

55-30 MUSIL, A., and THEIS, M., Die direkte komplexometrische Bestimmung des Zirkoniums. *Z. analyt. Chem.*, **144**, 427 (1955).

55-31 THEIS, M., Die komplexometrische Bestimmung des Mg, Ca und Ba mit Chromazurol S als Indicator sowie die Anwendung auf die Analyse des Magnesites und die Bestimmung der Wasserhärte. *Radex-Rundsch.*, 333 (1955).

55-32 POLYAK, L. Y., Analyse von Elektrolytbädern nach einer potentiometrischen Methode mit Trilon B. *Zavodskaya Lab.*, **21**, 1300 (1955).

55-33 MINAMI, E., and SATO, G., Titration von Mikromengen Blei mit AeDTE unter Verwendung von Dithizon als Extraktionsindicator. *Japan Analyst*, **4**, 579 (1955).

55-34 PINKSTON, J. L., and KENNER, C. T., Determination of lead in lead drosses and lead base alloys. Applications of EDTA-method. *Analyt. Chem.*, **27**, 446 (1955).

55-35 POKORNÝ, J., and PRIBYL, J., Bestimmung von Co in Kolophonium. *Chem. Zvesti*, **9**, 20 (1955).

55-36 DOLEŽAL, J., PATROVSKÝ, V., SULČEK, Z., and SVASTA, J., Analytische Chemie des Galliums. *Chem. Listy*, **49**, 1517 (1955).

55-37 BUDĚŠINSKÝ, B., Komplexometrie in der pharmazeutischen Analyse. X. Indirekte Bestimmung von Dimethylaminoantipyrin. *Cesk. Farm.*, **4**, 71 (1955).

55-38 BUDĚŠINSKÝ, B., Komplexometrie in der pharmazeutischen Analyse. XI. Bestimmung von Hexamethylentetramin und Isonicotinsäurehydrazid. *Cesk. Farm.*, **4**, 185 (1955).

55-39 BUDĚŠINSKÝ, B., Komplexometrie in der pharmazeutischen Analyse. XII. Indirekte Bestimmung von 8-Oxychinolin. *Cesk. Farm.*, **4**, 221 (1955).

55-40 BUDĚŠINSKÝ, B., Indirekte komplexometrische Bestimmung des Amidopyrins und des Isonicotinsäurehydrazids. *Pharmazie*, **10**, 597 (1955).

55-41 HARA, R., and WEST, P. W., High frequency titrations involving chelation with EDTA. II. Quantitative determination of some divalent metals. *Analyt. Chim. Acta*, **12**, 72 (1955).

55-42 HARA, R., and WEST, P. W., High frequency titrations involving chelation with EDTA. III. Determination of uranyl ion. *Analyt. Chim. Acta*, **12**, 285 (1955).

55-43 HARA, R., and WEST, P. W., High frequency titrations involving chelation with EDTA. IV. Complexation of thorium nitrate. *Analyt. Chim. Acta*, **13**, 189 (1955).

55-44 SWANN, M. H., and ADAMS, M. L., Direct volumetric determination of total zinc in mixed paint pigments with EDTA. *Analyt. Chem.*, **27**, 2005 (1955).

55-45 GREEN, H., Rapid method for determination of Mg in nodular cast iron using EDTA. *J. Brit. Cast Iron Res.*, **6**, 20 (1955).

55-46 JANKOVITS, L., and ERDEY, L., Calciumbestimmung in Tonerde. *Acta Chim. Acad. Sci. Hung.*, **7**, 155 (1955).

55-47 LOVE, S. K., and THATCHER, L. L., Water analysis. *Analyt. Chem.*, **27**, 680 (1955).

55-48 GROEBEL, W., and SCHNEIDER, E., Komplexometrische Bestimmung von Pyramidon in pharmazeutischen Zubereitungen. *Z. analyt. Chem.*, **146**, 191 (1955).

55-49 DIGGINS, F. W., Stabilization of murexide for colorimetric use. *Analyst*, **80**, 401 (1955).

55-50 DAVIS, I., and HOPKINSON, G., Determination of calcium in rain water. *Analyst*, **81**, 551 (1955).

55-51 CHERNIKOV, Y. A., DOBKINA, B. M., and KHERSONSKAYA, L. M., Komplexometrische Bestimmung von Aluminium in Silicaten und Schlacken. *Zavodskaya Lab.*, **21**, 638 (1955).

55-52 CAMPEN, W. A. C., NIJST, L. J. H., and NEIS, P. J., Het Bepalen van in 'Water oplosbar Ca' in Gypsum. *Chem. Weekblad*, **51**, 945 (1955).

55-53 ELDJARN, L., NYGAARD, O., and SVEINSSON, S. L., Determination of serum-Ca. A comparison of the methods of Clark and Collip and the titration with EDTA. *Scand. J. Clin. Lab. Invest.*, **7**, 92 (1955).

55-54 DOERING, H., Zur titrimetrischen Bestimmung von Ca und Mg in Gegenwart von Kupferspuren. *Das Papier*, **9**, 58 (1955).

55-55 ABBOTT, D. D., and REBER, L. A., An improved assay for magnesium-citrate solutions. *J. Am. Pharmaceut. Assoc. (Sci. Edn.)*, **44**, 287 (1955).

55-56 CHAPMAN, D., The infrared spectra of EDTA and its di- and tetrasodium salts. *J. Chem. Soc.*, 1766 (1955).

55-57 GERLACH, K., Mischindikator für komplexometrische Titrationen. *Angew. Chemie*, **67**, 178 (1955).

55-58 COLLIER, R. E., Examination of the interference of phosphate ion in the titrimetric determination of Ca and Mg with EDTA. *Chem. and Ind.*, 587 (1955).

55-59 MACNEVIN, W. M., and KRIEGE, O. H., Chelation of platinum group metals. *Analyt. Chem.*, **27**, 535 (1955).

55-60   MACNEVIN, W. M., and KRIEGE, O. H., Reactions of divalent Pd with EDTA. *J. Amer. Chem. Soc.*, **77**, 6149 (1955).

55-61   BUTT, L. T., and STRAFFORD, N., Determination of Fe in presence of large amounts of phosphate by titration with dihydrogen EDTA. *Analyt. Chim. Acta*, **12**, 124 (1955).

55-62   FLASCHKA, H., and ABDINE, H., Thioacetamid in the separation and complexometric determination of Mn. *Chem. Analyst*, **44**, 8 (1955).

55-63   FLASCHKA, H., and ABDINE, H., Thioacetamid in the separation and complexometric determination of Co and Ni. *Chem. Analyst*, **44**, 30 (1955).

55-64   FLASCHKA, H., and ABDINE, H., Mikrotitrationen mit AeDTE. XIII. Eine neue Methode zur Bestimmung von Al. *Mikrochim. Acta*, 37 (1955).

55-65   FLASCHKA, H., Mikrotitrationen mit AeDTE. XIV. Die Bestimmung einiger Seltener Erden. *Mikrochim. Acta*, 55 (1955).

55-66   FLASCHKA, H., and PÜSCHEL, R., Die komplexometrische Titration von Ni neben Co und einigen anderen Metallen. *Z. analyt. Chem.*, **147**, 353 (1955).

55-67   FLASCHKA, H., and FRANSCHITZ, W., Über die Anwendung des Indikator-systems nach Brown und Hayes bei komplexometrischen Titrationen. *Z. analyt. Chem.*, **144**, 421 (1955).

55-68   FLASCHKA, H., and PÜSCHEL, R., Complexometric determination of Mn in spiegeleisen and other ferromanganese alloys. *Chem. Analyst*, **44**, 71 (1955).

55-69   ACONSKY, L., and MORI, M., Spectrophotometric technique for calcium titration. *Analyt. Chem.*, **27**, 1001 (1955).

55-70   CHIARIONI, T., and BONATI, F., Determinazione del ferro emoglobinico mediante titolazione con acido EDTA. *L'Atteneo Parmese*, **26**, 40 (1955).

55-71   BANKS, C. V., and EDWARDS, R. E., Separation and determination of Th and Al. *Analyt. Chem.*, **27**, 947 (1955).

55-72   BROOKS, H. E., and JOHNSON, C. A., The separation and volumetric determination of Al, Bi, Ca and Mg in pharmaceutical preparations. *J. Pharm. Pharmacol.*, **7**, 836 (1955).

55-73   BUDANOVA, L. M., and PLATANOVA, O. P., Potentiometrische Titration einer Reihe von Metallen mit Trilon B. *Zavodskaya Lab.*, **21**, 1294 (1955).

55-74   GOETZ, C. A., and DEBBRECHT, F. J., Determination of lead in lead sulphide ores and concentrates. *Analyt. Chem.*, **27**, 1972 (1955).

55-75   REILLEY, C. N., and SCRIBNER, W. G., Chronopotentiometric titrations. *Analyt. Chem.*, **27**, 1210 (1955).

55-76   AMIN, A. M., Microvolumetric determination of Ag and Cu in coinage. *Chem. Analyst*, **44**, 17 (1955).

55-77   AMIN, A. M., and FARAH, M. Y., Complexometric analysis of zinc-lead ores using thioacetamide and ion-exchange resin. *Chem. Analyst*, **44**, 62 (1955).

55-78   AMIN, A. M., Thioacetamide in the separation and complexometric determination of copper and aluminium in alloys. *Chem. Analyst*, **44**, 66 (1955).

55-79   MILNER, G. W. C., The determination of Ga. *Analyst*, **80**, 77 (1955).

55-80   MILNER, G. W. C., and EDWARDS, J. W., An improved volumetric method for the determination of Zr. *Analyst*, **80**, 879 (1955).

55-81   MILNER, G. W. C., and WOODHEAD, J. L., The determination of Al in silicates (rocks and refractories). *Analyt. Chim. Acta*, **12**, 127 (1955).

55-82   ZENIN, A. A., The trilonometric method in the determination of aggressive carbon dioxide. *Gidrokhim. Materialy*, **23**, 165 (1955).

55-83   YATSIMIRSKIJ, K. B., Quantitative characteristics determining the use of complex compounds in volumetric analysis. *Zhur. analit. Khim.*, **10**, 94 (1955).

55-84   RITCHIE, J. A., Titration of Mg in presence of Al. *Analyst*, **80**, 402 (1955).

55-85   YOUNG, A., and SWEET, T. R., Complexes of Eriochrome Black T with calcium and magnesium. *Analyt. Chem.*, **27**, 418 (1955).

55-86   WISE, W. S., and SCHMIDT, N. O., Amperometric determination of EDTA with zinc ions. *Analyt. Chem.*, **27**, 1469 (1955).

55-87   SHAPIRO, L., and BRANNOCK, W. W., Automatic photometric titrations of calcium and magnesium in carbonate rocks. *Analyt. Chem.*, **27**, 725 (1955).

55-88   ADAMOVA, E., and ZÝKÁ, J., Komplexometrie in der pharmazeutischen Analyse. IX. Die Bestimmung von Codeinphosphat. *Cesk. Farm.*, **4**, 9 (1955).

55-89   PŘIBIL, R., Komplexometrische Titrationen (Chelatometrie). XI. 1,2-Diaminocyclohexan-N,N,N′,N′-tetraessigsäure als Maßreagens. Stufenweise Bestimmung von Fe und Mn (Mg, Ca); Bestimmung von Cu neben Fe, Ni, Co und Mn. *Coll. Czech. Chem. Commun.*, **20**, 162 (1955); *Chem. Listy*, **49**, 179 (1955).

55-90   ŠIR, Z., and PŘIBIL, R., Komplexometrische Titration (Chelatometrie). XII. DCyTE als Maßreagens. Die Bestimmung des Fe, Al. und Ti. *Coll. Czech. Chem. Commun.*, **20**, 871 (1955).

55-91   WÜNSCH, L., Komplexometrische Titrationen (Chelatometrie). XIII. Die Bestimmung des Scandiums. *Coll. Czech. Chem. Commun.*, **20**, 1107 (1955); *Chem. Listy*, **49**, 843 (1955).

55-92   BANERJEE, G., Direct complexometric titration of zirconium using SPADNS as indicator. *Z. analyt. Chem.*, **147**, 105 (1955).

55-93   BANERJEE, G., Rapid titrimetric determination of Th with fluoride using SPADNS. *Z. analyt. Chem.*, **146**, 417 (1955).

55-94   BANERJEE, G., Direct complexometric titration of thorium with versene, using SPADNS. *Analyt. Chem.*, **148**, 349 (1955).

55-95   BOND, R. D., Determination of low concentrations of sulphate using barium chloride and EDTA. *Chem. and Ind.*, 941 (1955).

55-96   SAJÓ, I., Eine neue Methode zur Schnellanalyse der Silikate, Gesteine, Erze, Schlacken, feuerfesten Stoffe usw. *Acta Chim. Acad. Sci. Hung.*, **6**, 233, 243, 251 (1955).

55-97   WILHITE, R. N., and UNDERWOOD, A. L., Ultraviolet photometric titrations of Bi and Pb with EDTA. *Analyt. Chem.*, **27**, 1334 (1955).

55-98   WÄNNINEN, E., and RINGBOM, A., Complexometric titration of Al. *Analyt. Chim. Acta*, **12**, 308 (1955).
55-99   DE SOUSA, A., Le dosage indirect du molybdène par voie complexométrique. *Analyt. Chim. Acta*, **12**, 215 (1955).
55-100 TAKAMOTO, S., Determination of heavy metals by EDTA. *Japan Analyst*, **4**, 178 (1955).
55-101 TAKAMOTO, S., Determination of heavy metals by means of EDTA. I. Titration of EDTA with cobaltous solution. *J. Chem. Soc. Japan, Pure Chem. Sec.*, **76**, 1339 (1955).
55-102 TAKAMOTO, S., Determination of heavy metals by means of EDTA. II. Titration of various metals using KCNS-acetone indicator. *J. Chem. Soc. Japan, Pure Chem. Sec.*, **76**, 1342 (1955).
55-103 TAKAMOTO, S., Determination of heavy metals by means of EDTA. III. Titration of various rare metals. *J. Chem. Soc. Japan, Pure Chem. Sec.*, **76**, 1344 (1955).
55-104 BELCHER, R., GIBBONS, D., and WEST, T. S., The effect of EDTA on $Fe(II)/Fe(III)$ and $Cu(I)/Cu(II)$ redox systems. *Analyt. Chim. Acta*, **12**, 107 (1955).
55-105 BELCHER, R., GIBBONS, D., and WEST, T. S., The determination of Cu by complexometric titration with EDTA. *Analyt. Chim. Acta*, **13**, 226 (1955).
55-106 BELCHER, R., GIBBONS, D., and WEST, T. S., The determination of sulphur in plain carbon steel. *Analyst*, **80**, 751 (1955).
55-107 FABER, J. S., De Complexometrische Titratiemethode bij het Onderzoek van Geneesmiddelen. *Dissertation*, Groningen (Holland) (1955).
55-108 FRITZ, J. S., and JOHNSON, M., Volumetric determination of Zr. *Analyt. Chem.*, **27**, 1653 (1955).
55-109 NIELSCH, W., and GIEFER, L., Komplexometrische Bestimmung von Pyrophosphorsäure. *Z. analyt. Chem.*, **142**, 323 (1955).
55-110 AKIYAMA, T., FUJIWARA, M., OKAMOTO, H., and FUSAKA, K., Determination of pyrophosphate and tripolyphosphate. *Bull. Kyoto Coll. Pharm.*, 19 (1955).
55-111 WEBSTER, H. L., and ROBERTSON, A., The rapid determination of total fatty acid in unbuilt soap products. *Analyst*, **80**, 616 (1955).
55-112 NOGAMI, H., and NAKAGOWA, F., High frequency titrations with EDTA. *J. Pharm. Soc. Japan*, **75**, 1289 (1955).
55-113 ŠARŠÚNÓVÁ, M., and ČIČMANČOVÁ, L., Komplexometrische Bestimmung von Bromoform in Syrup. *Cesk. Farm.*, **4**, 187 (1955).
55-114 VÝDRÁ, F., and PŘIBIL, R., Chelatometrie. XVIII. Die Bestimmung von Ni und Cu in Co und seinen Salzen. *Coll. Czech. Chem. Commun.*, **21**, 1147 (1956).
55-115 SEN SARMA, R. N., Complexometric titrations with versene. *Sci. & Culture (India)*, **20**, 448 (1955).
55-116 LEGGIERI, G., Analyse von Metallnaphthenaten. *Chimica (Mailand)*, **10**, 287 (1955).
55-117 NOGAMI, H., and NAKAGOWA, F., High frequency titrations with

EDTA. Det. of Co, Cu, Ni, Fe, Pb and Cd. *J. Pharm. Soc. Japan*, **75**, 1289 (1955).

55-118 SCHWARZENBACH, G., The complexones and their analytical application. *Analyst*, **80**, 713 (1955).

1956

56-1 KOVÁRIK, M., and MOUČKA, M., Die Anwendung der Pyrogal-lolcarbonsäure als komplexometrischer Indicator. *Z. analyt. Chem.*, **150**, 416 (1956).

56-2 YAKIMETS, E. M., and CHERNAVINA, N. M., Murexide as an indi-cator for the complexometric determination of Cu. *Trudy Uralsk Politekh. Inst.*, **57**, 106 (1956).

56-3 ZHDANOV, A. K., CHADEJEV, V. A., and MAKRICKAJA, E. K., Amperometrische Titration von Zn in Nichteisenlegierungen. *Zavodskaya Lab.*, **22**, 1286 (1956).

56-4 ZAK, B., HINDMAN, W. H., and BAGINSKI, E. S., Spectrophoto-metric titration of spinal fluid Ca and Mg. *Analyt. Chem.*, **28**, 1661 (1956).

56-5 TER HAAR, K., and BAZEN, J., The titration of Ni with EDTA at pH 2·8. *Analyt. Chim. Acta*, **14**, 209 (1956).

56-6 KABANOV, B. N., and POLYAK, L. Y., Elektrochemisches Verhalten von Elektroden in potentiometrischen Titrationen mit Trilon B. *Zhur. analit. Khim.*, **11**, 678 (1956).

56-7 REICHERT, R., Die Bestimmung von Mg in Eisensorten. *Z. analyt. Chem.*, **150**, 250 (1956); *Gießerei*, **44**, 51 (1957).

56-8 LACOURT, A., and HEYNDRICKX, P., Ultramicrométhode nou-velle de dosage direct du Zn sur papier. *Mikrochim. Acta*, 1621 (1956); . . . du Cu sur papier. *ibid.*, 1685 (1956).

56-9 SJÖSTRÖM, E., and RITTNER, W., Eine Methode zur quantitativen Bestimmung der Alkaloidsalze durch Kationenaustausch und nachfolgende komplexometrische Titration. *Z. analyt. Chem.*, **153**, 321 (1956).

56-10 ROWLEY, K., STOENNER, R. W., and GORDON, L., Spectrophoto-metric titration of milligram quantities of Ba. *Analyt. Chem.*, **28**, 136 (1956).

56-11 SPECHT, F., Über die Untersuchung des Flußspats und über einen Vorschlag zur Aufstellung einer deutschen Standardmethode für die Flußspatanalyse. *Z. analyt. Chem.*, **149**, 85 (1956).

56-12 HINSVARK, O. N., and O'HARA, F. J., Combustion and EDTA titrimetric determination of total sulphur in petroleum products. *Analyt. Chem.*, **28**, 919 (1956).

56-13 VLADIMIROVA, V. M., Komplexometrische Titration von Zr mit amperometrischer Anzeige des Äquivalenzpunktes. *Zavodskaya Lab.*, **22**, 529 (1956).

56-14 ZAK, B., HINDMAN, W. H., and BAGINSKI, E. S., Spectrophoto-metric titration of spinal fluid Ca and Mg. *Analyt. Chem.*, **28**, 1661 (1956).

56-15 TREINDL, L., Polarimetrische Bestimmung von In in Cd und In und Cd in Zn. *Coll. Czech. Chem. Commun.*, **21**, 1300 (1956).

348   Bibliography

56-16   SOCHEVANOVA, M. M., Murexid, seine Indicatoreigenschaften
        und eine Methode zu seiner Herstellung. *Zhur. analit. Khim.*, **11**,
        219 (1956).
56-17   KANENIWA, N., Complexes of Eriochrome Black T with some
        metals. *J. Pharm. Soc. Japan*, **76**, 136 (1956).
56-18   IRITANI, N., TANAKA, T., and SAKAI, E., Volumetric analysis of
        mercury salts by EDTA. I. Determination of $HgCl_2$, Hg, HgO,
        Hg_2Cl_2 and $HgNH_2Cl$. *J. Pharm. Soc. Japan.*, **76**, 1068 (1956).
56-19   KATO, T., HOGIWARA, Z., and SASAKI, I., Spectrophotometric
        determination of minute amounts of Th and its photometric
        titration. *Technol. Reports Tôhoku Univ.*, **21**, 15 (1956).
56-20   SCHÖNIGER, W., Die mikroanalytische Schnellbestimmung von
        Halogenen und Schwefel in organischen Verbindungen. *Mikro-
        chim. Acta*, 869 (1956).
56-21   GUREVICH, A. B., and KALINA-ZHIKAREVA, V. I., The use of
        cationite and of trilonometric titration for the determination of
        arsenic in high arsenic alloys. *Trudy Nauch. Tekh. Obshchest.
        Chernoi Met. Ukr. Resp. Pravlen*, **4**, 127 (1956).
56-22   NIKELLY, J. G., and COOKE, W. D., Amperometric titration of
        micromolecular solutions. *Analyt. Chem.*, **28**, 243 (1956).
56-23   PHILLIP, B., and HOYME, H., Zur komplexometrischen Bestim-
        mung von Sulfat mit Phthaleinkomplexon. *Faserforschung und
        Textiltechnik*, **7**, 525 (1956).
56-24   STYUNKEL, T. B., and YAKIMETS, E. M., Determination of sul-
        phate. *Zavodskaya Lab.*, **12**, 653 (1956).
56-25   HUNTER, J. A., and MILLER, C. C., The separation of Zn from some
        other elements by means of anion exchange and solvent extraction
        and its titrimetric determination with EDTA. *Analyst*, **81**, 79
        (1956).
56-26   PATTON, J., and REEDER, W., A new indicator for the titration of
        Ca with EDTA. *Analyt. Chem.*, **28**, 1026 (1956).
56-27   RAO, G. G., RAO, V. N., and SOMIDEVAMMA, G., Volumetric
        determination of Fe(III)-salts with EDTA using Fe(II)-cacotheline
        as indicator. *Z. analyt. Chem.*, **152**, 346 (1956).
56-28   PAUL, R. M., Titrimetric determination of Al with EDTA. *Biochem.
        J.*, **62**, 38 (1956).
56-29   PAULSEN, A., Bestimmung von Leptazol. *Medd. Norsk. Farm.
        Selskap*, **18**, 139 (1956).
56-30   PECZOK, R. L., and SAWYER, D. T., Molybdenum(V) and Molyb-
        denum(VI) complexes with EDTA. *J. Amer. Chem. Soc.*, **78**,
        5496 (1956).
56-31   MALINEK, M., and REHAK, B., Use of complexones in chemical
        analysis. XLVI. Gravimetric and volumetric determination of As.
        Coll. Czech. Chem. Commun., **21**, 777 (1956).
56-32   BUDĚŠINSKÝ, B., Komplexometrie in der organischen Analyse.
        I. Fällungen mit $KBiJ_4$. *Coll. Czech. Chem. Commun.*, **21**, 146
        (1956).
56-33   KÖRBL, J., and PŘIBIL, R., Anwendung des Silberpermanganates
        in der Analyse. V. Komplexometrische Bestimmung von verbrenn-

barem Schwefel als Mangan-(II)-Sulfat. *Coll. Czech. Chem. Commun.*, **21**, 322 (1956).

56-34 BUDĚŠINSKÝ, B., and VANÍČKOVÁ, E., Komplexometrie in der pharmazeutischen Analyse. XIV. Indirekte Bestimmung von Theophylin, Chinin, Analergen und Divascol. *Cesk. Farm.*, **5**, 77 (1956).

56-35 BUDĚŠINSKÝ, B., and VANÍČKOVÁ, E., Komplexometrie in der pharmazeutischen Analyse. XV. Die Bestimmung von Chinin. *Cesk. Farm.*, **5**, 277 (1956).

56-36 BUDĚŠINSKÝ, B., Komplexometrie in der pharmazeutischen Analyse. XVI. Halbmikrobestimmung von Papaverin, Narkotin, Kodein, Strychnin und Brucin. *Cesk. Farm.*, **5**, 579 (1956).

56-37 RYBA, O., CÍFKA, J., MALAT, M., and SUK, V., Chemische Indikatoren. II. Die Dissoziationskonstanten des Brenzcatechinvioletts. *Coll. Czech. Chem. Commun.*, **21**, 349 (1956).

56-38 CÍFKA, J., RYBA, O., SUK, V., and MALÁT, M., Chemische Indikatoren. III. Metallkomplexe von Brenzcatechinviolett. *Chem. Listy*, **50**, 888 (1956).

56-39 KÖRBL, J., and PŘIBIL, R., Xylenole orange: New indicator for EDTA titrations. *Chem. Analyst*, **45**, 102 (1956); **46**, 28 (1957).

56-40 MALÁT, M., PELIKAN, J., and SUK, V., EDTA titration of Th in gas mantels and glasses with pyrocatechol violet as indicator. *Chem. Analyst*, **45**, 61 (1956).

56-41 MALÁT, M., and SUK, V., Pyrocatechol violet: indicator for chelatometric titrations. *Chem. Analyst*, **45**, 30 (1956).

56-42 FLASCHKA, H., and SADEK, F., Komplexometrische Titrationen in stärker saurem Medium. Rücktitration mit Wismutnitrat gegen Brenzcatechinviolett. *Z. analyt. Chem.*, **149**, 345 (1956).

56-43 FLASCHKA, H., and ABDINE, H., Complexometric microtitration using PAN as indicator. *Chem. Analyst*, **45**, 2 (1956).

56-44 FLASCHKA, H., and ABDINE, H., EDTA titrations using Cu–PAN as indicator. *Chem. Analyst*, **45**, 58 (1956).

56-45 FLASCHKA, H., and ABDINE, H., Zur komplexometrischen Titration von Fe und Al und der Summe beider. *Z. analyt. Chem.*, **152**, 77 (1956).

56-46 FLASCHKA, H., and ABDINE, H., Mikrotitrationen mit AeDTE. XV. 1-(2-Pyridylazo)-2-naphthol als komplexometrischer Indikator. *Mikrochim. Acta*, 770 (1956).

56-47 FLASCHKA, H., and PÜSCHEL, R., Die komplexometrische Bestimmung von Zn und Cd in Anwesenheit größerer Mengen von Fe. *Z. analyt. Chem.*, **149**, 185 (1956).

56-48 LHEUREUX, M., HENRY, S., and HANISET, P., Titrage photocolorimétrique des ions Ca et Mg par complexométrie. *Ind. chim. belge.*, **21**, 695 (1956).

56-49 CHERKASHINA, T. V., Komplexometrische Bestimmung von Ga. *Zavodskaya Lab.*, **22**, 276 (1956).

56-50 MORGAN, L. O., and JUSTUS, N. L., Complex compounds of Zr and Hf with EDTA. *J. Amer. Chem. Soc.*, **78**, 38 (1956).

56-51   GUTMAN, F. W., Determination of Cu in plating baths. *Plating*, **43**, 345 (1956).

56-52   MACNEVIN, W. M., and KRIEGE, O. H., EDTA chelation of platinum group metals. Spectrophotometric determination of iridium. *Analyt. Chem.*, **28**, 16 (1956).

56-53   DATTA, S. K., Analytical aspects of some azo-dyes from chromotropic acid. II. Titrimetric determination of Th with EDTA using SNADNS. *Z. analyt. Chem.*, **149**, 328 (1956).

56-54   MCCALLUM, J. R., Determination of small amounts of Ca, Mg, Ba and sulphate. *Canad. J. Chem.*, **34**, 921 (1956).

56-55   ESCHMANN, H., and BROCHON, R., Chelatometric determination of H₃PO₄ precipitated by ethanolamine after separation of Fe and Al with cupferron. *Chem. Analyst*, **45**, 38 (1956).

56-56   CHERNUKHA, G. N., and GURKIN, K. M., Komplexometrische Bestimmung von Zn und Cd in Cyanidbädern. *Zavodskaya Lab.*, **22**, 656 (1956).

56-57   CONAGHAN, H. F., Investigation of some *o*-nitrophenols as indicators for the complexometric titration of Cu and Ni ions. *N.S. Wales Dept. Mines Tech. Rep.*, **4**, 104 (1956).

56-58   GHOSH, A. K., and RAY, K. L., Modified EDTA method for direct estimation of Mg. *Analyt. Chim. Acta*, **14**, 504 (1956).

56-59   HARA, R., and WEST, P. W., High frequency titrations involving chelation with EDTA. V. Complexation with rare earths. *Analyt. Chim. Acta*, **14**, 280 (1956).

56-60   CHENG, K. L., and LOTT, P. F., Reaction of hydrogen peroxide with complexes of EDTA and NTA. *Analyt. Chem.*, **28**, 462 (1956).

56-61   CHENG, K. L., Zinc complex of Eriochrome Blue Black R (zinchrome R) as a stable indicator for EDTA titrations. *Chem. Analyst*, **45**, 79 (1956).

56-62   CHMELAR, V., Bestimmung von Chloriden in biologischem Material durch komplexometrische Titration. *Chem. Listy*, **50**, 1326 (1956).

56-63   ABDINE, H., Die Analyse von Präparaten auf Basis AeDTE. *Arzneimittelforschung*, **6**, 698 (1956).

56-64   CHEW, B., and LINDLEY, G., Determination of Zn in Cu-alloys. *Metallurgia*, **53**, 45 (1956).

56-65   ALIMARIN, I. P., GOLOVINA, A. P., and GIBALO, I. M., Eine Studie der Absorptionsspektren der Komplexonate einiger Metalle. *Vestn. Moskau Univ. Ser. Math. Mech. Astron. Fis. i Khim.*, 135 (1956).

56-66   ALLSOPP, H. J., Determination of MgO in Mg-metal. *Analyst*, **81**, 469 (1956).

56-67   BELAVSKAYA, Y. I., Determinations with EDTA of ZnO in zinc white and in the catalyst for methanol syntheses. *Zavodskaya Lab.*, **22**, 422 (1956).

56-68   BÖHLER, K., Erfahrungen mit einer neuen Methode zur Bestimmung des Ca im Serum. *Schweiz. med. Woch.*, **86**, 68 (1956). Cf. also *Labor Praxis*, **10**, 5 (1957).

56-69   MICHAL, J., JANKOVSKÝ, J., and PAVLIKOVA, E., Die Bestimmung von Hg in Erzen und Konzentraten. *Z. analyt. Chem.*, **153**, 83 (1956).

56-70 BRÄUNINGER, H., and MENGERING, S., Bestimmung von Schwefel in pharmazeutischen Präparaten. *Pharmazie*, **11**, 574 (1956).

56-71 COHEN, A. I., and GORDON, L., Photometric titration of small amounts of Ba with EDTA. *Analyt. Chem.*, **28**, 1445 (1956).

56-72 DIEHL, H., and ELLINBOE, J. L., Indicator for titration of Ca in the presence of magnesium using EDTA. *Analyt. Chem.*, **28**, 882 (1956).

56-73 RINGBOM, A., Application of chelate complexes in analytical chemistry. (In Swedish.) *Finska Kemistsamfundets Medd.*, **65**, 82 (1956).

56-74 RÁDY, G., and ERDEY, L., Komplexometrische Bi-Bestimmung. *Z. analyt. Chem.*, **152**, 253 (1956).

56-75 MILNER, G. W. C., and WOODHEAD, J. L., The volumetric determination of plutonium(III) with EDTA. *Analyst*, **81**, 427 (1956).

56-76 SCHMID, R. W., and REILLEY, C. N., Rapid electrochemical method for the determination of metal chelate stability constants. *J. Amer. Chem. Soc.*, **78**, 2910, 5513 (1956).

56-77 REILLEY, C. N., and PORTERFIELD, W. W., Coulometric titrations with electrically released EDTA. Titrations of Ca, Cu, Zn and Pb. *Analyt. Chem.*, **28**, 443 (1956).

56-78 REILLEY, C. N., SCRIBNER, W. G., and TEMPLE, C., Amperometric titration of two- and three-component mixtures of metal ions with EDTA. *Analyt. Chem.*, **28**, 450 (1956).

56-79 MAURICE, M. J., Bestimmung von Schwefel in Rayon via Cu. N. V. Onderzoekingsinst. *Research* (1956).

56-80 MAURICE, M. J., Complexometrisch Bepaling van H₂S. *Chem. Weekblad*, **52**, 122 (1956).

56-81 AMIN, A. A. M., Thioacetamide in the separation and EDTA titration of Zn in Al-alloys. *Chem. Analyst*, **45**, 95 (1956).

56-82 SOMMER, L., and KOLÁŘIK, Z., Kojsäure als komplexometrischer Indikator für die Fe-Bestimmung. *Coll. Czech. Chem. Commun.*, **21**, 1645 (1956).

56-83 BUDĚŠINSKÝ, B., Komplexometrische Titrationen (Chelatometrie). XIV. Dipyridinzinkrhodanid als Urtitersubstanz in der Komplexometrie. *Coll. Czech. Chem. Commun.*, **21**, 255 (1956).

56-84 CÍFKA, J., MALÁT, M., and SUK, V., Komplexometrische Titrationen (Chelatometrie). XV. Brenzcatechinviolett als neuer spezifischer Indicator. Die Bestimmung des Bi(Th) neben Fe und Hg. *Coll. Czech. Chem. Commun.*, **21**, 412 (1956).

56-85 SUK, V., MALÁT, M., and JENIČKOVÁ, A., Komplexometrische Titrationen (Chelatometrie). XVI. Die Bestimmung von Bi, Ni und Co gegen Pyrogallolrot. *Coll. Czech. Chem. Commun.*, **21**, 422 (1956).

56-86 ŠIR, Z., and PŘIBIL, R., Komplexometrische Titrationen (Chelatometrie). XVII. Ein Beitrag zur Bestimmung von Cu, Fe, Al und Ti. *Coll. Czech. Chem. Commun.*, **21**, 872 (1950).

352   *Bibliography*

56-87   VÝDRÁ, F., and PŘIBIL, R., Komplexometrische Titrationen
         (Chelatometrie). XVIII. Die Bestimmung von Ni und Cu in Co
         und seinen Salzen. *Coll. Czech. Chem. Commun.*, **21**, 1149 (1956).
56-88   JENIČKOVÁ, A., SUK, J., and MALÁT, M., Komplexometrische
         Titrationen (Chelatometrie). XIX. Brompyrogallolrot als
         komplexometrischer Indicator. *Coll. Czech. Chem. Commun.*, **21**,
         1257 (1956).
56-89   DOLEŽAL, J., ŠIR, Z., and JANÁCEK, K., Komplexometrische
         Titrationen (Chelatometrie). XX. Die Bestimmung von In. *Coll.*
         Czech. Chem. Commun., **21**, 1300 (1956).
56-90   JENIČKOVÁ, A., MALÁT, M., and SUK, V., Komplexometrische
         Titrationen (Chelatometrie). XXI. Die komplexometrische Bestim-
         mung von Pb gegen Pyrogallol- und Brompyrogallolrot. *Coll.*
         Czech. Chem. Commun., **21**, 1599 (1956).
56-91   VODÁK, Z., and LEMINGER, O., Pyrogallolsulfonphthalein und
         sein Dibromderivat. Reaktionen mit Metallionen. *Chem. Listy*, **50**,
         2028 (1956).
56-92   VODÁK, Z., and LEMINGER, O., Sulfonphthaleine. II. Darstellung
         und Eigenschaften von Pyrogallolsulfonphthalein (Pyrogallolrot)
         eines neuen chelatometrischen Indikators. *Coll. Czech. Chem.*
         Commun., **21**, 1522 (1956).
56-93   VŘEŠTÁL, J., HAVÍR, J., and JILEK, A., Mg-Bestimmung in der
         Asche von rauchlosem Schießpulver. *Chem. Zvesti*, **10**, 188 (1956).
56-94   TANAKA, N., OIWA, I. T., and KODAMA, M., Amperometric and
         potentiometric titrations of Cd with EDTA using dropping Hg-
         electrodes. *Analyt. Chem.*, **28**, 1555 (1956).
56-95   BRUNISHOLZ, G., and CAHEN, R., Sur le dosage complexomé-
         trique des terres rares. *Helv. Chim. Acta*, **39**, 324, 2136 (1956).
56-96   ERDEY, L., and RÁDY, G., Komplexometrische Fe-Bestimmung in
         Anwesenheit von Variaminblau als Indikator. *Z. analyt. Chem.*,
         149, 250 (1956).
56-97   BALLCZO, H., and DOPPLER, G., Die mikroanalytische Barium-
         sulfatbestimmung. *Mikrochim. Acta*, 734 (1956).
56-98   SAJÓ, I., Maskierung von Fe bei AeDTE Titrationen. *Magyar Kém.*
         Folyóirat, **62**, 37 (1956).
56-99   SAJÓ, I., Bestimmungen mit Komplexon III in Gegenwart von
         Eisen(II)-Eisen(III)-cyanid und Benzidin als Indikator. *Magyar*
         Kém. Folyóirat, **62**, 56 (1956).
56-100   SAJÓ, I., Eine neue komplexometrische Indikationsmethode.
         Magyar Kém. Folyóirat, **62**, 176 (1956).
56-101   SCHNEYDER, J., Maßanalytische Bestimmung von Sulfat in Wein.
         Mittl. Landw. Bundesanst. Austria, Ser. A. Rebe und Wein, **6**, 155
         (1956).
56-102   FABER, J. S., Over Chelatometrie een farmaceutisch Belangrijke
         Titriermethode. *Pharm. Tijdschr. Belgie*, **33**, 25 (1956).
56-103   FABER, J. S., Farmaceutische Toepassingen der Chelatometrie.
         Pharm. Tijdschr. Belgie, **33**, 73 (1956).
56-104   FABER, J. S., De complexometrische Titratie van Ca, Mg, Sr en
         Ba in Genesmiddelen. *Pharm. Weekblad*, **91**, 145 (1956).

56-105 FABER, J. S., De complexometrische Titratie van Zn, Hg, Pb, Al en Bi in Geneesmiddelen. *Pharm. Weekblad*, **91**, 177 (1956).

56-106 JOHANNSEN, W., BOBOWSKI, E., and WEHBER, P., Chelatometrie. III. Zur Maßanalyse des Al mit AeDTE. *Metall*, **10**, 211 (1956).

56-107 WEHBER, P., Chelatometrie. IV. Das Zersetzungsprodukt des Bindschedler Grün als Redoxindicator. I. Die komplexometrische Bestimmung von Eisen(III)-salzen. *Z. analyt. Chem.*, **149**, 161 (1956).

56-108 WEHBER, P., Chelatometrie. V. Das Zersetzungsprodukt des Bindschedler Grün als Redoxindicator. 2. Die Bestimmung von AeDTA mit Eisen(III)-salzen. *Z. analyt. Chem.*, **149**, 241 (1956).

56-109 WEHBER, P., Chelatometrie. VI. Zur Bestimmung größerer Kupfer(II)-konzentrationen mit AeDTA. *Z. analyt. Chem.*, **149**, 244 (1956).

56-110 WEHBER, P., Chelatometrie. VII. Das Problem der Pufferung. Die Verwendung von Urotropin und Monochloressigsäure für komplexometrische Titrationen. *Z. analyt. Chem.*, **149**, 419 (1956).

56-111 WEHBER, P., Chelatometrie. VIII. Das Zersetzungsprodukt des Bindschedler Grün als Redoxindicator. 3. Komplexometrische Bestimmung von Chrom(III)-salzen. *Z. analyt. Chem.*, **150**, 186 (1956).

56-112 WEHBER, P., Chelatometrie. IX. Die Berechnung scheinbarer Stabilitätskonstanten. Ein Beitrag zur Yatsimirskij-Theorie. *Z. analyt. Chem.*, **153**, 249 (1956).

56-113 WEHBER, P., Die Leukobase des Bindschedler Grün als Redoxindicator in der Chelatometrie. *Z. analyt. Chem.*, **151**, 276 (1956).

56-114 WEHBER, P., Chelatometrie. X. Eriochromrot B ein neuer pM-Indikator. *Z. analyt. Chem.*, **153**, 253 (1956).

56-115 WEHBER, P., and JOHANNSEN, W., Chelatometrie. XI. Nitrilotriessigsäure als Maßreagenz. *Z. analyt. Chem.*, **195**, 324 (1956).

56-116 BARNARD, JR, A. J., BROAD, W. C., and FLASCHKA, H., The EDTA titration: nature and methods of end-point detection. *Chem. Analyst*, **45**, 86, 11 (1956); **46**, 18, 46, 76 (1957).

56-117 RAEMAEKERS, R., Die Bestimmung von Tri- und Pyrophosphat in Tripolyphosphat. *Coll. Czech. Chem. Commun.*, **21**, 1430 (1956). Cf. also *Compt. rend. 27th. Congr. Int. Chim. Ind.*, Brussels 1954; *Ind. chim. belge*, **20**, Spec. No. 603 (1955).

56-118 FOLEY, W. T., and POTTIE, R. F., Electrolytic separation and volumetric, absorptiometric and coulometric estimation of Tl. *Analyt. Chem.*, **28**, 1101 (1956).

56-119 SJÖSTRÖM, E., and RITTNER, W., Eine Methode zur quantitativen Bestimmung von Alkaloidsalzen durch Kationenaustausch und nachfolgende komplexometrische Titration. *Z. analyt. Chem.*, **153**, 321 (1956).

56-120 DOERING, H., Carboxylgruppenbestimmung in Zellstoffen mit Komplexon. *Papier*, **10**, 140 (1956).

56-121 ŠARŠUNÓVÁ, M., Bestimmung von Cyanid in Bittermandelöl. *Cesk. Farm.*, **7**, 196 (1956).

354   *Bibliography*

56-122   FLASCHKA, H., and HOLASEK, A., Über die komplexometrische Bestimmung des K. im Blutserum. *Z. physiol. Chem.*, **303**, 9 (1956).

56-123   FERRETT, D. J., and MILNER, G. W. C., The polarography of niobium. *J. Chem. Soc.*, 1186 (1956); see also *Nature*, **175**, 477 (1955).

56-124   SCHMID, R. W., and REILLEY, C. N., Coulometric titration of iron(III) with electrolytically generated iron(II)–EDTA. *Analyt. Chem.*, **28**, 520 (1956).

56-125   SCHWARZENBACH, G., and GUT, R., Die Komplexe der Seltenen Erdkationen und die Gadoliniumecke. *Helv. Chim. Acta*, **39**, 1589 (1956).

56-126   BRENNECKE, ERNA, Oxidations-Reduktions-Indikatoren. *Neuere Maßanalytische Methoden*, Ferd. Encke, Stuttgart (1956).

1957

57-1   BORCHERT, O., Die komplexometrische Maßanalyse im Betriebs-laboratorium. *Fertigungstechnik*, **7**, 389 (1957).

57-2   WEHBER, P., Chelatometrie. XII. Entametrische Mn-Bestimmung. *Z. analyt. Chem.*, **154**, 122 (1957).

57-3   WEHBER, P., Chelatometrie. XIII. Zur Maskierung von Ti(IV)-ionen. *Z. analyt. Chem.*, **154**, 182 (1957).

57-4   WEHBER, P., and JOHANNSEN, W., Chelatometrie. XIV. Der Fe(III)-rhodanindicator. *Z. analyt. Chem.*, **158**, 7 (1957).

57-5   WEHBER, P., Chelatometrie. XV. Über neue pM-Indikatoren. (PAR). *Z. analyt. Chem.*, **158**, 10 (1957).

57-6   WEHBER, P., Chelatometrie. XVI. Nebeneinanderbestimmung von Fe und Al. *Z. analyt. Chem.*, **158**, 321 (1957).

57-7   ABD-EL RAHEEM, A. A., Determination of Ca in milk. *Netherl. Milk Dairy J.*, **11**, 122 (1957).

57-8   SADEK, F. S., and REILLEY, C. N., Ultramicro chelometric titrations with potentiometric end-point detection. *Microchem. J.*, **1**, 183 (1957).

57-9   REILLEY, C. N., and SHELDON, M. V., Selective chelometric titrations of metal ions with triethylenetetramine. *Chem. Analyst*, **46**, 59 (1957).

57-10   SCHMID, R. W., and REILLEY, C. N., New complexone for titration of Ca in the presence of Mg. *Analyt. Chem.*, **29**, 264 (1957).

57-11   RAO, V. N., and RAO, G. G., Volumetric determination of Th with EDTA using Fe(II)–cacotheline as indicator. *Z. analyt. Chem.*, **155**, 334 (1957).

57-12   RAO, G. G., and SOMIDEVAMMA, G., Studies in U(VI)-complexes with organic ligands. Spectrophotometric study of the U(VI)–EDTA complex. *Z. analyt. Chem.*, **157**, 27 (1957).

57-13   RAMAIAH, N. A., and VISHNU, A new spectrophotometric method for the determination of Ca with EDTA. *Analyt. Chim. Acta*, **16**, 569 (1957).

57-14   RAMAIAH, N. A., Use of versene in sugar industry. *Indian Sugar*, **7**, 1 (1957).

57-15   RINGBOM, A., SIITONEN, S., and SKRIFVARS, B., The EDTA complexes of vanadium(V). *Acta Chim. Scand.*, **11**, 551 (1957).

57-16   WEINER, R., and NEY, E., Komplexometrische Titration von Cr(III)-Ion. *Z. analyt. Chem.*, **157**, 105 (1957).

57-17   WALLRAF, M., Schnellbestimmung von Ca und Mg in Zementen. *Tonindustr. Ztg. u. Keram. Rdsch.*, **81**, 41 (1957).

57-18   WALLRAF, M., Gerät zur optischen Endpunktsanzeige bei der Titration von Ca und Mg mit AeDTE. *Z. analyt. Chem.*, **156**, 332 (1957).

57-19   MUSHA, S., MUNEMORI, M., and OGAWA, K., Photometric determination of indicator end-points in complexometric titrations. *Bull. Chem. Soc. Japan*, **30**, 675 (1957).

57-20   MUSHA, S., and OGAWA, K., Determination of Ti with EDTA and H_2O_2. *J. Chem. Soc. Japan, Pure Chem. Section*, **78**, 1686 (1957).

57-21   LASSNER, E., and SCHLESINGER, H., Zur indirekten maßanalytischen Bestimmung des Mo mit AeDTE. *Z. analyt. Chem.*, **158**, 195 (1957).

57-22   POLCIN, J., Potentiometrische Bestimmung von Komplexon III mit Hg(II)-Salzen. *Coll. Czech. Chem. Commun.*, **22**, 1057 (1957).

57-23   RUSS, J. J., and REEDER, W., Determination of tetraethyllead in gasoline. *Analyt. Chem.*, **29**, 1331 (1957).

57-24   SJÖSTED, G., and GRINGRAS, L., Determination of Ca and Ag in photographic material via EDTA titration. *Chem. Analyst*, **46**, 58 (1957).

57-25   LANE, W. J., and FRITZ, J. S., The photometric titration of Nd. *U.S. A.E.C. Report*, ISC-945 (1957).

57-26   POLYAK, L. Y., Potentiometrische Bestimmung kleiner Ba-Mengen in Ni-Legierungen mittels AeDTA. *Zhur. analit. Khim.*, **12**, 224 (1957).

57-27   PHILLIPP, B., and HOYME, H., Zur komplexometrischen Fe-Bestimmung in Zellstoffen. *Faserforschung und Textiltechnik*, **8**, 34 (1957).

57-28   PASOVKAYA, G. B., Konduktometrische Schnellmethode für die Bestimmung der Wasserhärte. *Zhur. analit. Khim.*, **12**, 523 (1957).

57-29   PATZAK, R., and DOPPLER, G., Die Bestimmung von Cr, Fe und Al mit AeDTE bei gleichzeitiger Anwesenheit aller drei Kationen. *Z. analyt. Chem.*, **156**, 248 (1957).

57-30   HARA, S., and KATIHARA, S., Determination of Th in monazites. *Report Sci. Res. Inst.*, **33**, 343 (1957).

57-31   NEBBIA, L., and GUERRIERI, F., Bestimmung von Hexamethylenimin in Hexamethylendiamin. *Chim. e. Industr.*, **39**, 672 (1957).

57-32   MUSTAFIN, I. S., and KASHKOVSKAYA, E. A., Chromoxan Farbstoffe als komplexometrische Indikatoren. *Zavodskaya Lab.*, **23**, 519 (1957).

57-33   REISHAKHRIT, L. S., and SUKHOBOKOVA, N. S., Amperometric titration of ferric iron. *Zhur. analit. Khim.*, **12**, 146 (1957).

57-34   VÝDRÁ, F., and KARLIK, M., Konduktometrische Titrationen in der Chelatometrie. I. Einführungsmitteilung. *Coll. Czech. Chem. Commun.*, **22**, 401 (1957).

57-35   VÝDRÁ, F., and KARLIK, M., Konduktometrische Titrationen in

356   *Bibliography*

der Chelatometrie. II. Bestimmung der Gesamthärte des Wassers. *Coll. Czech. Chem. Commun.*, **22**, 979 (1957).

57-36   ASHBY, R. O., and ROBERTS, M., Determination of Ca in serum. *J. Lab. Clin. Med.*, **49**, 958 (1957).

57-37   BARON, D. N., and BELL, U. L., Determination of serum Ca. *Clinica Chim. Acta*, **2**, 327 (1957).

57-38   FABER, J. S., A note on the assay of Zn-stearate. *J. Amer. Pharm. Assoc.*, **46**, 512 (1957).

57-39   KÖRÖS, E., and REMPORT-HORVATH, Z., EDTA titration of Zn in brass and bronze and of Co in the presence of Cu. *Chem. Analyst*, **46**, 91 (1957).

57-40   EMI, K., TOEI, K., and MIYATA, H., Behaviour of o,o'-Hydroxyazo compounds towards Ca and Mg. *J. Chem. Soc. Japan, Pure Chem. Sect.*, **78**, 736 (1957).

57-41   EMI, K., TOEI, K., and TAKAMOTO, S., Behaviour of o-Carboxy-o'-hydroxyazo compounds towards Ca. *J. Chem. Soc. Japan, Pure Chem. Sect.*, **78**, 741 (1957).

57-42   BELCHER, R., CLOSE, R. A., and WEST, T. S., Fast Sulphone Black F as an indicator for the EDTA titration of Cu. *Chem. & Ind.*, 1647 (1957).

57-43   BELCHER, R., CLOSE, R. A., and WEST, T. S., Some o,o'-Dihydroxy-azo indicator dyes for EDTA titrations. *Chem. Analyst*, **46**, 86 (1957); **47**, 2 (1958).

57-44   GAGLIARDI, E., and REIMERS, H., Untersuchung und analytische Verwendung von Tripelsalzen vom Typus $NaMe(UO_2)_3(Acetat)_9$. H_2O. *Mikrochim. Acta*, 784 (1957).

57-45   GABRIELSON, G., Determination of Pb in Pb-fluoroborate solutions. *Metal Finishing*, 56 (1957).

57-46   ANDERSCH, M. A., Titration method for the determination of Ca in serum using a new indicator. *J. Lab. Clin. Med.*, **49**, 496 (1957).

57-47   DOBRYNINA, O. N., and BOGAREVA, K. G., Volumetrische Bestimmung von Ni in Katalysatorpulvern. *Maslob. Zhur. Prom.*, 40 (1957).

57-48   CASTIGLIONI, A., Komplexometrische Bestimmung von Pyridin. *Z. analyt. Chem.*, **156**, 426 (1957).

57-49   DETTNER, H. W., Bestimmung von Eisen in Chrombädern. *Metalloberfläche*, **11**, 12 (1957).

57-50   EMI, K., TOEI, K., and WADA, T., Behaviour of phthalein complexone reagent towards alkaline earth ions. *J. Chem. Soc. Japan, Pure Chem. Sect.*, **78**, 974 (1957).

57-51   DATTA, S. K., Thorium complexes of 2,7-dinitroso-chromotropic acid. III. Complexometric determination of Th with EDTA. *Analyt. Chim. Acta*, **16**, 115 (1957).

57-52   DELGA, J., and STORCK, J., Dosage complexométrique du Bi. Applications au contrôle des médicaments. *Ann. pharm. franç.*, **15**, 299 (1957).

57-53   BOUTEN, J., VERBEEK, F., and EECKHAUT, J., Détermination des constantes de stabilité du Tl^+ avec les acide EDTA et NTA. *Analyt. Chim. Acta*, **17**, 339 (1957).

57-54   CARPENTER, H., Determination of Ca in Natural Waters. *Limnol. and Oceanogr.*, **2**, 271 (1957).

57-55   BUDĚŠINSKÝ, B., Komplexometrie in der organischen Analyse. II. Bestimmung von Aminosäuren und Peptiden. *Coll. Czech. Chem. Commun.*, **2**, 230 (1957).

57-56   BUDĚŠINSKÝ, B., and VANÍČKOVÁ, E., Komplexometrie in der organischen Analyse. III. Bestimmung von quarternären Ammoniumsalzen. *Coll. Czech. Chem. Commun.*, **22**, 236 (1957).

57-57   BUDĚŠINSKÝ, B., Komplexometrie in der organischen Analyse. IV. Halbmikrobestimmung von Nitro- und Nitrosogruppen. *Coll. Czech. Chem. Commun.*, **22**, 1141 (1957).

57-58   BUDĚŠINSKÝ, B., Komplexometrie in der organischen Analyse. V. Bestimmung von C=C Bindungen mittels Hg(II)-Acetat. *Coll. Czech. Chem. Commun.*, **22**, 1147 (1957).

57-59   KÖRBL, J., and PŘIBIL, R., Metallochromic indicators. I. Introduction. *Coll. Czech. Chem. Commun.*, **22**, 1122 (1957).

57-60   KÖRBL, J., KRAUS, E., JANIČEK, F., and PŘIBIL, R., Metallochromic indicators. II. 3,4-Dihydroxy-4-nitrobenzene and 3,4-dihydroxybenzene-4′-sulphonic acid as simple metallochromic models of pyrocatechol violet. *Coll. Czech. Chem. Commun.*, **22**, 1416 (1957).

57-61   KÖRBL, J., Metallochromic indicators. III. Preparation of 3,3′-bis-N,N-di-(carboxymethyl)aminomethyl-thymolsulphophthalein (Methylthymol blue). *Coll. Czech. Chem. Commun.*, **22**, 1789 (1957).

57-62   BRAUN, K. H., Chelatometrische Bestimmung von Cu und Ni in galvanischen Bädern. *Chem. Techn.*, **9**, 541 (1957).

57-63   BRAKE, L. D., MCNABB, W. M., and HAZEL, J. F., The photometric titration of Ni in the presence of Co. *Analyt. Chim. Acta*, **17**, 314 (1957).

57-64   METZ, C. F., The analytical chemistry of plutonium. *Analyt. Chem.*, **29**, 1748 (1957).

57-65   UENO, K., Simultaneous complexometric titration of Cu and Hg. *Analyt. Chem.*, **29**, 1668 (1957).

57-66   GORYUSHINA, V. G., and ROMANOVA, E. V., Zr-Bestimmung mit Erio T und Carminsäure als Indikatoren. *Zavodskaya Lab.*, **23**, 781 (1957).

57-67   EK, C., Die Bestimmung von metallischem Zn in Gegenwart von ZnO. *Rev. universelle mines*, **13**, 249 (1957).

57-68   PILLERI, R., Komplexometrische Bestimmung von K. *Z. analyt. Chem.*, **157**, 1 (1957).

57-69   GREENE, JR, A., Sucrose as a stable carrier for murexide indicator. *Chem. Analyst*, **46**, 104 (1957).

57-70   MARTENS, G., and SCHWARZ, K., Die komplexometrische Bestimmung von Cu und Mn nebeneinander. *Z. analyt. Chem.*, **159**, 22 (1957).

57-71   MASHALL, J., and GEYER, L., The titration of Ca with EDTA in the presence of limited amounts of fluoride. *Bull. Res. Council Israel*, **6**, 74 (1957).

57-72   MANOLIU, C., Über die Verwendung des Eriochromcyanins als Indikator in der Chelatometrie. *Revista Chim.*, **8**, 716 (1957).

57-73   GENGE, J. A. R., and SALMON, J. E., The determination of phosphate and of metals in the presence of phosphate. *Lab. Practice*, **6**, 695 (1957).

57-74   DUVAL, C., Sur la stabilité thermique des étalons analytiques. V. *Analyt. Chim. Acta*, **16**, 545 (1957).

57-75   GEDANSKY, S. J., and GORDON, L., Indirect photometric titration of milligram quantities of Ag with EDTA. *Analyt. Chem.*, **29**, 566 (1957).

57-76   PATROVSKÝ, V., and HUKA, M., Komplexometrische Titrationen (Chelatometrie). XXII. Maßanalytische Bestimmung von Fe, Al und Ti in Silikaten. Bemerkung zur chelatometrischen Bestimmung von Ca und Mg. *Coll. Czech. Chem. Commun.*, **21**, 37 (1957).

57-77   VŘEŠTÁL, J., and KOTRLÝ, S., Chelatometrie. XXIII. Die Pb-Bestimmung gegen Brillantkongoblau in nichtgepuffertem Medium. *Coll. Czech. Chem. Commun.*, **22**, 1775 (1957).

57-78   VŘEŠTÁL, J., and HAVÍR, J., Komplexometrische Titrationen (Chelatometrie). XXIV. Die Pb-Bestimmung gegen Brenzcatechinviolett als Indikator. *Coll. Czech. Chem. Commun.*, **22**, 316 (1957).

57-79   KÖRBL, J., PŘIBIL, R., and EMR, A., Komplexometrische Titrationen (Chelatometrie). XXV. Xylenolorange als neuer spezifischer Indikator. *Coll. Czech. Chem. Commun.*, **22**, 961 (1957).

57-80   JANKOVSKÝ, J., Komplexometrische Titrationen (Chelatometrie). XXVI. Komplexometrische Bestimmung des Sn. *Coll. Czech. Chem. Commun.*, **22**, 1052 (1957).

57-81   BUDĚŠINSKÝ, B., Komplexometrische Titrationen (Chelatometrie). XXVII. Naphtholviolett, ein neuer einfacher komplexometrischer Indikator. *Coll. Czech. Chem. Commun.*, **22**, 1579 (1957).

57-82   KOTRLÝ, S., Komplexometrische Titrationen (Chelatometrie). XXVIII. Mikrobestimmung von Pb gegen Dithizon als Indikator. *Coll. Czech. Chem. Commun.*, **22**, 1765 (1957).

57-83   KÖRBL, J., and PŘIBIL, R., Komplexometrische Titrationen (Chelatometrie). XXIX. Die spezifische Maskierung und Bestimmung von Hg. *Coll. Czech. Chem. Commun.*, **22**, 1771 (1957).

57-84   LUKASZEWSKI, G. M., REDFERN, J. P., and SALMON, J. E., The volumetric determination of phosphate and of metals in the presence of phosphates. II. The complexometric determination of several metals in the presence of phosphate with dithizone as the indicator. *Lab. Practice*, **6**, 389 (1957).

57-85   BUDĚŠINSKÝ, B., and VANÍČKOVÁ, E., Die Bestimmung von Thiaminen und Mercaptothiaminen. *Cesk. Farm.*, **6**, 308 (1957).

57-86   FLASCHKA, H., HOLASEK, A., and ROSENTHAL, M., Über eine verbesserte komplexometrische Bestimmung des K im Blutserum. *Z. physiol. Chem.*, **308**, 183 (1957).

57-87   FLASCHKA, H., and SOLIMAN, A., Trien als analytisches Reagenz.

I. Allgemeines, Theorie und die photometrische Titration reiner Cu-Lösungen. *Z. analyt. Chem.*, **158**, 253 (1957).

57-88 FLASCHKA, H., and SOLIMAN, A., Trien als analytisches Reagenz. II. Photometrische Titration von Cu neben anderen Metallen. *Z. analyt. Chem.*, **159**, 29 (1957).

57-89 FLASCHKA, H., and KHALAFALLAH, S., Zur Theorie der visuellen Endpunktsbestimmung bei komplexometrischen Titrationen. *Z. analyt. Chem.*, **156**, 401 (1957).

57-90 FLASCHKA, H., BARNARD, A. J., and BROAD, W. C., EDTA titrations: Applications. *Chem. Analyst*, **46**, 106 (1957), **47**, 22, 52, 78, 109 (1958).

57-91 FLASCHKA, H., and BARAKAT, M. F., Über die Verwendung komplexchemischer Verdrängungsreaktionen bei polarographischen Analysen. II. Die amperometrische Titration von Th. *Z. analyt. Chem.*, **156**, 321 (1957).

57-92 FLASCHKA, H., and SADEK, F., Die Titerkonstanz stark verdünnter AeDTE-Maßlösungen. *Z. analyt. Chem.*, **156**, 23 (1957).

57-93 MAJUMDAR, S. K., and BANERJEE, N. G., A Volumetric method for the determination of sulphur in coal. *J. Industr. Chem. India*, **29**, 213 (1957).

57-94 BUDANOVA, L. M., and VOLODARSKAYA, R. S., Mg-Bestimmung in Al-Legierungen. *Zavodskaya Lab.*, **23**, 797 (1957).

57-95 RENTSCHLER, H., TANNER, H., and DJUNG, P., Komplexometrische Bestimmung von Zucker in gesüßten und fermentierten Getränken nach der Methode von Potterat und Eschmann. *Mitt. Lebensm. Hygiene, Bern*, **48**, 238 (1957).

57-96 SCHNEYDER, J., Schnellanalyse von Tabakaschen. *Fachl. Mittl. Österr. Tabakregie*, 25 (1957).

57-97 JORDAN, J., and ALLEMAN, T. G., Thermochemical titrations. Enthalpy titrations. *Analyt. Chem.*, **29**, 9 (1957).

57-98 KALEIS, O. YU., Komplexometrische Kontrolle der Qualität von destilliertem Wasser. *Aptechnoe Delo*, **6**, 29 (1957).

57-99 SOMMER, L., and KOLÁŘIK, Z., Triphenylmethanfarbstoffe als komplexometrische Indikatoren zur Fe-Bestimmung. *Coll. Czech. Chem. Commun.*, **50**, 203 (1957).

57-100 IRITANI, N., and TANAKA, T., Volumetric analysis of Hg-salts by EDTA. II. Determination of HgS, HgOHCN, Hg-salicylate, Phenyl-Hg-acetate and sodium ethyl-Hg-thiosalicylate. *J. Pharm. Soc. Japan*, **77**, 106 (1957).

57-101 KAO, H.-H., and YUN, W., Amperometric titration of small amounts of Zn in Co-plating bath. *Pei Ching Ta Hsueh Pao Tzu Ian K'o Hsueh.*, **3**, 217 (1957).

57-102 KANIE, T., Determination of Mg in Al-alloys by photometric titration. *Japan Analyst*, **6**, 711 (1957).

57-103 TUCKER, B. M., Calcein as indicator for the titration of Ca with EDTA. *Analyst*, **82**, 284 (1957).

57-104 TOLMACHEV, C. N., and VESTFRIED, T. Y., Spectrophotometric study on interaction of ions of Zn with purpuric acid. *Zhur. neorg. Khim.*, **2**, 60 (1957).

360 Bibliography

57-105 TICHOMIROVA, C. T., and SIMAČKOVA, O., Die Bestimmung von Ca in Gegenwart eines großen Überschusses Mg. *Coll. Czech. Chem. Commun.*, **22**, 982 (1957).

57-106 WAKAMATSU, S., Schnellbestimmung von Phosphor in Ferrophosphor. *Japan Analyst*, **6**, 579 (1957).

57-107 KINNUNEN, J., and WENNERSTRAND, B., EDTA titration of Sn in Cu base alloys. *Chem. Analyst*, **46**, 34 (1957).

57-108 SEN, B., Indirect complexometric titration of Na and K with EDTA. *Z. analyt. Chem.*, **157**, 2 (1957).

57-109 KUTZNETZOV, V. I., and MIKHAILOV, V. A., Laquer Scarlet C as complexometric indicator for Ca and Mg. *Zhur. analit. Khim.*, **12**, 59 (1957).

57-110 KUTZNETSOVA, E. T., TALALAEVA, O. D., and THIKONOV, A. S., Rapid analysis of Cd-alloys by EDTA. *Sb. Trud. Voran. Otd. Vses. Khim. Obshch. Mendeleev*, 151 (1957).

57-111 WÜNSCH, L., Kombinierte Anwendung von Chelaten und Ionenaustauschern. *Coll. Czech. Chem. Commun.*, **22**, 1339 (1957).

57-112 ZAITSEV, A. A., Use of EDTA in the determination of alkaloids. I. Determination of quinine hydrochloride. *Aptechnoe Delo*, **6**, 48 (1957).

57-113 SCHMITZ, B., Komplexometrische Titration von Zn, Pb und Hg in offizinellen Salben. *Dtsch. Apoth. Ztg.*, **97**, 399 (1957).

57-114 KRYLOVA, A. N., Use of EDTA in the determination of Ba in biological material. *Aptechnoe Delo*, **6**, 28 (1957).

57-115 SATO, T., and IKEGAMI, A., A rapid routine analysis of limestone. *Japan Analyst*, **6**, 706 (1957).

57-116 SCHOLZ, L., Die komplexometrische Bestimmung von Sulfaten in Wässern. *Mitt. Inst. Wasserwirtsch.*, **2**, 60 (1957).

57-117 ERDEY, L., and PÓLOS, L., Chelatometrische Bestimmung von Zn, Cd und Pb in Anwesenheit von Variaminblau als Indikator. *Analyt. Chim. Acta.*, **17**, 458 (1957).

57-118 OKADA, K., and SUGIYAMA, T., Bestimmung von Mikromengen Fluorid durch chelatometrische Titration. *Ann. Rep. Takamine Lab.*, **9** (1957).

57-119 LASZLOVSZKY, J., Analyse organischer Bi-Präparate. *Acta Pharm. Hung.*, **3**, 125 (1957).

57-120 VERMA, M. R., and BHUCHAR, V. M., Determination of Ba and Zn in lithopone. *Paint Manuf.*, **27**, 384 (1957).

57-121 VERMA, M. R., and PAUL, S. D., A screened indicator for the complexometric determination of Th. *Current Sci.*, **26**, 178 (1957).

57-122 FRITZ, J. S., LANE, W. J., and BYSTROFF, A. S., Complexometric titrations using azoxine indicator. *Analyt. Chem.*, **29**, 821 (1957).

57-123 HILDEBRAND, G. P., and REILLEY, C. N., New indicator for complexometric titration of Ca in presence of Mg. *Analyt. Chem.*, **29**, 258 (1957).

57-124 LOTT, P. F., and CHENG, K. L., Stepwise EDTA titration of Ca and Mg with CI 202 and CI 203 as indicators. *Chem. Analyst*, **46**, 30 (1957).

57-125 FLASCHKA, H., and SADEK, F., Mikrotitrationen mit AeDTE.

XVI. Metallbestimmungen im Ultramikrobereich. *Mikrochim. Acta*, 1 (1957).

57-126 RENAULT, J., and GAUTIER, J. A., Sur un nouveau procédé de dosage des alcoylsulfates alcalins à longues chaînes dérivés d'alcools primaires. *Bull. Soc. chim. France*, 208 (1957).

57-127 HENNART, C., and MERLIN, E., Applications de la chélatométrie. II. Dosage des esters et chlorures de monoesters oxaliques. *Analyt. Chim. Acta*, 17, 534 (1957).

57-128 GAUTIER, J. A., and RENAULT, J., Sur l'analyse des détergents; nouvelle methode applicable aux alcoylsulfates alcalins à longues chaînes, dérivés d'alcools primaires. *Chim. analyt.*, 39, 189 (1957).

57-129 HENNART, C., and MERLIN, E., Chelatométrie. I. Dosage du fluor dans les composés organiques. *Analyt. Chim. Acta*, 17, 463 (1957).

57-130 FOREMAN, J. K., and SMITH, T. D., The nature and stability of the complex ions formed by ter-, quadri-, and hexavalent plutonium ions with EDTA. (a) Part I. pH-titrations and ion-exchange studies. *J. Chem. Soc.*, 1752 (1957); (b) Part II. Spectrophotometric studies. *J. Chem. Soc.*, 1758 (1957).

57-131 BUSEV, A. I., Bi-Bestimmung mit PAN als Indikator. *Zhur. analit. Khim.*, 12, 386 (1957).

57-132 MILNER, G. W. C., and NUNN, J. H., Ion exchange separation in the analysis of bismuth base alloys. Part III. Binary alloys containing U and Th. *Analyt. Chim. Acta*, 17, 259 (1957).

57-133 MILNER, G. W. C., and NUNN, J. H., Ion exchange separation in the analysis of bismuth base alloys. Part II. Ternary alloys containing U and Th. *Analyt. Chim. Acta*, 17, 494 (1957).

57-134 FLASCHKA, H., and SADEK, F., Mikrotitrationen mit AeDTE. XVI. Metallbestimmungen im Ultramikrobereich. *Mikrochim. Acta*, 1 (1957).

57-135 PŘIBIL, R., and KÖRÖS, E., Komplexometrische Bestimmung des Hg-Gehaltes von Hg-Verbindungen. I. *Acta Pharm. Hung.*, 27, 1 (1957).

57-136 PŘIBIL, R., KÖRÖS, E., and BARCZA, L., Komplexometrische Bestimmung des Hg-Gehaltes von Hg-Verbindungen. II. *Acta Pharm. Hung.*, 27, 145 (1957).

57-137 KINNUNEN, J., and WENNERSTRAND, B., Some further applications of xylenole orange as indicator in EDTA titrations. *Chem. Analyst*, 46, 92 (1957).

57-138 BJERRUM, J., SCHWARZENBACH, G., and SILLÉN, L. G., Stability constants of metal complexes. *Chem. Soc. Special Publ. No. 6*, London 1957. Part I.

57-139 SCHWARZENBACH, G., SENN, H., and ANDEREGG, G., Komplexone. XXIX. Eine großer Chelateffekt besonderer Art. *Helv. Chim. Acta*, 40, 1886 (1957).

57-140 BAYER, E., Metallkomplexe von Schiffschen Basen aus *o*-Aminophenol und Dioxoverbindungen. *Ber.*, 90, 2325 (1957).

57-141 MORRISON, G. H., and FREISER, H., *Solvent Extraction in Analytical Chemistry*. John Wiley & Sons Inc., New York (1957).

1958

58-1 BELCHER, R., CLOSE, R. A., and WEST, T. S., The complexometric titration of Ca in the presence of Mg. A critical study. *Talanta*, 1, 238 (1958).

58-2 BELCHER, R., LEONARD, M. A., and WEST, T. S., New colour reactions of phthaleincomplexone. *Chem. & Ind.*, 128 (1958).

58-3 BELCHER, R., LEONARD, M. A., and WEST, T. S., The preparation and analytical properties of N,N-di-(carboxymethyl)aminomethyl derivatives of some hydroxyanthraquinones. *J. Chem. Soc.*, 2390 (1958).

58-4 RINGBOM, A., PENSAR, G., and WÄNNINEN, E., Complexometric titration method for determining Ca in the presence of Mg. *Analyt. Chim. Acta*, 19, 525 (1958).

58-5 CIMERMAN, CH., ALON, A., and MASHALL, J., Titrimetric determination of Al with EDTA in the presence of Fe, Cu, Ti, Mn, Ca, Mg and Phosphate. *Talanta*, 1, 314 (1958).

58-6 CIMERMAN, CH., ALON, A., and MASHALL, J., Volumetric determination of Ca and Mg with EDTA in the presence of phosphate. *Analyt. Chim. Acta*, 19, 461 (1958).

58-7 ABD-EL RAHEEM, A. A., and AMIN, A. A. M., Eriochrome Black A as indicator in chelatometric titrations. *Analyt. Chim. Acta*, 19, 327 (1958).

58-8 ABD-EL RAHEEM, A. A., and AMIN, A. A. M., Erio SE as indicator in chelatometric titrations. *Z. analyt. Chem.*, 163, 340 (1958).

58-9 STYUNKEL, T. B., and YAKIMETS, E. M., Acid Chrome Dark Blue and Acid Chrome Blue K as indicators in the complexometric determination of Ca. *Zavodskaya Lab.*, 24, 23 (1958).

58-10 ČERNY, A., Indirekte polarographische und komplexometrische Bestimmung kleiner Mengen von As in Gußeisen und Erzen. *Hutnicke Listy*, 13, 715 (1958).

58-11 REILLEY, C. N., and SCHMID, R. W., Chelometric titrations with potentiometric end-point detection. Hg as pM indicator electrode. *Analyt. Chem.*, 30, 947 (1958).

58-12 REILLEY, C. N., SCHMID, R. W., and LAMSON, D. W., Chelometric titrations of metal ions with potentiometric end-point detection. *Analyt. Chem.*, 30, 953 (1958).

58-13 SCRIBNER, W. G., and REILLEY, C. N., Indirect complexometric analysis with aid of liquid amalgams. *Analyt. Chem.*, 30, 1452 (1958).

58-14 REILLEY, C. N., and HOLLOWAY, J. H., The stability of metal-tetraethylenepentamine complexes. *J. Amer. Chem. Soc.*, 80, 2917 (1958).

58-15 WILLARD, H. H., MOSEN, A. W., and GARDNER, R. D., Volumetric determination of Th in U-alloys. *Analyt. Chem.*, 30, 1614 (1958).

58-16 KALINICHENKO, I. I., Complexometric determination of Ni without separation in alloys of Cu. *Zavodskaya Lab.*, 24, 266 (1958).

58-17 LIEBERMAN, M. V., Bestimmung von Mg in Gegenwart von Al. *Zavodskaya Lab.*, 24, 147 (1958).

58-18 BELLOMO, A., and BRUNO, E., Bestimmungen von Kationen mittels Hochfrequenzkomplexometrie. I. *Atti soc. peloritana sci. fis. math. e nat.*, **5**, 245 (1958–1959).

58-19 BADRINAS, A., Mikrobestimmung von Zn in Pflanzenmaterial durch AeDTA-Titration unter Verwendung eines Fluoreszenzindicators. *Publ. del Instit. de Biolog. Applicada*, **28**, 75 (1958).

58-20 BACHRA, B. N., DAUER, A., and SOBEL, A. E., Determination of serum Ca. *Clin. Chem.*, **4**, 107 (1958).

58-21 BAKER, R. A., Determination of Zn and Cd in plating solutions. *Metal Ind.*, **92**, 491 (1958).

58-22 HEADRIDGE, J. B., *Photometric Titrations*. Pergamon Press, London (1958).

58-23 HEADRIDGE, J. B., Photometric titrations. *Talanta*, **1**, 293 (1958).

58-24 FLEISCHER, K. D., SOUTHWORTH, B. C., HODECKER, J. H., and TUCKERMANN, M. M., Determination of phosphorus in organic compounds. *Analyt. Chem.*, **30**, 152 (1958).

58-25 LUCCHESI, C. A., and HIRN, C. F., Determination of the metal content of paint driers. EDTA titration in alcohol–benzene solutions. *Analyt. Chem.*, **30**, 1877 (1958).

58-26 COSTA, A. C., Dithizone as an indicator for the EDTA titration of various metals. *Chem. Analyst*, **47**, 39 (1958).

58-27 BRUNO, E., and BELLOMO, A., Bestimmungen von Kationen mittels Hochfrequenzkomplexometrie. II. Al-Mg- und Ni-Cr-Legierungen. *Atti. soc. peloritana sci. fis. math. e nat.*, **5**, 327 (1958/1959).

58-28 SOMMER, L., and HNILIČKOVÁ, M., Neue Pyridinazofarbstoffe als chelatometrische Indikatoren. *Naturwiss.*, **45**, 544 (1958).

58-29 FRANZON, O., IVARSSON, G., and SAMUELSON, O., Determination of Ca in sulphite spent liquor. *Svensk. Pappertidning*, **61**, 165 (1958).

58-30 LONGFORD, K. L., New approach to the analysis of brass, Cd and Zn plating solutions by using EDTA and Fast Sulphone Black F. *Electroplating*, **11**, 439 (1958).

58-31 LIU, S., and CHANG, M., Komplexometrische Titration von Se. *Acta Chim. Sinica*, **24**, 306 (1958).

58-32 BRAUN, T., MAXIM, I., and GALATEANU, I., Radiometric titrations with complexones. *Nature*, **182**, 936 (1958).

58-33 BENNET, M. C., and SCHMIDT, N. O., Determination of Ca and Mg in cane juice. *Int. Sugar J.*, **60**, 225 (1958).

58-34 KÖRBL, J., and PŘIBIL, R., Komplexometrische Titrationen (Chelatometrie). XXX. Methylthymolblau, ein neuer metallochromer Indikator vom Komplexontypus. *Coll. Czech. Chem. Commun.*, **22**, 873 (1958).

58-35 FLASCHKA, H., Zur Theorie der visuellen Indication und Selektivität komplexometrischer Titrationen. *Talanta*, **1**, 60 (1958).

58-36 FLASCHKA, H., ABD-EL RAHEEM, A. A., and SADEK, F., Erio SE als neuer Indikator zur komplexometrischen Bestimmung von Ca (und Mg) im Serum. *Z. physiol. Chem.*, **310**, 97 (1958).

58-37 FLASCHKA, H., and SADEK, F., Application of EDTA titration to

N

364 *Bibliography*

the determination of the tetraphenylboron and its salts. *Chem. Analyst*, **47**, 30 (1958).

58-38 DEWALD, A., Komplexometrische Bestimmung von Ni und Fe in der gleichen Probe durch photometrische Titration. *Acad. Rep. Popul. Romine, Baza Cercetari Stiint, Timisoara*, **5**, 125 (1958).

58-39 KENNY, A. D., and COHN, V. C., Complexometric titration of Ca in the presence of Mg. *Analyt. Chem.*, **30**, 1366 (1958).

58-40 CRAWLEY, R. H. A., The complexometric titration of Ga with fluorescent indicators. *Analyt. Chim. Acta*, **19**, 540 (1958).

58-41 ZHDANOV, A. K., KHADEEV, V. A., and KATS, A. L., Amperometrische Titration von Fe(III) mit Ascorbinsäure oder AeDTE. *Uzb. Khim. Zhur.*, 27 (1958).

58-42 WILKINS, D. H., and HIBBS, L. E., Determination of Al, Ni, Co, Cu and Fe in Alnico. *Analyt. Chim. Acta*, **18**, 372 (1958).

58-43 WILKINS, D. H., Determination of ThO_2 in thoriated tungsten wire. *Analyt. Chim. Acta*, **19**, 441 (1958).

58-44 KIES, H. L., Ampérométrie avec deux électrodes indicatrices. *Analyt. Chim. Acta*, **18**, 14 (1958).

58-45 FOSS, O. P., and ANDERSEN, B., An apparatus for rapid microphotometric titration of serum Ca with EDTA. *Scand. J. clin. Lab. Invest.*, **10**, 437 (1958).

58-46 VŘEŠTÁL, J., HAVÍR, J., BRANDŠTETTER, J., and KOTRLÝ, S., Komplexometrische Titrationen (Chelatometrie). XXXII. Indirekte komplexometrische Fluorbestimmung mittels Pb-Salzen. *Coll. Czech. Chem. Commun.*, **23**, 886 (1958).

58-47 KÖRBL, J., and VÝDRÁ, F., Metallochromic indicators. IV. A note on the preparation and properties of calcein. *Coll. Czech. Chem. Commun.*, **23**, 622 (1958).

58-48 KÖRBL, J., and KAKÁC, B., Metallochromic indicators. V. Acid–base properties of methylthymol blue. *Coll. Czech. Chem. Commun.*, **23**, 889 (1958).

58-49 KÖRBL, J., and PŘIBIL, R., Metallochromic indicators. VI. Analogues of *o*-cresolphthalein complexone. *Coll. Czech. Chem. Commun.*, **23**, 1213 (1958).

58-50 KÖRBL, J., KRAUS, E., and PŘIBIL, R., Metallochromic indicators. VII. Glycinethymol blue. *Coll. Czech. Chem. Commun.*, **23**, 1219 (1958).

58-51 BUDĚŠINSKÝ, B., and KÖRBL, J., Komplexometrie in der pharmazeutischen Analyse. XVII. Die Bestimmung von Bi. *Cesk. Farm.*, **7**, 78 (1958).

58-52 KÖRBL, J., VÝDRÁ, F., and PŘIBIL, R., The use of fluoresceine complexone. *Talanta*, **1**, 281 (1958).

58-53 BUDĚŠINSKÝ, B., Simple metallochromic indicators of the Erio T type. *Coll. Czech. Chem. Commun.*, **23**, 895 (1958).

58-54 KÖRBL, J., VÝDRÁ, F., and PŘIBIL, R., Ein Beitrag zur Charakterisierung des Fluoresceinkomplexons. *Talanta*, **1**, 138 (1958).

58-55 FAINBERG, S. YU., BLYAKHMAN, A. A., and FILATOVA, L. N., Rapid method of determining Cu, Pb and Zn in polymetallic ores and concentrates. *Zavodskaya Lab.*, **24**, 18 (1958).

58-56   KHALIFA, H., Analysis of binary mixtures of Ba, Sr or Mg together with Pb, Co, Ni or Cu. *Z. analyt. Chem.*, **159**, 410 (1958).

58-57   KHALIFA, H., PATZAK, R., and DOPPLER, G., Potentiometrische Titration von AeDTE im pH-Bereich 8—12. *Z. analyt. Chem.*, **161**, 264 (1958).

58-58   KHALIFA, H., Potentiometric estimation of Th. *Z. analyt. Chem.*, **161**, 401 (1958).

58-59   KHALIFA, H., Potentiometric estimation of Al and Mn. *Z. analyt. Chem.*, **163**, 81 (1958).

58-60   KHALIFA, H., Volumetric analysis of binary mixtures of the alkaline earths, Mg, Zn and Cd. *Analyt. Chim. Acta*, **18**, 310 (1958).

58-61   FRITZ, J. S., OLIVER, R. T., and PIETRZYK, D. J., Chelometric titrations using azoarsonic acid as indicator. *Analyt. Chem.*, **30**, 1111 (1958).

58-62   FRITZ, J. S., RICHARD, M. J., and KARRAKER, S. K., Potentio-metric titrations with EDTA. Use of masking agents to improve selectivity. *Analyt. Chem.*, **30**, 1347 (1958).

58-63   CHENG, K. L., Complexometric titrations of Cu and other metals in mixtures. Masking with thiosulphate. *Analyt. Chem.*, **30**, 243 (1958).

58-64   CHENG, K. L., EDTA titration of micro quantities of rare earths. *Chem. Analyst*, **47**, 93 (1958).

58-65   BIEBER, B., and VEČEŘA, Z., Vollständige Schnellanalyse von Kupolofenschlacken. *Przeglad Odlewictwa*, **3**, 66 (1958).

58-66   SAJÓ, I., Der Komplex des fünfwertigen Vanadiums mit AeDTE. *Acta Chim. Acad. Sci. Hung.*, **16**, 115 (1958).

58-67   BUDEVSKII, O. B., Complexometric determination of Cu. *Zavod-skaya Lab.*, **24**, 535 (1958).

58-68   MUSTAFIN, I. S., and KASHKOVSKAYA, E. A., Ein neuer Indikator zur komplexometrischen Mg-Bestimmung. *Zavodskaya Lab.*, **24**, 1060 (1958).

58-69   MILES, M. J., MESIMER, W. J., and ATKIN, M., Volumetric deter-mination in Mg in Ti. *Analyt. Chem.*, **30**, 361 (1958).

58-70   MENIS, O., MANNINGER, D. L., and BALL, R. G., Automatic spectrophotometric titration of fluoride, sulphate, U and Th. *Analyt. Chem.*, **30**, 1772 (1958).

58-71   DURCHAM, E. J., and RYSKIEWICH, S. P., The acid dissociation constants of diethylenetriamine penta-acetic acid and the stability constants of some of its metal complexes. *J. Amer. Chem. Soc.*, **80**, 4812 (1958).

58-72   SAITO, M., NAGAMURA, S., and UENO, K., Rapid determination of Zn in viscose spinning liquors. *Chem. Analyst*, **47**, 67 (1958).

58-73   MERZ, K. W., and LEHMANN, H., Bestimmung von Zn in Insulin. *Dtsch. Apoth. Ztg.*, **98**, 391 (1958).

58-74   TANAKA, N., KOIZUMI, T., MARAYAMA, T., KODAMA, M., and SAKUMA, Y., Use of rotated dropping mercury electrode. *Analyt. Chim. Acta*, **18**, 97 (1958).

58-75   GRÜNER, K., SOUKUP, M., and DRAHOTSKÁ, B., Komplexometrische Titration von Al in Salzen vom Kryolithtypus. *Hutnicke Listy*, **13**, 153 (1958).

58-76 HAHN, F. L., Härtebestimmung. *Angew. Chem.*, **70**, 712 (1958).

58-77 LUPAN, S., Komplexometrische Bestimmung von Th in Gegenwart von Ce. *Rev. Chim. Bucharest*, **9**, 101 (1958).

58-78 MACNEVIN, M., MCBRIDE, H. D., and HAKKILA, E. A., Evidence for the existence of a Rh complex with EDTA. *Chem. & Ind.*, 104 (1958).

58-79 DOBROWOLSKI, J., Komplexometrische Titration von Th, La und Ce(III) unter Verwendung von Carminsäure als Indikator. *Chem. Analit.* (*Warsaw*), **3**, 609 (1968).

58-80 KINNUNEN, J., and WENNERSTRAND, B., Contribution to the EDTA titration of Mo and W. *Chem. Analyst*, **47**, 38 (1958).

58-81 KINNUNEN, J., and MERIKANTO, B., EDTA titration of Pd. *Chem. Analyst*, **47**, 11 (1958).

58-82 LASSNER, E., and SCHARF, R., Die Maskierung von U(VI) bei komplexometrischem Titration bei pH 10. *Z. analyt. Chem.*, **159**, 212 (1958).

58-83 LASSNER, E., and SCHARF, R., Zur komplexometrischen Bestimmung von U(VI) mit AeDTE und PAN. *Z. analyt. Chem.*, **164**, 398 (1958).

58-84 STENGEL, E., and RIEMER, G., Beitrag zur photometrischen Endpunktsbestimmung bei der komplexometrischen Titration des Ca und Mg. *Z. analyt. Chem.*, **167**, 118 (1958).

58-85 RIVA, B., Bestimmung von Ca, Mg und Sulfat in Steinsalz. *Ann. Chim.* (*Roma*), **48**, 950 (1958).

58-86 SOUTHWORTH, B. C., HODECKER, J. H., and FLEISCHER, K. D., Determination of Hg in organic compounds. *Analyt. Chem.*, **30**, 1152 (1958).

58-87 OTTENDORFER, L. J., Rapid analysis of white metal. *Metallurgia*, **58**, 105 (1958). *Chem. Analyst*, **47**, 96 (1958).

58-88 PAVELKINA, V. P., and DOUSKAI, B. M., Determination of Zn in water. *Zavodskaya Lab.*, **24**, 548 (1958).

58-89 MUSTAFIN, I. S., and KASHKOVSKAYA, E. A., Schnellbestimmung von Ca und Mg in Gestein. *Zhur. Khim* (1958), Abstr. No. 70523.

58-90 SEN, B., EDTA titrations without metal indicators. *Analyt. Chim. Acta*, **19**, 551 (1958).

58-91 KORKISCH, J., and FARAG, A., Beiträge zur analytischen Chemie des Zr. I. Solochrome Violett R als Indikator zur Mikro- und Makrotitration des Zr mit AeDTE. *Z. analyt. Chem.*, **165**, 6 (1958).

58-92 WÄNNINEN, E., A nomogram for the evaluation of conditional constants for a number of metal–EDTA complexes. *Suomen Kemistelehti*, B **31**, 303.

58-93 YABRO, C. L., and GOLBY, R. L., Complexometric titration of urinary Ca and Mg. *Analyt. Chem.*, **30**, 505 (1958).

58-94 GAGLIARDI, E., and REIMERS, H., Komplexometrische Bestimmung von Na und K und deren Summe. *Z. analyt. Chem.*, **160**, 1 (1958).

58-95 VERMA, M. R., BHUCHAR, V. M., THERATTIL, K. J., and SHARMA, S. S., Determination of free lime in lime and silica products. *Analyst*, **83**, 160 (1958).

58-96 KNAPPE, E., and BÖCKEL, V., Über die komplexometrische Titration des Urinkalziums mit photometrischer Endpunktsanzeige. *Pharmazie*, **13**, 610 (1958).

58-97 WEINER, R., and NEY, E., Komplexometrische Titration von Cr(III) in Chromsäurelösungen. *Z. analyt. Chem.*, **161**, 432 (1958).

58-98 STREET, H. V., Determination of glucose in biological fluids with EDTA. *Analyst*, **83**, 628 (1958).

58-99 SEN, B., Indirect complexometric titration of K, Rb, Cs, Tl and NH$_4^+$. *Analyt. Chim. Acta*, **19**, 320 (1958).

58-100 TSYVINA, B. S., and VLADIMIROVA, V. M., Bestimmung von In in Sphalerit durch amperometrische Titration mit AeDTA. *Zavodskaya Lab.*, **24**, 278 (1958).

58-101 ŠTUDLAR, K., and JANOUŠEK, I., Komplexometrische Bestimmung von Pb in Lager- und Lötmetall. *Hutnicke Listy*, **13**, 805 (1958).

58-102 KÖRÖS, E., Some new possibilities in EDTA titrations using methyl thymol blue as indicator. *Proc. Intern. Sympos. Microchem.*, *Birmingham*, 474 (1958).

58-103 KÖRÖS, E., and POCZOK, I., Komplexon IV in der Komplexometrie. *Magyar Kém. Folyóirat*, **64**, 250 (1958).

58-104 SPOREK, K. F., Complexometric determination of sulphate. *Analyt. Chem.*, **30**, 1032 (1958).

58-105 SPOREK, K. F., Determination of phosphate in U ores, concentrates and liquors via an EDTA titration. *Chem. Analyst*, **47**, 12 (1958).

58-106 SÁRA, J., and BERNDT, W., Komplexometrische Cu-Bestimmung in Cupriäthylendiaminlösungen. *Svensk. Papperstidn.*, **61**, 353 (1958).

58-107 BUSEV, A. I., and TIPTSOVA, V. G., Komplexometrische Bestimmung von Tl. *Zhur. analit. Khim.*, **13**, 180 (1958).

58-108 BUSEV, A. I., KISELEVA, L. V., and CHERKESOV, A. I., Komplexometrische Bestimmung von Th. *Zavodskaya Lab.*, **24**, 13 (1958).

58-109 BUSEV, A. I., and PETRENKO, A. G., Complexometric indicators. *Zavodskaya Lab.*, **24**, 1449 (1958).

58-110 MALMSTADT, H. V., and HADJIIOANNOU, T. P., Rapid and accurate automatic titration of Ca and Mg in dolomites and limestones. *Analyt. Chim. Acta*, **19**, 563 (1958).

58-111 BERAL, H., DEMETRESCU, E., STOICESCU, V., and GRINTESCU, P., Beiträge zu den Untersuchungsmethoden von pharmazeutischen Zubereitungen. *Zentralhalle*, **97**, 524 (1958). *Farmacia*, 357 (1958).

58-112 PILZ, W., Die Bestimmung des Pflanzenschutzmittels Metasystox-i. *Z. analyt. Chem.*, **164**, 241 (1958).

58-113 MILNER, G. W. C., and EDWARDS, J. W., Analysis of Bi-base alloys. III. Ternary alloys containing U and Nd or Pr. *Analyt. Chim. Acta*, **18**, 513 (1958).

58-114 MARTYNENKO, L. I., *Nauchn. Doklad. Vyssh. Shkoly, Khim. i Khim. Tekhn.*, 718 (1958).

58-115 BOGNÁR, J., and JELLINEK, O., Technische Wasseranalyse. *Kohaszati Lapok*, **13**, 508 (1958).

58-116 IWANTSCHEFF, G., Das Dithizon und seine Anwendung in der Mikro- und Spurenanalyse. Verlag Chemie, Weinheim, 1958.

58-117 IRITANI, N., and TANAKA, T., Complexometric determination of sulphate. *Japan Analyst*, **7**, 42 (1958).

58-118 RYBA, O., CÍFKA, J., JEŽHOVA, D., MALÁT, M., and SUK, V., Chemische Indikatoren. IV. Komplexe des Brenzcatechin Violetts mit drei- und vierwertigen Metallen. *Coll. Czech. Chem. Commun.*, **23**, 71 (1958).

1959

59-1 LASSNER, E., SCHARF, R., and REISER, P. L., Zur komplexometrischen Bestimmung von Cu, Ni und Fe in W-Legierungen. *Z. analyt. Chem.*, **165**, 88 (1959).

59-2 LASSNER, E., and SCHARF, R., Komplexometrische Schnellbestimmung von Co in Hartmetallen. *Planseeber. Pulvermet.*, **7**, 129 (1959).

59-3 LASSNER, E., and SCHARF, R., Zur komplexometrischen Titration von Mo. *Z. analyt. Chem.*, **167**, 114 (1959); **168**, 429 (1959).

59-4 WILKINS, D. H., The determination of Ti with EDTA. *Analyt. Chim. Acta*, **20**, 113 (1959).

59-5 WILKINS, D. H., The determination of Ni, Co, Fe and Zn in ferrites. *Analyt. Chim. Acta*, **20**, 271 (1959).

59-6 WILKINS, D. H., and HIBBS, L. E., Determination of Ni in Au-Ni-alloys. *Analyt. Chim. Acta*, **20**, 273 (1959).

59-7 WILKINS, D. H., The chelometric determination of Cr(III), Co(III) and Cu with a metalfluorechromic indicator. *Analyt. Chim. Acta*, **20**, 324 (1959).

59-8 WILKINS, D. H., and HIBBS, L. E., The determination of Co, Fe, V and Mn in soft magnetic alloys. *Analyt. Chim. Acta*, **20**, 427 (1959).

59-9 HIBBS, L. E., and WILKINS, D. H., The determination of Al, Ti and Ni in their alloys. *Talanta*, **2**, 16 (1959).

59-10 WILKINS, D. H., The chelometric determination of Co and Fe using a fluorescent indicator. *Talanta*, **2**, 12 (1959).

59-11 WILKINS, D. H., and HIBBS, L. E., The determination of Cu with TRIEN using a metalfluorechromic indicator. *Talanta*, **2**, 201 (1959).

59-12 WILKINS, D. H., Metalfluorechromic indicators. *Talanta*, **2**, 277 (1959).

59-13 VŘEŠTÁL, J., HAVÍR, J., BRANDŠTETTER, J., and KOTRLÝ, S., Komplexometrische Titrationen (Chelatometrie). XXXIII. Urstoffe in der Chelatometrie. *Coll. Czech. Chem. Commun.*, **24**, 360 (1959).

59-14 MALÁT, M., and TENOROVÁ, M., Komplexometrische Titrationen (Chelatometrie). XXXIV. Chromazurol S als Indikator bei Th-,

Ni-, Ce-, und La-Bestimmung. *Coll. Czech. Chem. Commun.*, **24**, 632 (1959).

59-15 HOUDA, M., KÖRBL, J., BAZANT, V., and PŘIBIL, R., Komplexometrische Titrationen (Chelatometrie). XXXV. Über die indirekte Bestimmung von Al gegen Xylenolorange. *Coll. Czech. Chem. Commun.*, **24**, 700 (1959).

59-16 PŘIBIL, R., KÖRBL, J., KYSIL, B., and VOBORA, J., Komplexometrische Titrationen (Chelatometrie). XXXVI. Ein Beitrag zur Maskierung von Fe mit Triäthanolamin; Bestimmung von Ca mit Thymolphthaleinkomplexon. *Coll. Czech. Chem. Commun.*, **24**, 1799 (1959).

59-17 BUDĚŠINSKÝ, B., Komplexometrische Titrationen (Chelatometrie). XXXVII. Glycinnaphtholviolett als neuer chelatometrischer Indikator. *Coll. Czech. Chem. Commun.*, **23**, 1804 (1959).

59-18 KÖRBL, J., and PŘIBIL, R., Komplexometrische Titrationen (Chelatometrie). XXXVIII. Über die Indikatorempfindlichkeit und den Titrationsfehler in der Komplexometrie. *Coll. Czech. Chem. Commun.*, **24**, 2266 (1959).

59-19 BRHÁČEK, L., Komplexometrische Titrationen (Chelatometrie). XXXIX. Schnellbestimmung des Al in Ferrosilizium. *Coll. Czech. Chem. Commun.*, **24**, 2811 (1959).

59-20 MALÁT, M., SUK, V., and TENOROVÁ, M., Komplexometrische Titrationen (Chelatometrie). XL. Rücktitrationen gegen Pyrogallol- und Brompyrogallolrot als Indikatoren. *Coll. Czech. Chem. Commun.*, **24**, 2815 (1959).

59-21 SUK, V., and MIKETUKOVÁ, V., Komplexometrische Titrationen (Chelatometrie). XLI. Photometrische Titration von Bi und Cu gegen Brenzcatechinviolett. *Coll. Czech. Chem. Commun.*, **24**, 2818 (1959).

59-22 VÝDRÁ, F., PŘIBIL, R., and KÖRBL, J., Komplexometrische Titrationen (Chelatometrie). XLII. Einige weitere Reaktionen des Fluoreszein-Komplexons. Die Bestimmung des Cu und Mn. *Coll. Czech. Chem. Commun.*, **24**, 2623 (1959).

59-23 PŘIBIL, R., and VÝDRÁ, F., Komplexometrische Titrationen (Chelatometrie). XLIII. Maskierung einiger zweiwertiger Metalle mit o-Phenanthrolin. Selektive Bestimmung von Pb und Al. *Coll. Czech. Chem. Commun.*, **24**, 3103 (1959).

59-24 PATROVSKÝ, V., Komplexometrische Titrationen (Chelatometrie). XLIV. Yttriumbestimmung in Gemischen von Yttriumerden. *Coll. Czech. Chem. Commun.*, **24**, 3305 (1959).

59-25 DUBSKY, I., Komplexometrische Titrationen (Chelatometrie). XLVI. Direkte Bestimmung von zweiwertigem Sn. (Orig. russ.). *Coll. Czech. Chem. Commun.*, **24**, 4045 (1959).

59-26 JANOUŠEK, I., and ŠTUDLAR, K., Komplexometrische Titrationen (Chelatometrie). XLV. Selektive Zn-Bestimmung im schwachsauren Medium mit Xylenolorange und Methylthymolblau. *Coll. Czech. Chem. Commun.*, **24**, 3799 (1959).

59-27 VÝDRÁ, F., PŘIBIL, R., and KÖRBL, J., Direkte komplexometrische

370 *Bibliography*

Bestimmung von Fe mit Xylenolorange als Indikator. *Talanta*, 2 311 (1959).

59-28 SUK, V., and MIKETUKOVÁ, V., Chemische Indikatoren. V. Chelatometrischer Indikator Eriochromcyanin R; seine azido-basischen Eigenschaften und Bildung von Metallkomplexen. *Coll. Czech. Chem. Commun.*, **24**, 3629 (1959).

59-29 BUDEŠINSKÝ, B., and KÖRBL, J., Komplexometrie in der organischen Analyse. VI. Neue Möglichkeiten zur Bestimmung organischer Basen. *Coll. Czech. Chem. Commun.*, **24**, 1791 (1959).

59-30 PŘIBIL, R., and KOPANICA, M., Chelometric methods in rapid applied analysis. I. Chelometric titration of Mn in ferromanganese. *Chem. Analyst*, **48**, 35 (1959).

59-31 PŘIBIL, R., and KOPANICA, M., Chelometric methods in rapid applied analysis. II. Chelometric titration of Ni in Ni pellets and Ni–Fe alloys. *Chem. Analyst*, **48**, 66 (1959).

59-32 PŘIBIL, R., and KOPANICA, M., Chelometric methods in rapid applied analysis. III. Analysis of low-melting Bi–Pb–Cd–Sn-alloys. *Chem. Analyst*, **48**, 87, 89 (1959).

59-33 PŘIBIL, R., Contributions to the basic problems of complexometry. I. The blocking of indicators and its elimination. *Talanta*, **3**, 91 (1959).

59-34 PŘIBIL, R., Contributions to the basic problems of complexometry. II. Decomposition of xylenol orange in aqueous solutions. *Talanta*, **3**, 200 (1959).

59-35 HOLASEK, A., LIEB, H., and WINSAUER, K., Ein Mikrophotometer für colorimetrische Titrationen. *Mikrochim. Acta*, 402 (1959).

59-36 HOLASEK, A., and DUGANDŽIČ, M., Zur komplexometrischen Bestimmung des Na. *Mikrochim. Acta*, 488 (1959).

59-37 DUGANDŽIČ, M., FLASCHKA, H., and HOLASEK, A., Die komplexometrische Bestimmung des Na im Blutserum. *Clinica Chim. Acta*, **4**, 819 (1959).

59-38 BARNARD, JR., A. J., BROAD, W. C., and FLASCHKA, H., EDTA as a micro analytical reagent. *Microchem. J.*, **3**, 43 (1959).

59-39 SHAPIRO, M. YA., Neue photometrische Methode zur Bestimmung von Ni. *Zhur. analit. Khim.*, **14**, 365 (1959).

59-40 LACOURT, A., and HEYNDRICKX, P., Quantitative paper chromatography and direct titrimetric finish of determination of Zn spot concentrations. *Microchem. J.*, **3**, 181 (1959).

59-41 IRITANI, N., TANAKA, T., and OISHI, H., Chelometric determination of sulphate ion as $PbSO_4$. *Japan Analyst*, **8**, 30 (1959).

59-42 ITO, T., and Abe, M., Indirect determination of sulphate ions by spectrophotometric titration of Ba in the presence of barium sulphate. *J. Chem. Soc. Japan, Ind. Chem. Sec.*, **62**, 1801 (1959).

59-43 KANIE, T., Chelometric titrations with a mixture of murexide and 2-hydroxy-1-(2-hydroxy-4-sulpho-1-naphthylazo)-3-napthoic acid as an indicator. *Nagoyashi Kogyo Kenkynjo Hokoku*, **20**, 12 (1959).

59-44 USATENKO, J. I., and VITKINA, M. A., Die amperometrische Bestimmung von Fe mit EDTA. *Zavodskaya Lab.*, **24**, 1058 (1959).

59-45 TSYVINA, B. S., and KONKOVA, O. V., Komplexometrische Bestimmung von Sc. *Zavodskaya Lab.*, **25**, 1430 (1959).

59-46 WRONSKI, M., Synthese neuer Phthalein-Indikatoren. *Chem. Analit.* (Warsaw), **4**, 641 (1959).

59-47 ISHIDATE, M., and KIMURA, E., Micro determination of halogens in organic compounds. *Japan Analyst*, **8**, 739 (1959).

59-48 VORIŠEK, J., Über die Mg-Komplexe von AeDTE. *Coll. Czech. Chem. Commun.*, **24**, 3921 (1959).

59-49 KAROLEV, A., and BUDEVSKII, O., Schnellmethode zur Bestimmung von Cu und Zn in Messing. *Godishnik Nauch. Inst. Met. i Polezni Iskopaemi*, **1**, 163 (1959).

59-50 ERNSBERGER, F. M., Attack of glass by chelating agents. *J. Amer. Ceramic Soc.*, **42**, 373 (1959).

59-51 RUMLER, F., HERBOLSHEIMER, R., and WOLF, G., Ein Beitrag zur komplexometrischen Sulfatbestimmung. *Z. analyt. Chem.*, **166**, 23 (1959).

59-52 IWASE, A., Estimation of Zr by polarography. *Nippon Kagaku Zasshi*, **80**, 1142 (1959).

59-53 BORCHERT, O., Maßanalyse unter Verwendung von Komplexonen im Lacklaboratorium. *Plaste und Kautschuk*, **6**, 562 (1959).

59-54 BORCHERT, O., Komplexometrische Bestimmung von B neben Ba. *Talanta*, **2**, 387 (1959).

59-55 BARCZA, L., and KÖRÖS, E., Determination of Bi and divalent heavy metals in presence of each other. *Chem. Analyst*, **48**, 94 (1959).

59-56 BARCZA, L., Titerstellung von Komplexonlösungen. *Acta Pharm. Hung.*, **29**, 11 (1959).

59-57 SEDLÁČEK, B. A. J., Eine neue semimikro-komplexometrische Methode zur Bestimmung von Propylgallat in Fetten. *Z. Lebensm.- Unters. u. Forsch.*, **111**, 108 (1959).

59-58 KÖRÖS, E., and BARCZA, L., Some chelometric titrations of Fe(II). *Chem. Analyst*, **48**, 69 (1959).

59-59 ERDEY, L., Variamine blue: a versatile redox indicator. *Chem. Analyst*, **48**, 106 (1959).

59-60 BRAUN, T., MAXIM, I., and GALATEANU, I., Die Bestimmung von Ca, Sr und Mg mittels radiometrischer Titration durch Komplexon. *Zhur. analit. Khim.*, **14**, 542 (1959).

59-61 BUSEV, A. I., and CHAN, F., Direkte komplexometrische Bestimmung von Mo(VI). *Vest. Moskau Univ. Ser. Math. Mekh. Astr. Fiz i Khim.*, **14**, 203 (1959).

59-62 BUSEV, A. I., and CHAN, F., Chelometric determination of Mo after its reduction to the quinquevalent state. *Zhur. analit. Khim.*, **14**, 445 (1959).

59-63 ROBINSON, H. M. C., and RATHBUN, J. C., Determination of calcium and magnesium in serum. *Can. J. Biochem. Physiol.*, **37**, 225 (1959).

59-64 PETRIKOVA, M. N., Ultramicro determination of Na, K, Ca, Mg etc. in 0·5 microlitre cell juice of *ethmodiscus rex* (plancton diatomae). *Zhur. analit. Khim.*, **14**, 239 (1959).

372 Bibliography

59-65 KOMAR, N. P., The theory of complexometric titrations with metal indicators. *Zhur. analit. Khim.*, **14**, 152 (1959).

59-66 ISHIDATE, M., and KIMURA, E., Micro determination of sulphur in organic compounds. *Japan Analyst*, **8**, 733 (1959).

59-67 KRISTIANSEN, H., and LANGMYHR, F. J., Composition and stability constants of the complexes between Pb and PAR. *Acta Chim. Scand.*, **13**, 1473 (1959).

59-68 JARDIN, CL., Mise en évidence d'un complexe Sb–EDTA. *Médecine Tropicale*, **19**, 703 (1959).

59-69 JARDIN, CL., Complexométrie de Sb. *Médecine Tropicale*, **19**, 708 (1959).

59-70 KAROLEV, A. N., and KOJCEV, M. K., Complexometric determination of Pb in Pb-concentrates. *Zavodskaya Lab.*, **25**, 546 (1959).

59-71 MISUMI, S., and TAKETATSU, T., Indirect complexometric titration of Be with EDTA. *Bull. Chem. Soc. Japan*, **32**, 593 (1959).

59-72 MORRIS, A. G. C., Dissolution of $BaSO_4$ by EDTA in NaOH. *Chem. Analyst*, **48**, 76 (1959).

59-73 KNIGHT, W. S., and OSTERYOUNG, K. A., The amperometric and constant current potentiometric titration of EDTA with Cu. *Analyt. Chim. Acta*, **20**, 481 (1959).

59-74 RIEDEL, K., Eine neue maßanalytische Methode zur indirekten Bestimmung von Phosphat mit AeDTE. *Z. analyt. Chem.*, **168**, 106 (1959).

59-75 KORENMAN, I. M., GANINA, V. G., and LEIFER, E. I., A new indicator for the determination of Ca. *Trudy Khim. i Khim. Tekhnol.*, **2**, 108 (1959).

59-76 DRAGULESCU, C., SIMONESCU, T., and MENESSY, I., Photocolorimetrische Titration von Th mit Brenzcatechinviolett. *Acad. R. P. R. Baza cercet, stiint. Timisoara, Ser. stiint. chim.*, **6**, 21 (1959).

59-77 OCHYNSKA, J., and KROWCZYNSKI, L., Komplexometrische Bestimmung der Alkaloide in Chelidonium Majus. I. Bestimmung von Chelidoninhydrochlorid und der Alkaloide in den Wurzeln. *Chem. Analit.* (Warsaw), **4**, 309 (1959).

59-78 BRANDT, M., and VAN DEN BERG, R. H., Determination of tetraethyllead in gasoline by titration with EDTA. *Analyt. Chem.*, **31**, 1921 (1959).

59-79 LANGMYHR, F. J., and KRISTIANSEN, H., Indirect EDTA titration of Al with Pb solution and PAR as indicator. *Analyt. Chim. Acta*, **20**, 524 (1959).

59-80 TANAKA, N., and SAKUMA, Y., Constant current potentiometric titration of Ca, Pb and Cd at a semi-convection Hg-electrode. *Japan Analyst*, **7**, 223 (1959).

59-81 PHILLIPS, J. P., *Automatic Titrators*. Academic Press, New York (1959).

59-82 PEASE, B. F., and WILLIAMS, M. B., Spectrophotometric investigation of PAN and its Cu-complexes. *Analyt. Chem.*, **31**, 1044 (1959).

59-83 MUSTAFIN, I. S., and MOLOT, L. A., Mischindicatoren. II. Trilonometrische Bestimmung von Sulfat mit einem Indikator auf der

Basis von Natriumrhodizonat. *Izvest. Vysshikh. Ucheb. Zavedenii Khim. i Khim. Tekhnol.*, **2**, 293 (1959).

59-84 PANG, K., HSU, T. M., and WU, C. H., Spectrophotometric determination of the formation constants of ThY and CeY. *J. Chinese. Chem. Soc. (Taiwan)*, **6**, 12 (1959).

59-85 NIKITINA, E. I., Determination of Zr in borides and nitrides. *Zavodskaya Lab.*, **25**, 142 (1959).

59-86 OTTENDORFER, L. J., Indirect titrimetric determination of Zr and Hf in their oxide mixtures. *Chem. Analyst*, **48**, 97 (1959).

59-87 MISUMI, S., and TAKETATSU, T., Complexometric titration of rare earths elements. Dissolution of R. E. oxalates in EDTA and back titration with Mg. *Bull. Chem. Soc. Japan*, **32**, 973 (1959).

59-88 HARDER, R., and CHABEREK, S., Interaction of rare earths with diethylenetriamine pentaacetic acid. *J. Inorg. Nuclear Chem.*, **11**, 197 (1959).

59-89 LUCCHESI, C. A., STEARNS, J. A., and HIRN, C. F., EDTA titration of Cu in Cu-linoleate. *Chem. Analyst*, **48**, 9 (1959).

59-90 LITEANU, C., CRISAN, I., and CALU, C., Beiträge zur komplexometrischen Bestimmung von Kationengemischen. I. Bestimmung von Al und Cr. *Stud. Univ. Babes-Bolyai*, **1**, 105 (1959).

59-91 CARTIER, P., CLÉMENT-MÉTRAL, J., Ultra-microdosage automatique du Ca serique. *Clin. Chim. Acta*, **4**, 357 (1959).

59-92 GELMAN, A. D., ARTYNKHIN, P. I., and MOSKVIN, A. I., Complex formation of plutonium (V) in EDTA solutions. *Zhur. neorg. Khim.*, **4**, 1332 (1959).

59-93 LEWANDOWSKI, A., and WITKOWSKI, H., Titrimetrische Bestimmung von Phosphor in Apatit. *Prace. Kom. Mat. Przyr. Poznan Tow. Przyj. Nauk.*, **7**, 3 (1959).

59-94 CHABEREK, S., FROST, A. E., DORAN, M. A., and BICKWELL, N. J., Interaction of some bivalent metal ions with diethylene triamine pentaacetic acid. *J. Inorg. Nuclear Chem.*, **11**, 184 (1959).

59-95 BRIL, K. Y., HOLZER, S., and RETHY, B., Photometric titrations of Th and the rare earths with EDTA. *Analyt. Chem.*, **31**, 1353 (1959).

59-96 FRITZ, J. S., and PIETRZYK, D. J., Photometric titration of Sc. *Analyt. Chem.*, **31**, 1157 (1959).

59-97 BOGAREVA, K. G., Komplexometrische Bestimmung des Fe-Gehaltes in Ni- und Kupfersulfatlösungen nach Regeneration von Katalysatoren. *Masl. Zhur. Prom.*, **25**, 44 (1959).

59-98 ABD-EL RAHEEM, A. A., Omega Chrome Fast Blue 2 G. A new metal indicator for EDTA titrations. *Z. analyt. Chem.*, **167**, 98 (1959).

59-99 ABD-EL RAHEEM, A. A., and DOKHANA, M. M., The use of Metomega Chrome Blue BBL as a metal indicator and its application for the determination of Ca and Mg in serum. *Z. analyt. Chem.*, **168**, 165 (1959).

59-100 ABD-EL RAHEEM, A. A., and DOKHANA, M. M., Metomega Chrome Cyanine BLL as metal indicator in EDTA titrations. *Analyt. Chim. Acta*, **20**, 133 (1959).

59-101 ABD-EL RAHEEM, A. A., and MOUSTAFA, A. S., Omega Chrome Blue Green BL as an analytical reagent for Ca and Mg. *Analyt. Chim. Acta*, **21**, 379 (1959).

59-102 LOTT, P. F., and CHENG, K. L., Improved end point by addition of polyvinyl alcohol in the EDTA titration of Ca with calcon as indicator. *Chem. Analyst*, **48**, 13 (1959).

59-103 CHENG, K. L., and WARMUTH, F. J., Determination of Al in high temperature alloys. *Chem. Analyst*, **48**, 74 (1959).

59-104 HALL, J. L., GIBSON, J. A., and WILKINSON, P. R., Conductometric titration of EDTA with Ba. Effect of dilution and ammonia concentration on the curve form. *Proc. West Va. Acad. Sci.*, **31–2**, 143 (1959–60).

59-105 BERNDT, W., and SÁRA, J., Komplexometrische Cu-Bestimmung in Gegenwart einiger Aminoverbindungen. *Coll. Czech. Chem. Commun.*, **24**, 3181 (1959).

59-106 BUDEVSKII, O. B., KAROLEV, A. N., and KARANOV, R. A., Complexometric determination of Zn in concentrates with xylenole orange and methyl thymol blue as indicators. *Zavodskaya Lab.*, **25**, 1439 (1959).

59-107 CLAES, H. J., DE DONKER, K., and ROSSELLE, N., Routinebestimmung von Mg (und Ca) in Serum. *Chem. Weekblad*, **55**, 39 (1959).

59-108 BOOS, R. N., A volumetric microdetermination of organically bound sulphur and inorganic sulphates. *Analyst*, **84**, 633 (1959).

59-109 BENNEWITZ, R., and TANZER, I., Die komplexometrische Bestimmung von Phosphat im Anschluß an die Verbrennung nach Schöniger. *Mikrochim. Acta*, 836 (1959).

59-110 MILNER, G. W. C., and EDWARDS, J. W., The analysis of Zr–Th binary alloys. *Analyt. Chim. Acta*, **20**, 31 (1959).

59-111 SCRIBNER, W. G., Chelometric analysis of Mg–Mn and Mg–Mn–Zn mixtures. Fluoride ion as a demasking agent. *Analyt. Chem.*, **31**, 273 (1959).

59-112 RILEY, J. P., and WILLIAMS, H. P., Microanalysis of silicate and carbonate minerals. III. Determination of silica, phosphorus and metal oxides. *Mikrochim. Acta*, 804 (1959).

59-113 REILLEY, C. N., and VAVOULIS, A., Tetraethylenepentamine, a selective titrant. Potentiometric end-point detection. *Analyt. Chem.*, **31**, 243 (1959).

59-114 REILLEY, C. N., and SCHMID, R. W., Principles of endpoint detection in chelometric titrations using metallchromic indicators. Characterization of end-point sharpness. *Analyt. Chem.*, **31**, 887 (1959).

59-115 HOYLE, W., and WEST, T. S., Polarographic examination of the chelating power of EDTA and some closely related chelating agents. *Talanta*, **2**, 158 (1959).

59-116 HIRN, C. F., and LUCCHESI, C. A., Determination of Zr in Zr-driers. *Analyt. Chem.*, **31**, 1417 (1959).

59-117 MALMSTADT, H. V., and HADJIIOANNOU, T. P., Titration of Ca and Mg in plant material. *J. Agric. Food Chem.*, **7**, 418 (1959).

59-118 MALMSTADT, H. V., and HADJIIOANNOU, T. P., Automatic

titration of Ca and Mg in blood serum. *Clin. Chem.*, **5**, 50 (1959).

59-119 MALMSTADT, H. V., and HADJIIOANNOU, T. P., Determination of Ca, Mg and total hardness by automatic spectrophotometric titration. *J. Amer. Water Works Assoc.*, **51**, 411 (1959).

59-120 MALMSTADT, H. V., and HADJIIOANNOU, T. P., Determination of Cu and Zn in metallic products by automatic derivative spectrophotometric titration. *Analyt. Chim. Acta*, **21**, 41 (1959).

59-121 HADJIIOANNOU, T. P., EDTA titrations with automatic derivative spectrophotometric end-point detection. *Ph.D. Thesis*, Urbana Ill. 1959. Univ. Microfilm (Ann Arbor Mich.) L. C. Card No. Mic 60–1849. 133 pp.

59-122 MUSHA, S., MUNEMORI, M., and OGAWA, K., Photometric determination of indicator end-points in complexometric titrations. *Bull. Chem. Soc. Japan*, **32**, 132 (1959).

59-123 MUSHA, S., and OGAWA, K., Ferron as an indicator in the complexometric titration of Fe(III). *Japan Analyst*, **8**, 161 (1959).

59-124 KHALIFA, H., and SOLIMAN, A., Estimation of small amounts of Bi and the analysis of its binary mixtures with some other metals. *Z. analyt. Chem.*, **169**, 109 (1959).

59-125 KHALIFA, H., HAMDY, M., and SOLIMAN, A., Estimation of small amounts of La and analysis of its binary mixtures with some other metals. *Z. analyt. Chem.*, **171**, 178 (1959).

59-126 KLYGIN, A. E., SMIRNOVA, I. D., and NIKOLSKAYA, N. A., Solubility of EDTA in NH_3, HCl and its reaction with U(IV) and Pu(IV). *Zhur. neorg. Khim.*, **4**, 2766 (1959).

59-127 DIEHL, H., and LINDSTROM, F., Eriochrome Black T and its Ca and Mg derivatives. *Analyt. Chem.*, **31**, 414 (1959).

59-128 BARTELS, U., and HOYME, A., Komplexometrische Bestimmung von Schwefel in organischen Substanzen nach verschiedenen Aufschlußmethoden. *Chem. Tech. (Berlin)*, **11**, 600 (1959).

59-129 BETT, I. M., and FRASER, G. P., Rapid micromethod for the determination of serum Ca. *Clin. Chim. Acta*, **4**, 346 (1959).

59-130 CHUDINOVA, N. N., Amperometric titration of Ga with Complexone III in the presence of phosphate. *Zhur. analit. Khim.*, **14**, 636 (1959).

59-131 CHU, Y.-C., and CHI, J.-A., Antimonial chelates from some complexones. *Yao Hsüeh Pao*, **7**, 136 (1959).

59-132 HENNART, C., and MERLIN, E., Applications de la chélatométrie. III. Dosage volumétrique du diacetyle. *Chim. analyt.*, **41**, 287 (1959).

59-133 HELMSTÄDTER, G., Anwendung der Chelatometrie auf die Bestimmung stickstoffhaltiger Drogen. *Mitt. dtsch. pharm. Ges.*, **29**, 91 (1959).

59-134 HELMSTÄDTER, G., Analyse medizinischer Mandelsäureverbindungen. *Dtsch. Apoth. Ztschr.*, **99**, 589 (1959).

59-135 BERAL, H., DEMETRESCU, E., STOICESCU, V., and GRINTESCU, P., Beiträge zu den Untersuchungsmethoden von pharmazeutischen Zubereitungen. *Zentralhalle*, **98**, 49 (1959).

59-136 CHERKESOV, A. I., and MELNIKOVA, A. S., Trilonometrische Methode zur Bestimmung von Bi in Legierungen. *Zavodskaya Lab.*, **25**, 140 (1959).

59-137 CROUCH, E. A. C., and CRAWFORD, C. M., Titration of rare earths. *A.E.R.E. Report* C/R 2843 (1959).

59-138 ZAIKOVSKY, F. V., Chelometric and photometric determination of Th in minerals and ores. *Zhur. analit. Khim.*, **14**, 440 (1959).

59-139 SOCHEVANOVA, M. M., and SOCHEVANOV, V. G., Komplexometrische Bestimmung der Härte natürlicher Wässer in Gegenwart von Cu, Mn und anderen Störelementen. *Bull. Nauch. Tekchn. Inform. Min. Geol. i Okhrany Nedr. SSSR*, 98 (1959).

59-140 BJERRUM, J., SCHWARZENBACH, G., and SILLÉN, L. G., Stability constants of metal complexes. Part II. *Chem. Soc. Special Publ. No. 7.*

59-141 BOND, J., and JONES, T. I., Fe-chelates of polyaminocarboxylic acids. *Trans. Faraday Soc.*, **55**, 1310 (1959).

59-142 ANDEREGG, G., NÄGELI, P., MÜLLER, F., and SCHWARZENBACH, G., Diäthylentriaminpentaessigsäure. *Helv. Chim. Acta*, **42**, 827 (1959).

59-143 RINGBOM, A., Complexation Reactions. Chapter 14, Vol. 1 of *Treatise on Analytical Chemistry*, ed. by I. M. Kolthoff and P. J. Elving. The Interscience Encyclopedia Inc. New York, 1959.

59-144 HOELZL-WALLACH, O. F., SURGENOR, D. M., SODERBERG, J., and DELANO, E., Preparation and properties of 3,6-dihydroxy-2,4-bis-[*N,N'*-di-(carboxymethyl)-aminomethyl] fluoron. *Analyt. Chem.*, **31**, 456 (1959).

59-145 DICK, J., and RISTICI, J., Die Bestimmung von Pyramidon. *Acad. rep. pup. Romîne, Studii cercet. Stiinte chim.*, **6**, 47 (1959).

59-146 FLASCHKA, H., *EDTA-Titrations*. Pergamon Press, London (1959).

59-147 TSERKASEVICH, K. V., A new complexometric indicator for the determination of Ca. *Issled. Oblasti Ferm. Zapor. Gosudarst Farm. Ind.*, 120 (1959).

59-148 MARTIN, A. E., and REILLEY, C. N., EDTA titration of metal ions. Polarized Hg electrodes. *Analyt. Chem.*, **31**, 992 (1959).

59-149 WILKINS, D. H., Apparatus for titrations using ultraviolet light. *Talanta*, **2**, 88 (1959).

59-150 WALDVOGEL, P., *Dissertation*, ETH, Zürich, 1959.

59-151 GOLDSTEIN, D., A new indicator for the complexometric determination of calcium. *Analyt. Chim. Acta*, **21**, 339 (1959).

1960

60-1 PEĆAR, M., Zur komplexometrischen Bestimmung des K. *Mikrochim. Acta*, 567 (1960).

60-2 DUGANDŽIČ, M., and HOLASEK, A., Komplexometrische Bestimmung von saurer und alkalischer Phosphatase. *Wiener med. Wochensch.*, **110**, 460 (1960).

60-3 LASSNER, E., and SCHARF, R., Die chelatometrische Titration von Ti in Gegenwart von Nb und Ta. *Talanta*, **7**, 12 (1960).

60-4 LASSNER, E., and SCHARF, R., EDTA titration of Zr, Hf and Th in their alloys with W and in W-metal. *Chem. Analyst*, **49**, 22 (1960).

60-5 LASSNER, E., and SCHARF, R., EDTA titration of Co in cemented carbides. *Chem. Analyst*, **49**, 44 (1960).

60-6 WÜNSCH, L., Direct micro titration of serum Ca using fluorexone as indicator. *Casopis Lekarnů Czech.*, **99**, 754 (1960).

60-7 KOBAYASHI, M., and IWASE, A., Amperometric titration of europium with Na–DTPA. *Yamagata Daigaku Kiyo*, **5**, 301 (1060).

60-8 VAN SCHOUWENBURG, J. CH., Micro EDTA titration of Ca. Mg-interference. *Analyt. Chem.*, **32**, 709 (1960).

60-9 SOCHEVANOVA, M. M., and SOCHEVANOV, V. G., Komplexometrische Analyse von Eisenkarbonaterzen. *Zavodskaya Lab.*, **26**, 543 (1960).

60-10 KABANOVA, O. L., DANUSCHENKOVA, M. A., and PALEY, P. N., Sur les réactions des ions du plutonium avec EDTA. *Analyt. Chim. Acta*, **22**, 66 (1960).

60-11 ZAIKOVSKII, F. L., and GERKHARDT, L. I., A novel complexometric indicator, hydroxyhydroquinone pink, and its analytical utilization. *Trudy Kom. Analit. Khim. Akad. Nauk. SSSR Inst. Geokhim. i Analit. Khim.*, **11**, 346 (1960).

60-12 SUH, CH. C., Chelometric determination of As as $ZnNH_4AsO_4$. *J. Pharm. Soc. Korea*, **5**, 16 (1960).

60-13 SUN, P. J., Direct complexometric titration of Zr with versene using gallein as indicator. *J. Chinese Chem. Soc. Formosa.*, **7**, 143 (1960).

60-14 PANDE, C. S., and SRIVASTAVA, T. S., Analytical aspects of some organic acids. IV. Direct complexometric titration of Fe(III) with EDTA using 2-hydroxy-3-naphthoic acid as indicator. *Z. analyt. Chem.*, **172**, 356 (1960).

60-15 PANDE, C. S., and SRIVASTAVA, T. S., Analytical aspects of some organic acids. V. Back-titration procedure for determining Al, Zr and Th with Fe(III)-solution and 2-hydroxy-3-naphthoic acid as indicator. *Z. analyt. Chem.*, **173**, 195 (1960).

60-16 PANDE, C. S., and SRIVASTAVA, T. S., Analytical aspects of some organic acids. VI. Direct complexometric titration of Fe(III) with EDTA and the indirect titration of Zr and Th using *o*- and *p*-cresotic acids as indicators. *Z. analyt. Chem.*, **175**, 29 (1960).

60-17 JOHNSTON, M. B., BARNARD, JR., A. J., and BROAD, W. C., Thorin. Chromogenic agent and chelometric indicator. *Rev. Univ. Ind. Satander*, **2**, 137 (1960).

60-18 JENSEN, B. S., 1-(2-Thiazolylazo)-2-hydroxyaryl compounds as complexometric metal indicators. *Acta Chim. Scand.*, **14**, 927 (1960).

60-19 IWATA, S., Volumetrische Bestimmung von Na in Wässern verbunden mit der Produktion von Naturgas. *Japan Analyst*, **9**, 178 (1960).

60-20 HNILIČKOVÁ, M., and SOMMER, L., 5-(2-Pyridylazo)-6,7-dihydroxy-2-naphthalinsulfonsaure und 4-(2-Pyridylazo)-orcinol.

378   *Bibliography*

Zwei neue chelatometrische Indikatoren. *Z. analyt. Chem.*, **177**, 425 (1960).

60-21   DE SOUSA, A., Determination of chlorate and perchlorate in the presence of each other. *Chem. Analyst*, **49**, 18 (1960).

60-22   DE SOUSA, A., Determination of chloride and iodide in the presence of each other. *Chem. Analyst*, **49**, 45 (1960).

60-23   DE SOUSA, A., Determination of Ba in minerals. *Chem. Analyst*, **49**, 75 (1960).

60-24   DE SOUSA, A., Dosage complexométrique de Br⁻ et I⁻ en presence l'un de l'autre. *Analyt. Chim. Acta*, **22**, 520 (1960).

60-25   DE SOUSA, A., Dosage indirect du K par complexométrie. *Analyt. Chim. Acta*, **22**, 522 (1960).

60-26   DE SOUSA, A., The complexometric determination of bromide and bromate in the presence of each other. *Z. analyt. Chem.*, **174**, 337 (1960).

60-27   KRIJN, G. C., and DEN BOEF, G., Determination of metal ions in solution by means of thioacetamide and EDTA. I. General introduction and the determination of Zn. *Analyt. Chim. Acta*, **23**, 35 (1960).

60-28   KRIJN, G. C., and DEN BOEF, G., Determination of metal ions in solutions by means of thioacetamide and EDTA. II. Determination of Cu. *Analyt. Chim. Acta*, **23**, 186 (1960).

60-29   KRIJN, G. C., KOSTER, A. S., and DEN BOEF, G., Determination of metal ions in solution by means of thioacetamide and EDTA. III. Determination of Fe. *Analyt. Chim. Acta*, **23**, 240 (1960).

60-30   JABLONSKI, W. Z., and JOHNSON, E. A., Specific masking with acetylacetone in titrations with EDTA. *Analyst*, **85**, 297 (1960).

60-31   HOL, P. J., Indicators used in complexometric titrations. *Chem. and Tech.* (*Amsterdam*), **16**, 645 (1961); **17**, 133 (1962).

60-32   KATIGAWA, T., AC polarization titrations with two hanging Hg-drop-electrodes. Application to the chelatometric titration of Cd. *Bull. Chem. Soc. Japan*, **33**, 1124 (1960).

60-33   LEWIS, L. L., and MELNICK, L. M., Determination of Ca and Mg with EDTA. Studies of accuracy. *Analyt. Chem.*, **32**, 39 (1960).

60-34   LEWIS, L. L., and STRAUB, W. A., Determination of Ni and Co in high alloys and stainless steel. *Analyt. Chem.*, **32**, 96 (1960).

60-35   ŠTRÁFELDA, F., and RIHOVÁ, J., Komplexometrische Bestimmung von Ca und Mg mit potentiometrischer Endpunktsanzeige. *Coll. Czech. Chem. Commun.*, **25**, 1444 (1960).

60-36   DIEHL, H., and ELLINBOE, J. L., Azo-dyes as indicators for Ca and Mg. *Analyt. Chem.*, **32**, 1120 (1960).

60-37   VOLF, L. A., Masking of Zn, Cd, Hg, Pb, Sn with Unithiol in the complexometric determination of Sr and Ba. *Zavodskaya Lab.*, **26**, 1353 (1960).

60-38   VOLF, L. A., Application of unithiol in analytical chemistry. *Zhur. Khim. Obshchest. Mendeleeva*, **5**, 232 (1960).

60-39   SEDLÁČEK, B. A. J., Komplexometrische Bestimmung von Antioxydation. II. Eine neue semimikro-komplexometrische

Methode zur Bestimmung von Nordihydroguajaretsäure in Schweinefett. *Fette & Seifen*, **62**, 669 (1960).

60-40 SEDLÁČEK, B. A. J., Komplexometrische Bestimmung von Antioxydatien. III. Semimikrobestimmung von Butylhydroxylanisol in Schweinefett und Trennung von NDGA und BHA. *Fette & Seifen*, **62**, 1041 (1960).

60-41 MAJUMDAR, A. K., and SAVARIAR, C.P., *o*-Carboxyphenylazochromotropic acid as an analytical reagent. I. Chelometric titration of Th, Zr and Fe. *Z. analyt. Chem.*, **174**, 197 (1960).

60-42 ZHDANOV, A. K., KHADEEV, V. A., and ISHANKODZAYEV, S. D., Amperometrische Titration von Bi. *Uzb. Khim. Zhur.*, 29 (1960).

60-43 BUDĚŠINSKÝ, B., and KÖRBL, J., Komplexometrie in der organischen Analyse. VII. Bestimmung von organischen Basen mit Hilfe des Cd-Komplexonates. *Coll. Czech. Chem. Commun.*, **25**, 76 (1960).

60-44 BUDĚŠINSKÝ, B., VANIČKOVÁ, E., and KÖRBL, J., Komplexometrie in der organischen Analyse. VIII. Bestimmung einiger Derivate des Thioharnstoffes (Bestimmung von Sulfiden). *Coll. Czech. Chem. Commun.*, **25**, 456 (1960).

60-45 KÖRBL, J., and SVOBODA, V., Metalfluorechromic indicators. *Talanta*, **3**, 370 (1960).

60-46 PŘIBIL, R., and BURGER, K., Eine neue Schnellmethode zur Bestimmung von Th in Gegenwart von Zr, Fe, La, U und sonstigen Schwermetallen. *Talanta*, **4**, 8 (1960).

60-47 KOTRLÝ, S., and VRESTÁL, J., Komplexometrische Titrationen (Chelatometrie). XLVII. Mikrobestimmung von Bi und Pb durch stufenweise Titration gegen Xylenolorange. Untersuchung der Indikatorumschläge mit Hilfe von photometrischer Titration. *Coll. Czech. Chem. Commun.*, **25**, 1148 (1960).

60-48 KOPANICA, M., and PŘIBIL, R., Komplexometrische Titrationen (Chelatometrie). XLVIII. Komplexometrische In-Bestimmung. *Coll. Czech. Chem. Commun.*, **25**, 2230 (1960).

60-49 POVONDRA, P., and PŘIBIL, R., Chelometric method in rapid applied analysis. IV. Determination of Ca and Mg in high-Mn-welding fluxes. *Chem. Analyst*, **49**, 109 (1960).

60-50 SVOBODA, V., Thiazolyl analogues of PAN. *Talanta*, **4**, 201 (1960).

60-51 REHAK, B., and KÖRBL, J., Metallochrome Indikatoren. VIII. Physikalisch-chemische Untersuchung von Xylenolorange und einiger seiner Chelate. I. Dissoziationskonstanten v. Xylenolorange. *Coll. Czech. Chem. Commun.*, **25**, 797 (1960).

60-52 SVOBODA, V., DORAZIL, L., and KÖRBL, J., Metallochrome Indikatoren. IX. Indophenolkomplexone. *Coll. Czech. Chem. Commun.*, **25**, 1037 (1960).

60-53 WÄNNINEN, E., Complexometric titrations with diethylene-triamine-pentaacetic acid. *Acta Acad. Aboensis Math. Phys.*, **21**, 110 pp. (1960).

60-54 ZAKI, M. R., and SHAKIR, K., Organic azodyes in quantitative analysis. I. Complexometric titration of Th. *Z. analyt. Chem.*, **174**, 274 (1960).

380   *Bibliography*

60-55   ZAKI, M. R., and SHAKIR, K., Organic azodyes in quantitative analysis. II. Selective complexometric indicator for Th. *Z. analyt. Chem.*, **177**, 196 (1960).

60-56   ZAPP, E. E., and PUNGOR, E., Untersuchung der Bildung von Metallkomplexen mit Hilfe der Hochfrequenzmethode. *Ann. Univ. Sci. Budapest, Sec. Chim.*, **2**, 341 (1960).

60-57   PALEY, P. N., and UDALTSOVA, N. I., Amperometrische Titration von kleinen Th-Mengen mit AeDTE. *Trudy Kom. Analit. Khim. Akad. Nauk SSSR Inst. Geokhim i Analit. Khim.*, **11**, 299 (1960).

60-58   PALEY, P. N., and CHZHAN, V. T., Komplexometrische Bestimmung von Pu (IV) mit Arsenazo Indicator. *Zhur. analit. Khim.*, **15**, 598 (1960).

60-59   MORACHEVSKII, YU. V., and VOLF, A. L., Complexometric determination of Ca and Mg in the presence of cations of the As subgroup and Ge and Zn subgroup. *Uch. Zapiski Leningrad Gosudarst. Univ. A. A. Zhdanova Ser. Khim. Nauk.*, 144 (1960).

60-60   MORACHEVSKII, YU. V., and VOLF, L. A., Masking of cations with Unithiol in complexometric titrations. *Zhur. analit. Khim.*, **15**, 656 (1960).

60-61   LUKYANOV, V. F., and KNYAZEVA, E. M., Komplexometrische Bestimmung von Zr. *Zhur. analit. Khim.*, **15**, 69 (1960).

60-62   LUKYANOV, V. F., and SEDINA, L. I., Komplexometrische Bestimmung von Bi und Pb in Bi-Pb-Legierungen. *Zhur. analit. Khim.*, **15**, 595 (1960).

60-63   WILKINS, D. H., Further comments on metalfluorechromic indicators. *Talanta*, **4**, 79 (1960).

60-64   WILKINS, D. H., Calcein Blue. A new metalfluorechromic indicator for chelometric titrations. *Talanta*, **4**, 182 (1965).

60-65   WILKINS, D. H., The chelometric determination of Al, Ni and Mn without prior separation. *Analyt. Chim. Acta*, **23**, 309 (1960).

60-66   GUERRIN, G., DESBARRES, J., and TRÉMILLON, B., Titrages potentiométriques et ampérométriques par l'EDTA. Electrodes indicatrices de Hg et de Cu. *J. Electroanal. Chem.*, **1**, 226 (1960).

60-67   GUERRIN, G., SHELDON, M. V., and REILLEY, C. N., EDTA titrations employing SNAZOXS as the indicator. *Analyt. Chem.*, **49**, 36 (1960).

60-68   REILLEY, C. N., FLASCHKA, H., LAURENT, S., and LAURENT, B., Characterization of the colour quality of indicator transitions. Complementary tristimulus colorimetry. *Analyt. Chem.*, **32**, 1218 (1960).

60-69   MCKEND, J., Analysis of slags from the manufacture of U-metal. Determination of MgO and Mg-metal. *Analyt. Chem.*, **32**, 1193 (1960).

60-70   CHERNIKOV, Y. A., TRAMM, R. S., and PEVZNER, K. S., Folgetitration von Th und den Seltenen Erden. *Zavodskaya Lab.*, **26**, 921 (1960).

60-71   CHERNIKOV, Y. A., and VLADIMIROVA, V. M., Bestimmung von Zr in Nb-Legierung (durch amperometrische Rücktitration mit Bi bei −0·3 Volt). *Zavodskaya Lab.*, **26**, 1207 (1960).

60-72   CHERNIKOV, Y. A., LUKYANOV, V. F., and KOZLOVA, A. B., Komplexometrische Bestimmung von Th in Monazitkonzentraten. *Zavodskaya Lab.*, **14**, 567 (1959); *Zhur. analit. Khim.*, **15**, 452 (1960).

60-73   LASZLOVSZKY, J., BARCZA, L., and KÖRÖS, E., Anwendung der Komplexone in der pharmazeutischen Analyse. *Ann. Univ. Sci. Budapest, Sec. Chim.*, **2**, 277 (1960).

60-74   KHADEEV, V. A., and BAZARBAEV, A. T., Amperometrische Titration von Hg mit AeDTE und einer Ta-Elektrode. *Uzb. Khim. Zhur.*, 38 (1960).

60-75   KHADEEV, V. A., and KVASHINA, F. F., Direkte amperometrische Titration von Zr mit Komplexon III und einer rotierenden Ta-Mikroelektrode. *Izvest. Vysschikh. Uchb. Zavedenii Khim. i Khim. Teckhnol.*, **3**, 251 (1960).

60-76   RADKO, V. A., and YAKIMETS, E. M., Trilonometric determination of Mn in the systems Mn–Fe and Mn–Al. *Trudy Uralsk Politekh. Inst.*, 166 (1960).

60-77   RADKO, V. A., and YAKIMETS, E. M., Trilonometric determination of Ca, Mg and Mn in the presence of each other. *Trudy Uralsk Politekh. Inst.*, 176 (1960).

60-78   MALMSTADT, H. V., and HADJIIOANNOU, T. P., Automatic derivative titration of excess EDTA in the determination of Co, Cu and Fe. *Analyt. Chim. Acta*, **23**, 288 (1960).

60-79   HERBINGER, W., and HUBMAYER, W., Eine genaue und schnelle komplexometrische Methode zur Bestimmung von Serum-K. *Klin. Wochenschr.*, **38**, 822 (1960).

60-80   VERMA, M. R., and AGARWAL, K. C., Use of metalfluorchromic indicators for the complexometric titration of metal ions. *J. Sci. Indust. Res. (India)*, **19 B**, 319 (1960).

60-81   VICHEV, E. P., and KARAKASHOV, A. V., A micromethod for the direct complexometric titration of serum Ca. *Voprosy Med. Khim.*, **6**, 435 (1960).

60-82   CHATURVEDI, R. K., Spectrophotometric investigation of Cu-murexide. *Current Sci. (India)*, **29**, 128 (1960).

60-83   HEADRIDGE, J. B., The complexometric determination of Mo. *Analyst*, **85**, 379 (1960).

60-84   IRITANI, N., and TANAKA, T., Determination of phosphate and sulphate as $MgNH_4PO_4$ and $PbSO_4$. *Japan Analyst*, **9**, 1 (1960).

60-85   WENDLANDT, W. W., Thermogravimetric and differential thermal analysis of EDTA and its derivatives. *Analyt. Chem.*, **32**, 848 (1960).

60-86   WARD, G. M., and HEENEY, H. B., Collaborative study of methods for the determination of K, Ca and Mg in plant material. *Can. J. Plant Sci.*, **40**, 589 (1960).

60-87   STYUNKEL, T. B., and MIKHALEVA, Z. A., Acid Chrome Dark Blue and Acid Chrome Blue K as indicators in the trilonometric determination of Pb. *Trudy Uralsk Politekh. Inst.*, 159 (1960).

60-88   ROUBALOVA, D., and DOLEŽAL, J., The amperometric titration of EDTA and DCyTA. *Chem. Analyst*, **47**, 76 (1960).

382   *Bibliography*

60-89   ERDEY, L., and PÓLOS, L., Komplexometrische Bestimmung von Al und Pb in Anwensenheit von Variaminblau als Indikator. *Z. analyt. Chem.*, **174**, 333 (1960).

60-90   ERDEY, L., and BUZÁS, I., Complexometric and argentometric titrations using chemiluminescent indicators. *Analyt. Chim. Acta*, **22**, 524 (1960).

60-91   HISADA, M., and KASHIKAWA, K., Determination of Pb by titration with EDTA. *Japan Analyst*, **9**, 87 (1960).

60-92   ŠTUDLAR, K., Chelometric titration of Mn in ferromanganese. *Chem. Analyst*, **49**, 106 (1960).

60-93   VOLODARSKAYA, R. S., Complexometrische Bestimmung von Th und Zr in Mg-Legierungen. *Zavodskaya Lab.*, **26**, 925 (1960).

60-94   VASILIEV, R., and ANASTASESCU, G., Komplexometrische Bestimmung von Arsenaten. *Rev. Chim. Bukarest*, **11**, 298 (1960).

60-95   USATENKO, J. I., and VITKINA, M. A., Amperometrische Bestimmung von Bi in niedrig schmelzenden Legierungen mit Trilon B. *Zavodskaya Lab.*, **26**, 542 (1960).

60-96   POPOV, N. A., Complexometric titration of Ca in the presence of large amounts of Al. *Zavodskaya Lab.*, **26**, 540 (1960).

60-97   UMEDA, M., Indirect chelatometric determination of Mo as Pb molybdate. *Japan Analyst*, **9**, 172 (1960).

60-98   TERPILOWSKI, J., and MANCZYK, R., Komplexometrische Bestimmung von Zn in pharmazeutischen Zubereitungen. *Acta Polon. Pharm.*, **17**, 267 (1960).

60-99   SINGH, B., SAHOTA, S. S., and MANKOTA, M. S., Complexometric estimation of Fe(III). *Z. analyt. Chem.*, **173**, 275 (1960).

60-100   POWELL, J. E., FRITZ, J. S., and JAMES, D. B., A primary standard for alkalimetry and chelometry. *Analyt. Chem.*, **32**, 954 (1960).

60-101   PHILIPP, B., BARTELS, U., and HOYME, H., Untersuchungen zur komplexometrischen Endbestimmung des Schwefels als Sulfat oder Sulfid. *Mittls. Blatt Chem. Ges. DDR.*, *Sonderheft*, 359 (1960).

60-102   STANKOVIANSKY, S., PODANY, V., JASINGER, F., and MAJER, P., (N,N-bis-[carboxymethyl]-aminomethyl)-quinizarin als neuer metallochromer Indicator. *Chem. Zvesti*, **14**, 265 (1960).

60-103   RAMAIAH, N. A., and CHATURVEDI, R. K., Murexide as indicator in the estimation of Ca using EDTA. *Current Sci. (India)*, **29**, 305 (1960).

60-104   RALEA, R., and SIMININC, E., Indirekte komplexometrische Bestimmung von Thiocyanat mit AeDTE. *An. Stiint Univ. Al I. Cuza*, **6**, 171 (1960).

60-105   PAVLOV, N. N., KUZNETSOV, A. R., and ARBUZOV, G. A., Complexometry of Cr(III). I. Trilonometry of Cr-chloride. *Izv. Vysshikh. Ucheb. Zavedenii Technol. Legkai Prom.*, No. 1, 54 (1960); II. Effect of Cr ligands on the formation of trilon complexes of Cr. *ibid.*, Nr. 2, 55 (1960); III. Determination of Cr in high polymer films. *ibid.*, No. 3, 28 (1960).

60-106   OWENS, E. G., and YOE, J. H., Derivatives of 1,4-dihydroxyanthrachinone as metal indicators in the titration of Th with EDTA. *Analyt. Chim. Acta*, **23**, 321 (1960).

60-107 MERRILL, J. R., HONDA, M., and ARNOLD, J. R., Methods for separation and determination of Be in sediments and natural waters. *Analyt. Chem.*, **32**, 1420 (1960).

60-108 MUSTAFIN, I. S., and KRUCHKOVA, E. S., Hydron-II eine neuer Indikator zur komplexometrischen Bestimmung von Ca in Gegenwart von Mg. *Zhur. analit. Khim.*, **15**, 20 (1960), cf. also *Zavodskaya Lab.*, **27**, 1668 (1961).

60-109 OGAWA, K., and MUSHA, S., Photometric titration of Sr with EDTA and phthalein complexone as indicator. *Bull. Univ. Osaka, Prefect. Ser. A*, **8**, 63 (1960).

60-110 NYDHAL, F., The indirect complexometric determination of Al. A study of the Wänninen-Ringbom method. *Talanta*, **4**, 141 (1960).

60-111 HASLAM, J., SQUIRRELL, D. C. M., and BLACKWELL, I. G., Determination of Ca and Mg in waters by automatic titration. *Analyst*, **85**, 27 (1960).

60-112 TAKAGI, T., and IMOTO, H., Determination of Ca and Mg in steel. *Japan Analyst*, **8**, 782 (1959).

60-113 MORRIS, A. G. C., Chelometric titration of Mn in ferromanganese. *Chem. Analyst*, **49**, 105 (1960).

60-114 COSTA, A. C., 8-Quinolinol as an indicator for metal titrations with EDTA. *Anais. Assoc. brazil Quim*, **19**, 21 (1960).

60-115 COSTA, A. C., Indicators for metal titrations with EDTA. 1-Nitroso-2-naphthol; 2-nitroso-1-naphthol and sodium 1-nitroso-2-naphthol-3,6-disulfonate. *Anais. Assoc. brazil Quim.*, **19**, 29 (1960).

60-116 COSTA, A. C., Rapid determination of Pb in its alloys especially in type metal. *Anais. Assoc. brazil Quim.*, **19**, 145 (1960).

60-117 COSTA, A. C., Hematoxyline as an indicator in the chelatometric titration of metals. *Analyt. Chim. Acta*, **23**, 126 (1960).

60-118 MALOWAN, L. S., and ALEGRE, M., Salarcid. A new reagent for the complexometric determination of Fe. *Ciencia (Mexico)*, **20**, 205 (1960).

60-119 MINAMI, E., and WATANUKI, K., Determination of hydrogen sulphide in natural water by EDTA titration. *Japan Analyst*, **9**, 958 (1960).

60-120 MALKUS, Z., and HORAČEK, J., Komplexometrische Bestimmung von Propylgallat. *Prumysl potravni*, **11**, 43 (1960).

60-121 DUTT, N. K., Volumetric determination of Bi with Dalzin as indicator. *Sci. and Culture (India)*, **25**, 695 (1960).

60-122 MATSUO, T., and SUGAWARA, M., Determination of Al with EDTA. Back-titration with Fe(III). *Japan Analyst*, **9**, 706 (1960).

60-123 HANSELMAN, R. B., and ROGERS, L. R., Coulometric passage of reagents through ion exchange membranes. *Analyt. Chem.*, **32**, 1240 (1960).

60-124 HAHN, F. L., Zur komplexometrischen Titration von Ca und Mg. Die Vorteile der inversen Titration. *Z. analyt. Chem.*, **174**, 121 (1960).

60-125 DWYER, F. P., and GARVAN, F. L., Rhodium (III) complexes with EDTA. *J. Amer. Chem. Soc.*, **82**, 4823 (1960).

384 Bibliography

60-126 EGGERS, J. H., Umbellikomplexon und Xanthokomplexon. *Talanta*, **4**, 38 (1960).

60-127 GOTTSCHALK, G., Die komplexometrische Bestimmung von Al mit Dithizon als Indikator. *Z. analyt. Chem.*, **172**, 192 (1960).

60-128 GJESSING, L., Mg-Bestimmung im Serum. *Tidsk. Norske Loege-forening*, **80**, 494 (1960).

60-129 GJEMS, O., Stoichiometry of titrations of Ca, Mg and Mn at low concentrations with EDTA and the indicators murexide and Eriochrome Black T. *Analyst*, **85**, 738 (1960).

60-130 EFIMOV, I. P., and IVANOV, V. M., Spectrophotometric titration fo erbium with EDTA and PAR as indicator. *Zhur. analit. Khim.*, **15**, 750 (1960).

60-131 LUCCHESI, C. A., and HIRN, C. F., EDTA titration of total iron in iron(II)-Fe(III) mixtures. Application to iron driers. *Analyt. Chem.*, **32**, 1191 (1960).

60-132 LITEANU, C., MURGU, G., and MARINESCU, L., Beiträge zur komplexometrischen Bestimmung von Kationengemischen. II. Über die komplexometrische Titration des Zn und Cu in Gegenwart von Murexid als Indikator. *Z. analyt. Chem.*, **175**, 1 (1960).

60-133 GALATEANU, I., MAXIM, I., and BRAUN, T., Die radiometrische Titration von Zn und Cu mit AeDTE. *Z. analyt. Chem.*, **172**, 274 (1960).

60-134 LINDSTROM, F., and DIEHL, H., Indicator for the titration of Ca plus Mg with EDTA. *Analyt. Chem.*, **32**, 1123 (1960).

60-135 GIUFFRÉ, L., and CAPIZZI, F. M., Potentiometrische Bestimmung von Ti in Gegenwart von Al. *Ann. Chim. (Roma)*, **50**, 405 (1960).

60-136 LIEBER, F., Bestimmung von Titan mit AeDTE. *Z. analyt. Chem.*, **177**, 429 (1960).

60-137 LEMBREZ, Y., STORCK, J., and TERLAIN, B., Sur la détection et le dosage de l'EDTA et de ses dérivés dans les milieux complexes. *Ann. pharm. franç.*, **18**, 285 (1960).

60-138 ENDO, Y., and KOROKI, C., Bestimmung von Mn in Manganerzen. *Japan Analyst*, **9**, 993 (1960).

60-139 ELO, A., and POLKY, J. R., Rapid determination of Al in cracking catalysts by a modified Wänninen-Ringbom method. *Analyt. Chem.*, **32**, 294 (1960).

60-140 BURG, R. A., and CONAGHAN, H. F., Chelometric determination of Ca and Mg in minerals. *Chem. Analyst*, **49**, 100 (1960).

60-141 ENDO, Y., and TAKAGI, H., Bestimmung von Zr nach Abtrennung als Zr–Mg–Salz. *Japan Analyst*, **9**, 503 (1960).

60-142 BERBENNI, P., Bestimmung der Gesamtkohlensäure in Wasser mit Komplexon. *Boll. Lab. Chim. Provinciali*, **11**, 249 (1960).

60-143 DUNSTONE, J. R., MADSEN, N. P., and BELL, H. R., Determination of Ca and Mg in rat liver. *Analyst*, **85**, 519 (1960).

60-144 DAVIS, D. G., and JACOBSON, W. R., The determination Fe in Fe–Al mixtures by titration with EDTA. *Analyt. Chem.*, **32**, 215 (1960).

60-145 DATTA, S. K., Complexometric titration of Zr with 2-(4-sulpho-2-nitrosonaphthylazo)-1,8-dihydroxy-3,6-naphthalene disulphonic acid. *Chim. analyt.*, **42**, 562 (1960).

60-146 BOXER, J., Evaluation of a micro method for serum Ca determination with Calcon as indicator. *Clin. Chim. Acta*, **5**, 82 (1960).

60-147 BHATTACHARYYA, B. N., and GUPTA, N., Complexometric titration of Pb in polymetallic ores. *Indian J. Appl. Chem.*, **23**, 164 (1960).

60-148 BERNDT, W., and SÁRA, J., Contributions to the basic problems of complexometry. III. Some advantages of acetate buffers. *Talanta*, **5**, 281 (1960).

60-149 BRUNO, E., Analyse von Kationen mittels Hochfrequenzkomplexometrie. *Rass. Chim.*, **12**, 31 (1960).

60-150 BANYAI, E., GERE, E. B., and ERDEY, L., Complexometric determination of Hg(II)- and Al-ions. *Talanta*, **4**, 133 (1960).

60-151 BABENYSHEV, V. M., and KUZNETZOVA, O. M., Complexometric determination of Al in Mg alloys with amperometric indication of the end-point. *Zhur. analit. Khim.*, **15**, 568 (1960).

60-152 APPLE, R. F., and WHITE, J. C., Determination of Th present in fluoride salt mixtures. *Chem. Analyst*, **49**, 42 (1960).

60-153 BELCHER, R., REES, D. I., and STEPHEN, W. I., *N,N,N',N'*-tetracarboxymethyl derivatives of some benzidines as metallofluorescent indicators. *Talanta*, **4**, 78 (1960).

60-154 CLOSE, R. A., and WEST, T. S., A new selective metallochromic reagent for the detection and chelometric determination of Ca. *Talanta*, **5**, 221 (1960).

60-155 CLOSE, R. A., and WEST, T. S., Acid Alizarin Black SN as metallochromic indicator for Ca. *Analyt. Chim. Acta*, **23**, 361 (1960).

60-156 CLOSE, R. A., and WEST, T. S., Photometric titration of Ca in blood serum with Acid Alizarin Black SN as metallochromic indicator. *Analyt. Chim. Acta*, **23**, 370 (1960).

60-157 ANTONACCI, M., Volumetrische Schnellbestimmung von freier und gebundener Oleinsaure. *Chim. e Industr.*, **42**, 375 (1960).

60-158 FURLANI, A. D., Simultane Bestimmung von Al und Cr mit AeDTE. *Gazzetta*, **90**, 1380 (1960).

60-159 AOKI, M., KIMURA, H., and YMAUCHI, T., Chelometric titrations. I. Titrimetric determination of Zn and Cu with EDTA. *Bull. Nagoya Inst. Technol.*, **13**, 260 (1960).

60-160 BASHKIRTSEVA, A. A., and PRUDNIKOVA, L. D., Neue volumetrische Methode zur Analyse von Aluminatlösungen. *Zavodskaya Lab.*, **26**, 1107 (1960).

60-161 BASHKIRTSEVA, A. A., and YAKIMETS, E. M., Kalium- und Ammoniumrhodanid als Indikatoren in der komplexometrischen Fe-Titration. *Trudy Uralsk Politekh. Inst.*, 110 (1960).

60-162 BASHKIRTSEVA, A. A., and YAKIMETS, E. M., Sulfosalicylsäure als Indikator für die komplexometrische Titration von Fe. *Trudy Uralsk Politekh. Inst.*, 117 (1960).

60-163 KHALIFA, H., and ALLAM, M. G., Potentiometric determination of Hg(II) with EDTA. Analysis of binary mixtures. *Analyt. Chim. Acta*, **22**, 421 (1960).

60-164 KHALIFA, H., and OSMAN, F. A., Analysis of ternary mixtures of Hg with some other metals. *Z. analyt. Chem.*, **178**, 116 (1960).

60-165 COCOZZA, E. P., Determination of Ga and In in glycerol plating baths. *Chem. Analyst*, **49**, 46 (1960).

60-166 COCOZZA, E. P., Determination of In and Sn in glycerol plating baths. *Chem. Analyst*, **49**, 124 (1960).

60-167 BUSEV, A. I., and TIPTSOVA, V. G., Analytical chemistry of Tl. V. Complexometric indicators for Tl(III). *Zhur. analit. Khim.*, **15**, 573 (1960).

60-168 BUSEV, A. I., and TIPTSOVA, V. G., Determination of Tl in its alloys with Sb, As and P. *Izv. Vyssh. Ucheb. Zavedenii Khim. i Khim. Tekhnol.*, **3**, 69 (1960).

60-169 WAKAMATSU, S., Folgebestimmung von Fe, Al, Mn, Ca und Mg mit AeDTE. *Japan Analyst*, **9**, 238 (1960).

60-170 WAKAMATSU, S., Determination of Nb in Fe and steel by EDTA titrations. *Japan Analyst*, **9**, 587 (1960).

60-171 SEN SARMA, R. N., Complexometric estimation of micro amounts of Co using EDTA. Thiocyanate end-point. *Analyt. Chem.*, **32**, 717 (1960).

60-172 KOZLOV, A. G., and KROT, N. N., Spectrophotometric investigation of complexformation between UO_2^{2+} and EDTA. *Zhur. neorg. Khim.*, **5**, 1959 (1960).

60-173 SCHMITZ, B., Komplexometrische Gehaltsbestimmung offizineller Arzneistoffe. *Dtsch. Apoth. Ztg.*, **100**, 693 (1960).

60-174 SIERRA, F., and SANCHEZ, C., New observations on the indicator murexide. *Anal. Real. Soc. Espan. Fis. y Quim.*, **56**, 5 (1960).

60-175 KORTÜM, G., Photoelektrische, komplexometrische Titration von Ca und Mg in Serum. *Klin. Wochenschr.*, **38**, 452 (1960).

60-176 YURIST, I. M., and SHAKHANOVA, P. G., Complexometric determination of Pb in fluoboric acid electrolyte. *Zavodskaya Lab.*, **26**, 1354 (1960).

60-177 KORKISCH, J., Beiträge zur analytischen Chemie des Zr. VI. Bestimmung von Zr in Legierungen. *Z. analyt. Chem.*, **176**, 403 (1960).

60-178 KORKISCH, J., Beiträge zur analytischen Chemie des Zr. VII. Chelometrische Titration von Zr und Hf mit Solochromschwarz. 6 BN als Indikator. *Z. analyt. Chem.*, **176**, 42 (1960).

60-179 INCE, A. D., and FORSTER, W. A., Rapid titration of Ca in tricalciumphosphate. *Analyst*, **85**, 608 (1960).

60-180 BUSEV, A. I., TIPTSOVA, V. G., and SOKOLOVA, T. A., Die Reaktion von Tl mit AeDTE. *Zhur. neorg. Khim.*, **5**, 249 (1960).

60-181 BUSEV, A. I., and TIPTSOVA, V. G., Die Bestimmung von Tl in seinen Legierungen mit Sb, As und P. *Izv. Vissh. Ucheb. Zavedenii Khim. i Khim. Tekchn.*, **3**, 69 (1960).

60-182 HOLASEK, A., and DUGANDŽIČ, M., Die komplexometrische Bestimmung von Eilweiß. *Ärztl. Lab.*, **6**, 21 (1960).

60-183 DUGANDŽIČ, M., and HOLASEK, A., Komplexometrische Bestimmung der sauren und alkalischen Phosphatase. *Wiener Med. Wochenschr.*, **110**, 460 (1960).

60-184 DRAGUSIN, I., Schnellbestimmung von Ba und Sulfat. *Rev. Chim. (Bucharest)*, **11**, 110 (1960).

60-185 HOLASEK, A., LIEB, H., and PEČAR, M., Die komplexometrische Bestimmung von K neben Ammonium. *Mikrochim. Acta*, 750 (1960).

60-186 TSERKASEVICH, K. W., Complexometric determination of Pd. *Aptechnoe Delo*, **9**, 32 (1960).

60-187 ASENI, G., Nuevas quelometrias de tierras raras. *Inf. Quim. Anal.*, **14**, 121 (1960).

60-188 RAHEEM, A. A. ABD EL, MOUSTAFA, A. S., and AMIN, A. A., Omega Chrome green BLL: A new analytical reagent. *Z. analyt. Chem.*, **175**, 19 (1960).

60-189 TSERKOVNITSA, I. A., and KUSTOVA, N. A., Amperometric determination of V(IV) and V(III) when present together. *Vestn. Leningr. Univ. Ser. Fiz. i Khim.*, 148 (1960).

60-190 DRAGUSIN, I., Schnellbestimmung von Ba und Sulfat. *Rev. Chim. (Bucharest)*, **11**, 110 (1960).

60-191 KORENMAN, I. M., SHEYANOVA, F. R., and KUSHIN, S. D., Colorimetric and fluorescence reactions of Ga. *Zhur. analit. Khim.*, **15**, 36 (1960).

60-192 DATTA, S. K., Complexometric titration of Th by using some azo dyes from chromotropic acid. *J. Inst. Ind. Research (India)*, **19 B**, 168 (1960).

60-193 BECK, T., and GÖRÖG, S., Amphoteric properties of EDTA and the stability of its metal complexes. *Acta Chim. Acad. Sci. Hung.*, **22**, 159 (1960).

60-194 LEONARD, M. A., and WEST, T. S., Chelating reactions of 1,2-dihydroxyanthraquinone-3-methylamine-N,N'-diacetic acid with metal cations in aqueous medium. *J. Chem. Soc.*, 866, 4477 (1960).

60-195 KORENMAN, I. M., SHEYANOVA, F. R., and KUSHIN, S. D., Colorimetric and fluorescence reactions of Ga. *Zhur. analit. Khim.*, **15**, 35 (1960).

60-196 CIOGOLEA, G., MORAIT, G., and TEODORESCU, N., Eine neue Methode zur Bestimmung von Caesium. *Farmacia (Bucharest)*, **8**, 105 (1960).

60-197 ANDERSON, E., FORS, M., and LINDGREN, J. E., Quantitative determination of leptazole. *Acta Chim. Scand.*, **14**, 1957 (1960).

60-198 HOLLOWAY, J. H., and REILLEY, C. N., Metal chelate stability constants of aminopolycarboxylate ligands. *Analyt. Chem.*, **32**, 249 (1960).

60-199 PETROVA, G. S., A new reagent, Sulfarsazen (Plumbon). *Zavodskaya Lab.*, **26**, 1162 (1960).

60-200 BECK, M. T., Catalysis of complexformation reactions. *J. Inorg. Nuclear Chem.*, **15**, 250 (1960).

60-201 BARBIERI, R., BELLUCO, U., and TAGLIAVINI, G., Quantitative analysis of mixtures of organometallic compounds of lead. *Ricerca Sci.*, **30**, 1671 (1960).

1961

61-1 KUNDU, P. C., Komplexometric Titration of Fe(III) with EDTA using β-diketones (acetylacetone and acetoacetic ester) as indicators. *Z. analyt. Chem.*, **184**, 255 (1961).

61-2 KUNDU, P. C., Aluminon as metal indicator in complexometric titration. *Naturwiss.*, **18**, 644 (1961).

61-3 LADA, Z., and MLODEČKA, J., Indirekte komplexometrische Bestimmung von Brenzcatechin. *Chem. analit. (Warsaw)*, **6**, 95 (1961).

61-4 IVANOVA, I., and SIMOVA, L., Silikatanalyse von Ton. *Godishnik Nauchnoizseled. Inst. Metal. Obogat.*, **2**, 181 (1961).

61-5 KRUCHKOVA, E. S., and MUSTAFIN, I. S., Komplexometrische Bestimmung von Ca in Mineralen mit Hydron-II als Indikator. *Zavodskaya Lab.*, **27**, 668 (1961).

61-6 SCHMITZ, B., Komplexometrische Titration von pharmazeutischen Pb- und Zn-Präparaten. *Dtsch. Apoth. Ztg.*, **101**, 1673 (1961).

61-7 OLSEN, R. L., DIEHL, H., COLLINS, P. F., and ELLESTAD, R. B., Determination of Ca in Li-salts. *Talanta*, **7**, 187 (1961).

61-8 SCHEIDHAUER, G., Photometrische Endpunktsbestimmung in Titrationen mit Farbindikatoren. *Chem. Techn. (Berlin)*, **13**, 93 (1961).

61-9 SCHEIDHAUER, G., Photometrische Titration von Ca und Mg in Glassanden. *Jena Rundsch.*, **4**, 147 (1961).

61-10 SOLOMIN, G. A., and FESENKO, N. G., Komplexometrische Bestimmung von Ca, Mg, Fe(III) und Al in Kohlenminenwässern. *Gidrokhim. Materialy*, **33**, 128 (1961).

61-11 SOFFER, N., Determination of dithionate, sulphite and sulphate in manganese leach liquor. *Analyst*, **86**, 843 (1961).

61-12 WÜNSCH, L., Komplexe von Se mit Eriochromschwarz T. *Coll. Czech. Chem. Commun.*, **26**, 1886 (1961).

61-13 KHALIFA, H., and KHATER, M. M., Estimation of small amounts of Cr(III) and the analysis of binary mixtures with other metals. *Z. analyt. Chem.*, **178**, 260 (1961).

61-14 KHALIFA, H., and KHATER, M. M., Microestimation of Ga and In and the analysis of their mixtures with some other metals. *Z. analyt. Chem.*, **184**, 92 (1961).

61-15 SAWYER, W., and HAYES, J. F., Estimation of Ca in caseine. *Australian J. Dairy Technol.*, **16**, 108 (1961).

61-16 KAKABADSE, G., and WILSON, H. J., The complexes of V(V) with EDTA. *Analyst*, **86**, 402 (1961).

61-17 KOMARITSKAYA, I. D., Determination of Larusan in powders and tablets. *Farm. Zhur. (Kiev).*, **16**, 31 (1961).

61-18 KABANOVA, O. I., Komplexformation zwischen Plutonium(V) und AeDTE. *Zhur. neorg. Khim.*, **6**, 786 (1961).

61-19 PANDE, C. S., and SRIVASTAVA, T. S., Complexometric titration of Cu with nitrosochromotropic acid as indicator. *Z. analyt. Chem.*, **184**, 248 (1961).

61-20 STRELOW, F. W. E., Determination of Th in low-grade ores using a cation exchange separation–EDTA titration method. *Analyt. Chem.*, **33**, 1648 (1961).

61-21 PŘIBIL, R., KÖRÖS, E., and BARCZA, L., Die komplexometrische

Bestimmung des Hg-Gehaltes von Quecksilberverbindungen. *Acta Pharm. Hung.*, **100**, 522 (1961).

61-22 JENIK, J., Komplexometrische Mikrobestimmung von Kieselsäure. *Chem. Prumysl.*, **11**, 189 (1961).

61-23 TANANAEV, I. V., and SHEVCHENKO, G. V., Reaction of samarium with EDTA. *Zhur. neorg. Khim.*, **6**, 1909 (1961).

61-24 VEDA, S., YAMAMOTO, Y., and WAKIZAKA, H., Chelatometric determination of phosphate after precipitation with Bi. *Nippon Kagaku Zasshi*, **82**, 873 (1961).

61-25 DE SOUSA, A., Determination of Cr in minerals. *Chem. Analyst*, **50**, 9 (1961).

61-26 DE SOUSA, A., Determination of thiosulphate and sulphide content of samples. *Chem. Analyst*, **50**, 76 (1961).

61-27 DE SOUSA, A., Chelometrische Bestimmung von Anionen des Schwefels. I. Bestimmung von Persulfat und Sulfat, wenn gleichzeitig anwesend. *Inf. Quim. Anal.*, **15**, 91 (1961).

61-28 DE SOUSA, A., Chelometric determination of cesium. *Talanta*, **8**, 686 (1961).

61-29 DE SOUSA, A., Chelometric determination of CN⁻, CNS⁻ and Cl⁻ in presence of one another. *Talanta*, **8**, 782 (1961).

61-30 SUN, P. J., Chelatometric titration of Bi with gallein as indicator. *J. Chinese Chem. Soc. (Formosa)*, **8**, 71 (1961).

61-31 SUN, P. J., Chelatometric titration of Pb with gallein as indicator. *J. Chinese Chem. Soc. (Formosa)*, **8**, 181 (1961).

61-32 SUN, P. J., Consecutive chelatometric determination of Bi and Pb using gallein as indicator. *J. Chinese Chem. Soc. (Formosa)*, **8**, 314 (1961).

61-33 HOLASEK, A., and PEČAR, M., Die komplexometrische Bestimmung des K im Harn. *Clin. Chim. Acta*, **6**, 125 (1961).

61-34 MORRIS, A. G. C., Complexometric determination of calcium and magnesium in ferromanganese slags. *Analyt. Chem.*, **33**, 599 (1961).

61-35 HALL, L. C., and FLANIGAN, D. A., Stepwise titrations of Fe(III) and Fe(II) with EDTA and ferricyanide. Square wave titrimetric end-point. *Analyt. Chem.*, **33**, 1495 (1961).

61-36 KORKISCH, J., Komplexometrische Titration von U(IV). *Analyt. Chim. Acta*, **24**, 306 (1961).

61-37 SANGAL, S. P., and DEY, A. K., Complexometric determination of Th with Chromotrope 2 R. *Z. analyt. Chem.*, **183**, 178 (1961).

61-38 OCHYNSKA, J., Komplexometrische Bestimmung der Alkaloide in Chelidonium majus. II. Bestimmung der Gesamtalkaloide in Pflanze und Tinktur. *Chem. Analit. (Warsaw)*, **6**, 261 (1961).

61-39 SMITH, T. D., Chelates formed by tin(II) and certain aminopolycarboxylic acids. *J. Chem. Soc.*, 2554 (1961).

61-40 MELNIKOVA, A. S., and CHERKESOV, A. I., Trilonometric determination of Th with hematoxiline as indicator. *Zhur. Vses. Khim. Obchchestva Mendeleeva*, **6**, 469 (1961).

61-41 KALYNZHNAYA, G. A., and KHALININ, A. S., Analysis of In–As–Se alloys. *Zavodskaya Lab.*, **27**, 261 (1961).

61-42 WALLRAF, M., Volumetrische Bestimmung von Al, Fe, und Ti mit AeDTE. *Zement-Kalk-Gips*, **14**, 504 (1961).

61-43 MICHOD, J., Schnellbestimmung von Ca in Weißwein. *Ann. Agric. Suisse*, **62**, 273 (1961).

61-44 KRISTIANSEN, H., The direct titration of Al and stepwise titration of Fe and Al with EDTA and 3-hydroxy-2-naphthoic acid as indicator. *Analyt. Chim. Acta*, **25**, 513 (1961).

61-45 SHRIVANEK, V., and KLEIN, P., Komplexometrische Bestimmung von Al in Erzen. *Z. analyt. Chem.*, **184**, 360 (1961).

61-46 WAKAMATSU, S., Chemische Analyse von basischen Schlacken. VI. Bestimmung von metallischem Fe. *Tetso to Hayane*, **42**, 612 (1961).

61-47 MATSUI, Y., Thymolphthaleincomplexone in the rapid analysis of silicate rocks. *Japan Analyst*, **10**, 183 (1961).

61-48 FLASCHKA, H., Photometrische Titrationen. I. Allgemeine Betrachtungen und die Theorie chelometrischer Titrationen ohne Anwendung eines Indikators. *Talanta*, **8**, 381 (1961).

61-49 FLASCHKA, H., and SAWYER, P., Photometric titrations. II. Design and construction of a photometric titrator. *Talanta*, **8**, 521 (1961).

61-50 FLASCHKA, H., and GANCHOFF, J., Photometric titrations. III. The consecutive titration of Ca and Mg. *Talanta*, **8**, 720 (1961).

61-51 FLASCHKA, H., and GANCHOFF, J., Photometric titrations. V. A selective chelometric determination of Co. *Talanta*, **8**, 885 (1961).

61-52 ODLER, I., and GEBAUER, J., Komplexometrische Bestimmung von Sulfat. *Chem. Zvesti*, **15**, 563 (1961).

61-53 MIKLOS, I., and SZEGEDI, R., Polarographische Bestimmung der Stabilitätskonstante des Bi–AeDTE-Komplexes und seine analytische Verwendung. *Acta Chim. Acad. Sci. Hung.*, **26**, 365 (1961).

61-54 PESHKOVA, V. M., GROMOVA, M. I., EFIMOV, I. P., and ISA-CHENKO, A. V., Spectrophotometric titration of the rare earths elements. *Vestn. Moskau Univ. Ser. Khim.*, 59 (1961).

61-55 SAND, H. F., The dissociation of EDTA and EDTA-sodium salts. *Acta Odontol. Scand.*, **19**, 469 (1961).

61-56 SIMOVA, L., and IVANOVA, I., Schnellanalyse von Dolomiten und Magnesiten. *Godishnik Nauchnoizsled. Inst. Metal. Obogat.*, **2**, 177 (1961).

61-57 ROUSHDI, I. M., ABDINE, H., and AYAD, A., New method for the determination of barbiturates. *J. Pharm. Pharmacol.*, **13**, Suppl. 1535 (1961).

61-58 PERESHIN, G. S., and TANANAEV, I. V., The solubility product of EDTA. *Zhur. analit. Khim.*, **16**, 523 (1961).

61-59 VERMA, M. R., AGARWAL, K. C., and BAHL, J. S., Complexometric method for the determination of the thickness of chromplate on steel. *Electroplating*, **14**, 119 (1961).

61-60 RÁDY, G., GIIMESI, O., and ERDEY, L., Determination of total Pb and PbO in $PbCrO_4$. *Acta Chim. Acad. Sci. Hung.*, **28**, 237 (1961).

61-61 HARA, S., Dithizone and Zn dithizonate as indicator in EDTA titrations. *Japan Analyst*, **10**, 629 (1961).

61-62 HARA, S., EDTA titration of Zn and Pb with dithizone as indicator and masking of Pb with 3-mercapto propionic acid. *Japan Analyst*, **10**, 633 (1961).

61-63 LITEANU, C., LUKACS, I., and STRUSIVICI, C., Beiträge zur komplexometrischen Bestimmung von Kationengemischen. III. Simultane Bestimmung von Fe und Al mit NH₄SCN als Indikator. *Analyt. Chim. Acta*, **24**, 200 (1961).

61-64 HALASZ, A., JANOSI, A., and VILLANYI, K., Schnellbestimmung von Al und Mg in Elektronmetall. *Vesz. Vegyip. Egyet. Kozlemen*, **5**, 151 (1961).

61-65 PALATÝ, V., Hydroxyhydroquinonephthalein as a metallo-chromic indicator. *Chem. and Ind.*, 211 (1961).

61-66 IWAMOTO, T., Potassium acid tartrate as a supporting chelating agent and buffer in the chelatometric determination of Fe and Al. *Japan Analyst*, **10**, 190 (1961).

61-67 LOVASI, J., Komplexometrische Schnellbestimmung von Uran in Gegenwart von Verunreinigungen. *Femipari Kutató Intézet Közleményei*, **4**, 385 (1961).

61-68 LEWIS, L. L., NARDOZZI, M. J., and MELNICK, L. M., Rapid chemical determination of Al, Ca and Mg in raw materials, sinters and slags. *Analyt. Chem.*, **33**, 1351 (1961).

61-69 BÜRGER, K., Analyse organischer Sn-Verbindungen. *Z. Lebensm. Unters. und Forsch.*, **114**, 1 (1961).

61-70 KEMULA, W., BRAJTER, K., and RUBEL, S., Analyse von Ferriten. I. Polarographische und komplexometrische Bestimmung von Ni und Zn. *Chem. Analit.* (*Warsaw*), **6**, 331 (1961).

61-71 KEMULA, W., and BRAJTER, K., Analyse von Ferriten. II. Bestimmung von Ba in Ba-Ferriten. *Chem. Analit.* (*Warsaw*), **6**, 343 (1961).

61-72 TAKAO, H., and MUSHA, S., Amperometric titration of Ti with EDTA. *Japan Analyst*, **10**, 160 (1961).

61-73 SANGAL, S. P., and DEY, A. K., Dimethylaminohydroxyphenolazo-carboxylic acid as an indicator in the complexometric determination of Th. *J. Indian Chem. Soc.*, **38**, 75 (1961).

61-74 SANGAL, S. P., and DEY, A. K., Complexometric determination of Th using sulphodichlorohydroxo-dimethylfuchson-dicarboxylic acid as indicator. *Z. analyt. Chem.*, **178**, 415 (1961).

61-75 ARBUZOV, G. A., KUZNETSOV, A. R., and PAVLOV, N. N., Device for the titration of dark solutions. *Zavodskaya Lab.*, **27**, 225 (1961).

61-76 RÉPÁS, P., Schnellanalyse von Ferritmagneten. *Acta. Chim. Acad. Sci. Hung.*, **28**, 243 (1961).

61-77 RADKO, V. A., and YAKIMETS, E. M., Determination of Fe, Al and Mn in metallurgical slags. *Zavodskaya Lab.*, **27**, 1464 (1961).

61-78 MEE, J. E., and CORBETT, J. D., Titrimetric determination of Ga and of Ga in the presence of Al. *Chem. Analyst*, **50**, 74 (1961).

61-79 PALEY, P. N., and HSÜ, L.-Y., Komplexometrische Titration von U(IV) mit Thoron als Indikator. *Zhur. analit. Khim.*, **16**, 61 (1961).

61-80 PAVLOV, N. N., KUZNETSOV, A. R., and ARBUZOV, G. A., Study of the stability of Fe(III) complexes by the complexometric method. *Izv. Vissh. Ucheb. Zavedenii Techno. Legkai Prom.*, 85 (1961).

61-81 POVONDRA, P., and PŘIBIL, R., Komplexometrische Titrationen (Chelatometrie). XLIX. Bestimmung von Ca und Mg bei Gegenwart größerer Mengen Mn. *Coll. Czech. Chem. Commun.*, **26**, 311 (1961).

61-82 BIEBER, B., and VEČERA, Z., Komplexometrische Titrationen (Chelatometrie). L. Ti-Bestimmung gegen Xylenolorange in Gegenwart von H_2O_2. *Coll. Czech. Chem. Commun.*, **26**, 2081 (1961).

61-83 BUDĚŠINSKÝ, B., Komplexometrie in der organischen Analyse IX. Bestimmung von Aldehyden, Hydroxylamin, Hydrazin und dessen Derivaten durch Oxydation mit Hg(II)-Komplexonat. *Coll. Czech. Chem. Commun.*, **26**, 781 (1961).

61-84 POVONDRA, P., and PŘIBIL, R., Chelometric determination of Mn in ores and slags. *Coll. Czech. Chem. Commun.*, **26**, 2164 (1961).

61-85 SVOBODA, V., CHROMÝ, V., KÖRBL, J., and DORAZIL, L., Metallochromic indicators. X. A new mixed indicator for chelatometric determination of Ca. *Talanta*, **8**, 249 (1961).

61-86 HNILIČKOVÁ, M., and SOMMER, L., PAR as chelometric indicator. *Coll. Czech. Chem. Commun.*, **26**, 2189 (1961).

61-87 BIEBER, B., and VEČERA, Z., Trennung von Ca und Mg in chelatometrischen Bestimmung. *Coll. Czech. Chem. Commun.*, **26**, 59 (1961).

61-88 BIEBER, B., and VEČERA, Z., Komplexometrische Simultanbestimmung von Ca und Mg. *Analyt. Chem. (Czechosl.)*, **6**, 17 (1961).

61-89 KLYGIN, A. E., NIKOLSKAYA, N. A., KOLYADA, N. S., and ZAVRAZHNOVA, D. M., Complexometric determination of U(IV) with arsenazo I as indicator. *Zhur. analit. Khim.*, **16**, 110 (1961).

61-90 KLYGIN, A. E., KOLYADA, N. S., and ZAVRAZHNOVA, D. M., The interaction of Mo(V) with EDTA. *Zhur. analit. Khim.*, **16**, 442 (1961).

61-91 KLYGIN, A. E., and KOLYADA, N. S., Complexometric determination of Zr with xylenole orange as indicator. *Zavodskaya Lab.*, **27**, 23 (1961).

61-92 PŘIBIL, R., VESELÝ, V., and KRATOCHVIL, M., Contributions to the basic problems of complexometry. IV. Determination of Tl. *Talanta*, **8**, 52 (1961).

61-93 PŘIBIL, R., and VESELÝ, V., Contributions to the basic problems of complexometry. V. Mutual masking of Fe and Mn. *Talanta*, **8**, 271 (1961).

61-94 PŘIBIL, R., and VESELÝ, V., Contributions to the basic problems of complexometry. VI. The masking of Cr(III). *Talanta*, **8**, 565 (1961).

61-95 PŘIBIL, R., and VESELÝ, V., Contribution to the basic problems of complexometry. VII. Determination of Cu and Fe. *Talanta*, **8**, 743 (1961).

61-96   PŘIBIL, R., and VESELÝ, V., Contributions to the basic problems of complexometry. VIII. Thioglycollic acid as a masking reagent. *Talanta*, **9**, 880 (1961).

61-97   KRATOCHVIL, M., and BLECHA, J., Chelometric methods in rapid applied analysis. V. EDTA titration of Tl in monocrystals. *Chem. Analyst*, **50**, 11 (1961).

61-98   PŘIBIL, R., and VESELÝ, V., Chelometric methods in rapid applied analysis. VI. Analysis of Mg–Pb–Ni-ferrites. *Chem. Analyst*, **50**, 73 (1961).

61-99   PŘIBIL, R., and VESELÝ, V., Chelometric methods in rapid applied analysis. VII. Analysis of Cr–Ni–Fe-alloys. *Chem. Analyst*, **50**, 100 (1961).

61-100   PŘIBIL, R., and VESELÝ, V., Chelometric methods in rapid applied analysis. VIII. Analysis of Co–Mn–Mg-ferrites. *Chem. Analyst*, **50**, 108 (1961).

61-101   MALÁT, M., and MÚČKA, V., Chelometric methods in rapid applied analysis. IX. Determination of Th, La, Ba and B in a rare-element, non-silica optical glass. *Chem. Analyst*, **50**, 110 (1961).

61-102   MALÁT, M., and HOLEČEK, K., Chelometric methods in rapid applied analysis. X. Determination of Ni in cyanide silvering baths. *Chem. Analyst*, **50**, 115 (1961).

61-103   ALIMARIN, I. P., and TSE, Y.-H., Komplexometrische Fe-Bestimmung. *Vestn. Moskau Univ. Ser. Khim.*, 59 (1961).

61-104   ANDREW, T. R., and NICHOLS, P. N. R., The determination of Zn in Zn–Mn-ferrites and other ferrites. *Analyst*, **86**, 676 (1961).

61-105   BACHMANN, C., Komplexometrische Bestimmung von Liquor Alumin. Acet. empfohlen für DAB. VII. *Pharm. Praxis, Beilage Pharmazie*, 112 (1961).

61-106   ASHBROOK, A. W., and RITCEY, G. M., Volumetric determination of sulphate. *Analyst*, **86**, 740 (1961).

61-107   BAUGH, C. A., DECKER, K. H., and PALMER, J. W., Determination of small amounts of Ca in MgO. *Analyt. Chem.*, **33**, 1804 (1961).

61-108   CHETKOWSKA, M., ZOSIN, Z., and STRZESZEWSKA, I., Komplexometrische Bestimmung von Zink in der analytischen Kontrolle von Zn-Verbindungen. Anwendung auf die Analyse von Zn-Salzen. Bestimmung von Pb, Fe, Al und Zn. *Chem. Analit. (Warsaw)*, **6**, 309 (1961).

61-109   LYSTSOVA, G. G., Volumetric determination of copper in the presence of Ni and Fe. *Zavodskaya Lab.*, **27**, 964 (1961).

61-110   CICKERS, C., and WILKINSON, J. V., Flask combustion technique in pharmaceutical analysis: mercury-containing substances. *J. Pharm. Pharmacol.*, **13**, 156 Suppl. (1961).

61-111   ZEMLYANSKI, N. I., and DRACK, B. S., Complexometric determination of phosphorus. *Zhur. analit. Khim.*, **16**, 653 (1961).

61-112   FRITZ, J. S., ABBINK, J. E., and PAYNE, M. A., Naphthyl-azoxine S as a complexometric indicator. *Analyt. Chem.*, **33**, 1381 (1961).

61-113   AIKENS, D. A., SCHMUCKLER, G., SADEK, F. S., and REILLEY C. N., Increased selectivity in chelometric titrations through end-point location by linear extrapolation. Cu^{2+} as a photometric indicator. *Analyt. Chem.*, **33**, 1664 (1961).

61-114 BERNDT, W., and SÁRA, J., Über die Anwendung von Cystein als Maskierungsmittel bei komplexometrischen Titrationen. *Talanta*, **8**, 653 (1961).

61-115 SINGH, B., SAHOTA, S. S., and NARANG, A. S., Complexometric estimation of Cu. *Z. analyt. Chem.*, **182**, 241 (1961).

61-116 SINGH, B., SAHOTA, S. S., and GUPTA, M. P., Complexometric estimation of Fe(III). Acetylacetone as indicator. *Indian J. Appl. Chem.*, **24**, 78 (1961).

61-117 THOMANN, H., Photometrische Komplextitrationen. *Z. analyt. Chem.*, **184**, 241 (1961).

61-118 HEGEMANN, F., and THOMANN, H., Bestimmung von Ca und Mg in Silikaten mittels photometrisch-komplexometrischer Titration. *Ber. dtsch. keram. Ges.*, **38**, 345 (1961).

61-119 BEYERMANN, K., and CRETIUS, K., Mikromethoden zur Bestimmung einiger anorganischer Bestandteile in biologischem Material. *Clin. Chim. Acta*, **6**, 113 (1961).

61-120 BRUSH, J. S., Interaction of Ca with the dye Erio SE in highly alkaline solutions. Direct determination of Ca in serum. *Analyt. Chem.*, **33**, 798 (1961).

61-121 DENISOVA, N. E., and TSVETKOVA, E. V., Analysis of Al–Sb–Ga alloys. *Zavodskaya Lab.*, **27**, 656 (1961).

61-122 BUDANOVA, L. M., and MATRASOVA, T. V., Komplexometrische Bestimmung von Zn in Aluminiumlegierungen. *Zavodskaya Lab.*, **27**, 661 (1961).

61-123 BRUNO, M., and BARBIERI, R., Chromatographische Untersuchung der Komplexbilding zwischen Prometium und AeDTE. *Gazzetta*, **91**, 1055 (1961).

61-124 BORCHERT, O., Die AeDTE als Maßlösung in der pharmazeutischen Analyse. *Pharmazie*, **16**, 181 (1961).

61-125 BACHMANN, C., Komplexometrische Bestimmung von Bi in Noviform. *Pharm. Praxis, Beilage Pharmazie*, 178 (1961).

61-126 ABD-EL RAHEEM, A. A., and DOKHANA, M. M., Fast Navy 2 R as indicator for EDTA titration and its application for the determination of Hg in some pharmaceutical preparations. *Z. analyt. Chem.*, **180**, 339 (1961).

61-127 CHENG, K. L., Analysis of lead tellurite with an accuracy to better than 0.1%. *Analyt. Chem.*, **33**, 761 (1961).

61-128 SAJÓ, I., Bestimmung mehrerer Substanzen bei gleichzeitiger Anwesenheit durch Titration mit AeDTE. *Acta Chim. Acad. Sci. Hung.*, **28**, 253 (1961).

61-129 SAJÓ, I., Neue Ergebnisse in der Schnellanalyse von Erzen, Mineralen und Silicaten. *Acta Chim. Acad. Sci. Hung.*, **28**, 259 (1961).

61-130 BUSEV, A. I., and TIPTSOVA, V. G., Neue komplexometrische Methode zur Bestimmung von Tl. *Zhur. analit. Khim.*, **16**, 275 (1961).

61-131 BUSEV, A. I., PETRENKO, A. G., and BYKHOVSKAYA, I. A.,

Xylenolorange in der komplexometrischen Bestimmung von Al. *Zavodskaya Lab.*, **27**, 659 (1961).

61-132 BUSEV, A. I., TALIPOVA, L. L., and IVANOW, V. M., Direkte komplexometrische Titration von Tl mit 7-(2-Pyridylazo)-8-hydroxychinolin als Indikator. *Zhur. Vsesynz. Khim. Obshschestva Un. D. I. Mendeleeva*, **6**, 598 (1961).

61-133 SEDLÁČEK, B. A. J., Komplexometrische Bestimmung von Antioxydantien. IV. Semimikrobestimmung von Butylhydroxytoluol, Ascorbinpalmitat und von einigen Gallaten in Schweinefett. *Fette & Seifen*, **63**, 1053 (1961).

61-134 SEDLÁČEK, B. A. J., Komplexometrische Bestimmung von Antioxydantien. V. Semimikrobestimmung von Äthylgallat in Schweinfett. *Z. Lebensm. Unters. u. Forsch.*, **114**, 127 (1961).

61-135 CAMERON, A. J., and GIBSON, N. A., EDTA titrations with extractive end-points. I. Determination of Co. *Analyt. Chim. Acta*, **25**, 24 (1961).

61-136 CAMERON, A. J., and GIBSON, N. A., EDTA titrations with extractive end-points. II. Determination of Cu, Ni, Fe, Cr and V. *Analyt. Chim. Acta*, **25**, 429 (1961).

61-137 IWAMOTO, T., Acid-base properties and metal chelate formation of PAR. *Bull. Chem. Soc. Japan*, **34**, 605 (1961).

61-138 LASSNER, E., and SCHARF, R., Zur komplexometrischen Bestimmung des Mo. *Z. analyt. Chem.*, **183**, 187 (1961).

61-139 LASSNER, E., and SCHARF, R., Chelometric titrations in the presence of W and Mo. *Chem. Analyst*, **50**, 6 (1961).

61-140 FÜSTI-MOLNAR, S., and LEGRADI, L., Komplexometrische Kontrolle von Sulfonierungsreaktionen. *Magyar Kém. Folyóirat*, **67**, 455 (1961).

61-141 LEGRADI, L., Determination of alkali polysulphides. *Analyst*, **86**, 54 (1961).

61-142 VRCHLABSKÝ, M., and OKÁČ, A., Glyoxal-bis-(2-hydroxanil) für den empfindlichen Nachweis und die komplexometrische Bestimmung von Ca in Gegenwart von Mg. *Coll. Czech. Chem. Commun.*, **26**, 246 (1961).

61-143 GRASKE, A., Determination of Co, Zn, Mn and Pb in mixed driers. *Off. Dig. Fed. Soc. Paint Technol.*, **33**, 855 (1961).

61-144 LAUX, P. G., The determination of carbon dioxide in samples from ethanolamine gas treatment plants. *Chem. Analyst*, **50**, 49 (1961).

61-145 DUBSKÝ, I., Schnellbestimmung von Cu in alkalischen Cu- und Messingbädern. *Metalloberfläche*, **15**, 13 (1961).

61-146 DAVIS, D. G., Polarography of U(VI)–EDTA complexes. *Analyt. Chem.*, **33**, 492 (1961).

61-147 DITZ, J., Stilbazo als metallochromer Indikator. *Z. analyt. Chem.*. **178**, 274 (1961).

61-148 DRAGUSIN, I., Rapid complexometric determination of $BaSO_4$ in barite mineral. *Zhur. analit. Khim.*, **16**, 611 (1961).

61-149 ELOFSSON, A., Metod för komplexometrisk tetrering av Al. *Glastech. Tidskr.*, **16**, 171 (1961).

61-150 CALLEJA, J., and FERNANDEZ-PARIS, J. M., Komplexometrische

Bestimmung von SO_3 in Portland-Zement. *Rev. Cienc. Appl.*, **15**, 120 (1961).

61-151 wänninen, e., Selection of complexing agents for complexometric titrations. *Talanta*, **8**, 355 (1961).

61-152 ivanova, i., Komplexometrische Schnellanalyse von Chromiten und feuerfestem Material auf Basis Chrommagnesit. *Godishnik Nauchnoizsled. Inst. Metal Obogat.*, **2**, 167 (1961).

61-153 hozumi, k., and mizuno, k., Organic microanalysis. XXVII. Microdetermination of organic phosphorus by EDTA titration. *Japan Analyst*, **10**, 453 (1961).

61-154 aseni-mora, g., Die Verwendung flüssiger Amalgame in der Bestimmung von Ag-Ionen und seiner Mischungen. *Quim. e Ind. (Bilbao)*, **8**, 132 (1961).

61-155 wakamatsu, s., Determination of Pb in free-cutting steel. *J. Japan Inst. Metal. Sendai*, **25**, 265 (1961).

61-156 roushdi, i. m., abdine, h., and el-sheltawy, a. m., A complexometric determination for chloroform. *Egypt. Pharm. Bull.*, **42**, 277 (1961).

61-157 busev, a. i., talipova, l. l., and ivanov, v. m., Die direkte komplexometrische Titration von Tl mit 7-(2-Pyridylazo)-8-chinolinol als Indikator. *Zhur. Vsesyn z. Khim. Obshchestva Inst. D. I. Mendeleeva*, **6**, 598 (1961).

61-158 hennart, c., and merlin, e., Applications de la chélatométrie. V. Dosage volumétrique de la diacétyl-monoxime. *Chim. analyt.*, **43**, 28 (1961).

61-159 hennart, c., Chélatométrie. VI. Dosage volumétrique des vicdioximes. *Analyt. Chim. Acta*, **25**, 150 (1961).

61-160 hennart, c., and lefevre, y., Chélatométrie. VII. Dosage volumétrique des ortho-acylphenols. *Talanta*, **8**, 273 (1961).

61-161 hennart, c., Chélatométrie. XIV. Dosage volumétrique des vicdicétones. *Analyt. Chim. Acta*, **25**, 201 (1961).

61-162 hennart, c., Les apports de la chélométrie à l'analyse quantitative organique. Part I. Les dosages élémentaires. *Chim. analyt.*, **43**, 402 (1961). Part II. Les dosages fonctionelles et particulaires. *Chim. analyt.*, **43**, 447 (1961).

61-163 kurpiel, i., mojejko, j., and przyborowski, l., Determination of barbituric acid derivatives with Hg-salts. *Acta Polon. Pharm.*, **18**, 221 (1961).

61-164 holasek, a., and flaschka, h., *Komplexometrische und andere titrimetrische Methoden des klinischen Laboratoriums.* Springer-Verlag, Vienna (1961).

61-165 adamovich, l. p., and napadailo, i. n., Determination of the instability constant of the complex of Be with EDTA. *Zhur. analit. Khim.*, **16**, 158 (1961).

61-166 beck, m. t., and bàrdi, i., Reduction of Cr(VI) in the presence of complex-forming agents. I. Reduction in the presence of EDTA. *Acta Chim. Acad. Sci. Hung.*, **29**, 283 (1961).

61-167 holasek, a., and pećar, m., Die komplexometrische Bestimmung des K im Harn. *Clin. Chim. Acta*, **6**, 125 (1961).

61-168 LASSNER, E., and SCHARF, R., Determination of Ti, Nb and Ta in cemented carbides. *Chem. Analyst*, **50**, 69 (1961).

61-169 KENNEDY, J. H., Polarography of Nb(V) in EDTA and citric acid media. *Analyt. Chem.*, **33**, 943 (1961).

61-170 SPAUSZUS, S., and HUPFER, J., Die polarographische Bestimmung von Nb in Gegenwart von AeDTE. *Chem. Techn. (Berlin)*, **13**, 750 (1961).

61-171 KIRBY, R. E., and FREISER, H., Polarography of Ta–EDTA complexes. *J. Amer. Chem. Soc.*, **65**, 191 (1961).

61-172 HAMAGUCHI, H., and SUGISITA, R., Indirect determination of rhenium by EDTA titration. *Japan Analyst*, **10**, 1256 (1961).

61-173 DE SOUSA, A., La micro-détermination du Li avec EDTA. *Mikrochim. Acta*, 732 (1961).

61-174 SOLOMIN, G. A., and FESENKO, N. G., Komplexometrische Bestimmung von Ca, Mg, Fe(II), Fe(III) und Al in sauren Kohlenminenwässern. *Gidrokhim. Materialy*, **33**, 128 (1961).

61-175 DIMITROVA, M., Complexometric determination of Al in Zn-alloys and of Zn in Al-alloys, brass and bronze. *Mashinostroenie*, **10**, 23 (1961).

61-176 PESHKOVA, V. M., GROMOVA, M. I., EFIMOV, I. P., and ISA-CHENKO, A. V., Photometric titration of some rare earths using arsenazo as indicator. *Vestn. MGU. Seriya II, Khim.*, 59 (1961).

61-177 HOARD, J. L., LIND, M., and SILVERTON, J. V., The stereochemistry of the EDTA-aquoferrate(III)-ion. *J. Amer. Chem. Soc.*, **83**, 2771 (1961).

61-178 YOSHIWAKA, K., and MITSUI, T., Simultaneous micro determination of halogens and sulphur in organic compounds. *Japan Analyst*, **10**, 723 (1961).

61-179 DE SOUSA, A., Chelometrische Bestimmung von Schwefelanionen. II. Bestimmung von Mischungen von Sulfat und Sulfid. *Inf. Quim. Anal.*, **15**, 121 (1961).

61-180 HEGEMANN, F., and THOMANN, H., Bestimmung von Ca und Mg in Silikaten mittels photometrisch-komplexometrischer Titration. *Ber. dtsch. keram. Ges.*, **38**, 345 (1961).

61-181 POLVOLOTSKAYA, G. L., Complexometric determination of Fe in oxidized ferrochrome and chromites in the presence of Cr. *Met. i Khim. Prom. Kazakhstan. Nauchn.-Tekchn. Sb.*, 64 (1961).

61-182 HOFMAN, J., Komplexometrische Bestimmung von Ca in Serum und Urin unter Verwendung von Fluorexan. *Časopis lékařů českých*, **100**, 1171 (1961).

61-183 CHENG, K. L., Increasing selectivity of analytical reactions by masking. *Analyt. Chem.*, **33**, 783 (1961).

61-184 FLASCHKA, H., Application of complementary tri-stimulus colorimetry. III. The determination of indicator constants and their use in the calculation of screening conditions. *Talanta*, **8**, 342 (1961).

61-185 MALÁT, M., Dissoziationskonstanten des Chromazurol S. *Analyt. Chim. Acta*, **25**, 289 (1961).

61-186 COATES, E., and RIGG, B., Complex formation: data for Solochrome Violet R., I. Ionization constants. *Trans. Faraday Soc.,* 57, 1088 (1961).

61-187　PUNGOR, E., and ZAPP, E. E., Kinetic investigation on Al complexes by H.F. titrations. *Acta Chim. Acad. Sci. Hung.,* 27, 69 (1961).

1962

62-1　PHILLIPS, J. P., and CROWLEY, R. C., Photometric titrations. *Talanta,* 9, 178 (1962).

62-2　MORI, K., Acetylacetone as indicator. I. Titration of Fe(III). *Japan Analyst,* 11, 689 (1962).

62-3　MORI, K., Acetylacetone as indicator. II. Titrations of Al. *Japan Analyst,* 11, 690 (1962).

62-4　WILLS, V., STAMBAUGH, O. F., and PROCTOR, Z. G., New indicator for the volumetric determination of Cu. *Analyt. Chem.,* 34, 225 (1926).

62-5　DATTA, S. K., and SAHA, S. N., Photometric EDTA titration of Th with β-SNADNS-6 as indicator. *Chem. Analyst,* 51, 49 (1962).

62-6　ONOSOVA, S. P., Complexometric determination of rare earths and Th. *Zavodskaya Lab.,* 28, 271 (1962).

62-7　BRADY, G. W. F., and GWILT, J. R., The chelometric determination of Al. *J. Applied Chem. (London),* 12, 75 (1962).

62-8　BOZHEVOLNOV, E. A., and KREINGOLD, S. U., Fluoreszenzeigenschaften von Fluoresceinkomplexon. *Zhur. analit. Khim.,* 17, 291 (1962).

62-9　MALISSA, H., and KOTZIAN, H., Vergleich von drei komplexometrischen Methoden zur Bestimmung von Al. *Analyt. Chim. Acta,* 26, 128 (1962).

62-10　NOVAKOVSKAYA, E. G., Determination of Cd and Ni in the active mass of alkaline accumulators by amperometric titration. *Zavodskaya Lab.,* 28 (1962).

62-11　CORSINI, A., YIH, I. M. L., FERNANDO, Q., and FREISER, H., Potentiometric investigation of metal complexes of PAN and PAR. *Analyt. Chem.,* 34, 1090 (1962).

62-12　BELIKOV, V. V., and SCHRAIBAR, M. S., Complexometric determination in the analysis of medical preparations. *Farm. Zhur. (Kiev),* 15, 10, 25 (1961); 17, 7 (1962).

62-13　BEGELFER, K. I., SZONOVE, P. A., and FUNTIKOV, K. M., Komplexometrische Schnellmethode zur getrennten Bestimmung von Fe und Al in Al-haltigem Material. *Steklo i Keram.,* 19, 30 (1962).

62-14　SIERRA, F., and SANCHEZ, C., Disodium mono barium EDTA in chelometry. *Inf. Quim. Anal.,* 14, 91 (1960).

62-15　SIERRA, F., and SANCHEZ, C., Weitere Bestimmungen unter Verwendung von Ba–AeDTE: Zn–Cr(III) und Zn–Al Mischungen. *Anal. Real. Soc. Fis. y Quim. Ser. B.,* 58, 223 (1962).

62-16　SILVERSTONE, N. M., The potentiometric determination of Ni. *Metallurgia,* 65, 99 (1962).

62-17　STRÁFELDA, F., Eine Silberelektrode als Indicator für potentiometrische komplexometrische Titrationen. *Coll. Czech. Chem. Commun.,* 27, 343 (1962).

62-18 OLSON, D. C., and MARGERUM, D. W., Semixylenole Orange: a sensitive reagent for Zr. *Analyt. Chem.*, **34**, 1299 (1962).

62-19 PESHKOVA, V. M., GROMOVA, M. I., and ALEKSANDROVA, N. M., The successive spectrophotometric titration of Th and total rare earths. *Zhur. analit. Khim.*, **17**, 218 (1962).

62-20 GREGORY, G. R. E. C., and JEFFEREY, P. G., Determination of Ga in pure solutions. *Talanta*, **9**, 800 (1962).

62-21 GEARY, W. J., NICKLESS, G., and POLLARD, F. H., The metal complexes of some azomethine dyestuffs. *Analyt. Chim. Acta*, **26**, 575 (1962); **27**, 71 (1962).

62-22 FOLEY, W. T., and OSYANY, J. M., Determination of Cd and Zn in amalgams. *Chem. Analyst.* **51**, 16 (1962).

62-23 DOORNBOS, D. A., and FABER, J. S., Die komplexometrische Titration von Cu in Gegenwart von Fe und Mn. *Pharm. Weekblad*, **97**, 257 (1962).

62-24 LINDSTROM, F., and STEPHENS, B. G., $Mg(IO_3)_2.4H_2O$ as primary standard for EDTA. *Analyt. Chem.*, **34**, 993 (1962).

62-25 MONK, R. G., and STEED, K. C., Microchemical methods in radiometric analysis. II. Determination of chemical yields by microcoulometry. *Analyt. Chim. Acta*, **26**, 305 (1962).

62-26 FLASCHKA, H., and GANCHOFF, J., Photometric titrations. IV. The chelometric titration of Cd in the presence of Zn. *Talanta*, **9**, 76 (1962).

62-27 FLASCHKA, H., and SAWYER, P., Photometric Titrations. VI. The determination of submicrogram quantities of Ca and Mg. *Talanta*, **9**, 249 (1962).

62-28 DIXON, M., HEINLE, P. J., HUMLICEK, D. E., and MILLER, R. L., Determination of Sn and Pb in simple Sn–Pb solders via EDTA titrations. *Chem. Analyst*, **51**, 42 (1962).

62-29 SCHMID, R. W., Application of the Hg-indicator electrode in potentiometry. *Chem. Analyst*, **51**, 56 (1962).

62-30 SUN, P. J., Chelatometric titration of Cd with gallein as indicator. *J. Chinese Chem. Soc.* (*Formosa*), **9**, 37 (1962).

62-31 SUN, P. J., Chelatometric titration of La with gallein as indicator. *J. Chinese Chem. Soc.* (*Formosa*), **9**, 41 (1962).

62-32 GOLDSTEIN, G., MANNING, D. L., and ZITTEL, H. E., Fe(II) as indicator for amperometric titrations with EDTA. Application to the determination of Th. *Analyt. Chem.*, **34**, 358 (1962).

62-33 CAMPBELL, R. T., and REILLEY, C. N., Chelometric titrations with amperometric end-point detection. *Talanta*, **9**, 153 (1962).

62-34 SPAUSZUS, S., and SCHWARZ, C., Direkte komplexometrische Bestimmung von Al in Stahl. *Neue Hütte*, **7**, 180 (1962).

62-35 SANDERSON, I. P., and WEST, T. S., Selective chelatometry of Cu(II) with ethylenediamine-*N,N,N',N'*-*n*-propionic acid. *Talanta*, **9**, 71 (1962).

62-36 HAMM, R. E., and FURSE, C. T., Square-wave polarography. Amperometric titrations of In with EDTA. *Analyt. Chem.*, **34**, 219 (1962).

400   *Bibliography*

62-37   KRIJN, G. C., Quantitative determination of metals by means of EDTA and thioacetamide. *Chem. Weekblad*, **58**, 127 (1962).

62-38   KRAFT, G., Elektrometrische Endpunktsbestimmung in komplexometrischen Titrationen. *Z. analyt. Chem.*, **186**, 187 (1962).

62-39   KLASS, C. S., Use of the indicator Calceine and its fluorescence in a rapid ultramicrotitration of serum Ca. *Techn. Bull. Registry Med. Technologists*, **32**, 77 (1962).

62-40   SNESHNOVA, L. P., Determination of Ca in streptomycine-calcium chloride by complexometric titration. *Med. Prom. SSSR*, **16**, 49 (1962).

62-41   KAWAHATA, M., MICHIZUKI, H., KAJIYAMA, R., and ISHII, M., Analysis of steel-making materials. I. A rapid titrimetric determination of Mo in ferromolybdenum and MoO_3 with EDTA. *Japan Analyst*, **11**, 748 (1962).

62-42   KAWAHATA, M., MICHIZUKI, H., KAJIYAMA, R., WATANAKE, M., ISHII, M., and KASAKI, K., Determination of low concentration of carbon in steel by EDTA titration. *Japan Analyst*, **11**, 192 (1962).

62-43   IWAMOTO, T., and KANAMORI, K., Metallic Cu for the standardization of EDTA solutions. *Analyt. Chim. Acta*, **26**, 167 (1962).

62-44   HARMELIN, M., Thermometric titrations. *Chim. analyt.*, **44**, 153 (1962).

62-45   KITAGAWA, H., and SHIBATA, N., Determination of Ni in ferrites by EDTA titration. *Japan Analyst*, **11**, 358 (1962).

62-46   VRCHLABSKÝ, M., and OKÁČ, A., Glyoxal-bis-(-2-hydroxyanil) for the detection and complexometric determination of Cd. *Coll. Czech. Chem. Commun.*, **27**, 492 (1962).

62-47   JUNEK, H., and WITTMANN, H., Neue Indikatoren für die komplexometrische Titration von Pb. *Mikrochim. Acta*, 114 (1962).

62-48   BUSEV, A. I., TALIPOVA, L. L., and SKREBKOVA, L. M., Direct complexometric titration of Ga with 7-(1-naphthylazo)-8-hydroxyquinoline-5-sulphonic acid as indicator. *Zhur. analit. Khim.*, **17**, 180 (1962).

62-49   BUSEV, A. I., and TALIPOVA, L. L., Komplexometrische Titration von In mit 7-(Naphthylazo)-8-oxychinolin-5-sulfonat und 7-(4-Sulfo-1-naphthylazo)-8-oxychinolin-5-sulfonat als Indikator. *Vestn. Moskau Univ. Ser. II. Khim.*, **17**, 73 (1962).

62-50   LASSNER, E., and SCHARF, R., Improved chelometric titration of Ti in the presence of Nb and Ta. *Chem. Analyst*, **51**, 49 (1962).

62-51   CHENG, K. L., Stepwise determination of Cd, Hg and Te. *Talanta*. **9**, 501 (1962).

62-52   CHENG, K. L., and GOYDISH, B. L., Determination of Fe and Ni in electroplating solutions. *Chem. Analyst*, **51**, 45 (1962).

62-53   PŘIBIL, R., and VESELÝ, V., Contributions to the basic problems of complexometry. IX. Determination and masking of Al. *Talanta*. **9**, 23 (1962).

62-54   ARIKAWA, Y., and KATO, T., Complexometric titration of Fe(III) using α,5,5′-trimethylformaurin-3,3′-dicarboxylic acid as indicator. *Technol. Reports Tôhoku Univ.*, **26**, 205 (1962).

62-55 SOCOLOVSCHI, R., Indirekte Titration von Phosphat. *Rev. Chim.* (*Bucharest*), **13**, 306 (1962).

62-56 CIOGOLEA, G., MORAIT, G., and NGUYEN, H. B., Komplexometrische Bestimmung von Alkalihypophosphiten. *Farmacia* (*Bucharest*), **10**, 331 (1962).

62-57 FÜLÖP, L., and BLAZSEK, A., Komplexometrische Bestimmung von Hypophosphiten. I. Komplexometrische Bestimmung von Alkalihypophosphiten. *Farmacia* (*Bucharest*), **10**, 525 (1962).

62-58 STEFANAC, Z., Chelometrische As-Bestimmung und ihre Anwendung auf die Mikroelementaranalyse. *Mikrochim. Acta*, 1115 (1962).

62-59 TSUCHIYA, Y., EDTA titration of Zr and Hf and their practical application. *Japan Analyst*, **11**, 1176 (1962).

62-60 KONONENKO, L. I., and POLUETKOV, N. S., Komplexometrische Bestimmung von Hf in Mischung mit Zr. *Zavodskaya Lab.*, **28**, 794 (1962).

62-61 LUKYANOV, V. F., and KNYAZEVA, E. M., Direkte komplexometrische Bestimmung von Zr in Zr–Cu Legierungen. *Zavodskaya Lab.*, **28**, 804 (1962).

62-62 YURIST, I. M., and KOROTKOVA, O. I., Komplexometrische Bestimmung von Sn, Pb, Cd, Ni in Legierungen. *Zavodskaya Lab.*, **28**, 660 (1962).

62-63 ERDEY, L., RADY, C., and GIMESI, O., Analyse von Pb–Ag Legierungen. *Acta Chim. Acad. Sci. Hung.*, **32**, 151 (1962).

62-64 TOROPOVA, V. F., and BATYRSHINA, F. M., Amperometric titration with indicators. *Teoriya i Praktika Polyarogr. Anal. Kishinev Sb.*, 352 (1962).

62-65 BOASE, D. G., FOREMAN, J. K., and DRUMMOND, J. L., Complexometric determination of plutonium in reactor fuel processing plant solutions. I. Nitric acid solutions of irradiated U. *Talanta*, **9**, 53 (1962).

62-66 SZELAG, M., and KOZLICKA, M., Bestimmung von Cu in Erzen mittels Komplexon III. *Chem. Analit.* (*Warsaw*), **7**, 815 (1962).

62-67 MAJER, J., and SPRINGER, V., Zimthydroxamsäure, ein neuer komplexometrischer Indikator für Fe(III)-Ionen. *Chem. Zvesti*, **16**, 633 (1962).

62-68 BUSEV, A. I., and TIPTSOVA, V. G., Die komplexometrische Titration von Tl mit 7-(5,7-Disulfo-2-naphthylazo)-8-hydroxychinolin-5-sulfonat und 7-(1-Naphthylazo)-8-hydroxychinolin-5-sulfonat als Indikator. *Uzbekch. Khim. Zhur.*, **6**, 24 (1962).

62-69 HENNART, C., Chélatométrie. XII. Dosage volumétrique des halogénures d'alkyl-magnesium. *Chim. analyt.*, **44**, 7 (1962).

62-70 HENNART, C., Chélatométrie. XIII. Dosage volumétrique des sulfamides. *Chim. analyt.*, **44**, 8 (1962).

62-71 HENNART, C., Chélatométrie. XV. Dosage volumétrique du 6-purinthiol. *Talanta*, **9**, 97 (1962).

62-72 BERKA, A., and BUSEV, A. I., Reaction of Tl(III) with hydrazin sulphate. *Analyt. Chim. Acta*, **27**, 493 (1962).

402   Bibliography

62-73   CEGLARSKI, R., and ROMAN, J., Komplexometrische Bestimmung einiger Histamine. *Farm. Polska*, **18**, 327 (1962).

62-74   AIKENS, D. A., and REILLEY, C. N., Rapid chelometric determination of chromate as Cr(III)–EDTA by reduction in the presence of EDTA. *Analyt. Chem.*, **34**, 1707 (1962).

62-75   KHALIFA, H., ROBERTS, J. E., and KHATER, M. M., Spectrophotometric determination and titration of Cr(III) with DCTA. *Z. analyt. Chem.*, **156**, 428 (1962).

62-76   DATSENKO, O. V., Indirekte Schnellmethode zur Bestimmung von Kieselsäure in Ofenschlacken. *Zavodskaya Lab.*, **28**, 279 (1962).

62-77   LASSNER, E., and SCHARF, R., Selective detection test for Nb with methylthymol blue. *Chem. Analyst*, **51**, 14 (1962).

62-78   YURIST, I. M., and TYUKOVA, Z. V., The complexometric determination of Pd. *Zavodskaya Lab.*, **28**, 798 (1962).

62-79   GIMESI, O., RÁDY, G., and ERDEY, L., Die komplexometrische Bestimmung von elementarem Schwefel und Selen. *Acta Chim. Acad. Sci. Hung.*, **33**, 381 (1962).

62-80   GIUFFRÉ, L., and CAPIZZI, F. M., Determinazione spettrofotometrica del Ti in presenza di Al. *Ann. Chim.* (*Roma*), **52**, 398 (1962).

62-81   KROT, N. N., ERMOLAEV, N. P., and GELMAN, A. D., Behaviour of EDTA in acid solution and its reaction with U(IV). *Zhur. neorg. Khim.*, **7**, 2054 (1962).

62-82   GATTOW, G., and SCHOTT, D., Zur komplexometrischen Bestimmung des Bi. I. Indikatoren. *Z. analyt. Chem.*, **188**, 10 (1962).

62-83   GATTOW, G., and SCHOTT, D., Zur komplexometrischen Bestimmung des Bi. II. Bildung von Hydrolyseprodukten (Polykationen) bei der Zugabe von basischen Reagentien. *Z. analyt. Chem.*, **188**, 81 (1962).

62-84   BELKOVSKAYA, Y. I., Indirekte Bestimmung von W mit Trilon B. *Tr. Inst. Met. A. A. Baikova*, 244 (1962).

62-85   ELWELL, W. T., and WOOD, D. F., Analysis of Nb and Nb-alloys. *Metallurgia*, **65**, 308 (1962).

62-86   YURIST, I. M., and KOROTKOVA, O. I., Complexometric determination of Sn, Pb, Cd and Ni in alloys. *Zavodskaya Lab.*, **28**, 660 (1962).

62-87   PŘIBIL, R., and VESELÝ, V., Contributions to the basic problems of complexometry. IX. Determination and masking of Al. *Talanta*, **9**, 23 (1962).

62-88   TIKHONOV, V. N., Complexometric determination of Al in Mg-alloys. *Zhur. analit. Khim.*, **17**, 422 (1962).

62-89   KUTEINIKOV, A. F., and BRODSKAY, V. M., Complexometric determination of rare earths in the presence of Al, Fe, Ca, Th and F. *Zavodskaya Lab.*, **22**, 792 (1962).

62-90   GIMESI, O., RADÝ, G., and ERDEY, L., Anwendung von Arsenazo Indicatoren in der Chelatometrie. *Period. Polytech.*, **6**, 15 (1962).

62-91   TERESHIN, G. S., and TANANAEV, I. V., Determination of EDTA and rare earths when present together. *Zhur. analit. Khim.*, **17**, 526 (1962).

62-92   BUSEV, A. I., IVANOV, V. M., and TIPTSOVA, V. G., Complexometric determination of Th after separation as iodate. *Zavodskaya Lab.*, **28**, 799 (1962).

62-93 FRITZ, J. S., and ABBINK, J. E., Cation exchange separation of V from other metal ions. *Analyt. Chem.*, **34**, 1080 (1962).

62-94 SAJÓ, I., Volumetrische Bestimmung von V(V) mit AeDTE. *Z. analyt. Chem.*, **188**, 168 (1962).

62-95 KAIMAL, V. R. M., and SHOME, S. C., Direct titration of V(V) with EDTA using *N*-benzoyl-phenylhydroxylamin as metal indicator. *Analyt. Chim. Acta*, **27**, 594 (1962).

62-96 DAVIES, D. T., and WHITE, J. C. D., Determination of Ca and Mg in milk and milk diffusate. *J. Dairy Res.*, **29**, 285 (1962).

62-97 DE SOUSA, A., EDTA titration of micro amounts of V. *Microchem. J. Symposium Ser.*, **2**, 821 (1962).

62-98 FABREGAS, R., PRIETO, A., and GARCIA, C., The EDTA titration of Zn in the presence of Cd. *Chem. Analyst*, **51**, 77 (1962).

62-99 JOHNSON, N. C., Electrometric titration of Cu and Zn in biological material. *Clin. Chem.*, **8**, 497 (1962).

62-100 LANGER, H. G., and BOGUCKI, R. F., The chelates formed by Sn(II) with EDTA. *J. Chem. Soc.*, 375 (1962). Criticism of Smith's paper, (61–39).

62-101 MOELLER, T., and THOMPSON, L. C., Observations on the rare earths. LXXV. The stabilities of diethylentriaminpentaacetic acid chelates. *J. Inorg. Nucl. Chem.*, **24**, 499 (1962).

62-102 PŘIBIL, R., and VESELÝ, V., Triethylentetramine hexaacetic acid as a new titrimetric reagent. *Talanta*, **9**, 939 (1962).

62-103 PECHANEC, V., Elementaranalyse von Hg enthaltenden Substanzen. IV. Komplexometrische Semimikrobestimmung von S. *Coll. Czech. Chem. Commun.*, **27**, 1817 (1962).

62-104 DRAGUSIN, I., Schnellmethode zur Bestimmung von BaSO₄ in Baryt. *Rev. Chim. (Bucharest)*, **13**, 103 (1962).

62-105 DE SOUSA, A., Chelometrische Bestimmung von Schwefelanionen. III. Analyse von Thiocyanat-Cyanid-Mischungen. *Inf. Quim. Anal.*, **16**, 126 (1962).

62-106 DE SOUSA, A., Chelometrische Bestimmung von Schwefelanionen. IV. Analyse von Sulfat-Sulfit-Mischungen. *Inf. Quim. Anal.*, **16**, 177 (1962).

62-107 ANTONOVA, A. A., and TERENTEVA, T. A., Rapid complexometric determination of sulphate in potash. *Steklo i Keram.*, **19**, 23 (1962).

62-108 UEDA, S., YAMAMOTO, Y., and MURATA, K., Chelatometric determination of arsenates by precipitation with Bi-Nitrate. *Nippon Kagaku Zasshi*, **83**, 1301 (1962).

62-109 CIOGOLEA, G., MORAIT, G., TEODORESCU, N., CREANGA, S., and NICOARA, N., Titrimetrische Bestimmung von Pb. *Farmacia (Bucharest)*, **10**, 457 (1962).

62-110 KUSAKINA, N. P., and YAKIMETS, E. M., Trilonometric analysis of lead vanadate. *Tr. Uralsk. Politekhn. Inst.*, 91 (1962).

62-111 NARUSKEVICIUS, L., and DAUKSAS, K., A rapid method of

404   *Bibliography*

Babbitt-analysis. *Lietuvos TSR Aukstuju Mokyklu Mikslo Darbai Chem. i Chem. Techn.*, **2**, 5 (1962).

62-112 ALEKSANDROVICH-MELNIKOVA, A. S., Trilonometric determination of chloride. *Izv. Vys. Uchebn. Zaved. Pishchevaya Techn.*, 141 (1962).

62-113 GUSEV, S. I., SOKOLOVA, E. V., and KOZHEVNIKOVA, I. A., Determination of chlorides by means of 2-(2-hydroxy-1-naphthyl-methyleneamino)-pyridine. *Zhur. analit. Khim.*, **17**, 499 (1962).

62-114 SAYUN, M. G., and TIKHANINA, S. P., Complexometric determination of In in concentrates. *Zavodskaya Lab.*, **28**, 544 (1962).

62-115 BUSEV, A. I., SKREBKOVA, L. M., and TALIPOVA, L. L., Four 7-(sulphonaphthylazo)-8-hydroxy-5-quinoline-sulphonic acid dyes as indicators for the direct complexometric titration of Ga. *Zhur. analit. Khim.*, **17**, 831 (1962).

62-116 KRAFT, G., Elektrometrische Endpunktsbestimmung in komplexometrischen Titrationen. *Z. analyt. Chem.*, **186**, 187 (1962).

62-117 ERDEY, L., GEGUS, E., and VÁNDORFFY, M. T., Analysis of natural waters. *Magyar Kém. Lapja*, **17**, 277 (1962).

62-118 RINGBOM, A., EKLUND, B., and BERGMAN, L., Photometric titrations of dilute solutions. *Proc. Sympos. Microchem. Techniques 1961*. Interscience New York, 797 (1962).

62-119 VISHNU and SRIVASTAVA, V. K., The Cu–PAN-complex for the spectrophotometric determination of Ca. *Current Sci. (India)*, **31**, 330 (1962).

62-120 KHALIFA, H., and KHATER, M. M., Back-titrations with Hg-nitrate in alkaline medium. Micro estimation of Sc and Pd and analysis of their mixtures with some other metals. *Z. analyt. Chem.*, **191**, 339 (1962).

62-121 PALAČEK, M., Volumetrische Bestimmung von Na₂O in Glas. *Skalř. & Keram.*, **12**, 3 (1962).

62-122 DOLEŽAL, J., NOVOZÁMSKÝ, I., and ZÝKA, J., Indirekte komplexometrische Bestimmung von Na. *Coll. Czech. Chem. Commun.*, **27**, 1830 (1962).

62-123 KLEIN, P., and SKRIVÁNEK, V., Complexometric determination of Ag in AgNO₃. *Chem. Prumsyl*, **12**, 359 (1962).

62-124 BUDĚŠINSKÝ, B., BEZDEKOVÁ, A., and VRZALOVÁ, D., Bestimmung von U. Modifikation der Kinnunen-Wennerstrand Methode. *Coll. Czech. Chem. Commun.*, **27**, 1528 (1962).

62-125 ENDO, Y., and TOMORI, T., Rapid determination of Mo in Fe–Mo and MoO₃ clinker by using EDTA, xylenol orange and Pb solutions. *Japan Analyst*, **11**, 1310 (1962).

62-126 YURIST, I. M., and SHAKHOVA, P. G., Complexometric determination of Cu, Co and Mn. *Zavodskaya Lab.*, **28**, 1183 (1962).

62-127 WOLFSBERG, K., Determination of rare earths in fission products by ion exchange at room temperature. *Analyt. Chem.*, **34**, 518 (1962).

62-128 HOSHINO, Y., Separation of Zr and Hf and their micro determination. VII. Separation of Hf from Zr by extraction with cyclohexanone. *Japan Analyst*, **11**, 1032 (1962).

62-129 HAHN, R. B., and WARDI, A. H., Determination of Zr–Hf ratio by EDTA titration. *Abstracts* 10. *Anachem Conference*, Detroit, Oct. 1962.

62-130 KONONENKO, L. I., and POLUETKOV, N. S., Complexometric determination of Hf in mixtures with Zr. *Zavodskaya Lab.*, 28, 794 (1962).

62-131 VEREPH, J., and FÜLÖP, L., Komplexometrische Bestimmung von Hexaphosphat. *Orv. Szemle*, 8, 180 (1962).

62-132 OMBOLY, V., and DERZSI, E., Determination of 8-hydroxyquinoline and its derivatives. *Acta Pharm. Hung.*, 32, 246 (1962).

62-133 RADOVIĆ, G., and BLAGOJEVIĆ, Z., Dosage de l'acide L-ascorbique par complexometrie. *Farmakeftikon Deltion*, 2, 187 (1962).

62-134 DICK, J., and RISTICI, J., Eine neue, spezifische Methode zur komplexometrischen Bestimmung von Antipyrin. *Acad. rep. pupul. Romine, Stud. cercet. stiint chim.*, 269 (1962).

62-135 VANDAEL, C., Une électrode indicatrice pour les titrages potentiométriques au moyen de EDTA. Spirale de Pt garnie d'une couche de PbO_2 ou d'un oxyde de Bismuth (Bi_2O_4?–Bi_2O_5?). *Ind. chim. belge*, 27, 932 (1962).

62-136 SZEKERES, L., and KELLNER, A., Bestimmung von Alkalikarbonat in Gegenwart von Hydroxid. *Pharm. Zentralhalle*, 101, 700 (1962).

62-137 SEN, A. B., and SRIVASTAVA, T. S., Use of nitrosochromotropic acid as a metal indicator. *Z. analyt. Chem.*, 187, 401 (1962).

62-138 KISS, T. A., and CANIĆ, V. D., Komplexometrische Bestimmung von Metallen. I. Bestimmung von Cu. *Glasn. Hem. Drustva, Beograd*, 27, 5 (1962).

62-139 HABERCETL, M., Komplexometrische Bestimmung von Al in Stahl. *Hutn. Listy*, 18, 138 (1962).

62-140 LANGER, H. G., and BOGUCKI, R. F., The chelates formed by Sn(II) with EDTA. *J. Chem. Soc.*, 375 (1962).

62-141 SMITH, T. D., Chelates formed by Sn(IV) with EDTA. *Nature*, 196, 1092 (1962).

62-142 ONUKI, S., and UKIGAYA, K., Rapid determination of Sn in Sn–Pb alloys. *Japan Analyst*, 11, 1313 (1962).

62-143 KONISHI, S., Analysis of Sn plating solutions using EDTA. *J. Metal Finish. Soc. Japan*, 13, 159 (1962).

62-144 WAKAMATSU, S., Determination of Sn in steel. *Japan Analyst*, 11, 1151 (1962).

62-145 VAN DER REYDER, A. J., and VAN LINGEN, R. L. M., Determination of trace amounts of Ca in caustic soda. Complexometric titration after separation on Dowex A-1 chelating resin. *Z. analyt. Chem.*, 187, 241 (1962).

62-146 VLADIMIROV, L. V., and SHULGINA, M. N., Trilonometric determination of Ca in presence of Mg and phosphate. *Zavodskaya Lab.*, 28, 548 (1962).

62-147 VISHNU and STRIVASTAVA, V. K., Cu–PAN complex for the spectrophotometric titration of Ca. *Current Sci. (India)*, 31, 330 (1962).

406 *Bibliography*

62-148 KONRAD, H., Komplexometrische Bestimmung von Ca in Nahrungsmitteln, besonders Milch und Milchprodukten. *Z. Lebensm.-Unters.*, **118**, 35 (1962).

62-149 BOZHEVOLNOV, E. A., and KREINGOLD, S. U., Fluorescent properties of fluorescein complexone. *Zhur. analit. Khim.*, **17**, 291 (1962).

62-150 BOZHEVOLNOV, E. A., and KREINGOLD, S. U., Fluorescence-complexometric determination of micro amounts of Ca. *Zhur. analit. Khim.*, **17**, 560 (1962).

62-151 KIRKBRIGHT, G. F., and STEPHEN, W. I., The screening of metal-fluorescent indicators. *Analyt. Chim. Acta*, **27**, 294 (1962).

62-152 KARANOV, R., KAROLEV, A., and TOSCHEWA, D., Komplexometrische Titration von Cu in Cu–Erzen und Konzentraten. *Talanta*, **9**, 409 (1962).

62-153 CHENG, K. L., EDTA as masking agent in selective spectrophotometric determination of Cu with TRIEN. *Analyt. Chem.*, **34**, 1392 (1962).

62-154 KIM, J. H., and RIM, H. K., Determination of germanium by indirect titration with EDTA. *Chosum Kwahakwon Tongbo*, **3**, 13 (1962).

62-155 KIRKBRIGHT, G. F., REES, D. I., and STEPHEN, W. I., 4,4'-Diaminostilbenetetraacetic acid as metallofluorescent indicator. *Analyt. Chim. Acta*, **27**, 558 (1962).

62-156 BAETSLÉ, L., and BENGSCH, E., Ion exchange characteristic of Ra–EDTA complexes. *J. Chromatog.*, **8**, 265 (1962).

62-157 COATES, E., and RIGG, B., Complex formation. Solochrome Violet R. IV. Metal-dye stability constants. *Trans. Faraday Soc.*, **58**, 2058 (1962).

62-158 KÉKEDY, L., and BALOGH, G., Complexometric determination of Bi and Pb with gallein as indicator. *Stud. Univ. Babes-Bolyai Ser. Chem.* **1**, 109 (1962).

62-159 BAGDASARYAN, A. M., ABRAMYAN, S. A., and CHILINGARYAN, M. N., Complexometric determination of Pb, Cu, Fe and Zn in polymetallic ores. *Nauchn. Trudy Nauchn. Issled. Gorn.-Met. Inst. Sov. Nar. Khoz. Arm.*, **3**, 353 (1962).

1963

63-1 SAJÓ, I., Indikatorsysteme auf VY-Basis. *Talanta*, **10**, 493 (1963).

63-2 BIZON, P., Komplexometrische Bestimmung von Sulfat in Gips. *Stavino*, **41**, 223 (1963).

63-3 CHENG, K. L., Determination of metal content in phthalocyanine chelates. *Microchem. J.*, **7**, 23 (1963).

63-4 STAROSTIN, V. V., SPITSYN, V. I., and SILINA, G. F., Complex compounds of Be with EDTA. *Zhur. neorg. Khim.*, **8**, 660 (1963).

63-5 RYABCHIKOV, D. I., YAO, K. M., and ZARINSKII, V. A., Complex formation of In with some complexones. *Zhur. neorg. Khim.*, **8**, 388 (1963); cf. also *ibid.*, 641.

63-6 BROWN, W. B., ROGERS, D. R., MERSHAD, E. A., and AMOS, W. R., The determination of plutonium by EDTA titration. *Analyt. Chem.*, **35**, 1000 (1963).

63-7   BLOOD, F. H., and NEBERGALL, W. H., EDTA titration of Pb in the presence of orthophosphate and alkaline earths. *Analyt. Chem.*, **35**, 1089 (1963).

63-8   VOLODARSKAYA, R. S., and DEREVYANKO, G. N., Complexometric determination of Zr and Th with Xylenol Orange. *Zavodskaya Lab.*, **29**, 28 (1963).

63-9   FLASCHKA, H., and SPEIGHTS, R., EGTA in the amperometric titration of Cd in the presence of Zn. *Analyt. Chim. Acta*, **28**, 433 (1963).

63-10 FLASCHKA, H., and GARRETT, J., The determination of some organic compounds by oxidation with permanganate and subsequent EDTA titration. *Chem. Analyst*, **52**, 103 (1963).

63-11 FLASCHKA, H., and CARLEY, F. B., Photometric titrations. VII. The consecutive titration of Cd and Zn. *Talanta*, **11**, 423 (1963).

63-12 KIRBY, R. E., and FREISER, H., Polarography of Nb–EDTA complexes. *Analyt. Chem.*, **35**, 122 (1963).

63-13 GOLDSTEIN, G., MANNING, D. L., and ZITTEL, H. E., Vanadyl ion as back-titrant for indirect amperometric titrations with EDTA. Application to the determination of Al, Zr and Th in fluoride-bearing material. *Analyt. Chem.*, **35**, 17 (1963).

63-14 YAMAGUCHI, K., and UENO, K., Chelatometric titrations. I. Masking by acetylacetone. *Japan Analyst*, **12**, 55 (1963).

63-15 SHEKA, Z. A., and SINYAVSKAY, E. I., Complexometric determination of rare earths in the presence of dialkylphosphoric acid. *Zhur. analit. Khim.*, **18**, 460 (1963).

63-16 ROGERS, J. R., and BROWN, W. B., Indirect determination of Th with EDTA. *Analyt. Chem.*, **35**, 1261 (1963).

63-17 PŘIBIL, R., and VESELÝ, V., Contributions to the basic problems of complexometry. XI. Determination of Ti, Fe and Al in the presence of each other. *Talanta*, **10**, 383 (1963).

63-18 DATTA, S. K., ß-SNADNS-6 as chelometric indicator for Th. *Z. analyt. Chem.*, **195**, 22 (1963).

63-19 BIECHLER, D. G., Analysis of In–Zn–Sn alloys. *Chem. Analyst*, **52**, 48 (1963).

63-20 ASHTON, A. A., Direct EDTA titration of Zn with murexide. *Chem. Analyst*, **52**, 49 (1963).

63-21 ASHTON, A. A., Titrimetric micro determination of Zn and Cu with EDTA. *Analyt. Chim. Acta*, **28**, 296 (1963).

63-22 BERKA, A., Indirekte komplexometrische Bestimmung einiger mehrwertiger Alkohole und α-Hydroxysäuren. *Z. analyt. Chem.*, **195**, 263 (1963).

63-23 CHAPMAN, D., LLOYD, D. R., and PRINCE, R. H., An IR and NMR study of the nature of EDTA and some related substances in solution: hydrogen bonding in α-amino-polycarboxylic acid systems. *J. Chem. Soc.*, 3645 (1963).

63-24 KUDRUNOVSK-PAVLIKOVA, E., KVAPIL, M., and WEISS, D., Komplexchemische Analyse von Tetrahaedrit. *Rudy*, **11**, 9 (1963).

63-25 NAKAMOTO, K., MORIMOTO, Y., and MARTELL, A. E., Infrared spectra of aqueous solutions. III. EDTA, HDTA and DTPA. *J. Amer. Chem. Soc.*, **85**, 309 (1963).

63-26 SAWYER, D. T., and TACKETT, J. E., Properties and IR-spectra of EDTA complexes. IV. Structure of the ligand in solution. *J. Amer. Chem. Soc.*, **85**, 314 (1963).

63-27 PŘIBIL, R., and VESELÝ, V., Contributions to the basic problems of complexometry. XII. Successive determination of Th, rare earths and some other elements. *Talanta*, **10**, 899 (1963).

63-28 SCHWARZENBACH, G., and SCHWARZENBACH, K., Die Stabilität der Fe(III)-Komplexe einfacher Hydroxamsäuren und des Ferrioxamins B. *Helv. Chim. Acta*, **46**, 1390 (1963).

63-29 WILLS, V., STAMBAUGH, O. F., and PROCTOR, Z. G., *N,N'*-bis(2-hydroxyethyl)-dithiooxamid in volumetric determination of Ni and Ni–Cu mixtures. *Analyt. Chem.*, **35**, 257 (1963).

63-30 DE SOUSA, A., Chelometrische Bestimmung von Schwefelanionen. V. Analyse von Sulfit-Sulfid-Gemischen. *Inf. Quim. Anal.*, **17**, 51 (1963).

63-31 DE SOUSA, A., Chelometrische Bestimmung von Schwefelanionen. VI. Analyse von Thiosulfat-Sulfid-Gemischen. *Inf. Quim. Anal.*, **17**, (1963).

63-32 POZNANSKI, S., Kolorimetrische Bestimmung von H_2S in Käse. *Roczniki Technol. Chem. Zywnosci*, **9**, 17 (1963).

63-33 TURINA, S., and KRAJOVAN-MARJANOVIC, V., Komplexometrische Bestimmung von Schwefel in Fe und niedrig legierten Stählen. *Z. analyt. Chem.*, **196**, 32 (1963).

63-34 VETTER, G., Mikrobestimmung von S in flüssigen organischen Verbindungen mit bis zu 70% S Gehalten. *Chem. Techn. (Berlin)*, **15**, 43 (1963).

63-35 BERKA, A., Indirekte komplexometrische Bestimmung einiger Hydrazinderivate. *Z. analyt. Chem.*, **193**, 276 (1963).

63-36 HENSLEY, A. L., and BERGMANN, J. G., Rapid and accurate determination of metals in motor oils and additives with DTPA. *Analyt. Chem.*, **35**, 1318 (1963).

63-37 LASSNER, E., and PÜSCHEL, R., Über Peroxo-Mischkomplexe von Ti, Nb und Ta. II. Peroxo-Mischkomplexe von Nb und Ta mit Methylthymolblau und PAR und photometrische Untersuchung zum Nachweis der Bildung von Nb- und Ta-Chelonaten. *Mikrochim. Acta*, 950 (1963).

63-38 LASSNER, E., Über Peroxo-Mischkomplexe von Ti, Nb und Ta. III. Die chelometrische Titration von fünfwertigem Nb. *Talanta*, **10**, 1229 (1963).

63-39 CHENG, K. L., and GOYDISH, B. L., Analysis of mixtures of Sb and Bi tellurides containing Se and I. *Analyt. Chem.*, **35** 1965 (1963).

63-40 BLOOD, F. H., and NEBERGALL, W. H., EDTA titration of Pb in the presence of orthophosphate and alkaline earths. *Analyt. Chem.*, **35**, 1089 (1963).

63-41 GUSEV, S. I., and KOZHEVNIKOVA, I. A., Indirect titrimetric determination of bromides. *Zhur. analit. Khim.*, **18**, 366 (1963).

63-42 YAMADA, T., Rapid analysis of clay. *Japan Analyst*, **12**, 183 (1963).

63-43 KONKIN, V. D., and ZHIKHAREVA, V. I., Complexometric determination of Mo in Permalloy and ferromolybdenum. *Zavodskaya Lab.*, **29**, 791 (1963).

63-44 KEATTCH, C. J., The purification of Calcein. *Talanta*, **10**, 1303 (1963).

63-45 MILES, T. D., HOFFMAN, F. A., and DELASANTA, A. C., Direct photometric EDTA-titration of Cu-8-quinolinate. *Amer. Dyestuff Reporter*, **22** (1963).

63-46 LACY, J., Semiautomatic determination of Ca and Mg hardness in water. *Talanta*, **10**, 1031 (1963).

63-47 SCAIFE, D. B., Precise micro determination of Zn and Cd by photometric titration with EDTA. *Analyst*, **88**, 618 (1963).

63-48 ROZHDESTVENSKAYA, Z. B., SONGINA, O. A., and BARIKOV, V. G., Amperometric determination of U with graphite electrode. *Zavodskaya Lab.*, **29**, 30 (1963).

63-49 LYLE, S. J., and RAHMAN, M. M., Complexometric titration of yttrium and the lanthanons. I. A comparison of direct methods. *Talanta*, **10**, 1177 (1963).

63-50 LYLE, S. J., and RAHMAN, M. M., Complexometric titration of yttrium and the lanthanons. II. Methods for their determination in oxalates. *Talanta*, **10**, 1183 (1963).

63-51 KYŘŠ, M., and CALETKA, R., The stability constants of the complex of Zr with EDTA. *Talanta*, **10**, 1115 (1963).

63-52 SKRIVANEK, V., and KLEIN, P., Komplexe Analyse von Zirkonsanden. *Rudy (Prace)*, **11**, 89 (1963).

63-53 BUSS, H., KOHLSCHÜTTER, H. W., and PREISS, M., Komplexometrische Phosphatbestimmung. I. Das Verfahren und seine Leistungsfähigkeit. *Z. analyt. Chem.*, **193**, 264 (1963).

63-54 BUSS, H., KOHLSCHÜTTER, H. W., and PREISS, M., Komplexometrische Phosphatbestimmung. II. Anwendung des Verfahrens. *Z. analyt. Chem.*, **193**, 326 (1963).

63-55 FURUYA, M., and TAJIRI, M., Rapid determination of P in P–Cu metal and P–Cu brazing filler alloy. *Japan Analyst*, **12**, 288 (1963).

63-56 SOLODOVNIKOV, P. P., Violuric acid as an indicator for the complexometric titration of Cu. *Zhur. analit. Khim.*, **18**, 1026 (1963).

63-57 BOLTON, S., Chelatometric determination of Fe(II) with 2-pyridinaldoxime as an indicator. *J. Pharm. Sci.*, **52**, 858 (1963).

63-58 PARUSHEV, M., Komplexometrische Bestimmung von Captax. *Khim. Ind. (Sofia)*, **35**, 49 (1963).

63-59 TAMURA, Z., and MIYAZAKI, M., Metal complexes of glucosamine and its derivatives. II. Complexometric micro determination of D-glucosamine–HCl and N-methyl-D-glucosamine–HCl with Fehlings reagent. *Japan Analyst*, **12**, 561 (1963).

63-60 WENG, Y. K., and LAN, C. T., Amperometric titration of piperazine salts in their preparations. *Yao Hsueh Hsueh Pao*, **10**, 303 (1963).

63-61 PŘIBIL, R., and VESELÝ, V., Separation of Ti from Fe and Al. *Talanta*, **10**, 233 (1963).

410   Bibliography

63-62   PŘIBIL, R., and VESELÝ, V., Rapid analysis. XII. Determination of Ti in Ti–Al-alloys. *Chem. Analyst*, **52**, 43 (1963).

63-63   ELINSON, S. V., and POBEDINA, L. I., Komplexometrische Bestimmung von Ti in Legierungen. *Zavodskaya Lab.*, **29**, 139 (1963).

63-64   RINGBOM, A., *Complexation in Analytical Chemistry*. Interscience New York (1963).

63-65   THOMAS, R. S., Potentiometric determination of Ca in mixed feeds using EGTA. *Chem. Analyst*, **52**, 6 (1963).

63-66   PRIESTLEY, P. T., An automatic digital thermometric titrator. *Analyst*, **88**, 194 (1963).

63-67   PRIESTLEY, P. T., SEBBORN, W. S., and SELMAN, R. F., Automatic thermometric titrations. *Analyst*, **88**, 797 (1963).

63-68   DE SOUSA, A., Bestimmung der Anionen des Schwefels. VII. Bestimmung von Thiocyanat in Gegenwart von Halogenen. *Inf. Quim. Anal.*, **17** (1963).

63-69   DE SOUSA, A., Modifikation der Methode von Beck zur chelometrischen Bestimmung von Cl⁻, Br⁻ und J⁻ bei gleichzeitiger Anwesenheit. *Inf. Quim. Anal.*, **17** (1963).

63-70   BUSEV, A. I., IVANOV, V. M., and TALIPOVA, L. L., Complexometric determination of Cu with 7-(2-pyridylazo)-8-hydroxyquinoline. *Zhur. analit. Khim.*, **18**, 33 (1963).

63-71   ROSKAM, R. T., and DE LANGEN, D., Complexometric determination of dissolved oxygen in water. *Analyt. Chim. Acta*, **28**, 78 (1963).

63-72   FURUYA, M., and TAJIRI, M., Rapid analysis of phosphor bronzes by chelometric titration. I. The chelometric titration of tin. *Japan Analyst*, **12**, 59 (1963).

63-73   RICHARDSON, M. L., Determination of small amounts of Ca in the presence of large amounts of alkali metal salts by amperometric chelatometry. *Talanta*, **10**, 103 (1963).

63-74   TOFT, R. J., EASTMAN, H. W., and MCKENNY, D. B., An optical aid to end-point detection in Ca–EDTA titrations. *Chem. Ind.*, 1030 (1963).

63-75   YAMAGUCHI, K., and UENO, K., Use of masking agents in chelometric titrations. II. β-Aminoethymercaptane. *Talanta*, **10**, 1041 (1963).

63-76   YAMAGUCHI, K., and UENO, K., Use of masking agents in chelometric titrations. III. β-Mercaptopropionic acid. *Talanta*, **10**, 1195 (1963).

63-77   GALAKTIONOV, Y. G., LIDIN, R. A., and ASTAKHOV, K. V., Polarographic study of complex formation between Eu and EDTA. *Zhur. Fiz. Khim.*, **37**, 829 (1963).

63-78   JACOBSON, E., and KALLAND, G., Polarographic determination of Tl. *Analyt. Chim. Acta*, **29**, 215 (1963).

63-79   STRAFELDA, F., Potentiometrische Titrationen mit Komplexon III an einer Ag-Elektrode. *Coll. Czech. Chem. Commun.*, **28**, 3345 (1963).

63-80   ANDEREGG, G., Komplexone. XXXIII. Reaktionsenthalpie und -entropie bei der Bildung der Metallkomplexe von Äthylendiamin-

und Diaminocyclohexan-tetraessigsäure. *Helv. Chim. Acta,* **46,** 1833 (1963).

63-81 FLASCHKA, H., and BUTCHER, J., The EDTA titration of Zn in the presence of Cd. *Microchem. J.,* **7,** 407 (1963).

63-82 SAMUELSON, O., *Ion Exchange Separations in Analytical Chemistry.* John Wiley and Sons Inc., New York (1963).

63-83 KUAN, T.-Y., Indirect method of potassium determination using a chelating agent. *T'u Jang T'ung Pao,* **5,** 46 (1963).

63-84 TSIRUL, V. A., Complexometric determination of sugar products (for Ca). *Tr. Tsentr. Nauch.-Issled. Sakharn. Prom.,* 154 (1963).

63-85 BREKKE, J. E., TAYLOR, D. H., and STANLEY, W. L., Determination of Ca in cherry brines by versenate titration. Elimination of anthocyanin interference by means of carbonyl reagents. *J. Agric. Food Chem.,* **11,** 260 (1963).

63-86 ALFAROVA, V. N., and SERIKOVA, L. I., The determination of Ca and Sr in Ba(OH)$_2$. *Khim. Prom. Nauk. Tekhn. Zborn.,* 68 (1963).

63-87 DATE, Y., and TOEI, K., Selective titration of Ca in the presence of Mg. *Bull. Chem. Soc. Japan,* **36,** 518 (1963).

63-88 COPP, D. H., Simple and precise micro method for the EDTA titration of Ca. *J. Lab. Clin. Med.,* **60,** 1029 (1963).

63-89 DUNSBACH, F., Bestimmung von Ca in Serum und Urin mit DCyTA. *Clin. Chim. Acta,* **8,** 481 (1963).

63-90 SPACK, C., and BRAY, E., Complexometric determination of Ca with a substituted phenylazonaphthalene dye. Application to the measurement of Ca in human urine and faeces. *Bull. Soc. Chim. Biol.,* **45,** 1191 (1963).

63-91 PŘIBIL, R., and VESELÝ, V., Contributions to the basic problems of complexometry. XIII. Determination of Al and Cr(III) in the presence of chromate. *Talanta,* **10,** 1287 (1963).

63-92 TSUBOTA, H., Ion exchange method for the determination of Cs-137 and Sr-90 in fall-out. *Bull. Chem. Soc. Japan,* **36,** 1545 (1963).

63-93 WÜNSCHER, H. J., and HOFFMANN, H., Verbesserte Bestimmung der Resthärte. *Chem. Techn. (Berlin),* **15,** 492 (1963).

63-94 KUSAKINA ,N. P., Redox properties of Acid Chrome Dark Blue, Erio T and murexide. *Tr. Uralsk Politekhn. Inst.,* **130,** 42 (1963).

63-95 BUDĚŠINSKÝ, B., and HAAS, K., Spectrophotometric study of the reaction of Metalochrome Violet A with protons and various metal ions. *Acta Chim. Acad. Sci. Hung.,* **39,** 7 (1963).

63-96 VLADIMIRTSEV, I. F., Complexometric metallochromic indicators. *Tr. Uralsk Politekhn. Inst.,* **130,** 5, 15 (1963).

63-97 VLADIMIRTSEV, I. F., and STARTSEVA, Z. P., Complexometric metallochromic indicators. *Tr. Uralsk Politekhn. Inst.,* **130,** 35 (1963).

63-98 JENSEN, B. S., Complex compounds in analysis. *Dansk. Kemi.,* **44,** 35 (1963).

63-99 ZHIVOPITSEV, V. F., KALMYKOVA, I. S., and PYATOSIN, L. P., 4-(2-Hydroxyphenylazo)-1-(*p*-sulphophenyl)-3-methyl-5-hydroxypyrazole as a complexometric indicator. *Uch. Zap. Permsk. Gos. Univ.*, **25**, 108 (1963).

63-100 CANIĆ, V. D., and KISS, T. A., Complexometric determination of metals. IV. Determination of Al. *Bull. Soc. Chim. Beograde*, **28**, 143 (1963).

63-101 CANIĆ, V. D., and KISS, T. A., Complexometric determination of metals. III. Addition of ethanol in the EDTA titration of Ni. *Chem. Analyst*, **52**, 111 (1963).

63-102 KÉKEDY, L., and BALOGH, G., Gallein, a new indicator for the complexometric titration of Th. *Stud. Univ. Babes-Bolyai Ser. Chem.* No. 1, 205 (1963).

63-103 SHAFRAN, I. G., PARTASHNIKOVA, M. Z., MAKAROVA, N. I., SOLOV'EV, E. A., and ZELICHENOK, S. L., Analytical applications of Calcion for the complexometric and photocolorimetric determination of Ca. *Tr. Vses. Nauchn. Issled. Inst. Khim. Reaktivov*, **25**, 203 (1963).

63-104 KREINGOLD, S. U., and BOZHEVOLNOV, E. A., Analytical properties of Fluorexon. *Tr. Vses. Nauchn. Issled. Inst. Khim. Reaktivov*, **25**, 358 (1963).

63-105 SPRINGER, V., MAJER, J., and KARLICEK, R., Use of cinnamohydroxamic acid as a new complexometric indicator for Fe(III) in the control of drugs. *Cesk. Farm.*, **12**, 4 (1963).

63-106 GROMOVA, M. T., EFIMOV, I. P., and PESHKOVA, V. M., Spectrophotometric titrations of the rare earth elements. *Rekozm. Elementy Akad. Nauk. SSSR Inst. Geokhim. i Analit. Khim.*, 318 (1963).

63-107 BASHKIRTSEVA, A. A., Trilonometric determination of Fe and Ti in titanomagnetite ores. *Tr. Uralsk Polyt. Inst. No.* 130, 84 (1963).

63-108 NOVÁK, J., and AREND, H., Complexometric determination of Ti in Ba titanate in the presence of small amounts of Fe. *Silikáty*, **7**, 150 (1963).

63-110 DYATLOVA, N. M., and KOLESNIK, E. S., Complexometric determination of Th, Ga and Fe with a H.F. titrimeter. *Vestn. Tekh. i Ekon. Inform. Nauch.-Issled. Inst. Tekh. Ekan. Issled. Gos. Khim. i Neft. Prom. pri Gosplane SSSR*, 42 (1963).

63-111 TSERKOVNITSKAYA, I. A., and YEH, J.-C., Amperometric titration of Th with two Pt electrodes. Determination of Th in uraninite. *Vestn. Leningr. Univ. Ser. Fis. i Khim.*, 168 (1963).

63-112 PAVLOV, N. N., and KUZNETSOVA, A. R., Complexometry of mixtures of Cr(III) and Al. *Nauch. Trudy Mosk. Tekhnol. Inst. Legkoi Prom.*, 61 (1963).

63-113 NEMIROVSKAY, A. F., KEREMEDZHIDI, L. N., and YASHINA, N. I., Determination of W and Mo when present together. *Trudy Novocherk. Polit. Inst.*, **55**, 143 (1963).

63-114 ZHDANOV, A. K., and UMURZAKOV, I. A., Amperometric titration of Mo(VI) with Complexone III at a Ta rotating electrode. *Nekototye. Vopr. Khim. Tekhnol. i Fiz.-Khim. Analyza neorg. Sistem. Akad. Nauk Uz. SSR Otd. Khim. Nauk.*, 163 (1963).

63-115 KONKIN, V. D., and ZHIKAREVA, V. I., Complexometric determination of Mo in Permalloy and Fe–Mo. *Zavodskaya Lab.*, **29**, 791 (1963).

63-116 LITEANU, C., CRISAN, I., and GHEORGHE, F., Complexometric determination of Mo by back-titration of excess complexone with FeCl₃. *Stud. Univ. Babes-Bolyai, Ser. Chem.*, **8**, 107 (1963).

63-117 ASENI-MORA, G., Technical application of chelates. I. Determination of Mn in normal steels and ferromanganese. *Anales real Soc. espan. Fis. Quim.*, **B59**, 695 (1963).

63-118 ZHDANOV, A. K., and UMAROVA, M. M., Anodic amperometric titration of Fe with Complexone III at a rotating Ta electrode. *Nekatorye Vorp. Khim. Tekhnol. i Fiz.-Khim. Analiza neorg. System Akad. Nauk Uz. SSR Khim. Nauk*, 115 (1963).

63-119 RENGER, F., and JENIK, J., Volumetric micro-determination of Fe in ferrocene and its derivatives. *Sb. Ved. Prac. Vysoke Skoly, Chem. Techn. Pardubice*, **1**, 55 (1963).

63-120 KALINICHENKO, I. I., STYUNKEL, T. B., MIKHALEVA, Z. A., and MEKHANOSHINA, E. Y., Complexometric determination from one weighing of Zn and Ni in alloys of German silver type. *Trudy Uralsk. Polit. Inst.*, **130**, 54 (1963).

63-121 BRUNO, E., and CIURLO, R., Complexometric determination of Cu, Hg and Pb simultaneously. *Rass. Chim.*, **15**, 265 (1963).

63-122 FURUYA, M., and TAJIRI, M., Rapid chelometric analysis of phosphor bronzes. II. Rapid titration of Cu with Xylenol Orange as indicator. *Japan Analyst*, **12**, 389 (1963).

63-123 KRUYSSE, L., Volumetric determination of Zn in pigment mixture. *Verfkroniek*, **36**, 374 (1963).

63-124 KLEWSKA, A., and DOMINIK-SRYCHARSKA, M., Determination of small amounts of Zn and Cd in biological material. *Acta Polon. Pharm.*, **20**, 433 (1963).

63-125 RYABCHIKOV, D. I., MAROV, I. N., and YAO, K.-M., Studies of In complex formation with some complexones by ion exchange. *Zhur. neorg. Khim.*, **8**, 641 (1963).

63-126 LITEANU, C., CRISAN, I., and TRUTA, L., Complexometric determination of anions. I. Determination of phosphate by titration of excess Fe after precipitation as FePO₄. *Stud. Univ. Babes-Bolyai Ser. Chem.*, **8**, 31 (1963).

63-127 LITEANU, C., CRISAN, I., and TRUTA, L., Complexometric determination of anions. II. Determination of phosphate by titration of the Fe(III) in FePO₄. *Stud. Univ. Babes-Bolyai Ser. Chem.*, **8**, 39 (1963).

63-128 LITEANU, C., CRISAN, I., and TRUTA, L., Complexometric determination of anions. III. Determination of phosphate by precipitation of AlPO₄. *Stud. Univ. Babes-Bolyai Ser. Chem.*, **8**, 45 (1963).

63-129 POPOV, M. A., and BERSHADSKAYA, N. M., Complexometric determination of Bi in W-containing products. *Uch. Zap. Tsentr. Nauchn.-Issled. Inst. Olovyan. Prom.*, 47 (1963).

63-130 BARANENKO, S. E., and KRIVOSHEEVA, V. I., Trilonometric determination of H_2S in natural, accessory and other gases *Vopr. Razwitiya Gaz. Prom. Ukr. SSR Kiev Sb.*, 300 (1963).

63-131 LOUNAMAA, N., and FIGMAN, W., Bestimmung von S in Stahl. *Z. analyt. Chem.*, **199**, 98 (1963).

1964

64-1 CYGANSKI, A. D., Complexometric determination of rubidium. *Chem. Analit. (Warsaw)*, **9**, 749 (1964).

64-2 TIKHONOV, V. N., GRANKINA, M. Y., and KOROLEVA, V. I., Complexometric determination of Ca and Mg in products of the Ti industry. *Zhur. Analit. Khim.*, **19**, 59 (1964).

64-3 GOLDSTEIN, M., Rapid methods for the analysis of plating baths. *Chim. analyt. (Paris)*, **46**, 125 (1964).

64-4 PARRY, E. P., and DOLLMAN, G. W., Determination of traces of alkaline earths by spectrophotometric titration in the UV. *Analyt. Chem.*, **36**, 1783 (1964).

64-5 WHARTON, H. W., and CHAPMAN, L. R., Application of spectrophotometric titration to micromolecular solutions of Ca, Mg, oxalate and sulphate. *Analyt. Chem.*, **36**, 1679 (1964).

64-6 MONNIER, D., and ROUÉCHE, A., Méthode de dosage ampérométrique rapide de submicro-, micro- et macro quantités de Ca et de Mg en presence l'un de l'autre. *Helv. Chim. Acta*, **47**, 103 (1964).

64-7 ROUÉCHE, A., and MONNIER, D., Dosage ampérométrique rapide et directe de Ca et de Mg dans l'eau, les cements et le soil. *Analyt. Chim. Acta*, **31**, 429 (1964).

64-8 MONNIER, D., and ROUÉCHE, A., Dosage ampérométrique de Ca et de Mg dans le serum et l'urine. *Helv. Chim. Acta*, **47**, 869 (1964).

64-9 MAREVA, S., and BUDEVSKY, O., A complexometric method of determining Ca and Ba. *Mitt. Inst. allgem. anorg. Chem. Bulg. Akad. Wiss.*, **2**, 39 (1964).

64-10 DUNSBACH, F., Komplexometrische Bestimmung des Ca in Blutserum und Urin mit DGITA. *Ärztl. Labor.*, **10**, 153 (1964).

64-11 KOWALSKA, E., and SOLLORZ, J., Determination of calcium ortho-phosphoric acid solutions by anion exchange and complexometry. *Chem. Analit. (Warsaw)*, **9**, 349 (1964).

64-12 SZEKERES, L., Rapid titrimetric determination of Ca, sulphate and phosphate in crude phosphate ores and in superphosphate fertilizers. *Magyar Kem. Lapja*, **19**, 321 (1964).

64-13 BRAUN, T., and TÖLGYESSY, J., Application of precipitate membranes in analytical chemistry. *Talanta*, **11**, 1543 (1964).

64-14 BRAUN, T., and TÖLGYESSY, J., Radiometric titrations. A review. *Talanta*, **11**, 1277 (1964).

64-15 TÖLGYESSY, J., JESANAK, V., and BRAUN, T., Contributions to the theory and methods of radiometric titrations. *Radiochem. Methods Anal. Proc. Symp., Salzburg, Austria*, **2**, 199 (1964). [Published 1965.]

64-16 TIKHONOV, V. N., and MUSTAFIN, I. S., Complexometric determination of Ca and Mg without separation of Fe. *Zavodskaya Lab.*, **30**, 1448 (1964).

64-17 PRUMBAUM, R., and ORTHS, K., Beitrag zur Ausführung und Bewertung von Schlackenanalysen. *Giesserei*, **51**, 73 (1964).

64-18 SMART, A., Volumetric determination of Mg in Al-alloys. *Metallurgia*, **69**, 245 (1964).

64-19 HESSLER, W., Automatische Bestimmung von Mg und Ca in Kalilauge und Salzen. *Bergakademie*, **16**, 685 (1964).

64-20 PŘIBIL, R., and VESELÝ, V., Contributions to the basic problems of complexometry. XIV. Determinations of Zr, Th and Ti in the presence of each other. *Talanta*, **11**, 1197 (1964).

64-21 PŘIBIL, R., and VESELÝ, V., Contributions to the basic problems of complexometry. XV. Determination of In and Ga in the presence of each other without the use of a screening agent. *Talanta*, **11**, 1319 (1964).

64-22 PŘIBIL, R., and VESELÝ, V., Contributions to the basic problems of complexometry. XVI. Determination of Th and Sc in the presence of each other. *Talanta*, **11**, 1545 (1964).

64-23 PŘIBIL, R., and VESELÝ, V., Contributions to the basic problems of complexometry. XVII. Determination of Zn and Cd in the presence of each other. *Talanta*, **11**, 1613 (1964).

64-24 COMBS, H. F., and GROVE, E. L., EDTA titration of Ba in the presence of other alkaline earth elements. *Analyt. Chem.*, **36**, 400 (1964).

64-25 SZEKERES, L., KARDOS, E., and SZEKERES, G. L., Resolution of systems containing Ba and certain dipositive ions via EDTA titrations. *Chem. Analyst*, **53**, 72 (1964).

64-26 REISHAKRIT, L. S., PUSTOSHKINA, M. P., and TIKHONOVA, Z. I., Amperometric titration of cations with Trilon B with a rotating Pt electrode. *Vestn. Leningr. Univ. Ser. Fis. i Khim.*, **19**, 122 (1964).

64-27 FILIPOV, D., and KIRTSCHEVA, N., Complexometric determination of Al and V in the presence of each other. *Compt. rend. Acad. bulg. Sci.*, **17**, 467 (1964).

64-28 KÉKEDY, L., and BALOGH, G., Komplexometrische Bestimmung unter Verwendung von Gallein als metallochromer Indikator. *Stud. Univ. Babes-Bolyai*, 101 (1964).

64-29 MCGLYNN, J. A., Determination of Ca in the presence of Mg. *N.S.W. Department of Mines, Chem. Lab. Report No.* **10**, 17 (1962–1963–1964).

64-30 PETRENKO, A. G., KOROTKIKH, A. E., KUTZNEZOVA, L. A., and KUPERSHTEIN, M. A., Complexometric determination of Ga. *Khim. Analiz. Tsvetn. i Redikh Metal. Akad. Nauk. SSSR Sibirsk. Otd. Khim.-Met. Inst.*, 32 (1964).

64-31 NAKAGAWA, G., and TANAKA, M., The theory of the indicator transition in chelatometry. *Bull. Chem. Soc. Japan*, **37**, 27 (1964).

64-32 SPARE, P. D., Stable murexide reagent for the determination of Ca in microquantities of serum. *Clin. Chem.*, **10**, 726 (1964).

64-33 GÖRNER, F., and TÜRÄKOVA, J., Complexometric determination of Ca in milk. *Prumsyl Potravin*, **15**, 424 (1964).

416 *Bibliography*

64-34 DITZ, J., Systematic classification of complexometric indicators. *Chem. Listy*, **58**, 946 (1964).

64-35 CHERKESOV, A. I., and ALYKOV, N. M., Selection of complexometric indicators from a series of azo derivatives of chromotropic acid. *Zhur. analit. Khim.*, **19**, 943 (1964).

64-36 WADA, H., and NAKAGAWA, G., PAR and PAN as metal indicators for Cu. *Nippon Kagaku Zasshi*, **85**, 549 (1964).

64-37 BUDĚŠINSKÝ, B., Complexes of metallochromic substances. I. Some basic problems. *Z. analyt. Chem.*, **206**, 262 (1964).

64-38 TANAKA, H., and YAMAUCHI, O., Imidazolylazo compounds as metallochromic indicators. *Chem. Pharm. Bull.*, **12**, 1268 (1964).

64-39 DATTA, S. K., and SAHA, S. N., Metal complexes of mononitrosochromotropic acid. Cu complex and its complexometric titration. *Bull. Chem. Soc. Japan*, **37**, 1418 (1964).

64-40 KUTEINKOV, A. F., and LYSENKO, S. A., Composition of complex compounds by a method of masking agents. *Zhur. analit. Khim.*, **19**, 1289 (1964).

64-41 TALIPOV, SH.T., and NIGAI, K. G., Complexometric titration of Bi with 4-(2-N-methylanabasineazo)-resorcinol. *Zhur. analit. Khim.*, **19**, 851 (1964).

64-42 TALIPOV, SH.T., and NIGAI, K. G., Complexometric titration of Tl(III) with 4-(2-N-methylanabasineazo)-resorcinol. *Zhur. analit. Khim.*, **19**, 697 (1964).

64-43 TALIPOV, SH.T., and ABDULLAEVA, KH.S., Complexometric titration of In in the presence of 4-(N-methyl-2-anabasineazo)-resorcinol. *Usbeksk. Khim. Zhur.*, **6**, 18 (1964).

64-44 SANGAL, S. P., Quinizarinsulphonic acid as a chelatometric indicator for Th. *Chim. analyt. (Paris)*, **46**, 138 (1964).

64-45 NAKAGAWA, G., and WADA, H., Some o-(2-Thiazolylazo)-phenol derivatives as metal indicators for Ni. *Nippon Kagaku Zasshi*, **85**, 202 (1964).

64-46 MOMIKI, K., and SEKINO, J., Reaction mechanism of EDTA titration of Fe(III) in acetate buffered solution of pH 5 using Cu–PAN indicator. *Japan Analyst*, **13**, 213 (1964).

64-47 KATIYAR, S. S., and TANDON, S. N., 1-Isonicotinoyl-2-salicylidenehydrazine as a metallochromic indicator for Fe. *J. Indian Chem. Soc.*, **41**, 219 (1964).

64-48 KUROKAWA, K., EDTA titration of Ca with NN indicator in NaOH–KCN buffer. *Japan Analyst*, **13**, 464 (1964).

64-49 UHLIG, E., and BERNDT, H., 2,5-Bis-(β-hydroxyethylamino)-terephthalsäure als Indikator in Mercurometrie und Chelatometrie. *Z. analyt. Chem.*, **203**, 241 (1964).

64-51 DATTA, S. K., and SAHA, S. N., Chelometric determination of Cu, Zn and Sn in mixtures, alloys and minerals by using a new indicator: mononitrosochromotropic acid. *Indian J. Appl. Chem.*, **27**, 7 (1964).

64-52 ANDEREGG, G., and L'EPLATTENIER, F., Metall-Indikatoren. VII. Aciditäts-konstanten und der Fe-Komplex von N,N'-

Athylenbis-(*o*-hydroxyphenylglycin) (EHPG). *Helv. Chim. Acta*, **47**, 1067 (1964).

64-53 GUSEV, S. I., and SHCHUROVA, L. M., Pyridylazo compounds as metallochromic indicators. I. 5-(2-Pyridylazo)-2-monoethylamino-*p*-cresol as an analytical reagent. *Zhur. analit. Khim.*, **19**, 799 (1964).

64-54 MALEVANNYI, V. A., and SHUMINA, V. A., Complexometric determination of Pb in lead pigments. *Lekokrasochn. Materialy ikh Primenenie*, 57 (1964).

64-55 CHERNOVA, R. K., Use of phys.-chem. methods of analysis for certain complex compounds. *Materialy k Konf. Molodykh Nauchn. Rabotn. Saratovsk. Med. Inst. Sb.*, 37 (1964).

64-56 IDEMORI, M., UV spectrophotometric determination of the composition and stabilities of some aminopolycarboxylate complexes of Fe(III). *Nippon Kagaku Zasshi*, **85**, 311 (1964).

64-57 JOUSSOT-DUBIEN, J., and CORTRAIT, M., Determination of the dissociation constants of EDTA complexes of Ag. *J. Chim. phys.*, **61**, 1211 (1964).

64-58 KEMULA, W., BRACHACZYK, W., and HULANICKI, A., A direct photometric titration of Mg in Ni. *Chem. Analit (Warsaw)*, **9**, 43 (1964).

64-59 RICHARDS, C. S., and BOYMAN, E. C., Rapid analysis of refractory chrome ores and chrome-bearing refractories. *Analyt. Chem.*, **36**, 1790 (1964).

64-60 PETRENKO, A. G., and KUZNETSOVA, L. A., Xylenol orange as a complexometric indicator for In, Ce and La. *Khim. Analiz. Tsvetn. i Redkikh Metal. Akad. Nauk SSSR, Sibirsk. Otd. Khim.-Met.-Inst.*, 23 (1964).

64-61 PŘIBIL, R., and VESELÝ, V., Rapid analysis. XIV. Analysis of some Ni–Cu alloy systems. *Chem. Analyst*, **53**, 38 (1964).

64-62 PŘIBIL, R., and VESELÝ, V., Rapid analysis. XV. Determination of La, Al, Ca, Ba, Si and B in optical glass. *Chem. Analyst*, **53**, 43 (1964).

64-63 PŘIBIL, R., and VESELÝ, V., Rapid analysis. XVI. Resolution of Th–Fe, and Th–Al systems. *Chem. Analyst*, **53**, 77 (1964).

64-64 VÝDRÁ, F., and VORLIČEK, J., Rapid analysis. XVII. EDTA titration of Fe(III) in the presence of Fe(II) in ores and slags. *Chem. Analyst*, **53**, 103 (1964).

64-65 SILNICHENKO, V. G., and DMITRIEVA, F. I., Determination of Sm of various valency in fluorite. *Zhur. analit. Khim.*, **19**, 84 (1964).

64-66 MORACHEWSKI, YU.V., and VOLF, L. A., Complexometric determination of lanthanides in the presence of some metal complexes. *Izvest. Vysshikh. Uchebn. Zaved. Khim. i Khim. Teklnol.*, **7**, 513 (1964).

64-67 DOLEŽAL, J., and ZÝKA, J., Determination of inorganic and organic systems using EDTA after amalgam reduction. *Chem. Analyst*, **53**, 68 (1964).

418   *Bibliography*

64-68   SUZUKI, T., Determination of Fe, Al and Ti in silicates by EDTA titration. *Japan Analyst*, **13**, 524 (1964).

64-69   JENIK, J., and RENGER, F., Analyse der metallorganischen Verbindungen vom Sandwichtyp. III. Gleichzeitige Mikrobestimmung von Fe and Ti in Donor-Akzeptorkomplexen der Ferrocenbasen. *Coll. Czech. Chem. Commun.*, **29**, 2237 (1964).

64-70   LASSNER, E., and PÜSCHEL, R., Über Peroxokomplexe von Ti, Nb and Ta. IV. Reaktionen von Peroxo Ti, Nb und Ta mit verschiedenen metallochromen Indikatoren. *Mikrochim. Acta*, 753 (1964).

64-71   KONKIN, V. D., and ZHIKHAREVA, V. I., Complexometric method for the analysis of materials containing Zr, Ti and Fe. *Sb. Tr. Nauchn. Issled. Inst. Metal*, 442 (1964).

64-72   PŘIBIL, R., and VESELÝ, V., Beitrag zur komplexometrischen Bestimmung von Zr. *Z. analyt. Chem.*, **200**, 332 (1964).

64-73   CALETKA, R., KYŘŠ, M. M., and RAIS, J., Sorption of Zr by silica gel from nitrate medium in the presence of oxalate, EDTA and Arsenazo. I. *J. Inorg. Nuclear Chem.*, **26**, 1443 (1964); *ibid.*, p. 1760.

64-74   BARAN, V., and TYMPL, M., Direct complexometric determination of Zr with Catechol Violet. *Coll. Czech. Chem. Commun.*, **29**, 2250 (1964).

64-75   HUNG, S.-C., and CHANG, H.-S., Chelometric micro determination of Zr. *Hua Hsueh Pao*, **30**, 492 (1964).

64-76   KHADEEV, V. A., and TUYAKOV, N. B., Amperometric determination of Zr by using EDTA and a Tl(III) salt. *Nauch. Tr. Tashkentsh. Gosud. Univ.*, **264**, 33 (1964).

64-77   PANKRATOVA, L. N., VLASOV, L. G., and LAPITSKII, A. V., Reaction of Zr with some complexing agents. *Zhur. neorg. Khim.*, **9**, 1763 (1964).

64-78   GUPTA, A. K., and POWELL, J. E., Successive determination of Th and rare earths by complexometric titration. *Talanta*, **11**, 1339 (1964).

64-79   TSERKOVNITSKAYA, I. A., GRIGOR'EVA, M. F., and KUSTOVA, N. A., Organic reagents for the determination of different oxidation states of V. *Khim. Redkikh Elementov, Leningrad Gos. Univ.*, 453 (1964).

64-80   OGAWA, K., and MUSHA, S., Spectrophotometric study of the complex of vanadium (IV) with EDTA. *Bull. Univ. Osaka Prefect Ser.*, *A***13**, 113 (1964).

64-81   KONKIN, V. D., and ZHIKHAREVA, V. I., Use of trilon B for the analysis of alloys containing vanadium. *Sb. Tr. Ukr. Nauch. Issled. Inst. Metal*, 447 (1964).

64-82   UVAROVA, E. I., and RIK, V. M., Complexometric determination of Mo in catalysts. *Khim. i Tekhnol. Topliv. i Masel.*, **9**, 67 (1964).

64-83   CHANG, T.-H., TSAO, F.-Y., and SHU, S.-M., Amperometric titration of W with EDTA. *Hua Hsueh Pao*, **30**, 230 (1964).

64-84   BHAT, T. R., and KRISHNAMURTHY, M., Studies on EDTA complexes. II. The UO_2–Y system. *J. Inorg. Nuclear Chem.*, **26**, 587 (1964).

64-85 RAKLMILEVICH, N. M., Trilon B in the analysis of non-ferrous metals. *Zavodskaya Lab.*, **30**, 507 (1964).

64-86 YAMAMURA, S. S., Determination of Fe(III) by EDTA replacement titrimetry following selective separation by column extraction. *Analyt. Chem.*, **36**, 1858 (1964).

64-87 TACHIKAWA, T., Potentiometric titration of Fe with EDTA. *Japan Analyst.*, **13**, 456 (1964).

64-88 PROVAZNIK, J., and KNIZEK, M., Rapid complexometric determination of the metal component in Fe and Co silicides. *Chem. Listy*, **58**, 1158 (1964).

64-89 BRUILE, E. S., and MERKULOVA, K. S., Complexometric determination of Ni in alloy steel containing Co. *Zhur. priklad. Khim.*, **37**, 216 (1964).

64-90 AGASYAN, P. K., NIKOLAEVA, E. R., and DEMINA, L. A., Electrometric uranium determination by vanadate or complexometric titration. *Zavodskaya Lab.*, **30**, 1434 (1964).

64-91 FAERMAN, D. V., and CHIGRIN, D. S., Complexometric determination of Mn catalysts in paraffin hydrocarbons. *Zavodskaya Lab.*, **30**, 288 (1964).

64-92 NASSLER, J., and HOLEČEK, K., EDTA titration of Ni in electrolytic Ni plating-baths. *Chem. Analyst*, **53**, 76 (1964).

64-93 TANAKA, N., and OGINO, H., Substitution inert metal complex as indicator in a.c. polarographic titrations. A new type of metal indicator. *J. Electroanalyt. Chem.*, **7**, 332 (1964).

64-94 GREEN, H., and RICHARDS, P. J., Determination of Ni in cast iron using DMG precipitation and EDTA titration. *Brit. Cast Iron Res. Assoc. J.*, **12**, 578 (1964).

64-95 KOVACH, A., Complexometric determination of Zn in the presence of Cu. *Magyar Kém. Folyóirat*, **70**, 252 (1964).

64-96 BERMEJO, F., and MARGALET, A., Chelons. XXXVIII. Chelometric determination of Cu, Co, Ni, Fe, Cr, Al, Zn, Cd, V and Mn. *Inform. Quim. Anal.*, **18**, 35 (1964).

64-97 PORTIGLIA, E., and QUATTRONE, C., Complexometric determination of Cu in alkali cyanide plating-baths. *Galvanotecnica*, **15**, 237 (1964).

64-98 KOMATSU, S., KITAZAWA, C., and HATANAKA, T., Indirect chelatometric determination of Ag and Cl by using a suspension of Cu diethyldithiocarbamate. *Nippon Kagaku Zasshi*, **85**, 435 (1964).

64-99 KHALIFA, H., Back-titration with Hg(II) in alkaline medium. Estimation of some rare earths and analysis of mixtures with some other metals. *Z. analyt. Chem.*, **203**, 93 (1964).

64-100 JANKOVSKÝ, J., Zur Chelatometrische Bestimmung von Zn in Konzentraten. *Z. analyt. Chem.*, **201**, 330 (1963).

64-101 KANISHI, S., Zn and Cd thickness determination. Weight-area method employing EDTA. *Metal Finishing*, **62**, 84 (1964).

64-102 ONUKI, S., WATANUKI, K., and YOSHINO, Y., Determination of Zn in Zn base white metal by EDTA titration. *Japan Analyst*, **13**, 462 (1964).

420   *Bibliography*

64-103 PECHENTKOVSKAYA, L. E., and NAZARCHUK, T. N., Complexo-
        metric determination of Zn in ferro-alloys. *Zhur. analit. Khim.*, **19**,
        897 (1964).
64-104 KINOSHITA, S., and HOZUMI, K., Micro determination of Hg with
        a simple wet combustion method. *Microchem. J.*, **8**, 79 (1964).
64-105 KOMATSU, S., KITAZAWA, C., and NASU, S., Indirect chelometric
        determination of Hg(II) using a suspension of Cu(DITICA).
        Nippon Kagaku Zasshi, **85**, 598 (1964).
64-106 OLSEN, E. D., Convenient anion exchange resin preparation of MgY
        chelate for substitution titrations. *Analyt. Chem.*, **36**, 2461 (1964).
64-107 SOLOV'EVA, L. A., STOLYAROV, K. P., and GRIGOR'EV, N. N.,
        Determination of small amounts of Ga by microluminescent
        titration. *Vestn. Leningr. Univ. Ser. Fiz. i Khim.*, **3**, 134 (1964).
64-108 TEODORESCU, GR., NICULESCU, I., POPESCU, F., and IANESCU,
        A. M., Rapid method for the determination of Ga, In and Tl. *Bul.
        Inst. Polt. Bucaresti*, **26**, 67 (1964).
64-109 ZABOLOTSKII, T. V., and NAVROTSKAYA, V. A., Reactions of
        NTA with In, Ga and Al. *Khim. Anal. Tsvetn. i Redkikh. Metal
        Akad. Nauk SSSR Sibirsk. Otd. Met.-Inst.*, **38** (1964).
64-110 METCALFE, J., and KNOWLES, C. J., A rapid method for the
        determination of In in cyanide In-plating solutions. *Analyst*, **89**,
        293 (1964).
64-111 YURIST, I. M., Complexometric determination of In, Sn and Pb
        in solutions. *Zavodskaya Lab.*, **30**, 805 (1964).
64-112 CHENG, K. L., Bromine oxidation in a highly selective EDTA
        titration of Tl. *Microchem. J.*, **8**, 225 (1964).
64-113 BALLSCHMITER, K. H., and TÖLG, G., Bestimmung des Kohlen-
        stoffgehaltes von schwerflüchtigen Verbindungen unter Ver-
        wendung von 10–30 µg Proben. *Z. analyt. Chem.*, **203**, 20 (1964).
64-114 HENTRICH, K., and PFEIFER, S., Komplexometrische Bestimmung
        der Reineckate organischer Basen. *Pharmazie*, **19**, 630 (1964).
64-115 BACZYK, ST., and KACHELSKA, O., New complexometric deter-
        mination of L-ascorbic acid. *Mikrochim. Acta*, 873 (1964).
64-116 ECKSCHLAGER, K., and FIDLEROVA, J., Complexometric
        determination of nicotinic acid in Nicoflavine. *Cesk. Farm.*, **13**,
        371 (1964).
64-117 FLESZAR, B., Polarographic and complexometric determination
        of ethyl-*p*-nitrobenzoylacetate. *Chem. Analit. (Warsaw)*, **9**, 387
        (1964).
64-118 GITTNER-MINDERMANN, A., LEUPIN, K., and BAUMGARTNER,
        R., Komplexometrische Bestimmung von Theophyllin. *Pharm.
        Acta Helv.*, **39**, 390 (1964).
64-119 GRAMBERG, K., and BLAGOJEVIĆ, Z., Determination of Pro-
        methazine. *Acta Pharm. Jugosl.*, **14**, 17 (1964).
64-120 SAKAI, T., Chelatometric determination of hydroquinone using
        Tl(III) as oxidizing reagent. *Japan Analyst*, **85**, 30 (1964).
64-121 ŠARŠÚNOVÁ, M., and LUKÁČOVÁ, O., Complexometric deter-
        mination of some barbituric acid derivatives in mixtures and
        galenic preparations. *Cesk. Farm.*, **13**, 74 (1964).

64-122 NAZARENKO, V. A., LEBEDEVA, N. V., and VIBAROVA, L. I., Complexometric determination of Ge(IV). *Zhur. analit. Khim.*, **19**, 87 (1964).

64-123 LANGER, H. G., Chemistry of Sn(II) Chelates. I. Solid chelates of Sn(II) with EDTA. *J. Inorg. Nuclear Chem.*, **26**, 767 (1964).

64-124 ONUKI, S., WATANUKI, K., YOSHINO, Y., and NAGASHIMA, K., Some applications of masking in the analysis of non-ferrous metals and alloys. *Sci. Papers College Gen. Educ. Univ. Tokyo.*, **14**, 205 (1964).

64-125 MARCHENKO, P. V., Reactions of Pb with Xylenol Orange. *Ukr. Khim. Zhur.*, **30**, 224 (1964).

64-126 HARZDORF, C., Chelometrische Titration von Pb in Gegenwart störender Elemente. *Z. analyt. Chem.*, **203**, 101 (1964).

64-127 SHTEINMA, E. A., DOBRYNINA, Z. G., and MORDOVSKAYA, E. A., Complexometric determination of Bi and Pb in the presence of Sn. *Zavodskaya Lab.*, **30**, 1200 (1964).

64-128 BUDEVSKI, O., PENCHEVA, L., and RUSSINOVA, R. Indirect complexometric determination of phosphate by precipitation with Zr. *Talanta*, **11**, 1225 (1964).

64-129 VASILIEV, R., CHIALDA, I., and ANASTASESCU, G., Complexometric determination of phosphate in the product Fosfovit. *Farmacia (Bucharest)*, **12**, 238 (1964).

64-130 BUSS, H., KOHLSCHÜTTER, H. W., and RISCH, A., Bestimmung von löslichem Phosphat in Düngemitteln. *Z. analyt. Chem.*, **204**, 97 (1964).

64-131 FORMANEK, I., FÜLÖP, L., and SZANTHO, C., Rapid assay of Na methylarsenate. *Rev. Med. (Targu-Mures)*, **10**, 312 (1964).

64-132 AMSHEEVA, A. A., and BEZUGLYI, D. V., Photometric or complexometric determination of Bi in cast iron with Xylenol Orange. *Zhur. analit. Khim.*, **19**, 97 (1964).

64-133 GATTOW, G., and GOTTHARDT, B., Komplexometrische Titration von Bi. III. Die Bi Bestimmung in Gegenwart von Fremdionen. *Z. analyt. Chem.*, **206**, 331 (1964).

64-134 BALABANOFF, L., and SOTO, J., Rapid determination of H_2S. *Chim. analyt. (Paris)*, **46**, 90 (1964).

64-135 JANGIDA, B. L., VARDE, M. S., and VENKATASUBRAMANIAN, V., Microtitration of sulphate in $Th(NO_3)_4$ and reagent chemicals. *Indian J. Chem.*, **2**, 149 (1964).

64-136 BUDESVSKY, O., and PENTSHEVA, L., Complexometric determination of S in pyrites and sulphides. *Z. analyt. Chem.*, **203**, 14 (1964).

64-137 KERIN, D., Complexometric determination of S in soil. *Agrochimica*, **8**, 222 (1964).

64-138 BURRIEL-MARTI, F., and ALVAREZ-HERRERO, C., Complexometric determination of mixtures of sulphates, sulphides and sulphites in water. *Inf. Quim. Anal.*, **18**, 142 (1964).

64-139 KRAL, J., Field analysis of water. I. Determination of sulphate. *Sb. Vyokeho Ucein. Tech. Brno*, 37 (1964).

64-140 NONOVA, D., Complexometric determination of small amounts of Se. *Ann. Univ. Sofia Fac. Chim.*, **57**, 113 (1964).

64-141 HUBICKA, K., and SAWICKI, B., Complexometric determination of F with La(NO₃)₃. *Diss. pharm. Warsaw*, **16**, 317 (1964).

64-142 TIKHOMIROV, V. I., Approximative calculation of the best conditions for complexometric titrations. *Methody Kilichestv. Gred. Elementov, Leningr. Gos. Univ.*, 135 (1964).

64-143 GOLDMAN, J. A., and MEITES, L., Theory of titration curves. V. Location of extrema on conductometric, amperometric, spectrophotometric and other linear ('segmented') curves. *Analyt. Chim. Acta*, **30**, 280 (1964).

64-144 FLASCHKA, H., and GARRETT, J., Photometric titrations. X. The EDTA titration of Fe(III) in the presence of high concentrations of Bi. *Talanta*, **11**, 1651 (1964).

64-145 SCHONEBAUM, R. C., and BREEKLAND, E., Possibility of automation of potentiometric titrations. *Talanta*, **11**, 659 (1964).

64-146 BERKOVICH, M. T., SIRINA, A. M., and LAGUNOVA, N. L., Physico-chemical methods for the determination of Al and Fe in chromites. II. Potentiometric determination with complexone III. *Trudy Uralsk. Nauchn.-Issled. Khim. Inst.*, 31 (1964).

64-147 HALL, J. L., WILKINSON, P. R., and GIBSON, J. A., Equivalent conductance of some salts of EDTA. *Proc. West. Va. Acad. Sci.*, **36**, 87 (1964).

64-148 FLASCHKA, H., and BUTCHER, J., Photometric titrations. IX. DTPA titration of Zn in the presence of Cd and other metals. *Talanta*, **11**, 1067 (1964).

64-149 MEKAOTA, T., YAMAGUCHI, K., and UENO, K., Use of masking agents in chelatometric titrations. IV. Dimercapto succinic acid. *Talanta*, **11**, 1461 (1964).

64-150 ALEKSANDROVICH-MELNIKOVA, A. S., and AKCHURINA, G. S., Gallein as a metal indicator in the trilonometric determination of Pb. *Spektraln. i Khim. Metody Analiza Materialov Sb. Metodik*, 156 (1964).

64-151 SAJÓ, I., Direkte chelometrische Titration von Mo(VI). *Z. analyt. Chem.*, **199**, 16 (1964).

1965

65-1 CHRISTIAN, G. D., KNOBLOCK, E. C., and PURDY, W. C., Coulometric generation of EGTA. Titration of Ca in the presence of Mg. *Analyt. Chem.*, **37**, 292 (1965).

65-2 TIKHONOV, V. N., Methods for the complexometric determination of Mg and Ca. *Zhur. analit. Khim.*, **20**, 214 (1965).

65-3 KODAMA, K., MIZUNO, K., OGA, T., and MTSUNAMI, M., EDTA titration of Ca in the presence of large amounts of Mg. *Bunseki Kagaku*, **14**, 474 (1965).

65-4 STRAFELDA, F., Titrimetrische Bestimmung von Kationen mit EDTA unter Verwendung einer Silberelektrode. *Coll. Czech. Chem. Commun.*, **30**, 2320 (1964).

65-5 HESSLER, W., Automatische Bestimmung von Mg und Ca in Kalilauge und Salzen. *Bergakademie*, **17**, 223, 478 (1965).

65-6 FLEKSER, G. I., Complexometric determination of Ca with

Hydron II in the presence of large amounts of magnesium (in Mg metal). *Probl. Bol'shoe Met. i Fiz. Khim. Novykh Splavov Akad. Nauk. SSSR Inst. Met.*, 324 (1965).

65-7   PŘIBIL, R., and VESELÝ, V., Contributions to the basic problems of compleximetry. XVIII. Masking of Fe with fluoride. *Talanta*, **12**, 385 (1965).

65-8   PŘIBIL, R., and VESELÝ, V., Contributions to the basic problems of compleximetry. XIX. Determination of Zn and Cd. β-Mercaptoproprionic acid as a masking reagent for Cd. *Talanta*, **12**, 475 (1965).

65-9   YURASOVA, G. M., and ZINOV'EVA, L. D., Experimental complexometric determination of Ba in barite ores. *Sb. Nauchn. Tr. Vses. Nauch.-Issled. Gornp-Met. Inst. Tsvetn. Metal*, 53 (1965).

65-10 STEPANOV, A. V., and MAKAROVA, T. P., Complex formation of Pu(III) in EDTA solution by electromigration. *Radiokhimiya*, **7**, 664 (1965).

65-11 JURCZYK, J., Complexometric determination of Al in the presence of Ti. *Chem. Analit. (Warsaw)*, **10**, 441 (1965).

65-12 TIKHONOV, V. N., Complexometric methods for the Al determination. *Zhur. analit. Khim.*, **20**, 1219 (1965).

65-13 MYSHLYAEVA, L. V., and SHATUNOVA, T. G., Determination of Al in Al–Si organic compounds. *Trudy Mosk. Khim.-Tekhn. Inst.*, **48**, 48 (1965).

65-14 MILNER, O. I., and GEDANSKY, S. J., Chelometric determination of rare earths in the presence of Al. *Analyt. Chem.*, **37**, 931 (1965).

65-15 MINER, F. J., and DE GRAZIO, R. P., Ion exchange separation and volumetric determination of Ga in Ga–Pu alloys. *Analyt. Chem.*, **37**, 1071 (1965).

65-16 STILL, E., Theory of complexometric titrations based on extractive end-point detection. *Talanta*, **12**, 817 (1965).

65-17 BUDĚŠINSKÝ, B., Complexes of metallochromic substances. II. Complexes of La with Arsenazo I and Monoarsenazo III. *Z. analyt. Chem.*, **207**, 105 (1965).

65-18 BUDĚŠINSKÝ, B., Complexes of metallochromic substances. III. Classification of metal complexes with 1 : 1 molar ratio. *Z. analyt. Chem.*, **207**, 178 (1965).

65-19 BUDĚŠINSKÝ, B., Complexes of metallochromic substances. IV. Metallochromic properties of 2-(o-Carboxyphenylazo)-naphthionic acid and 2-(o-Arsonophenylazo)-naphthionic acid. *Z. analyt. Chem.*, **207**, 241 (1965).

65-20 BUDĚŠINSKÝ, B., Complexes of metallochromic substances. V. Optimum conditions for the metal complex formations and their relation to the complex composition. *Z. analyt. Chem.*, **207**, 247 (1965).

65-21 BEŽDEKOVÁ, A., and BUDĚŠINSKÝ, B., Spectrophotometric study of the reaction between Ca ions and calcion and calcichrome. *Coll. Czech. Chem. Commun.*, **30**, 811 (1965).

65-22 BUDĚŠINSKÝ, B., HAAS, K., and VRŽALOVÁ, D., Einige neue

Analoge von Arsenazo II und ihre analytischen Eigenschaften. *Coll. Czech. Chem. Commun.*, **30**, 2373 (1965).

65-23 ABDULLAEVA, KH.S., and TALIPOV, SH.T., Complexometric titration of Ga with the use of 4-(2-*N*-methylanabasineazo)-resorcinol. *Uzbek. Khim. Zhur.*, **9**, 25 (1965).

65-24 KISS, T. A., Deblockierung von Erio T bei Rücktitrationen von Co, Fe, Cu, Ni and Al unter Verwendung von organischen Lösungsmitteln. *Z. analyt. Chem.*, **208**, 334 (1965).

65-25 PARTASHNIKOVA, M. Z., and SHAFRAN, I. G., Sulfarzen as a complexometric indicator for Zn, Cd, Ni and Pb determinations. *Zhur. analit. Khim.*, **20**, 313 (1965).

65-26 HOSOKAWA, Y., and TOEI, K., Metallochromic indicator 'Mandelic-azo-chromotropic acid' for Th in chelatometry. *Bunseki Kagaku*, **14**, 161 (1965).

65-27 MALIK, A. U., Ferron as an indicator in the complexometric titration of vanadium(V). *Indian J. Chem.*, **3**, 316 (1965).

65-28 MUSTAFIN, I. S., GORYUNOVA, N. N., and FRUMINA, N. S., New complexometric indicator for Cu. *Zavodskaya Lab.*, **31**, 786 (1965).

65-29 RADKO, V. A., YAKIMETS, E. M., and VLADIMIRTSEV, I. F., Indicators for the complexometric determination of Mn. *Zhur. analit. Khim.*, **20**, 955 (1965).

65-30 PATROVSKÝ, V., Phthaleinviolett als Indikator in der Chelato-metrie. *Z. analyt. Chem.*, **214**, 321 (1965).

65-31 FISCHER, W., Mikrobestimmung von Fe(III) mit Variaminblau B als Redoxindikator. *Z. analyt. Chem.*, **215**, 251 (1965).

65-32 HEYER, F. C., Direct EDTA titration of Cu(II) with haematoxylin as indicator. *Chem. Analyst*, **54**, 10 (1965).

65-33 MORI, ITSUE, Hydroxyfluoran and its derivatives as organic reagents. V. Derivatives of 3′,4′,5′,6′-tetrahydroxyfluoran as metallochromic indicators. *Yakugaku Zasshi*, **85**, 686 (1965).

65-34 DRAGŬSIN, I., Use of bisazo derivatives of chromotropic acid in volumetric analysis. I. As indicators in complexometric analysis. *Rev. Chim. (Bucharest)*, **16**, 390 (1965).

65-35 BECK, M. T., and GERGELY, A., Stability constants of Bi–EDTA and Bi–DCTA complexes. *Magyar Kém. Folyóirat*, **71**, 163 (1965).

65-36 EGOROVA, L. G., Complexometric determination of Sc and Al without separation. *Trudy Inst. Met. Obogashch. Akad. Nauk. Kaz. SSR*, **12**, 151 (1965).

65-37 STRAFELDA, F., and MATOUSEK, J., Potentiometrische Titration mit Komplexon III gegen die Pt-Elektrode mittels eines Indikator-Redox Systems. *Coll. Czech. Chem. Commun.*, **30**, 2334 (1965).

65-38 BHAT, T. R., and RADHAMMA, D., EDTA complexes. III. Fe(III) and Ce(IV) EDTA complexes. *Indian J. Chem.*, **3**, 151 (1965).

65-39 GALAKTINOV, YU.P., ASTAKHOV, K. V., and ZHIRNOVA, N. M., Complex formation of Nd(III) with EDTA in aqueous medium. *Zhur. neorg. Khim.*, **10**, 2386 (1965).

65-40 PŘIBIL, R., and VESELÝ, V., Rapid analysis. XVIII. Analysis of low melting In-alloys. *Chem. Analyst*, **54**, 12 (1965).

65-41 PŘIBIL, R., and VESELÝ, V., Rapid analysis. XIX. Determination of Fe and Al in Cu alloys. *Chem. Analyst*, **54**, 46, 51 (1965).

65-42 VORLIČEK, J., and VÝDRÁ, F., Rapid analysis. XX. EDTA titration of Cu in ores and alloys with biamperometric indication. *Chem. Analyst*, **54**, 87 (1965).

65-43 PŘIBIL, R., and VESELÝ, V., Rapid analysis. XXI. Determination of Ni in alloys containing Fe, Co and Cu. *Chem. Analyst*, **54**, 114 (1965).

65-44 HUNG, S.-C., and JEN, H.-T., Chelometric microdetermination of rare earths. *Hua Hsueh Pao*, **31**, 91 (1965).

65-45 LYUBOMILOVA, G. V., Complexometric determination of Ti in microanalysis. *Mineral. Mikrovklyuch. Akad. Nauk. SSSR Inst. Min. Geokhim. i Kristallokhim. Redkikh Elementov*, 66 (1965).

65-46 CHEN, Y.-C., and LI, H.-J., Masking agents in chelatometry. II. Masking of Ti(IV) with lactic acid and the indirect titration of Ti. *Hua Hsueh Pao*, **31**, 391 (1965).

65-47 NAZARCHUK, T. N., and MEKHANOSHINA, L. N., Complexometric analysis of Zr–Ni–Al alloys. *Zhur. analit. Khim.*, **20**, 260 (1965).

65-48 CHERKESOV, A. I., PUSHINOV, YU.V., and TONKOSHUROV, V. S., Complexometric determination of Zr with Stilbazogall II as indicator. *Zhur. analit. Khim.*, **20**, 459 (1965).

65-49 PILKINGTON, E. S., and WILSON, W., Influence of polynuclear Zr species on the direct titration of Zr with EDTA. *Analyt. Chim. Acta*, **33**, 577 (1965).

65-50 WILSON, D. W., and GINGERICH, K. A., Chemical analysis of Th phosphides. *Analyt. Chem.*, **37**, 595 (1965).

65-51 DE HEER, B. H. J., VAN DER PLAS, TH., and HERMANS, M. E. A., The successive complexometric determination of Th and U in nitrate solution. *Analyt. Chim. Acta*, **32**, 292 (1965).

65-52 FARROW, R. N. P., and HILL, A. G., A modified method for determining traces of NTA in EDTA. *Analyst*, **90**, 241 (1965).

65-53 FARROW, R. N. P., and HILL, A. G., The effect of NTA impurities on the standardization of solutions of EDTA. *Analyst*, **90**, 210 (1965).

65-54 PODLAHOVA, J., Herstellung und Eigenschaften von AeDTE Komplexon. III. Verbindungen mit Vanadium(III). *Coll. Czech. Chem. Commun.*, **30**, 2000 (1965).

65-55 PODLAHOVA, J., Herstellung und Eigenschaften von AeDTE Komplexen. IV. Verbindungen mit Vanadium(IV). *Coll. Czech. Chem. Commun.*, **30**, 2012 (1965).

65-56 PODLAHOVA, J., V(II), V(III) and V(IV) EDTA systems in water solutions. *Chem. Zvesti*, **19**, 530 (1965).

65-57 BABKO, A. K., and LUKACHINA, V. V., Photometric titration (of Nb) with Complexone III. *Ukr. Khim. Zhur.*, **31**, 1313 (1965).

65-58 CRISAN, I., and LITEANU, C., Rapid formation of Cr(III) complexonate by a new reaction and its analytical significance. *Stud. Univ. Babes-Bolyai Ser. Chim.*, **10**, 113 (1965).

65-59 RAO, V. K., SUNDRA, D. S., and SASTRI, M. N., Photometric determination of Cr using EDTA. *Chem. Analyst*, **54**, 86 (1965).

65-60 NIKITINA, E. I., and ADRIANOVA, N. N., Titration of Mo in MoSi₂ and Fe–Mo. *Zavodskaya Lab.*, **31**, 654 (1965).

65-61 BUSEV, A. I., TIPTSOVA, V. G., and SOKOLOVA, T. A., Reactions of reduced forms of W with Complexone III. *Zhur. neorg. Khim.*, **10**, 1857 (2965).

65-62 YAGUCHI, H., and KAJIWARA, T., A new complexometric titration of Mo. *Bunseki Kagaku*, **14**, 785 (1965).

65-63 TANAKA, N., SHIRAKASHI, T., and OGINO, H., The redox potential on Mn(II) and Mn(III) EDTA complexes and the stability constants of Mn(III)Y. *Bull. Chem. Soc. Japan*, **38**, 1515 (1965).

65-64 VORLIČEK, J., and VÝDRÁ, F., Amperometry with two polarized electrodes. III. Chelatometric determination of Fe(II) using an indicator system of two carbon electrodes. *Talanta*, **12**, 377 (1965).

65-65 WAKAMATSU, S., Determination of metallic Fe, FeO and Fe₂O₃ in cutting Fe powder and spongy Fe. *Bunseki Kagaku*, **14**, 297 (1965).

65-66 TANAKA, N., and OGINO, H., Stability constants of some substitution inert Co(III) complexes. *Bull. Chem. Soc. Japan*, **38**, 1054 (1965).

65-67 STRAFELDA, F., KARLIK, M., and MATOUSEK, J., Potentiometrische Titrationen mit AeDTE und Verwendung einer Pt-Elektrode. *Coll. Czech. Chem. Commun.*, **30**, 2327 (1965).

65-68 WADA, H., and NAKAGAWA, G., Some *o*-(2-Thiazolylazo)-phenol derivatives as metal indicators for Cu. *Japan Analyst*, **14**, 28 (1965).

65-69 DANOWSKI, K., and LEWANDOWSKA, K., Rapid determination of Ni in catalysts. *Tluszcze i Srodki Piorace*, **9**, 160 (1965).

65-70 KOMATSU, S., Indirect spectrophotometric and chelatometric determination of Pd with diethyldithiocarbamate. *Nippon Kagaku Zasshi*, **86**, 827 (1965).

65-71 MUKAIDA, M., OKUNO, H., and ISHIMORI, T., Ru(III) complexes with EDTA. *Nippon Kagaku Zasshi*, **86**, 598 (1965).

65-72 EZERSKAYA, N. A., FILIMONOVA, V. N., and SOLOVYKH, T. P., Pd–EDTA complexes. *Zhur. neorg. Khim.*, **10**, 2657 (1965).

65-73 BORCHERT, O., Schnellbestimmung von Cu und Zn in Cu-Legierungen. *Neue Hütte*, **10**, 52 (1965).

65-74 ACKERMANN, G., and KADEN, W., Gravimetrische und Titrimetrische Bestimmung von Cu mit Reinecke Salz. *Z. analyt. Chem.*, **214**, 88 (1965).

65-75 PETRENKO, A. G., KHRISTOFOROV, B. S., and LEONARDOVA, L. A., Complexometric determination of Cu and Zn in the presence of each other. *Metody Izuch. Veshchestv. Sostava i ikh Primenenie Akad. Nauk. SSSR, Sibirsk Otd.*, **1**, 59 (1965).

65-76 HENNART, C., Application de la Chelones. XVII. Dosage volumetrique des ions argent par reduction a l'etat elementaire. *Talanta*, **12**, 694 (1965).

65-77 HUNTER, T. L., Determination of Zn in rubber products via EDTA titration. *Analyt. Chem.*, **37**, 1436 (1965).

65-78 CURRO, P., and CALABRO, G., Thiosulphate masking reaction for the complexometric determination of Hg in the presence of other cations. *Atti Soc. Pelor. Sci. Fis. Mat. Nat.*, **11**, 49 (1965).

65-79 BOGS, U., and BORMAN, G., Zur quantitativen Bestimmung von Ca in der Arzneimittelanalyse. *Pharmazie*, **20**, 21 (1965).

65-80 MIZUNO, K., Chelometric micro determination of Ga in red mud waste from Al industry. *Bunseki Kagaku*, **14**, 410 (1965).

65-81 MAKASHEVA, I. E., KIRIN, I. S., and MAKASHEV, YU. A., Ga determination in Nb–Ga alloys. *Zavodskaya Lab.*, **31**, 1192 (1965).

65-82 ZHIRNOVA, N. M., ASTAKHOV, K. V., and BARKOV, S. A., Spectrophotometric study of complex formation of In with EDTA and NTA. *Zhur. fiz. Khim.*, **39**, 1224 (1965).

65-83 AWE, W., and SCHILLING, D., Pyramidon-Bestimmung in DRF-Drogen. *Pharm. Ztg.*, **110**, 748 (1965).

65-84 GRECU, I., and CUREA, E., Determination of 1,4-Dihyrazinophthalzide. *Rev. Chim. (Bucharest)*, **16**, 348 (1965).

65-85 HENNART, C., Le récent apports de la chélométrie a l'analyse quantitative organique. *Chim. analyt.*, **47**, 249 (1965).

65-86 HENNART, C., Les methodes volumétriques de dosage des sucre réducteurs. *Bull. Soc. Chim. France*, 1585 (1965).

65-87 IOANID, N., ARMASASCU, L., and BEM, M., Complexometric method for the estimation of Luminal. *Farmacia Bucharest*), **13**, 459 (1965).

65-88 KELLNER, A., and SZEKERES, L., Chelometric methods in the determination of oxidants and reductants. I. Determination of some oxidants and reductants via EDTA titrations. *Chem. Analyst*, **54**, 75 (1965).

65-89 SZEKERES, L., KARDOS, E., and SZEKERES, G. L., Chelometric methods in the determination of oxidants and reductants. II. Determination of some inorganic ions and oxalate via oxidation with MnO_4^- and subsequent EDTA titration. *Chem. Analyst*, **54**, 116 (1965).

65-90 LAMBERT, R. S., and DU BOIS, R. J., Modification of Fe(II) reduction method for nitroglycerine. *Analyt. Chem.*, **37**, 427 (1965).

65-91 SASONGO, Chelometric assay of sodium citrate. *Suara Pharm. Madjalah*, **8**, 46 (1965).

65-92 PUKHONTO, A. N., ZHAVORONKOVA, A. YA., MOISEEVA, E. I., and SMIRNOV, V. F., Simultaneous determination of butyl hydrogen phosphates, tributyl phosphate and kerosine in aqueous solution. *Zhur. analit. Khim.*, **20**, 372 (1965).

65-93 CHUMACHENKO, M. N., and MUKHADSHINA, R. A., Determination of nitrogen by decomposition of organic substances with potassium metal. *Izvest. Akad. Nauk SSSR, Ser. Khim.*, 1262 (1965).

65-94 YURIST, I. M., Direct complexometric determination of Sn in solders. *Zavodskaya Lab.*, **37**, 267 (1965).

65-95 FILIPPOVA, N. A., and KOROSTELEVA, V. A., Trilonometric determination of Sn. *Sb. Nauchn. Tr. Gos. Nauchn.-Issled. Inst. Tsvet. Metal*, **23**, 352 (1965).

428   Bibliography

65-96   LYASHENKO, T. V., and MILAEV, S. M., Determination of Pb in metallic Bi. *Sb. Nauchn. Tr. Vses. Nauchn.-Issled. Gorno-Metal. Inst. Tsvet. Metal.*, 78 (1965).

65-97   KUAN, T. -Y., Indirect determination of P by complexing reagents in fertilizers. *T'u Jang T'ung Pao*, 46 (1965).

65-98   BLAZSEK, A., FÜLÖP, L., and VEREPH, J., Complexometric determination of arsenites and of As in organic compounds. *Farmacia (Bucharest)*, 73, 349 (1965).

65-99   BHAT, T. R., and IYER, R. K., EDTA complexes. V. Sb(III) and Bi(III) EDTA systems. *Z. anorg. Chem.*, 335, 331 (1965).

65-100   WANG, E. K., and CHANG, J. P., Polarographic study of Sb(III) complex with EDTA. *Hua Hsueh Hsueh Pao*, 31, 18 (1965).

65-101   T'AO, T. C., Complexometric determination of Bi with EDTA. *Hua Hsueh Tung Pao*, 51 (1965).

65-102   ZUSE, M., Schnellbestimmung von Schwefelsäure in Chromatbädern. *Galvanotechn.*, 56, 20 (1965).

65-103   CHANG, W. P., MIN, S. K., and CHUNG, I. B., Determination of F via an EDTA titration. *Chem. Analyst*, 54, 41 (1965).

65-104   JOHANNSON, A., and WÄNNINEN, E., Graphic representation of complexometric titrations. *Svensk. Kem. Tidskr.*, 77, 492 (1965).

65-105   VORLIČEK, J., and VÝDRÁ, F., Amperometry with two polarizable electrodes. *Coll. Czech. Chem. Commun.*, 30, 4272 (1965).

65-106   BLAEDEL, W. J., and LAESSIG, R. H., Continuous EDTA titration with a dropping mercury-electrode. Automated titrations based on non-symmetrical curves. *Analyt. Chem.*, 37, 1255 (1965).

65-107   FLASCHKA, H., Photometric titrations. *Pure and Applied Chem.*, 10, 165 (1965).

65-108   FLASCHKA, H., and BUTCHER, J., Photometric titrations. XI. Construction and evaluation of a semi-immersion phototitrator. *Talanta*, 12, 913 (1965).

65-109   STILL, E., and RINGBOM, A., Photometric titrations with indicators. *Analyt. Chim. Acta*, 33, 50 (1965).

65-110   BLAEDEL, W. J., and LAESSIG, R. H., Continuous back-titration with direct read-out. Application to EDTA systems. *Analyt. Chem.*, 37, 1650 (1965).

65-111   NOMURA, T., DONO, T., and NAKAGAWA, G., Potentiometric titration of metals with EDTA by the use of the Fe(III)–Fe(II) system as indicator. *Japan Analyst*, 14, 197 (1965).

65-112   TACHIKAWA, T., Bimetal electrodes in the potentiometric titration of Cu with EDTA. *Bunseki Kagaku*, 14, 697 (1965).

65-113   FLASCHKA, H., and BUTCHER, J., The determination of Zn and Cd in the same solution via EDTA titration. *Chem. Analyst*, 54, 36 (1965).

65-114   PATROVSKY, V., Möglichkeiten der Maskierung von Begleitelementen bei der Chelatometrischen Bestimmung von Ni und Cu mit Murexid als Indikator. *Z. analyt. Chem.*, 214, 261 (1965).

65-115   LUZINA, L. N., Rapid method of solder analysis. *Zavodskaya Lab.*, 31, 808 (1965).

65-116   BENEV, B., Complexometric determination of Fe, Al, Ca, Mg and

Ni in ammonia synthesis catalyst. *God. Nauchnoizsled. Inst. Koksokhim. Neftoprerab.*, **4**, 191 (1965).

65-117 DIAZ ROJAS, C. A., Spectrophotometric titrations using solutions of EDTA as titrating agent. *Rev. Fac. Quim. Univ. nac. mayor San Marcos*, **17**, 66 (1965).

65-118 KARADAKOV, B., Complexometric determination of iron(II) with thioglycollic acid as indicator. *God. Vissh. Khimikotekhnol. Inst.*, **12** (1), 61 (1965).

65-119 KARADAKOV, B., Complexometric determination of Mn in the presence of large amounts of Fe with Methylthymol Blue as indicator. *God. Vissh. Khimikotechnol. Inst.*, **12**, 101 (1965).

65-120 KOLOBOVA, K. K., Chromatographic and complexometric determination of the content of aluminium in zirconium concentrates. *Tr. Vses. Inst. Nauch.-Issled. Prockt. Rab. Ogneupor. Prom.*, **37**, 74 (1965).

65-121 KONISHI, S., Analysis of copper pyrophosphate baths with EDTA. *Metal Finish*, **63** (3), 58, 62 (1965).

65-122 KONISHI, S., Silver analysis with EDTA. *Metal Finish.*, **63**, 77 (1965).

65-123 PRZYBOROWSKI, L., SCHWARZENBACH, G., and ZIMMERMANN, TH., Komplexe XXXVII. Die EDTA-Komplexe des Vanadiums(V). *Helv. Chim. Acta*, **48**, 1556 (1965).

1966

66-1 BRAUN, T., TÖLGESSY, J., and KONECNY, J., New techniques in radiometric titrations. *Acta Chim. Acad. Sci. Hung.*, **49**, 131 (1966).

66-2 PŘIBIL, R., and VESELÝ, V., Contributions to the basic problems of complexometry. XX. Determination of Ca and Mg. *Talanta*, **13**, 233 (1966).

66-3 PŘIBIL, R., and VESELÝ, V., Contributions to the basic problems of complexometry. XXI. Determination of Ni in the presence of Co. *Talanta*, **13**, 515 (1966).

66-4 ESCARILLA, A. M., Fluorometric titration of Ca, Mg and Fe using Calcein Blue as indicator. *Talanta*, **13**, 363 (1966).

66-5 SAJÓ, I., Einstellen des pH-Wertes bei der chelatometrischen Bestimmung von Ca. *Z. analyt. Chem.*, **219**, 279 (1966).

66-6 ASHTON, A. A., Tetracycline as a fluorescent indicator in the complexometric micro determination of Group II cations. *Analyt. Chim. Acta*, **35**, 543 (1966).

66-7 OLSEN, E. D., and NOVAK, R. J., Convenient substitution titration of Ba and other alkaline earths with DTPA. *Analyt. Chem.*, **38**, 152 (1966).

66-8 CHENG, K. L., and GOYDISH, B. L., Selective EDTA titration of Ga or Al in the presence of other metals. *Talanta*, **13**, 1161 (1966).

66-9 ALIMARIN, I. A., and PETRIKOVA, M. N., Complexometric ultramicro titration by using a Hg-electrode. *Zhur. analit. Khim.*, **21**, 1257 (1966).

66-10 JAIMNI, J. P. C., PUROHIT, D. N., and SOGANI, N. C., Complexometric determination of Fe in natural samples using 3-hydroxy-1-*p*-sulphonatophenyl-3-phenyltriazene as metal indicator. *J. Proc. Inst. Chemists (India)*, **38**, 123 (1966).

66-11 JAIMNI, J. P. C., BHANDRI, M. R., and SOGANI, N. C., Na *p*-(nercaptoacetanilido)benzene sulphonate as a metal indicator for the complexometric determination of Ni with EDTA. *J. Proc. Inst. Chemists (India)*, **38**, 63 (1966).

66-12 TANDON, K. N., Complexometric determination of Hg(II) using Congo Red as indicator. *Talanta*, **13**, 161 (1966).

66-13 HAVÍR, J., and VŘEŠTÁL, J., 1-(Thiazolylazo)-2-naphthol as a metallochromic indicator in chelatometric titrations. *Chem. Listy*, **60**, 64 (1966).

66-14 LINDSTROM, F., and ISAAC, R., Crystalline calmagite and a study of sulphonation effects on azo dye metal indicators. *Talanta*, **13**, 1003 (1966).

66-15 MORI, ITSUO, Hydroxyfluoran and its derivatives as organic reagents. VI. 2′,3′,6′,7′-Tetrahydroxyfluoran derivatives as metallochromic indicators. *Yakugaku Zasshi*, **86**, 140 (1966).

66-16 GUSEV, S. I., and NIKOLAEVA, E. M., Pyridylazo compounds as reagents for In. II. Complexometric and photometric determination of In with pyridylazoaminophenols. *Zhur. analit. Khim.*, **21**, 281 (1966).

66-17 SRIVASTAVA, T. N., and SINGH, N., Complexometric determination of Ga, In and Tl(III) by using NaN₃ as metal indicator. *Z. analyt. Chem.*, **218**, 261 (1966).

66-18 DAS, H. R., and SHOME, S. C., Titration of Fe(III) and Cu with EDTA using *N*-benzoyl-*N*-phenylhydroxylamine as indicator. *Analyt. Chim. Acta*, **35**, 256 (1966).

66-19 ASHTON, A. A., Tetracycline as a fluorescent indicator in the complexometric microdetermination of Group II cations. *Analyt. Chim. Acta*, **35**, 543 (1966).

66-20 GUSEV, S. I., SHCHUROVA, L. M., and BITOVT, Z. A., Picramineazo compounds as metallochromic indicators for Bi. *Zhur. analit. Khim.*, **21**, 568 (1966).

66-21 SANGAL, G., SANGAL, S. P., and MUSHRAN, S. P., Complexometric determination of Th with PAR, alizarin-3-sulphonate and sulphohydroxydimethylfuchsone dicarboxylic acid as indicators. *Microchem. J.*, **11**, 513 (1966).

66-22 BETTERIDGE, D., Extractive indicators in complex formation titrations. Theory and practice. *Talanta*, **13**, 1497 (1966).

66-23 KRIEGE, O. H., and THEODORE, M. L., Analysis of rare earth sulphides, selenides and tellurides. *Talanta*, **13**, 265 (1966).

66-24 PŘIBIL, R., and VESELÝ, V., Rapid analysis. XXII. Determination of Cadmium in the presence of Zn and Pb via EGTA titration. *Chem. Analyst*, **55**, 4 (1966).

66-25 PŘIBIL, R., and VESELÝ, V., Rapid analysis. XXIII. Analysis of In–Al–Ni and In–Ni–Zn in gold plating baths. *Chem. Analyst*, **55**, 38 (1966).

66-26 PŘIBIL, R., and VESELÝ, V., Rapid analysis. XXIV. Determination of Fe, Al, Mg, Ca, Ti and Mn in iron ores and slags. *Chem. Analyst*, **55**, 68 (1966).

66-27 PŘIBIL, R., and VESELÝ, V., Rapid analysis. XXV. Determination of Mg (and Ca) in calcareous materials and silicates. *Chem. Analyst*, **55**, 82 (1966).

66-28 PODLAHOVA, J., and PODLAHA, J., Ti(III) complexes of EDTA. *J. Inorg. Nuclear Chem.*, **28**, 2267 (1966).

66-29 RYBININ, A. I., and AFANAS'EV, YU. A., Complexometric determination of Th in U in the presence of tributylphosphate. *Zhur. analit. Khim.*, **21**, 374 (1966).

66-30 FUHRMAN, D. L., LATIMER, G. W., and BISHOP, J., Determination and differentiation of NTA and EDTA. *Talanta*, **13**, 103 (1966).

66-31 MONK, R. G., Comments on the effect of NTA impurity on the standardization of solutions of EDTA. *Analyst*, **91**, 597 (1966).

66-32 HEYL, H. J., Metal specific indicator solutions. *Brit. Patent* 953 128, March 25, 1964.

66-33 KRTIL, J., Bestimmung von V in schwerlöslichen Vanadylhexacyanoferrate. *Z. analyt. Chem.*, **219**, 412 (1966).

66-34 TANAKA, M., and ISHIDA, A., Direct EDTA titration of V(V) using Variamine Blue as indicator in the presence of excess Fe(II). *Analyt. Chim. Acta*, **36**, 515 (1966).

66-35 IRVING, H. M. N. H., and TOMLINSON, W. R., Electron exchange catalysis of the formation of the EDTA complex of chromium (III) and its photometric determination. *Chem. Analyst*, **55**, 14 (1966).

66-36 FOUCHECOURT, P., and BELIN, P., Formation de l'ion complexe $[CrY(H_2O)]^-$ à partir de melange de solutions de Cr(VI) et EDTA en milieu acide. *Compt. rend. (Paris)*, **262**, 605 (1966).

66-37 RAO, V. K., SUNDRA, D. S., and SASTRI, M. N., Rapid chelatometric determination of Cr(III). *Z. analyt. Chem.*, **218**, 93 (1966).

66-38 KULA, R. J., Solution equilibriums and structures of Mo(IV) chelates. EDTA. *Analyt. Chem.*, **38**, 1581 (1966).

66-39 KULA, R. J., Potentiometric determination of stabilities of Mo(VI) and W(VI) chelates. *Analyt. Chem.*, **38**, 1934 (1966).

66-40 SHLENSKAYA, V. I., BIRYUKOV, A. A., and SHUMKOVA, N. G., Spectrophotometric study of the interaction of Pd(II) with complexone III in perchlorate and chloride media. *Zhur. analit. Khim.*, **21**, 702 (1966).

66-41 EZERSKAYA, N. A., and SOLOVYKH, T. P., The reaction of RuY with H_2O_2. *Zhur. neorg. Khim.*, **11**, 2179 (1966).

66-42 EZERSKAYA, N. A., and SOLOVYKH, T. P., Compounds of Ru(III) hydroxy EDTA with oxygen. *Zhur. neorg. Khim.*, **11**, 2569 (1966).

66-43 SHAKHOVA, G. P., Complexometric titration of Pt. *Zavodskaya Lab.*, **32**, 1201 (1966).

66-44 POPOVA, O. I., and GODOVANNAYA, I. N., Complexometric determination of Cu, Fe and Al in their alloys. *Ukrain. Khim. Zhur.*, **32**, 217 (1966).

66-45 STILES, R. E., and MUNNECKE, D. L., Determination of Cu with EDTA. *U.S. Bureau of Mines Rep. Invest.* No. 6852 (1966).

66-46 CHIACCHIERINI, E., Reaction between Cr(III) and some complexones at room temperature. Direct amperometric titration. *Ann. Chim. (Rome)*, **56**, 1405 (1966).

66-47   BUDEVSKY, O., RUSSEVA, E., and MESROB, B., Dithiocarbamino acetic acid as a masking agent in complexometry. *Talanta*, **13**, 277 (1966).

66-48   HICKEY, J. J., and OVERBECK, C. J., Chelometric titration for Zn in municipal and industrial water supplies. *Analyt. Chem.*, **38**, 932 (1966).

66-49   GUSEV, E. I., and NIKOLAEVA, E. M., Pyridylazo compounds as reagents for In. III. Complexometric and spectrophotometric determination of In with some derivatives of pyridylazoamino-cresol. *Zhur. analit. Khim.*, **21**, 1183 (1966).

66-50   STRELOW, F. W. E., and TOERIEN, F. V. S., Accurate determination of Tl by direct titration with EDTA using methyl thymol blue as indicator. *Analyt. Chim. Acta*, **36**, 189 (1966).

66-51   PANTANI, F., Complex formation of Tl(I) with CyDTA and EGTA. *Ricerca Sci.*, **36**, 702 (1966).

66-52   BALICA, G., and SAVU, M., Complexometric determination of CO_2 in monoethanolamine solutions. *Rev. Chim. (Bucharest)*, **17**, 565 (1966).

66-53   KOMATSU, S., and NOMURA, T., Indirect chelometric determination of CN^- with HgY. *J. Chem. Soc. Japan*, **87**, 1060 (1966).

66-54   PŘIBIL, R., Komplexometrie, Band IV. Pharmazeutische und organische Analyse. VEM Deutscher Verlag für Grundstoff industrie, Leipzig, 1966, p. 120.

66-55   SZEKERES, L., and KELLNER, A., Chelometric methods in the determination of oxidants and reductants via EDTA titration. *Chem. Analyst*, **55**, 77 (1966).

66-56   LUR'E, I. S., Complexometric determination of the sum of citric and oxalic acids in fermentation solutions. *Khlebopek. i Konditer. Prom.*, **10**, 18 (1966).

66-57   HENTRICH, K., and PFEIFER, S., Bestimmung von Pyridin in Komplexverbindungen. *Pharmazie*, **21**, 58 (1966).

66-58   YAO, M. S., and HUANG, C. S., Chelometric determination of Zn (Pb, Cu, Cd) with gallic acid as sequestering agent for Sn(IV). *Hua Hsueh Pao*, 48 (1966).

66-59   CHROMÝ, V., and VŘEŠTÁL, J., New chelatometric determination of Sn and its use for organo tin compounds. *Chem. Listy*, **60**, 1537 (1966).

66-60   KONISHI, S., and DOLI, N., Pb–Sn alloy bath control. *Metal Finish.*, **44**, 66 (1966).

66-61   TIWARI, G. D., and TREVEDI, S. R., Complexometric estimation of phosphate. *Current Sci. (India)*, **35**, 568 (1966).

66-62   YURIST, I. M., Complexometric determination of Sb. *Zavodskaya Lab.*, **32**, 1050 (1966).

66-63   FREESE, F., and DEN BOEF, G., Amperometric complex-formation titrations of traces of cations. *Talanta*, **13**, 865 (1966).

66-64   BLAEDEL, W. J., and LAESSIG, R. H., Continuous EDTA titrations at low concentrations. *Analyt. Chem.*, **38**, 186 (1966).

66-65   ISRAEL, Y., and VROMEN, A., Spontaneous voltammetry and voltammetric titrations. *J. Electroanalyt. Chem.*, **11**, 262 (1966).

66-66 VASSILIADES, C., KAWASSIADES, C. TH., HADJIIOANNOU, T. P., and COLOVOS, G., Automatic derivative spectrophotometric titration of Fe and (or) Al with EDTA. *Analyt. Chim. Acta*, **36**, 115 (1966).

66-67 FLASCHKA, H., and GARRET, J., Chelometric titration of Ga in the presence of In and other metals by chloride masking. *Z. analyt. Chem.*, **218**, 338 (1966).

66-68 GARDEL, M. C., and CORNWELL, J. C., Automatic potentiometric EDTA and redox titrations for detection of stoicheiometry. *Analyt. Chem.*, **38**, 774 (1966).

66-69 MARPLE, L. W., and SCHEPPERS, G. J., Acid–base equilibria in t-butyl alcohol. *Analyt. Chem.*, **38**, 553 (1966).

66-70 ISHIBASHI, N., KOHARE, H., and FUKUDA, M., Radiometric titration of small amounts of Co with EDTA. *Bunseki Kagaku*, **15**, 637 (1966).

66-71 MÜLLER, K., Rasche radiometrische Bestimmung durch Rücktitration mit AeDTE. *Analyt. Chim. Acta*, **35**, 162 (1966).

66-72 AUSTIN, J. H., and KLETT, C. A., EGTA titration of calcium in algae samples high in magnesium and phosphate. *Chem. Analyst*, **55**, 11 (1966).

66-73 BODEN, H., Determination of magnesium in aluminium alloys via a DCTA titration. *Chem. Analyst*, **55**, 75 (1966).

66-74 BECK, M. T., and GERGELY, A., Stability constant of Bi(III)–EDTA and Bi(III)–DCTA complexes. *Acta Chim. Acad, Sci. Hung.*, **50**, 155 (1966).

66-75 MONNIER, D., DELPHIN, G., and HAERDI, W., Rapid electrochemical determination of Ca and Mg in presence of each other in one operation without separation. *Analyt. Chim. Acta*, **35**, 231 (1966).

66-76 AGGARWAL, R. C., and SRIVASTAVA, A. K., Direct complexometric determination of Tl(III) using iron(III) sulphosalicylate complex as a metal indicator. *Bull. Chem. Soc. Japan*, **39**, 2178 (1966).

66-77 BACQUIAS, G., Rapid testing of sulphate in chromium solutions. *Galvano*, **35**, 203 (1966).

66-78 BOGATOVA, E. I., Complexometric determination of Ni in Cu–Ni catalysts. *Nauch. Tr. Krasnodar. Gos. Pedagog. Inst.*, No. 70, 40 (1966).

66-79 BALOGH, G., and FELSZEGHY, E., Complexometric determination of the phosphates in inorganic systems and biological materials. *Stud. Univ. Babes-Bolyai. Ser. Chem.*, No. 2, 111 (1966).

66-80 BONDAREVA, E. G., and KOBYAK, G. G., Complexometric determination of P in copper–phosphorus alloys by titration without a burette. *Uch. Zap. Perm. Gos. Univ.*, **141**, 276 (1966).

66-81 BONDAREVA, E. G., SHARDAKOVA, M. A., and KOBYAK, G. G., Complexometric determination of Al in titanium alloys by titration without a burette. *Uch. Zap. Perm. Gos. Univ.*, **141**, 279 (1966).

66-82 BRUILE, E. S., and MERKULOVA, K. S., Complexometric determination of Cu and Zn in brasses. *Tr. Vses. Nauch.-Issled. Konstr. Inst. Khim. Mashinostr.*, **51**, 174 (1966).

66-83 CEAUSESCU, D., PIRIRI, F., and PIRIRI, I., Conductometric titrations with the tetra sodium salt of ethylenediaminetetraacetic acid. *Proc. Conf. Appl. Phys.-Chem. Methods Chem. Anal.*, (*Budapest*), **1**, 314 (1966).

66-84 CHAUDHRI, R. S., Complexometric determination of iron in Dagshai rocks. *Indian Mineral.*, **7**, 71 (1966).

66-85 CIURLO, R., and BIINO, L., Chelatometric micro-determination of sulphate in small quantities. *Atti. Soc. Peloritana Sci. Fis. Mat. Natur.*, **12**, 537 (1966).

66-86 COPELLO, M. A., and DE DORFMAN, E. A., Chelatometric and potentiometric determinations. Use of a glass electrode as a reference electrode. *Ann. Soc. Cient. Argent.*, **182**, 3 (1966).

66-87 CRISAN, I. A., and LAKATOS, A., The complexometric analysis of Kaolin. *Rev. Chim.* (*Roumania*), **17**, 557 (1966).

66-88 GORBENKO, F. P., and DEGTYARENKO, L. I., Extraction complexometric determination of zinc in copper compounds. *Tr. Vses. Nauch.-Issled. Inst. Khim. Reaktivov Osobo Chist. Khim. Veshshestv.*, **29**, 73 (1966).

66-89 GORBENKO, F. P., TESLINSKII, YU. K., and NADEZHDO, A. A., Determination of traces of calcium in magnesium salts and oxide with Azo-Azoxy BN. *Tr. Vses. Nauch.-Issled. Inst. Khim. Reaktivov Osobo Chist. Khim. Veshchestv.*, **29**, 59 (1966).

66-90 HORIUCHI, T., and IIMURA, I., Chelatometric titration of chromate by reduction. *Technol. Reports. Iwate Univ.*, **2**, 49 (1966).

66-91 HORIUCHI, Y., and ICHIJYO, O., Chelatometric determination of Ti in the presence of Al. *Iwate Daigaku Kogakubu Kenkyu Hokoku*, **20**, 49 (1966).

66-92 HORIUCHI, Y., and SHIMOZI, M., Chelatometric titration of hydroxylamine with iron as oxidant. *Iwate Daigaku Kogakubu Kenkyu Hokoku*, **20**, 63 (1967).

66-93 IRITANI, N., TANAKA, T., and OGAWA, K., Chelatometric determination of pharmaceuticals. Part 5. Determination of zinc salts. *J. Pharm. Soc. Japan*, **86**, 1112 (1966).

66-94 KEMULA, W., BRAJTER, K., and BOGDANSKA, E., Determination of Ba in barium ferrites by using Dowex-50 cation exchanger. *Chem. analit.* (*Warsaw*), **11** (6), 1239 (1966).

66-95 KHLYSTOVA, K. B., Rapid complexometric determination of Ni and Fe in an alloy. *Uch. Zap. Yaroslav. Tekhnd. Inst.*, **9**, 121 (1966).

66-96 KIM, H. R., and KANG, K. H., Equations for chelatometric titration curve of a mixture of metals in aqueous solution. *Chosun Kwahakwon Tongbo*, 36 (1966).

66-97 KISS, T., RÁDY, G., and ERDEY, L., Complexometric determination of copper with EDTA by back-titration in the presence of Eriochrome Black as indicator. *Periodica Polytech.*, **10**, 303 (1966).

66-98 KLETENIK, YU. B., and BYKHOVSKAYA, I. A., Extraction of titanium with 2-ethylhexylphosphoric acids. Part 2. Complexonometric determination of Ti in strip solution. *Zhur. analit. Khim.*, **21**, 1499 (1966).

66-99 KONISHI, S., and DOLI, N., Control of lead–tin fluoroborate plating baths and analyses of the deposit from the bath. *Kinzoku Hyomen Gijutsu*, **17**, 343 (1966).

66-100 KOTRLY, S., Theory of colour change in metallochromic indicators. *Sci. Papers Univ. Chem. Technol. Pardubice*, **14**, 165 (1966).

66-101 LAESSIG, R. H., Continuous EDTA titrations with direct read-out. *Diss. Abs.*, **27**, V, 1052 (1966).

66-102 LAZAR-JUCU, D., DRAGULESCU, C., and KUZMAN-ANTON, R., New titrimetric determination of Be with anthranyldiacetic acid. *Bull. Stiint. Tek. Inst. Politek. Timisoara*, **11** (1), 69 (1966).

66-103 LENGYEL NADLER, V., Complexometric determination of sulphide and calcium content of lime water. *Bor Cipo-Tech.*, **17**, 144 (1966).

66-104 DE LOPIDANA., Metallofluorescent indicators. *Acta Cient. Compostilana*, **3**, 173 (1966).

66-105 LUKIN, A. M., SMIRNOVA, K. A., and PETROVA, G. S., Sulfarsazen as a new reagent for mercury. *Tr. Vses. Nauch.-Issled. Inst. Khim. Reaktivov Osob Chist. Khim. Veshchestov.*, **29**, 290 (1966).

66-106 LUKIN, A. M., SMIRNOVA, K. A., and ZAVARIKHINA, G. B., Determination of calcium with Calcion. *Metody. Anal. Khim. Reaktivov Prep. Moscow*, **12**, 57 (1966).

66-107 MEISSL, R. F., and GOMEZ, A. A., Complexometric determination of Al in natural sulphates of Calingasta. *Acta Cuyana Ing.*, **8** (1), 71 (1966).

66-108 MOIZHES, I. B., and FILYUZINA, V. S., Complexometric analysis of garnets. *Tr. Vses. Nauch.-Issled. Geol. Inst.*, **125**, 195 (1966).

66-109 MOLOT, L. A., PETRIKOVA, G., and SAMSONOVA, N. N., Rapid analysis of an Al–Si–Fe alloy. *Metody Kontr. Khim. Sostava Neorg. Orrg. Soedin*, 169 (1966).

66-110 NAKASUKA, N., and TANAKA, M., Theory of complexometric titrations with redox indicators. EDTA titration of iron(III) using Variamine Blue as indicator. *Analyt. Chim. Acta*, **36**, 422 (1966).

66-111 PANTANI, F., and CATANI, I., Complexometric determination of Tl(III) with potentiometric detection of the equivalence point. *Ricerca Sci.*, **36**, 829 (1966).

66-112 PODLAHA, J., and PODLAHOVA, J., Preparation and properties of EDTA complexes. Part 6. Halogeno- and thiocyanato-complexes of titanium(III) and vanadium(III). *Coll. Geol.*, **31**, 4467 (1966).

66-113 PŘIBIL, R., Determination of Th and rare earths with triethylenetetraminehexaacetic acid (TTHA). *Talanta*, **13**, 1711 (1966).

66-114 SHAFRAN, I. G., KON'HOVA, O. V., and RACHKEVICH, V. F., Unified complexometric determination of the main constituents in scandium compounds. *Tr. Vses. Nauch.-Issled. Inst. Khim. Reaktivov Osobo Chist. Khim. Veshchestv.*, **28**, 86 (1966).

66-115 SHAFRAN, I. G., PARTASHNIKOVA, M. Z., and PLETNEVA, T. I., Complexometric determination of In in In–Sb–As solid solution.

Tr. Vses. Nauch.-Issled. Inst. Khim. Reaktivov Osobo Chist. Khim. Veshchestv., **28**, 92 (1966).

66-116 SIERRA, F., and SANCHEZ-PEDRINO, C., Potentiometric and volumetric determination of Pb^{2+} and Zn^{2+} with $BaNa_2EDTA$ and a redox system. *Anales real Soc. espan. Fis. Quim.*, **62B**, 1149 (1966).

66-117 SOLODOVNIKOV, P. P., Determination of metallic aluminium in the presence of aluminium oxide. *Tr. Kazan. Aviats. Inst.*, No. **90**, 64 (1966).

66-118 SZARVAS, P., KORONDAN, I., and RAISZ, I., Use of the indicator Luminol (3-aminophthalichydrazide) for the complexometric determination of alkaline earth metals. *Magyar Kém. Folyóirat*, **72**, 441 (1966).

66-119 VITAL'SKAYA, N. N., GLAVINA, V. S., and KLEMINA, T. N., Rapid complexometric determination of $BaSO_4$ in lithopone and technical grade baryte, and zinc sulphide and zinc oxide in lithopone. *Proiz. Shin. Rezino-Tekh. Asbestotekh. Izdelii*, **4**, 27 (1966).

66-120 VOŘLIČEK, J., Biamperometric indication of EDTA titration in the determination of Ca, Mg, Sr and Ba. *Proc. Conf. Appl. Phys.-Chem. Methods Chem. Anal.*, *Budapest*, **1**, 357 (1966).

66-121 VOŘLIČEK, K., and VÝDRÁ, F., Determination of alkaline earths by EDTA titration and bi-amperometric indication. *Sb. Pr. Vyzk. Ustavu ZDHE*, **7**, 124 (1966).

66-122 VÝDRÁ, F., and VOŘLIČEK, J., Amperometry with two polarizable electrodes. XII. Chelometric determination of Ga and In. *Proc. Conf. Appl. Phys.-Chem. Methods Chem. Anal.*, *Budapest*, **1**, 233 (1966).

66-123 ZHIVOPISTSEV, V. P., SELEZNEVA, E. A., and CHEREPANOVA, T. B., Extraction complexometric determination of germanium with diantipyrylmethane. *Uch. Zap. Perm. Gos. Univ.*, **141**, 179 (1966).

66-124 ZHIVOPISTSEV, V. P., and KALMYKOVA, I. S., Complexometric determination of scandium after extraction of its diantipyryl-methane–nitrate complex. *Uch. Zap. Perm. Gos. Univ.*, **141**, 186 (1966).

1967

67-1 OLSEN, E. D., and ADAMO, F. S., Automatic chelometric titrations of metal ions using a Ag–DTPA electrode. *Analyt. Chem.*, **39**, 81 (1967).

67-2 PŘIBIL, R., and VESELÝ, V., Rapid analysis. XXVI. Determination of rare earths in the presence of phosphate. *Chem. Analyst*, **56**, 23 (1967).

67-3 KINNUNEN, J., and WENNERSTRAND, B., Rapid determination of rare earths in phosphate rock. *Chem. Analyst*, **56**, 24 (1967).

67-4 CULP, S. L., Determination of Ti and Al in their binary alloys. *Chem. Analyst*, **56**, 29 (1967).

67-5 AIKENS, D. A., and GAHBAH, F. J., Potentiometric characterization of Al aminopolycarboxylate complexes. *Analyt. Chem.*, **39**, 646 (1967).

67-6   ARTHUR, P., and HUNT, B. R., Amperometric titration of Pb, Cd and Zn in 2-propanol with DCyTA. *Analyt. Chem.*, 39, 95 (1967).

67-7   ELSHEIMER, H. N., Complexometric determination of Ga with Calcein Blue as indicator. *Talanta*, 14, 97 (1967).

67-8   KONOPIK, N., Reaktion von Germansäure mit AeDTE. *Z. analyt. Chem.*, 224, 107 (1967).

67-9   LANGER, H. G., and BOGUECKI, R. F., The chemistry of Sn(II) chelates. II. The Sn(II)–EDTA system in aqueous solutions. *J. Inorg. Nuclear Chem.*, 29, 495 (1967).

67-10   MEITES, L., and MEITES, T., Theory of titration curves. VI. Slopes and inflection points of potentiometric chelometric titrations. *Analyt. Chim. Acta*, 37, 1 (1967).

67-11   ABDURAKHMANOV, M., TALIPOV, SH.T., and DZHIYANBAEVA, R. KH., Complexometric determination of zirconium. *Tr. Tashkent. Gos. Univ.*, 288, 79 (1967).

67-12   AGULÁ, J. F., Fluorescent aluminium indicators derived from substituted coumarins. *Talanta*, 14, 1195 (1967).

67-13   AMIRKHANOVA, T. B., IVANOVA, I. YA., and PODGORNOVA, V. S., Complexometric titration of Ni with *N*-methylanabasine-α'-azo-α-naphthol as indicator. *Tr. Tashkent. Gos. Univ.*, 288, 106 (1967).

67-14   AMIRKHANOVA, T. B., PODGORNOVA, V. S., and SHESTEROVA, I. P., *N*-Methylanabasine-α'-azo-α-naphthol as a reagent for the determination of cadmium. *Tr. Tashkent. Gos. Univ.*, 288, 109 (1967).

67-15   BENSCH, H., HELMBOLDT, O., KOESTER, M., HUEBNER, K., and PROTZER, H., Determination of aluminium in bauxite. *Erzgebau Metalhüttenw.*, 20 (11), 522 (1967).

67-16   BRÜCK, A., and LAUER, K. F., Precise complexometric titration of uranium(VI). *Analyt. Chim. Acta*, 37, 325 (1967).

67-17   BUDĚŠINSKÝ, B., Xylenol Orange and Methylthymol Blue as chromogenic reagents. *Chelates Anal. Chem.*, 1, 15-47 (1967).

67-18   CHROMÝ, V., and SOMMER, L., 2-(2-Thiazolylazo)-4-methoxy-phenol and 2-(2-thiazolylazo)-5-methoxyphenol as metal-ion indicators. *Talanta*, 14, 393 (1967).

67-19   CONRADI, G., Determination of magnesium, calcium, sodium and potassium in tissue samples. *Chem. Analyst*, 56, 87 (1967).

67-20   COPELLO, M. A., DORFMAN, E. A., and AMAROS, J. L., Complexometric determination of phosphate. *Safybi*, 7, 10 (1967).

67-21   CRISAN, I. A., Complexometry. I. Determination of iron(III) using Complexon III. *Stud. Cercet. Chim.*, 15 (3), 233 (1967).

67-22   CRISAN, I. A., and BERCEA, E., Titration of a mixture of iron(III), cobalt(II) and aluminium(III) with Complexon III. *Rev. Chim. (Roumania)*, 18, 370 (1967).

67-23   CRISAN, I. A., and BOLOS, V., Determination of Sn(II) by back-titration of excess Complexon III with lead nitrate in the presence of Pyrocatechin Violet. *Rev. Chim. (Roumania)*, 18, 307 (1967).

67-24   CRISAN, I. A., and DANIEL, A., Determination of chromium(VI) by Complexon III. *Stud. Univ. Babes-Bolyai., Ser. Chem.*, 12 (1), 135 (1967).

67-25 CRISAN, I. A., and DOMOCOS, V. F., Determination of titanium by back-titration of excess of Complexon III with lead nitrate in the presence of Pyrocatechin Violet. *Stud. Univ. Babes-Bolyai, Ser. Chem.*, 12, 109 (1967).

67-26 CRISAN, I. A., ONISOR, M.-I., RUSU, R., GÖRBE, E., and OPRESCU, C.-L., Determination of indium(III) by back-titration of the excess of complexon III. *Stud. Univ. Babes-Bolyai, Ser. Chem.*, 12, 55 (1967).

67-27 CRISAN, I. A., and PANITI-CETEAN, M., Complexometric determination of thallium(I). *Stud. Univ. Babes-Bolyai, Ser. Chem.*, 12, 39 (1967).

67-28 CRISAN, I. A., and PFEIFFER, M. M., Complexometric analysis of a steel alloyed with nickel and chrome. *Rev. Chim. (Roumania)*, 18, 109 (1967).

67-29 CRISAN, I. A., and TIRA, L., Titration of mixtures of Zn(II) and Cr(III) with the aid of Complexon III. *Stud. Univ. Babes-Bolyai, Ser. Chem.*, 12, 139 (1967).

67-30 CYGANSKI, A., Analytical applications of thiocyanatobismuthates. Part 3. New complexometric and gravimetric determination of caesium as caesium tetrathiocyanatobismuthate. *Chem. Analit. (Warsaw)*, 12, 765 (1967).

67-31 DUBSKY, I., Copper determination in cyanide-containing baths. *Metalloberfläche*, 21, 5 (1967).

67-32 DZHIYANBAEVA, R. KH., TALIPOV, SH. T., and KIRIKA, A. L., *N*-Methylanabasine-(α'-azo-2)-1-naphthol-5-sulphonic acid as a metallochromic indicator for copper. *Tr. Tashkent Gos. Univ.*, 288, 69 (1967).

67-33 ELINSON, S. V., and NEZHNOVA, T. I., Complexometric determination of Ti and Zr in alloys with Nb. *Zavodskaya Lab.*, 33, 927 (1967).

67-34 EVANS, W. H., Rapid complexometric determination of Al and total iron in silicate and other rock material. *Analyst*, 92, 685 (1967).

67-35 EZERSKAYA, N. A., and SOLOVYKH, T. P., Compound formation in the reaction of ruthenium(III)–EDTA with hydrogen peroxide. *Zhur. neorg. Khim.*, 12, 2922 (1967).

67-36 FEDEROV, A. A., OZERSKAYA, F. A., and STREBULAEVA, E. N., Use of a photoelectric titrimeter for the trilonometric determination of cobalt oxide in chromium ores and concentrates. *Zavodskaya Lab.*, 33 (12), 1502 (1967).

67-37 FLASCHKA, H., and MANN, J., Some observations concerning the quality of murexide end-points. *Analyt. Letters*, 1, 19 (1967).

67-38 FOERSTER, W., ZIEGER, M., and RUEDIGER, H., Determination of aluminium in titaniferous alloys. *Neue Hütte*, 12 (3), 150 (1967).

67-39 GAUR, J. N., and JAIN, D. S., Determination of thallium(I) in presence of cadmium(II) by polarographic and amperometric methods using 1,2-diaminocyclohexane-tetra-acetic acid as complexing agent. *Z. analyt. Chem.*, 233, 109 (1967).

67-40 GERTNER, A., and GRDINIC, V., Ring oven methods. II. Complexometric ultramicrodetermination of iron(III). *Mikrochim. Acta*, 1048 (1967).

67-41 GEYER, R., and BORMANN, R., Xylene Orange and Methylthymol Blue as indicators for the direct chelatometric determination of aluminium. *Z. Chem.*, **7**, 30 (1967).

67-42 GLADYSHEVA, K. F., and TSAREVA, K. KH., Thioacetamide for separation of zinc and cadmium. *Zavodskaya Lab.*, **33**, 30 (1967).

67-43 GLAUSER, S. C., IFKOVITS, E., GLAUSER, E. M., and SEVY, R. W., Complexometric titrations using calcium-specific electrodes. *Proc. Soc. Exp. Biol. Med.*, **124**, 131 (1967).

67-44 GORBENKO, F. P., and LAPITSHAYA, E. V., Comparative study of the complexometric determination of calcium by means of various metallic indicators. *Ukrain. khim. Zhur.*, **33**, 1302 (1967).

67-45 GOYAL, S. S., and TANDON, J. P., Vanadium(IV) complexes of Azoxine S dyes (as indicators for the determination of V). *Talanta*, **14**, 1449 (1967).

67-46 GROSSKREUTZ, W., SCHULTZ, D., and WILKE, K. T., Complexometric determination of yttrium and aluminium in yttrium–aluminium–garnet crystals. *Z. analyt. Chem.*, **232**, 278 (1967).

67-47 GUSEV, S. I., KOZHEVNIKOVA, I. A., MAL'TSEVA, L. S., and SHCHUROVA, L. M., Pyridylazo-compounds as metallochromic indicators. Part 4. The reactions of pyridylazo-*p*-cresol, pyridylazo-monoethylamino-*p*-cresol and its bromo-derivatives with lead ions. *Zhur. analit. Khim.*, **22**, 1190 (1967).

67-48 GUSEV, S. I., and KUREPA, G. A., 5-(2-pyridylazo)-2-monoethyl-amino-*p*-cresol and its bromo derivatives as indicators for the complexometric determination of thallium. *Zhur. analit. Khim.*, **22**, 863 (1967).

67-49 HANNEMA, V., and DEN BOEF, G., Titration curves of complexometric titrations of mixtures of metal ions with one ligand. Part 1. Mathematical expressions. *Analyt. Chim. Acta*, **39**, 167 (1967).

67-50 HANNEMA, V., and DEN BOEF, G., Titration curves of complexometric titrations of mixtures of metal ions with one ligand. Part 2. Conditions for sharp end-points. *Analyt. Chim. Acta*, **39**, 479 (1967).

67-51 HANUS, J., Determination of Co, Mn, Pb, Zn and Ca in mixtures in naphthenates. *Chem. prumysl.*, **17** (4), 215 (1967).

67-52 HEYER, F. C., Stepwise EDTA titrations for zinc or cadmium and alkali cyanide in plating-baths. *Chem. Analyst*, **56**, 86 (1967).

67-53 HORIUCHI, Y., and ICHIJYO, O., Methylthymol Blue as an indicator in the chelatometric titration of chromium. *Iwate Daigaku Kogakubu Kenkyu Hokoku*, **19**, 91 (1966); *Amer. Chem. Abs.*, **67**, 87384 (1967).

67-54 HUITINK, G. M., Substituted coumarins as metallofluorochromic indicators. *Diss. Abs.*, **28 B**, 1386 (1967).

67-55 IRITANI, N., MIYAHARA, T., YANO, K., and KITANO, S., Chelatometric determination of pharmaceuticals. Part 6. Determination of aluminium salts. *J. Pharm. Soc. Japan*, **87**, 1427 (1967).

440   Bibliography

67-56   JOHRI, K. N., and SINGH, K., Potassium thiocarbonate (PTC) as a masking and demasking agent and as a metal-indicator in EDTA titrations. *Indian J. Appl. Chem.*, **30**, 1 (1967).

67-57   JOHRI, K. N., and SINGH, K., Potassium thiocarbonate as a masking agent and as a metal indicator in EDTA titrations. *Bull. Chem. Soc. Japan*, **40**, 990 (1967).

67-58   JORDAN, D. E., and MONN, D. E., Rapid determination of Mg in the presence of Ca and phosphate by titration with DCTA. *Analyt. Chim. Acta*, **37**, 42 (1967).

67-59   JORDAN, D. E., and MONN, D. E., Spectrophotometric end-point detection for the determination of Mg in the presence of Ca and phosphate by titration with DCTA. *Analyt. Chim. Acta.* **39**, 401 (1967).

67-60   KAMAEVA, G., TALIPOV, SH. T., and DZHIYANBAEVA, R. KH., Complexometric determination of Cu in the presence of 4-(2-N-methylanabasinoazo)-resorcinol (MAAR). *Uzbek. khim. Zhur.*, No. **4**, 14 (1967).

67-61   KAROLEV, A. N., Rapid complexometric determination of zinc. *Zavodskaya Lab.*, **33**, 811 (1967).

67-62   KARPOV, O. N., SAVOS'KIN, V. M., GARKUSHA-BOZHKO, I. P., and STETS, R. G., Complexometric determination of metals (e.g. Co, Cu, Mn) in naphthenates. *Zhur. analit. Khim.*, **22**, 169 (1967).

67-63   KAWABE, A., Colour reactions of analogues of 1-(2-pyridylazo)-2-naphthol (PAN) with metal ions. *Bunseki Kagaku*, **16** (6), 569 (1967).

67-64   KHADEEV, V. A., and AKHMEDZHANOVA, S. A., Amperometric titration of Ca and Mg with Complexon III on an apparatus with a solid rotating micro-electrode. *Nauch. Tr. Tashkent. Univ.*, **284**, 3 (1967).

67-65   KHADEEF, V. A., and KOCHERGINO, S. A., Amperometric determination of thorium in the presence of foreign metal ions forming stable complexonates. *Uzbeck. Khim. Zhur.*, **11**, 15 (1967).

67-66   KHADEEF, V. A., and MASALOVA, L. S., Amperometric determination of traces of thorium. II. Amperometric titration of Th with Complexon III by using two indicating electrodes. *Tr. Tashkent. Gos. Univ.*, **288**, 12 (1967).

67-67   KHALIZOVA, V. A., KRASYUKOVA, I. G., DONCHENKO, V. A., ALEKSEEVA, A. Y., and SMIRNOVA, E. P., Complexometric determination of lead in ores after concentration on an ion-exchanger. *Zavodskaya Lab.*, **33**, 1064 (1967).

67-68   KIDAMA, M., Spectrophotometric studies of the solution equilibria and the kinetics of the substitution reaction between Eriochrome Black T and cobalt(II)–EDTA chelate. *Bull. Soc. Chem. Japan*, **40**, 2575 (1967).

67-69   KIRIKA, A. L., DZHIYANBAEVA, R. KH., and TALIPOV, SH. T., Complexometric titration of Bi with N-methylanabasine-(α'-azo-2)-1,5-dihydroxynaphthalene as indicator. *Tr. Tashkent. Gos. Univ.*, **288**, 62 (1967).

67-70   KISELEVA, L. V., Complexometric method of determining all heavy-metal contamination in high-purity materials. *Zhur. analit. Khim.*, **22**, 557 (1967).

67-71 KLASSOVA, N. S., A rapid complexometric method for determining Al in microsample of minerals and rocks. *Zhur. analit. Khim.*, **22**, 810 (1967).

67-72 KRASYUKOVA, N. G., Analysis of iron ores based on anion-exchange and complexometry. *Mater. Geol. Polezn. Iskop. Buryat. ASSR, Buryat. Geol. Vpr.*, **10**, 238 (1967). *Amer. Chem. Abs.*, **69**, 8088e (1967).

67-73 KRLEŽA, F., Chelatometric determination of Ni, Co, Mn and Zn in iron systems previously treated with glycerine. *Croat. Chem. Acta*, **39**, 47 (1967).

67-74 KRUMINA, V. T., ASTAKHOV, K. V., BARKOV, S. A., and KORNEV, V. I., Spectrophotometry of complexes formed in a mercury(II)–diethylenetriaminepenta-acetic acid system. *Zhur. neorg. Khim.*, **12**, 3356 (1967).

67-75 KUTEINIKOV, A. F., and LYSENKO, S. A., Complexometric determination of niobium. *Zavodskaya Lab.*, **33**, 141 (1967).

67-76 KUTEINIKOV, A. F., and LYSENKO, S. A., Conditions for the complexometric determination of Nb. *Zhur. analit. Khim.*, **22**, 1366 (1967).

67-77 KUZ'MENKO, N. I., YAKIMETS, E. M., and VAINER, M. G., Some obstacles encountered during the trilonometric determination of Ca and Mg and methods for eliminating the hindrances. *Tr. Ural. Politekh. Inst.*, **163**, 94 (1967).

67-78 KYŘŠ, M., and SELUCKÝ, P., Indirect determination of caesium by complexometric titration of calcium after extraction with calcium dipicrylaminate in nitrobenzene. *Anal. Chim. Acta*, **38**, 460 (1967).

67-79 LASTOVSKII, R. P., DYATLOVA, N. M., and SELIVERSTOVA, I. A., Complexing of 2,3-dihydroxy-1,4-diaminobutanetetra-acetic acid with transition and rare earth elements. *Zhur. neorg. Khim.*, **12**, 3351 (1967).

67-80 LEBEDEV, I. A., MAKSIMOVA, A. M., STEPANOV, A. V., and SHALINETS, A. B., Determination of the stability constants for complexes of americium and curium with EDTA by means of electromigration. *Radiokhimiya*, **9**, 707 (1967).

67-81 LEVINE, S. L., and GOLDEN, H. J., Photometric end-point determination with EDTA titration of Zn in the presence of Mn. *Analyt. Letters*, **1**, 39 (1967).

67-82 LITEANU, C., and CRISAN, I. A., Rapid analytical methods. IX. Chromite analysis. *Stud. Univ. Babes-Bolyai, Ser. Chem.*, **12**, 7 (1967).

67-83 LITEANU, C., CRISAN, I. A., and POPESCU, M., Complexometric determination of iron(III) in the presence of large amounts of aluminium. *Stud. Univ. Babes-Bolyai, Ser. Chem.*, **2**, 25 (1967).

67-84 LITEANU, C., CRISAN, I. A., and THEISS, G., Determination of Ag(I) with Complexon III using a new substitution method and double titration. *Rev. Roumaine Chim.*, **12**, 569 (1967).

67-85 MAEKAWA, S., YONEYAMA, Y., and MORINAGA, H., Simultaneous determination of Mg and Ca in iron and steel using a photometric EDTA titration. *Japan Analyst*, **16**, 455 (1967).

67-86 MARTINY, E., and STRESKO, V., Improved end-point sharpness in the complexometric determination of iron(III) with biamperometric indication through the addition of manganese(II). *Z. analyt. Chem.*, **231**, 17 (1967).

67-87 MATTEUCCI, E., Complexometric titration of mercury in the presence of chloride ions. Part 2. Titration through substitution in acidic media. *Chimica e Industria*, **49**, 474 (1967).

67-88 MESZAROS, P., VINEK, G., and KONOPIK, N., Determination of dissociation constants of the forms hydrogen(5)–Y(IV), hydrogen(4)–Y(IV) and hydrogen(3)–Y(IV) of EDTA. *Monatsh.*, **98**, 1810 (1967).

67-89 MIHALESCU, M., Complexometric determination of Pb and Zn in bronzes. *Metalurgia (Bucharest)*, **19** (8), 443 (1967).

67-90 MOLOT, L. A., MUSTAFIN, I. S., and NEMKOVA, N. K., Comparative study of metallochromic indicators for direct complexometric determination of aluminium. *Izvest. Vysshikh. Uchebn. Zavenenni Khim. i Khim. Tekhnol.*, **10**, 1060 (1967).

67-91 MORI, S., HIRATA, H., and TONOOKA, N., Determination of iron(III) and iron(IV) in barium orthoferrate, $BaFeO_{2.5-3.0}$ by EDTA titrations. *Japan Analyst*, **16**, 1340 (1967).

67-92 MOTTOLA, H. A., and FREISER, H., Use of metal-ion catalysis in detection and determination of micro-amounts of complexing agents: determination of EDTA. *Analyt. Chem.*, **39**, 1294 (1967).

67-93 NEMODRUK, A. A., and GLUKHOVA, L. P., Complexometric determination of quadrivalent plutonium using Xylenol Orange as indicator. *Zhur. analit. Khim.*, **22**, 193 (1967).

67-94 NEVROZOV, A. N., and GANENKO, Z. G., Determination of zirconium in samples containing Nb and Ta by Xylenol Orange. *Zavodskaya Lab.*, **33**, 285 (1967).

67-95 NOMURA, T., and NAKAGAWA, G., Potentiometric determination of Ca with EGTA (glycoletherdiaminotetra-acetic acid) using silver as an indicator electrode. *Japan Analyst*, **16**, 1309 (1967).

67-96 NOMURA, T., and NAKAGAWA, G., Potentiometric titration with chelating agents using metal oxide electrode as an indicator electrode. *Japan Analyst*, **16**, 1314 (1967).

67-97 NOMURA, T., NAKAGAWA, G., and DONO, T., Potentiometric titration with chelating agents, with an indicator electrode of copper. *Japan Analyst*, **16** (3), 216 (1967).

67-98 NOVAK, V., KOTOUČEK, M., LUČANSKÝ, J., and MAJER, J., New complexones. Part 10. Polarographic investigation of chelates of ethylenediamine-*NN'*-diaceto-*NN'*-(α,α'-dipropionic)acid with lanthanides and some bivalent cations. *Chem. Zvesti*, **21**, 687 (1967).

67-99 ORLOVA, YU. YA., GUSENKO, T. V., and SVERZHINA, R. E., Rapid determination of Ca and Mg oxides in sintered agglomerates. *Zavodskaya Lab.*, **33**, 565 (1967).

67-100 PÁLYI, G., Complexometric determination of magnesium in the presence of Luminol (3-aminophthalic acid hydrazide) as indicator. *Magyar Kém. Folyóirat*, **73**, 320 (1967).

67-101 PASOVSKAYA, G. B., Determination of Ca, Fe and Al by conductometric titration with sodium nitrilotriacetate. *Izvest. V.U.Z.M.O.SSSR Khim. i khim. Tekhnol.*, **10**, 624 (1967).

67-102 PATROVSKY, V., Use of *p*-aminophenylazochromotropic acid (Victoria Violet) as chelatometric indicator and photometric reagent in the determination of magnesium. *Z. analyt. Chem.*, **230**, 428 (1967).

67-103 PAVLIKOVA, E., and WEISS, D., Complexometric determination of Ca and Mg in iron ores. *Sb.Pr.Ustavu Vyzk. Rud.*, **6**, 244 (1967).

67-104 PORTELL, R., and BUSQUETS, P., Complexometric analysis of carbonate rocks. *Breviora Geol. Asturica Univ. Oviedo, Fac. Cienco*, **9**, 83 (1967).

67-105 PISON, R. E., Potentiometric determination of lead in ethylated petrol using a disodium salt of EDTA. *Quimica e Industria*, **14**, 107 (1967).

67-106 PŘIBIL, R., Present state of complexometry. Part 3. Determination of univalent metals. Part 4. Determination of rare earths. *Talanta*, **14**, 613 (1967); **14**, 619 (1967).

67-107 PŘIBIL, R., and HORÁČEK, J., Complexometry. Part 23. Determination of Th, Sc and some lanthanides in the presence of each other. *Talanta*, **14**, 313 (1967).

67-108 PŘIBIL, R., and HORÁČEK, J., Determination of zirconium–yttrium mixtures employing EDTA. *Chem. Analyst*, **56**, 76 (1967).

67-109 PŘIBIL, R., and VESELÝ, V., Identification tests for DCTA, DPTA, EDTA, EGTA, MEDTA, NTA and TTHA. *Chem. Analyst*, **56**, 51 (1967).

67-110 PŘIBIL, R., and VESELÝ, V., Determination of EDTA, DPTA and TTHA in their mixtures. *Chem. Analyst*, **56**, 83 (1967).

67-111 PŘIBIL, R., VESELÝ, V., and HORÁČEK, J., Basic problems of complexometry. Part 22. Determination of Th and Sc in the presence of each other. *Talanta*, **14**, 266 (1967).

67-112 RAD'KO, V. A., Analysis of AZhMts bronzes for Al, Mn, Fe and Cu content by a complexometric method. *Tr. Ural. Politekh. Inst.*, **163**, 98 (1967).

67-113 ROCHE, M., Rapid titrimetric determination of Ca and Mg in limestone. *Sucr. Fr.*, **108** (10), 2 (1967).

67-114 ROLLINS, O. W., and HAYNES, B. J., Determination of gallium in molybdo- and tungsto-gallates via an EDTA titration. *Chem. Analyst*, **56**, 98 (1967).

67-115 ROZYCKI, C., Complexometric analysis of Mn–Zn–Mg ferrite. *Chem. analit. (Warsaw)*, **12**, 573 (1967).

67-116 SAPEK, A., and ZMINKOWSKA, T., Extractive removal of interfering Fe, Cu, Mn and Al ions in the complexometric determination of Ca and Mg in soil extracts. *Chem. analit. (Warsaw)*, **12**, 907 (1967).

67-117 SCHULTZ, F. A., and SAWYER, D. T., Electrochemical studies of

444   *Bibliography*

molybdenum–ethylenediaminetetra-acetic acid complexes. *J. Electroanalyt. Chem. Interfacial Electrochem.*, **17**, 207 (1967).

67-118   SELIVERSTOVA, I. A., SAMOILOVA, O. I., DYATLOVA, N. M., and YASHUNSKII, V. G., Substances with complexing capacity. Synthesis and properties of meso-2,3-dihydroxy-1,4-diaminobutanetetra-acetic acid. *Zhur. obsch. Khim.*, **37**, 2643 (1967).

67-119   SIERRA, F., and HERNÁNDEZ-CAÑAVATE, J., Potentiometric determination of Hg^{2+} with BaL^{2-} (barium ethylenediaminetetracetate anion) and a Pt electrode. *An. Real. Soc. Espan. Fis. Quim.*, *Ser. B.*, **63**, 817 (1967).

67-120   SILVER, G. L., and BOWMAN, R. C., Reactions of Eriochrome Blue Black R. *Talanta*, **14**, 893 (1967).

67-121   SINGHAL, G. K., and TANDON, K. N., Metallochromic indicators. Part 1. Pyrocatecholphthalein as a metallochromic indicator. *Talanta*, **14**, 1127 (1967).

67-122   SINGHAL, G. K., and TANDON, K. N., Direct EDTA titration of vanadium(IV) with haematoxylin as indicator. *Chem. Analyst*, **56**, 60 (1967).

67-123   SINGHAL, G. K., and TANDON, K. N., Studies on metallochromic indicators. Part 2. Zincon and its mercury and zinc complexes as indicators in EDTA titrations. *Talanta*, **14**, 1351 (1967).

67-124   SINHA, B. C., and DAS GUPTA, S., Direct complexometric determination of zirconium(IV) in relation to polymerization. *Analyst*, **92**, 558 (1967).

67-125   SKORIK, N. A., KUMOK, V. N., and SEREBRENNIKOV, V. V., Stability of complexes of La, Sc and Th with nitriloacetate ion. *Zhur. neorg. Khim.*, **12**, 3381 (1967).

67-126   STEPANOV, A. V., MAKAROVA, T. P., MAKSIMOVA, A. M., and SHALINETS, A. B., Complexing of cerium(III), americium(III) and curium(III) with 1,2-diaminocyclohexanetetra-acetic acid by means of electromigration. *Radiokhimiya*, **9**, 710 (1967).

67-127   SZARVAS, P., KORONDÁN, I., and RAISZ, I., Rapid determination of Al, Ca, Fe(III), Mg and Mn(II) content in cupole slags by means of Na_2EDTA reagent. *Magyar Kém. Lapja*, **22**, 149 (1967).

67-128   SZEKERES, L., KARDOS, E., and SZELERES, G. L., Chelatometric determination of chromium(III). *Microchem. J.*, **12**, 147 (1967).

67-129   TELESHOVA, R. L., Complexometric micro-determination of Ca and Mg in rocks and minerals. *Metody Khim. Anal. Khim. Sostav. Miner., Akad. Nauk. SSSR, Inst. Geol. Rud. Mestorozhd., Petrogr., Mineral. Geokhim.*, 49 (1967).

67-130   TEMKINA, V. YA., BOZHEVOL'NOV, E. A., DYATLOVA, N. M., KREINGOL'D, S. V., YAROSHENKO, G. F., ANTONOV, V. N., and LASTOVSKII, R. P., Triazinylstilbexone. A new luminescent reagent for Cr(III). *Zhur. analit. Khim.*, **22**, 1830 (1967).

67-131   TÔEI, K., and KOBATAKE, T., Successive chelatometric titration of Ca and Mg. *Talanta*, **14**, 1354 (1967).

67-132   TÖLGYESSY, J., KONECNY, J., and BRAUN, T., Ion-exchange membranes in radiochelatometric titrations of traces of metals. *Nucl. Appl.*, **3** (6), 383 (1967).

67-133 VAN-DER-REYDEN, A. J., and POLMAN, R. J., Determination of trace amounts of Ca and Ba in caustic soda. Complexometric titration after separation and concentration on Dowex A-1 chelating resin. *Zeit. analyt. Chem.*, **232**, 274 (1967).

67-134 VAZIRANI, V. A., Complexometric determination of yield of aluminium oxide from aluminium glycinate using Xylenol Orange as metal indicator. *J. Inst. Chemists (India)*, **39**, 150 (1967).

67-135 VDOVENKO, M. E., and LISICHENOK, S. L., Photometric determination of rare earth elements by Arsenazo III in the presence of Complexon III. *Zavodskaya Lab.*, **33**, 1372 (1967).

67-136 VERMA, M. R., AGRAWAL, K. C., and AMAR, V. K., Complexometric method for the determination of Th and Cr in a mixture. *India J. Chem.*, **5**, 79 (1967).

67-137 VINOGRADOZ, A. V., APIRINA, R. M., and PAVLOVA, I. V., Complexometric determination of Be and U by means of cobalt. *Zhur. analit. Khim.*, **22**, 1320 (1967).

67-138 VITKINA, M. A., and BEKLESHOVA, G. E., Amperometric titration of Ca by Complexon III in the presence of Mg. *Metody Khim. Spektral. Anal. Mater.*, 237 (1967).

67-139 VORLIČEK, J., Biamperometric titrations with chelatometric reagents. *Rudy (Prague)*, **15**, 373 (1967).

67-140 VORLIČEK, J., and PETÁK, P., Amperometry with two polarizable electrodes. Part 15. Chelometric determination of small amounts of bismuth(III). *Microchem. J.*, **12**, 466 (1967).

67-141 VORLIČEK, J., FARA, M., and VÝDRÁ, F., Amperometry with two polarizable electrodes. Part 14. Chelometric determination of very small amounts of calcium. *Microchem. J.*, **12**, 409 (1967).

67-142 VÝDRÁ, F., and HORÁČEK, J., Amperometry with two polarizable electrodes: chelometric determination of rare earths. *Analyt. Letters*, **1**, 31 (1967).

67-143 VÝDRÁ, F., and ŠTULÍK, K., Conductimetric and high-frequency impedimetric titrations involving chelates and chelating agents. *Chelates Anal. Chem.*, **1**, 81 (1967).

67-144 WASHIZUKA, S., Indirect chelatometric titration of thioacetamide with Hg(II)–EDTA. *Japan Analyst*, **16**, 963 (1967).

67-145 YAMAGUCHI, N., HATA, A., and HASEGAWA, M., Determination of calcium in ferrosilicon by a chelataometric titration method. *Japan Analyst*, **16**, 253 (1967).

67-146 YAMAGUCHI, H., IWASAKI, K., YAMAGUCHI, K., and UENO, K., Masking agents in chelatometric titrations. V. Masking properties of thio ethers and comparative study on the masking abilities of various thiol derivatives. *Bunseki Kagaku*, **16** (7), 703 (1967).

67-147 YOTSUYANAGI, T., YAMAGUCHI, T., GOTO, K., and NAGAYAMA, M., Consecutive chelatometric titration of sum of Ca and Mg, Mn, and Zn in acid mine drainage. *Japan Analyst*, **16**, 1056 (1967).

67-148 YURIST, I. M., Ethylenediamine as a masking agent in volumetric analysis. *Zavodskaya Lab.*, **33**, 680 (1967).

67-149 YURIST, I. M., *o*-Phenanthroline as a masking agent for the com-

446   *Bibliography*

plexonometric determination of manganese. *Zhur. analit. Khim.*, **22**, 442 (1967).

67-150   ZAHRADNÍČEK, M., BLEŠOVÁ, M., and ŠUBERT, J., Study of mixed indicators by complementary tristimulus colorimetry. Part 3. Stability of mixed indicators. *Cesk. Farm.*, **16**, 334 (1967).

67-151   ZHADANOV, A. K., AKENT'EVA, N. A., and KAPITSA, N. V., Sequential amperometric titration of mixtures containing Tl, Cd and Ag. *Uzh. Khim. Zhur.*, **11** (6), 12 (1967).

67-152   ZHADANOV, A. K., KAPITSA, N. V., and AKHMEDOV, G., Amperometric analysis of cation mixtures. IV. Successive amperometric titration of Bi, Cu and Ag in a single sample with a rotating tantalum electrode. *Tr. Tashkent Gos. Univ.*, **288**, 39 (1967).

67-153   GURN'EV, S. D., and LUTCHENKO, N. N., Determination of bismuth in products which contain large amounts of molybdenum, lead, zinc, iron and copper. *Sb. Nauch. Tr., Gos. Nauch.-Issled. Ist. Tsvest. Metal.*, **27**, 59 (1967).

67-154   SHEVCHUK, I. A., NIKOL'SKAYA, N. N., and SIMONOVA, T. N., Determination of nickel in nickel–zinc ferrite powders. *Metody Anal. Khim. Reaktivov. Prep.*, **14**, 63 (1967).

67-155   SHEVCHUK, I. A., NIKOL'SKAYA, N. N., and SIMONOVA, T. N., Determination of Mg and Mn in magnesium–manganese ferrites. *Metody Anal. Khim. Reaktivov Prep.*, **14**, 71 (1967).

67-156   SHEVCHUK, I. A., NIKOL'SKAYA, N. N., and SIMONOVA, T. N., Determination of Mn in manganese–zinc ferrites. *Metody Anal. Khim. Reaktivov Prep.*, **14**, 74 (1967).

67-157   DA SILVA, F. J. J. R., Theoretical basis of complexometry. *Technica (Lisbon)*, **42**, 53 (1967).

67-158   HENTRICH, K., and PFEIFER, S., Metal thiocyanate complexes of alkaloids and organic bases and their applications in chelatometric determination methods. *Pharm. Zentralhalle*, **106**, 735 (1967).

67-159   ORHANOVIC, M., and WILKINS, R. G., Kinetic studies of the formation of iron(III)–EDTA–hydrogen peroxide complex. *Croat. Chem. Acta*, **39**, 149 (1967).

67-160   PAWELCZAK, M., Complexometric determination of Al in Al phosphate in the presence of indicator 4-(2-pyridylazo)resorcinol. *Acta Polon. Pharm.*, **24**, 615 (1967).

67-161   PAWELCZAK, M., Titration of solutions of disodium EDTA by means of lead nitrate in the presence of the reagent, 4-(2-pyridylazo)resorcinol. *Acta Polon. Pharm.*, **24**, 661 (1967).

67-162   TSCHAN, D., and LEUPIN, K., Complexometric titration of barbituric acids. *Pharm. Acta Helv.*, **42**, 657 (1967).

67-163   TUKHONOVA, L. I., Dissociation constants of EDTA in 1·2M potassium chloride. *Zhur. Prikl. Khim.*, **40**, 1887 (1967).

1968

68-1   ALEKSANDROVICH-MEL'NIKOVA, A. S., and ZHIGALKINA, T. S., Use of the metal indicator gallein in the complexometric determination of copper. *Zavodskaya Lab.*, **34**, 16 (1968).

68-2   ANDERSEN, R. G., and NICKLESS, G., Heterocyclic azo dyestuffs in analytical chemistry. Part I. The ligand properties of 2-(2-pyridylazo)-1-naphthol and its sulphonated analogues. *Analyst*, **93**, 13 (1968).

68-3   ANDERSEN, R. G., and NICKLESS, G., Heterocyclic azo dyestuffs in analytical chemistry. Part II. The ligand properties of 2-(2-pyrimidylazo)-1-naphthol and its sulphonated analogues. *Analyst*, **93**, 20 (1968).

68-4   BAGROVA, R. K., and SMOLENSKAYA, A. M., Complexometric determination of sulphate ions in technological solutions of the sulphite-alcohol industry. *Gidroliz Lesokhim. Prom.*, **21**, 24 (1968).

68-5   BANDYOPADHYAY, S. S., Use of direct complexometric titration in the analysis of antacid tablet containing magnesium trisilicate, heavy magnesium carbonate, dried aluminium hydroxide gel, dried aluminium glycinate. *J. Inst. Chem. India*, **40**, 49 (1968).

68-6   BARNA, L., and UDVARDI, M., The analytical determination of magnesium in silicates by using ethyleneglycol-bis(β-amino-ethyl(ether-*NNN'N'*-tetra-acetic acid (EGTA) and EDTA as complexing agents. *Epitoanyag*, **20** (2), 64 (1968); *Amer. Chem. Abs.*, **69**, 15964r (1968).

68-7   BARTURA, T., and BODENHEIMER, W., Deficit of aluminium in the analysis of silicates with EDTA. *Israel J. Chem.*, **6**, 61 (1968)

68-8   BEIZEROV, E. M., and SHVETSOVA, T. I., Separation of phosphoric acid as iron phosphate during the complexometric determination of Ca and Mg in phosphorites. *Ref. Zh. Khim.* abstracted in *Amer. Chem. Abs.*, **69**, 32675k (1968).

68-9   BUSEV, A. I., and SHVEDOVA, N. V., Complexometric determination of bismuth in the presence of indium. *Zavodskaya Lab.*, **34**, 140 (1968).

68-10   BUSEV, A. I., SUKHORUKOVA, N. V., TERZEMAN, L. N., and BASENKO, N. L., Complexometric titration of gallium and indium. *Vestn. Mosk. Univ. Ser. II Khim.*, **23** (2), 119 (1968).

68-11   CAMARGO, R. S., Complexometry. *Rev. Brasil Quim. (Sao Paulo)*, **65**, 77, 82, 84, 86, 88 (1968).

68-12   CANDORI, R., and FURLANI, C., Luminescence of Cr(III) chelates with EDTA. *Atti accad. Naz. Lince. Rend. Cl. Sci. Fis. Mat. Nat.*, **44** (3) 415 (1968).

68-13   CORNWELL, J. C., and CHENG, K. L., Analysis of tellurides of Pb and Sn by automatic titrations. *Anal. Chim. Acta*, **42**, 189 (1968).

68-14   CRISAN, I., and TĂNASE, D., Complexometric analysis of a mixture of copper(II) and zinc(II). *Rev. Chim. (Roumania)*, **19**, 228 (1968).

68-15   DAS, H. R., and SHOME, S. C., Direct complexometric titration of In, Tl and Th with EDTA using iron-*N*-benzoyl-*N*-phenyl-hydroxylamine as indicator. *Anal. Chim. Acta*, **43**, 140 (1968).

68-16   DEAN, J. R., and HARRIS, W. E., Theoretical-slope method of end-point detection. *Analyt. Chem.*, **40**, 1213 (1968).

68-17 DERKO, V. I., and SAVCHENKOVA, V. P., Complexometric determination of zinc in the presence of ions of other metals. *Lakokrasoch. Mater. Ikl Primen.* (2), 48 (1968). *Amer. Chem. Abs.*, **69**, 24262n (1968).

68-18 FERNANDEZ, T., CASASSAS, E., and MONTELONGO, G., Non-aqueous complexometric determination of metals using a high-frequency technique. *Anales de Quim.*, **64**, 315 (1968).

68-19 FLEET, B., SOE-WIN, and WEST, T. S., Rapid visual complexometric titration of calcium in natural waters. *Talanta*, **15**, 333 (1968).

68-20 FOG, H. M., Complexometric determination of magnesium in manganese containing aluminium alloys. *Acta Chem. Scand.*, **22**, 791 (1968).

68-21 GIULIANI, A. M., GATTEGNO, D., and FURLANI, A., Polarography of copper(II)–cyclohexane-1,2-diaminetetra-acetic acid. *J. Electroanalyt. Chem. Interfacial Electrochem.*, **18**, 151 (1968).

68-22 GOLDMAN, J. A., Redox equilibria. V. The location of inflection points on titration curves for homogeneous reactions. *J. Electroanalyt. Chem. Interfacial Electrochem.*, **18**, 41 (1968).

68-23 GUREVICH, A. M., POLOZHENSKAYA, L. P., and SOLNTSEVA, L. F., Complexing in a uranyl(II)–hydrogen peroxide–EDTA water system. Equilibrium in solution. *Radiokhimiya*, **10**, 195 (1968).

68-24 HALL, L. H., and LAMBERT, J. L., Structure and photo-decomposition studies of the complex acid aquoethylenediaminetetra-acetatoferrate(III), and several of its metal(I) salts. *J. Amer. Chem. Soc.*, **90**, 2036 (1968).

68-25 HEFLY, A. J., Structure and uses of chelating derivatives of fluorescein. *Diss. Abs.*, **28**, 4035 (1968).

68-26 HUBER, C. O., and TALLANT, D. R., Electrometric titration of dilute EDTA solutions. *J. Electroanalyt. Chem. Interfacial Electrochem.*, **18**, 421 (1968).

68-27 IDENO, E., and HOZUMI, K., Photometric titration of Zn in a Cd and Zn mixture with EDTA. *Japan Analyst*, **17**, 727 (1968).

68-28 ITO, A., and UENO, K., Chelometric titrations with TTHA (tri-ethylenetetraminehexa-acetic acid). *Japan Analyst*, **71**, 327 (1968).

68-29 JURCZYK, J., Chelatometric determination of Mn in ferromanganese. *Neue Hütte*, **13**, 58 (1968).

68-30 KHALIFA, H., and EL-BARBARY, I., Application of back-titration of EDTA with mercury(II) to the analysis of alloys. Analysis of bearing metals, solders, type metals and stainless steels. *Microchem. J.*, **13**, 137 (1968).

68-31 KISS, T. A., and DORIĆ, T., EDTA titrimetry of aluminium in aluminium alloys. *Chem. and Ind.*, 1567 (1968).

68-32 KISS, T. A., GAÁL, F. F., SURÁNYI, T. M., and ZSIGRAI, I. J., Folgetitrimetrische Bestimmung von Calcium und Magnesium mit ÄDTA in Äthanolhaltiger Lösung. *Anal. Chim. Acta*, **43**, 340 (1968).

68-33 KONOPIK, N., Complexes of germanic acid with EDTA and EDTA analogues. EDTA. *Monatsh.*, **99**, 902 (1968).

68-34 KRAFT, G., Voltammetric indication of complexometric titrations. *Zeit. analyt. Chem.*, **238**, 321 (1968).

68-35 KRISHNAN, K., Raman spectra of EDTA and its metal complexes. *J. Amer. Chem. Soc.*, **90**, 3195 (1968).

68-36 LASSNER, E., and SCHEDLE, H., Molybdenum(VI)–hydroxylamine complexes. Chelometric determination of molybdenum as the ternary molybdenum(VI)–hydroxylamine–EDTA complex and comparison with the already known method of titration of Mo(VI) with EDTA. *Talanta*, **15**, 623 (1968).

68-37 LEBEDEV, I. A., SHALINETS, A. B., and YAKOVLEV, G. N., Complexing of cerium(III) with hexamethylenediaminetetra-acetic acid. *Radiokhimiya*, **10**, 98 (1968).

68-38 LEYDEN, D. E., and WHIDBY, J. F., Nuclear magnetic resonance studies of EDTA in the presence of excess calcium or strontium. *Anal. Chim. Acta*, **42**, 271 (1968).

68-39 MECHOS, K. Z., GUTNIKOVA, R. I., and TEODOROVICH, I. L., Determination of cobalt in steel using complexing agents. *Dokl. Akad. Nauch. Uzb.SSR*, **25** (2), 33 (1968).

68-40 MENDEZ-BEZERRA, A. E., and UDEN, P. C., Determination of sulphate by titrimetric and colorimetric measurements of equivalent displaced zinc ion. *Analyt. Letters*, **1**, 355 (1968).

68-41 NEUMANN, J., DITZ, J., and SUK, V., New metallochromic indicators of the 'calcon' type. Determination of calcium. *Z. analyt. Chem.*, **239**, 167 (1968).

68-42 NIGAM, R. C., Determination of Ca and Mg in magnesite by a complexometric method employing coprecipitation of Ca with strontium sulphate. *Z. analyt. Chem.*, **237**, 351 (1968).

68-43 NOMURA, T., TAKEUCHI, K., and KOMATSU, S., Volumetric determination of cyanide ion with mercury(II)–EDTA employing mercury(II)–cresolphthaleincomplexone as an indicator. *Nippon Kagaku Zasshi*, **89**, 291 (1968).

68-44 PAGE, J. O., Complexometric determination of calcium in impure calcium carbonate and limestone. *Anal. Chim. Acta*, **42**, 233 (1968).

68-45 PASOVSKAYA, G. B., Determination of calcium in the presence of Mg, Fe and Al by conductometric titration with sodium nitrilotriacetate. *Izvest. VUZMVOSSSR Khim. i khim. Takohl.*, **11**, 17 (1968).

68-46 PHATAK, G. M., and BHAT, T. R., Rapid chelatometric estimation of Cr, Fe and Al in mixtures of Cr and Fe or Cr and Al. *Z. analyt. Chem.*, **233**, 418 (1968).

68-47 RAMAIAH, N. A., TEWARI, G. D., TRIVEDI, S. R., and KATIYAR, S. S., Spectrophotometric titration of bismuth with EDTA. *Talanta*, **15**, 352 (1968).

68-48 ROSTROMINA, N. A., and KOMASHKO, G. A., Structure and stability of neodymium complexes with some polydentate complexones. *Zhur. neorg. Khim.*, **13**, 1041 (1968).

68-49 SAHA, N. N., MAZUMDAR, S. K., and HANDA, R., Crystal and molecular structure of calcium EDTA. *Sci. Cult. Calcutta*, **34**, 70 (1968).

68-50 SCHERZER, J., and CLAPP, L. B., Ruthenium complexes with ethylenediaminetetra-acetic acid. *J. Inorg. Nuclear Chem.*, **30**, 1107 (1968).

68-51 DA SILVA, J. J. R. F., and SIMOES, M. L. S., Uranyl complexes. Uranyl complexes of EDTA. *Talanta*, **15**, 609 (1968).

68-52 SINGH, R. P., CHICKERUR, N. S., and NARASARAJU, T. S. B., Complexometric determination of calcium and phosphorus in synthetic hydroxyapatite. *Zeit. analyt. Chem.*, **237**, 117 (1968).

68-53 SINGHAL, G. K., and TANDON, K. N., Metallochromic indicators. Part 3. Haematoxylin and haematein. *Talanta*, **15**, 707 (1968).

68-54 SINHA, B. C., DAS GUPTA, S., and KUMAR, S., A titrimetric method for the determination of phosphate. *Analyst*, **93**, 409 (1968).

68-55 SOLJIC, Z., and MARJANOVIC-KRAJOVAN, V., Rapid method for determining silica, ferric oxide, alumina, titania, calcium oxide and magnesia in bauxite. Rapid analysis of limestone and dolomite. *Chim. Anal. (Paris)*, **50**, 122 (1968).

68-56 STEFANOVIC, D., VAJGAND, V., and KISS, T., Oscillographic, chronopotentiometric titrations. *Magyar Kém. Folyóirat*, **74**, 307 (1968).

68-57 STOLYAROV, K. P., and VINOGRADOVA, N. I., Solubility of metal oxides and carbonates in sodium EDTA solutions. *Vestn. Leningrad. Univ. Ser. Fiz. i Khim.*, **23**, 115 (1968).

68-58 ŠTULÍK, K., and VÝDRÁ, F., Voltammetry of EDTA and related compounds on a rotating platinum electrode. *J. Electroanalyt. Chem. Interfacial Electrochem.*, **16**, 385 (1968).

68-59 SZEKERES, L., and KELLNER, A., Determination of formaldehyde via EDTA titration. *Microchem. J.*, **13**, 227 (1968).

68-60 TALLANT, D. R., Voltammetry at the lead dioxide electrode. Anodic EDTA currents. *J. Electroanalyt. Chem. Interfacial Chem.*, **18**, 413 (1968).

68-61 TANAKA, M., FUNAHASHI, S., and SHIRAI, K., Kinetics of the ligand substitution reaction of the zinc(II)-4-(2-pyridylazo)-resorcinol complex with ethyleneglycol-bis(2-aminoethylether)-NNN′N′-tetra-acetic acid. *Inorg. Chem.*, **7**, 573 (1968).

68-62 THORNELEY, R. N. F., and SYKES, A. G., Extent to chelation in some chromium(III)–EDTA complexes. *Chem. Commun.*, 340 (1968).

68-63 TSELINSKY, U. K., and LAPITSKAYA, E. V., Comparative study of metallochromic indicators for the complexometric determination of zirconium. *Ukr. Khim. Zhur.*, **34**, 189 (1968).

68-64 TSUNOGAI, S., NISHIMURA, M., and NAKAYA, S., Complexometric determination of calcium in the presence of larger amounts of magnesium. *Talanta*, **15**, 385 (1968).

68-65 UHLIG, E., and HERRMANN, D., Acidity and complex chemical behaviour of *p*- and *m*-phenylenediamine-NNN′N′-tetra-acetic acid towards ions of 30 elements. *Z. anorg. allg. Chem.*, **359**, 135; **360**, 158 (1968).

68-66 URBAIN, H., BACAUD, R., CHARCOSSET, H., and TOURNAYAN,

L., Determination of metallic nickel in nickel–silica–alumina catalysts. *Chim. Anal. (Paris),* **50,** 242 (1968).

68-67 VANDERDEELEN, J., and VAN-DER-HENDE, A., Titrimetric determination of EDTA. *Chim. Analyt. (Paris),* **50,** 237 (1968).

68-68 VASIL'EVA, V. F., LAVROVA, O. U., DYATLOVA, N. M., and YASHUNSKII, V. G., Substances with complexing capacity. (*N'*-Hydroxyethyl)-diethylenetriamine-*N,N*-tetra-acetic acid. *Zhur. obsch. Khim.,* **38,** 473 (1968).

68-69 VÝDRÁ, F., and ŠTULÍK, K., Biamperometric indication in chelatometric titrations in acidic solutions. *Electroanalyt. Chem. Interfacial Electrochem.,* **16,** 375 (1968).

68-70 WADA, H., and NAKAGAWA, G., 2-(2-Pyridylazo)-4-methylphenol (PAC) complexes with indium and the use of PAC as a metal indicator. *Nippon Kagaku Zasshi,* **89,** 499 (1968).

68-71 WATANABE, T., Determination of calcium sulphate by EDTA titration. *Fukushima-Ken Eisei Kenkyusho Kenkyu Hokoku,* **71,** 27 (1968).

68-72 YAMAGUCHI, O., TANAKA, H., and UNO, T., Imidazole derivatives as chelating agents. Part 5. Applicability of azoimidazoles as metallochromic indicators. *Talanta,* **15,** 459 (1968).

68-73 YATSYK, I. YA., Standardization of a working solution of Complexon III by using Mn. *Zavodskaya Lab.,* **34,** 281 (1968).

68-74 YOFÉ, J., and RAPPART, B. R., Complexometric titration of phosphate with lanthanum as precipitant. *Anal. Chim. Acta,* **43,** 346 (1968).

68-75 YURIST, I. M., Polyethylene polyamine as a masking reagent during the complexometric determination of lead. *Zavodskaya Lab.,* **34,** 539 (1968).

68-76 BABKO, A. K., and LORIYA, N. V., Ternary complex of iron–EDTA–hydrogen peroxide. *Zhur. neorg. khim.,* **13,** 506 (1968).

Author Index

A

Abbink, J. E.: 61-112; 62-93
Abbott, D. D.: 55-55
Abd-El Raheem, A. A.: 57-7; 58-7, 8, 36; 59-98, 99, 100, 101; 60-188; 61-126
Abdine, H.: 54-61; 55-62, 63, 64; 56-43, 44, 45, 46, 63; 61-57, 156
Abdullaeva, Kh. S.: 64-43; 65-23
Abdurakhmanov, M.: 67-11
Abe, M.: 59-42
Abramyan, S. A.: 62-159
Ackermann, G.: 65-74
Ackermann, H.: 47-2; 48-6; 49-2, 3; 52-18
Aconsky, L.: 55-69
Adamo, F. S.: 67-1
Adamova, E.: 55-88
Adamovich, L. P.: 61-165
Adams, M. L.: 55-44
Adams, R. N.: 54-22
Adrianova, N. N.: 65-60
Afanas'ev, Yu. A.: 66-29
Affsprung, H. E.: 54-5
Agarwal, K. C.: 60-80; 61-59; 67-136
Agasyan, P. K.: 64-90
Aggarwal, R. C.: 66-76
Ågren, A.: 54-72
Agulá, J. F.: 67-12
Aikens, D. A.: 61-113; 62-74; 67-5
Akchurina, G. S.: 64-150
Akent'eva, N. A.: 67-151
Akhmedov, G.: 67-152
Akhmedzhanova, S. A.: 67-64
Akiyama, T.: 55-110
Aldrovandi, R.: 52-10
Alegre, M.: 60-118
Aleksandrova, N. M.: 62-19
Aleksandrovich-Melnikova, A. S.: 62-112; 64-150; 68-1
Alekseeva, A. Y.: 67-67
Alfarova, V. N.: 63-86

Alimarin, I. A.: 66-9
Alimarin, I. P.: 56-65; 61-103
Allam, M. G.: 60-163
Alleman, T. G.: 57-97
Allsopp, H. J.: 56-66
Alon, A.: 58-5, 6
Alvarez-Herrero, C.: 64-138
Alykov, N. M.: 64-35
Amar, V. K.: 67-136
Amaros, J. L.: 67-20
Amin, A. A.: 60-188
Amin, A. A. M.: 56-81; 58-7, 8
Amin, A. M.: 53-24, 28; 54-59; 55-76, 77, 78
Amirkhanova, T. B.: 67-13, 14
Amos, W. R.: 63-6
Amsheeva, A. A.: 64-132
Anastasescu, G.: 60-94; 64-129
Anderegg, G.: A-7; 52-19, 21; 53-1; 54-27, 28; 57-139; 59-142; 63-80; 64-52
Andersch, M. A.: 57-46
Andersen, B.: 58-45
Andersen, R. G.: 68-2, 3
Anderson, E.: 60-197
Andrew, T. R.: 61-104
Antonacci, M.: 60-157
Antonov, V. N.: 67-130
Antonova, A. A.: 62-107
Aoki, M.: 60-159
Apirina, R. M.: 67-137
Apple, R. F.: 60-152
Arbuzov, G. A.: 60-105; 61-75, 80
Arend, H.: 63-108
Arikawa, Y.: 54-23; 62-54
Armasescu, L.: 65-87
Arnold, J. R.: 60-107
Arthur, P.: 67-6
Artynkhin, P. I.: 59-92
Aseni, G.: 60-187
Aseni-Mora, G.: 61-154; 63-117
Ashbrook, A. W.: 61-106
Ashby, R. O.: 57-36
Ashton, A. A.: 63-20, 21; 66-6, 19

Asperen van, K.: 54-4
Astakhov, K. V.: 63-77; 65-39, 82; 67-74
Atkin, M.: 58-69
Austin, J. H.: 66-72
Awe, W.: 65-83
Ayad, A.: 61-57

B

Babenyshev, V. M.: 60-151
Babko, A. K.: 65-57; 68-76
Bacaud, R.: 68-66
Bach, R. O.: 47-1
Bachmann, C.: 61-105, 125
Bachra, B. N.: 58-20
Bacquias, G.: 66-77
Baczyk, St.: 64-115
Badrinas, A.: 58-19
Baetslé, L.: 62-156
Bagdasaryan, A. M.: 62-159
Baginski, E. S.: 56-4, 14
Bagrova, R. K.: 68-4
Bahl, J. S.: 61-59
Baker, R. A.: 58-21
Balabanoff, L.: 64-134
Balica, G.: 66-52
Ball, R. G.: 58-70
Ballczo, H.: 54-94; 56-97
Ballschmiter, K. H.: 64-113
Balogh, G.: 62-158; 63-102; 64-28; 66-79
Bandyopadhyay, S. S.: 68-5
Banerjee, G.: 55-92, 93, 94
Banerjee, N. G.: 57-93
Banewitz, J. J.: 52-43
Bangerter, F.: 46-2
Banks, C. V.: 55-71
Banks, J.: 52-39
Banyai, E.: 60-150
Barakat, M. F.: 57-91
Baran, V.: 64-74
Baranenko, S. E.: 63-130
Barbieri, R.: 60-201; 61-123
Barcza, L.: 55-21; 57-136; 59-55, 56, 58; 60-73; 61-21
Bàrdi, I.: 61-166
Barikov, V. G.: 63-48
Barkov, S. A.: 65-82; 67-74
Barna, L.: 68-6

Barnard, Jr, A. J.: 56-116; 57-90; 59-38; 60-17
Baron, D. N.: 57-37
Bartels, U.: 59-128; 60-101
Bartolotti, T. R.: 51-9
Bartura, T.: 68-7
Basenko, N. L.: 68-10
Bashkirtseva, A. A.: 60-160, 161, 162; 63-107
Batyrshina, F. M.: 62-64
Baugh, C. A.: 61-107
Baumgartner, R.: 64-118
Bayer, E.: 57-140
Bazant, V.: 59-15
Bazarbaev, A. T.: 60-74
Bazen, J.: 53-9, 22; 54-17, 104; 56-5
Beck, M. T.: 60-193, 200; 61-166; 65-35; 66-74
Begelfer, K. I.: 62-13
Beizerov, E. M.: 68-8
Bekleshova, G. E.: 67-138
Belavskaya, Y. I.: 56-67
Belcher, R.: 53-14; 54-1; 55-104, 105, 106; 57-42, 43; 58-1, 2, 3; 60-153
Belikov, V. V.: 62-12
Belin, P.: 66-36
Belkovskaya, Y. I.: 62-84
Bell, H. R.: 60-143
Bell, R. P.: 53-15, 59
Bell, U. L.: 57-37
Bellomo, A.: 58-18, 27
Belluco, U.: 60-201
Bem, M.: 65-87
Benev, B.: 65-116
Bengsch, E.: 62-156
Bennet, M. C.: 58-33
Bennewitz, R.: 59-109
Bensch, H.: 67-15
Beral, H.: 58-111; 59-135
Berbenni, P.: 60-142
Bercea, E.: 67-22
Berg van den, R. H.: 59-78
Bergman, L.: 62-118
Bergmann, J. G.: 63-36
Berka, A.: 62-72; 63-22, 35
Berkovich, M. T.: 64-146
Bermejo, F.: 64-96

Berndt, H.: 64-50
Berndt, W.: 58-106; 59-105; 60-148; 61-114
Bershadskaya, N. M.: 63-129
Bertschinger, J. P.: 55-7
Bett, I. M.: 59-129
Betteridge, D.: 66-22
Betz, J. D.: 50-4
Beyermann, K.: 61-119
Bezděková, A.: 62-124; 65-21
Bezuglyi, D. V.: 64-132
Bhandri, M. R.: 66-11
Bhat, T. R.: 64-84; 65-38; 99; 68-46
Bhattacharyya, B. N.: 60-147
Bhuchar,V.M.: 55-25;57-120;58-95
Bickwell, N. J.: 59-94
Bieber, B.: 58-65; 61-82, 87, 88
Biechler, D. G.: 63-19
Biedermann, W.: 46-2; 47-4; 48-1, 2, 3, 4, 5
Biino, L.: 66-85
Bionda, G.: 53-7, 8
Biryukov, A. A.: 66-40
Bishop, J.: 66-30
Bitovt, Z. A.: 66-20
Bizon, P.: 63-2
Bjerrum, J.: A-3; 50-1; 57-138; 59-140
Blackwell, I. G.: 60-111
Blaedel, W. J.: 54-2, 3; 65-106, 110; 66-64
Blagojević, Z.: 62-133; 64-119
Blazsek, A.: 62-57; 65-98
Blecha, J.: 61-97
Blešová, M.: 67-150
Blood, F. H.: 63-7, 40
Blumer, M.: 52-11
Blyakhman, A. A.: 58-55
Boase, D. G.: 62-65
Bobowski, E.: 56-106
Böckel, V.: 58-96
Boden, H.: 66-73
Bodenheimer, W.: 68-7
Bodor, A.: 53-36
Bogareva, K. G.: 57-47; 59-97
Bogatova, E. I.: 66-78
Bogdanska, E. E.: 66-94
Bognár, J.: 58-115
Bogs, U.: 65-79

Boguecki, R. F.: 62-100, 140; 67-9
Böhler, K.: 56-68
Bolos, V.: 67-23
Bolton, S.: 63-57
Bonati, F.: 55-70
Bond, J.: 59-141
Bond, R. D.: 54-34; 55-95
Bondareva, E. G.: 66-80, 81
Boos, R. N.: 59-108
Borchert, O.: 57-1; 59-53, 54; 61-124; 65-73
Borman, G.: 65-79
Bormann, R.: 67-41
Botha, G. R.: 52-8
Boulanger, F.: 53-17
Bouten, J.: 57-53
Bovalini, E.: 53-16
Bowman, R. C.: 67-120
Boxer, J.: 60-146
Boyman, E. C.: 64-59
Bozhevolnov, E. A.: 62-8, 149, 150; 63-104; 67-130
Brachaczyk, W.: 64-58
Brady, G. W. F.: 62-7
Brajter, K.: 61-70, 71; 66-94
Brake, L. D.: 57-63
Brandštetter, J.: 58-46; 59-13
Brandt, M.: 59-78
Brannock, W. W.: 55-87
Braun, K. H.: 57-62
Braun, T.: 58-32; 59-60; 60-133; 64-13, 14, 15; 66-1; 67-132
Bräuninger, H.: 56-70
Bray, E.: 63-90
Bray, R. H.: 51-7; 52-46; 53-61; 55-18
Breekland, E.: 64-145
Brekke, J. E.: 63-85
Brennecke, Erna: 56-126
Brháček, L.: 59-19
Bricker, C. E.: 53-37; 54-14
Bright, H. A.: 53-56
Bril, K. Y.: 59-95
Brintzinger, H.: A-9
Britton, H. T. S.: A-2
Broad, W. C.: 56-116; 57-90; 59-38; 60-17
Brochon, R.: 56-55
Brodskay, V. M.: 62-89

Brooke, M.: 52-7
Brooks, H. E.: 55-72
Brown, E. G.: 53-33, 34, 35; 54-57, 58
Brown, W. B.: 63-6, 16
Brück, A.: 67-16
Bruile, E. S.: 64-89; 66-82
Brunisholz, G.: 52-12; 53-5; 54-33; 56-95
Bruno, E.: 58-18, 27; 60-149; 63-121
Bruno, M.: 61-123
Brush, J. S.: 61-120
Buckley, E. S. Jr: 51-9
Budanova, L. M.: 55-73; 57-94; 61-122
Buděšinský, B.: 55-37, 38, 39, 40; 56-32, 34, 35, 36, 83; 57-55, 56, 57, 58, 81, 85; 58-51, 53; 59-17, 29; 60-43, 44; 61-83; 62-124; 63-95; 64-37; 65-17, 18, 19, 20, 21, 22; 67-17
Budevskii, O. B.: 58-67; 59-49, 106
Budevsky, O.: 64-9, 128, 136; 66-47
Bultasova, H.: 54-7
Burg, R. A.: 60-140
Burger, K.: 60-46
Bürger, K.: 61-69
Buriel-Marti, F.: 64-138
Burtner, D. C.: 54-97
Busev, A. I.: 57-131; 58-107, 108, 109; 59-61, 62; 60-167, 168, 180, 181; 61-130, 131, 132, 157; 62-48, 49, 68, 72, 92, 115; 63-70; 65-61; 68-9, 10
Busquets, P.: 67-104
Buss, H.: 63-53, 54; 64-130
Butcher, J.: 63-81; 64-148; 65-108, 113; 66-64
Butt, L. T.: 55-61
Buzás, I.: 60-90
Bykhovskaya, I. A.: 61-131
Bystroff, A. S.: 57-122

C

Cabell, M. J.: 52-41
Cahen, R.: 56-95
Calabro, G.: 65-78
Caletka, R.: 63-51; 64-73

Calleja, J.: 61-150
Calu, C.: 59-90
Calvin, M.: 51-8
Camargo, R. S.: 68-11
Cameron, A. J.: 61-135, 136
Campbell, D. N.: 54-8
Campbell, R. T.: 62-33
Campen, W. A. C.: 55-52
Candori, R.: 68-12
Canić, V. D.: 62-138; 63-100, 101
Capizzi, F. M.: 60-135; 62-80
Carini, F. F.: 54-9
Carley, F. B.: 63-11
Carpenter, H.: 57-54
Cartier, P.: 59-91
Casassas, E.: 68-18
Casini, A.: 53-16
Castiglioni, A.: 57-48
Catani, I.: 66-111
Ceausescu, D.: 66-83
Ceglarski, R.: 62-73
Černy, A.: 58-10
Certa, A. J.: 52-49
Chaberek, S.: 54-13; 59-88, 94
Chadejev, V. A.: 56-3
Chalmers, R. A.: 54-10
Chan, F.: 59-61, 62
Chang, H.-S.: 64-75
Chang, J. P.: 65-100
Chang, M.: 58-31
Chang, T.-H.: 64-83
Chang, W. P.: 65-103
Chapman, D.: 55-56; 63-23
Chapman, L. R.: 64-5
Charcosset, H.: 68-66
Charles, R. G.: 54-63
Chaturvedi, R. K.: 60-82, 103
Chaudri, R. S.: 66-84
Chen, Y.-C.: 65-46
Cheng, K. L.: 51-7; 52-46; 53-61; 54-6; 55-18, 19, 20; 56-60, 61; 57-124; 58-63, 64; 59-102, 103; 61-127, 183; 62-51, 52, 153; 63-3, 39; 64-112; 66-8; 68-13
Cherepanova, T. B.: 66-123
Cherkashina, T. V.: 56-49
Cherkesov, A. I.: 58-108; 59-136; 61-40; 64-35; 65-48
Chernavina, N. M.: 56-2

Chernikov, Y. A.: 55-51; 60-70, 71, 72
Chernova, R. K.: 64-55
Chernukha, G. N.: 56-56
Chetkowska, M.: 61-108
Chew, B.: 56-64
Chi, J.-A.: 59-131
Chiacchierini, E.: 66-46
Chialda, I.: 64-129
Chiarioni, T.: 55-70
Chickerur, N. S.: 68-52
Chigrin, D. S.: 64-91
Chiligaryan, M. N.: 62-159
Chmelar, V.: 56-62
Christian, D. G.: 65-1
Christianson, G.: 54-21
Chromý, V.: 61-85; 66-59; 67-18
Chu, Y.-C.: 59-131
Chudinova, N. N.: 59-130
Chumachenko, M. N.: 65-93
Chung, I. B.: 65-103
Chzhan, V. T.: 60-58
Cickers, C.: 61-110
Čičmancová: 55-6, 113
Cífka, J.: 56-37, 38, 84; 58-118
Čihalik, J.: 53-12, 13; 54-82
Cimerman, Ch.: 58-5, 6
Ciogolea, G.: 60-196; 62-56, 109
Ciurlo, R.: 63-121; 66-85
Claes, H. J.: 59-107
Clapp, L. B.: 68-50
Clark, S. J.: 53-14
Clément-Métral, J.: 59-91
Close, R. A.: 57-42, 43; 58-1; 60-154, 155, 156
Cluley, H. J.: 54-20
Coates, E.: 61-186; 62-157
Cocozza, E. P.: 60-165, 166
Cohen, A. I.: 56-71
Cohn, V. C.: 58-39
Collier, R. E.: 54-25; 55-58
Collins, P. F.: 61-7
Colovos, G.: 66-66
Combs, H. F.: 64-24
Conaghan, H. F.: 56-57; 60-140
Conradi, G.: 67-19
Cooke, W. D.: 56-22
Copello, M. A.: 66-86; 67-20
Copp, D. H.: 63-88

Corbett, J. D.: 61-78
Cornwell, J. C.: 66-68; 68-13
Corsini, A.: 62-11
Cortrait, M.: 64-57
Costa, A. C.: 58-26; 60-114, 115, 116, 117
Coulter, S. T.: 54-21
Crawford, C. M.: 59-137
Crawley, R. H. A.: 58-40
Creanga, S.: 62-109
Cretius, K.: 61-119
Crisan, I.: 59-90; 63-116, 126, 127, 128; 65-58; 66-87; 67-21, 22, 23, 24, 25, 26, 27, 28, 29, 82, 83, 84; 68-14
Crouch, E. A. C.: 59-137
Crowley, R. C.: 62-1
Culp, S. L.: 67-4
Curea, E.: 65-84
Curro, P.: 65-78
Cyganski, A.: 64-1; 67-30

D

Daniel, A.: 67-24
Danowski, K.: 65-69
Danuschenkova, M. A.: 60-10
Das, H. R.: 66-18; 68-15
Das Gupta, S.: 67-124; 68-54
Date, Y.: 63-87
Datsenko, O. V.: 62-76
Datta, S. K.: 56-53; 57-51; 60-145, 192; 62-5; 63-18; 64-39, 51
Dauer, A.: 58-20
Dauksas, K.: 62-111
Davies, C. W.: 51-12
Davies, D. T.: 62-96
Davis, D. G.: 60-144; 61-146
Davis, I.: 55-50
Dean, J. R.: 68-16
Debbrecht, F. J.: 55-74
Debney, E. W.: 52-37
Decker, K. H.: 61-107
De Grazio, R. P.: 65-15
Degtyarenko, L. I.: 66-88
De Heer, B. H. J.: 65-51
Delano, E.: 59-144
Delasanta, A. C.: 63-45
Delga, J.: 57-52
Delphin, G.: 66-75

Demina, L. A.: 64-90
Demetrescu, E.: 58-111; 59-135
Den Boef, G.: 60-27, 28, 29; 66-63; 67-49, 50
Denisova, N. E.: 61-121
Derevyanko, G. N.: 63-8
Derko, V. I.: 68-17
Derzsi, E.: 62-132
Desbarres, J.: 60-66
Dettner, H. W.: 57-49
Dewald, A.: 58-38
Dey, A. K.: 61-37, 73, 74
Diaz Rojas, C. A.: 65-117
Dick, J.: 59-145; 62-134
Diehl, H.: 50-7, 8; 56-72; 59-127; 60-36, 134; 61-7
Diggins, F. W.: 55-49
Dimitrova, M.: 61-175
Diskant, E. M.: 52-47
Ditz, J.: 61-147; 64-34; 68-41
Dixon, M.: 62-28
Djung, P.: 57-95
Dmitrieva, F. I.: 64-65
Dobkina, B. M.: 55-51
Dobrowolski, J.: 58-79
Dobrynina, O. N.: 57-47
Dobrynina, Z. G.: 64-127
Dodd, E. N.: A-2
Doering, H.: 55-54; 56-120
Dokhana, M. M.: 59-99, 100; 61-126
Doležal, J.: 51-19; 52-14; 53-12, 13; 54-82; 55-36; 56-89; 60-88; 62-122; 64-67
Doli, N.: 66-60, 99
Dollman, G. W.: 64-4
Dominik-Srycharska, M.: 63-124
Domocos, V. F.: 67-25
Donchenko, V. A.: 67-67
Donker de, K.: 59-107
Dono, T.: 65-111; 67-97
Doornbos, D. A.: 62-23
Doppler, G.: 54-94; 56-97; 57-29; 58-57
Doran, M. A.: 59-94
Dorazil, L.: 60-52; 61-85
Dorfman, E. A.: 66-86; 67-20
Dorić, T.: 68-31
Douskai, B. M.: 58-88

Drack, B. S.: 61-111
Dragulescu, C.: 59-76; 66-102
Dragusin, I.: 60-184, 190; 61-148; 62-104; 65-34
Drahotská, B.: 58-75
Drummond, J. L.: 62-65
Du Bois, R. J.: 65-90
Dubský, I.: 59-25; 61-145; 67-31
Dugandžič, M.: 59-36, 37; 60-2, 182, 183
Dunsbach, F.: 63-89; 64-10
Dunstone, J. R.: 60-143
Durcham, E. J.: 58-71
Dutt, N. K.: 60-121
Duval, C.: 57-74
Duyckaerts, G.: 53-51
Dwyer, F. P.: 60-125
Dyatlova, N. M.: 63-110; 67-79, 118, 130; 68-68
Dzhiyanbaeva, R. Kh.: 67-11, 32, 60, 69

E

Eastman, H. W.: 63-74
Eckschlager, K.: 64-116
Edwards, J. W.: 55-80; 58-113; 59-110
Edwards, R. E.: 55-71
Eeckhaut, J.: 57-53
Efimov, I. P.: 60-130; 61-54, 176; 63-106
Eggers, J. H.: 60-126
Egorova, L. G.: 65-36
Eichlin, D. W.: 55-10
Ek, C.: 57-67
Eklund, B.: 62-118
El-Barbary, I.: 68-30
Eldjarn, L.: 55-53
Elinson, S. V.: 63-63; 67-33
Ellestad, R. B.: 61-7
Ellinboe, J. L.: 56-72; 60-36
Elliot, W. E.: 52-9
Elo, A.: 60-139
Elofsson, A.: 61-149
Elsheimer, H. N.: 67-7
El-Sheltawy, A. M.: 61-156
Elwell, W. T.: 62-85
Emi, K.: 57-40, 41, 50
Emmerich, A.: 51-18

Emr, A.: 57-79
Endo, Y.: 60-138, 141; 62-125
Engelen van, H. Th. J.: 55-12
Erdey, L.: 53-36; 55-46; 56-74, 96;
    57-117; 59-59; 60-89, 90, 150;
    61-60; 62-63, 79, 90, 117; 66-97
Ermolaev, N. P.: 62-81
Ernsberger, F. M.: 59-50
Escarilla, A. M.: 66-4
Esch van, J.: 54-4
Eschmann, H.: 54-93; 56-55
Evans, W. H.: 67-34
Ezerskaya, N. A.: 65-72; 66-41, 42;
    67-35

F

Faber, J. S.: 54-30; 55-107; 56-102,
    103, 104, 105; 57-38; 62-23
Fabregas, R.: 62-98
Faerman, D. V.: 64-91
Fainberg, S. Yu.: 58-55
Faller, E. F.: 53-6
Fara, M.: 67-141
Farag, A.: 58-91
Farah, M. Y.: 55-77
Farrow, R. N. P.: 65-52, 53
Federov, A. A.: 67-36
Felszeghy, E.: 66-79
Fernandez, T.: 68-18
Fernandez-Paris, J. M.: 61-150
Fernando, Q.: 62-11
Ferrett, D. J.: 54-55; 56-123
Fesenko, N. G.: 61-10, 174
Fidlerova, J.: 64-116
Figman, W.: 63-131
Filatova, L. N.: 58-55
Filimonova, V. N.: 65-72
Filipov, D.: 64-27
Filippova, N. A.: 65-95
Filyuzina, V. S.: 66-108
Fischer, W.: 65-31
Flanigan, D. A.: 61-35
Flaschka, H.: 51-20; 52-22, 23, 24,
    25, 26, 27, 28, 29, 30, 31, 32, 33,
    34, 35; 53-1, 21, 22, 23, 24, 25, 26,
    27, 28, 58, 62; 54-59, 60, 61, 62;
    55-62, 63, 64, 65, 66, 67, 68;
    56-42, 43, 44, 45, 46, 47, 116,
    122; 57-86, 87, 88, 89, 90, 91, 92,
    Q

125, 134; 58-35, 36, 37; 59-37,
    38, 146; 60-68; 61-48, 49, 50, 51,
    164, 184; 62-26, 27; 63-9, 10, 11,
    81; 64-144, 148; 65-107, 108,
    113; 66-67; 67-37
Fleet, B.: 68-19
Fleischer, K. D.: 58-24, 86
Flekser, G. I.: 65-6
Fleszar, B.: 64-117
Foerster, W. I.: 67-38
Fog, H. M.: 68-20
Foley, W. T.: 56-118; 62-22
Ford, J. J.: 53-19
Foreman, J. K.: 57-130; 62-65
Formanek, I.: 64-131
Fors, M.: 60-197
Forster, W. A.: 60-179
Fortuin, J. M. H.: 54-43
Foss, O. P.: 58-45
Fouchecourt, P.: 66-36
Franschitz, W.: 55-67
Franzon, O.: 58-29
Fraser, G. P.: 59-129
Freese, F.: 66-63
Freiser, H.: 57-141; 61-171; 62-11;
    63-12; 67-92
Freitag, Elsi: 51-2, 3
Fretzdorff, A. M.: 53-53
Fritz, J. J.: 53-19
Fritz, J. S.: 54-44, 45, 46; 55-108;
    57-25, 122; 58-61, 62; 59-96; 60-
    100; 61-112; 62-93
Frost, A. E.: 59-94
Frumina, N. S.: 65-28
Fuhrman, D. L.: 66-30
Fujiwara, M.: 55-110
Fukuda, M.: 66-70
Fulda, M. O.: 54-45, 46
Fülöp, L.: 62-57, 131; 64-131; 65-98
Funahashi, S.: 68-61
Funtikov, K. M.: 62-13
Furby, E.: 54-99
Furlani, A.: 68-21
Furlani, A. D.: 60-158
Furlani, C.: 68-12
Furse, C. T.: 62-36
Furuya, M.: 63-55, 72; 63-122
Fusaka, K.: 55-110
Füsti-Molnar, S.: 61-140

G

Gaál, F. F.: 68-32
Gabrielson, G.: 57-45
Gagliardi, E.: 57-44; 58-94
Gahbah, F. J.: 67-5
Galaktinov, Yu. P.: 65-39
Galaktionov, Y. G.: 63-77
Galateanu, I.: 58-32; 59-60; 60-133
Ganchoff, J.: 61-50, 51; 62-26
Ganenko, Z. G.: 67-94
Ganina, V. G.: 59-75
Garcia, C.: 62-98
Gardel, M. C.: 66-68
Gardner, R. D.: 58-15
Garkusha-Bozhko, I. P.: 67-62
Garrett, J.: 63-10; 64-144; 66-67
Garvan, F. L.: 60-125
Gattegno, D.: 68-21
Gattow, G.: 62-82, 83; 64-133
Gaugin, R.: A-5
Gaur, J. N.: 67-39
Gautier, J. A.: 57-126, 128
Geary, W. J.: 62-21
Gebauer, J.: 61-52
Gedansky, S. J.: 57-75; 65-14
Gegus, E.: 62-117
Gehrke, Ch. W.: 54-5
Gelman, A. D.: 59-92; 62-81
Genge, J. A. R.: 57-73
Genton, M.: 52-12; 53-5
George, J. H. B.: 53-15, 59
Gere, E. B.: 60-150
Gergely, A.: 65-35; 66-74
Gerkhardt, L. I.: 60-11
Gerlach, K.: 55-57
Gertner, A.: 67-40
Geyer, L.: 57-71
Geyer, R.: 67-41
Gheorghe, F.: 63-116
Ghosh, A. K.: 56-58
Gibalo, I. M.: 56-65
Gibbons, D.: 54-1; 55-104, 105, 106
Gibson, J. A.: 54-32; 59-104; 64-147
Gibson, J. G.: 51-9
Gibson, N. A.: 61-135, 136
Giefer, L.: 55-109
Gimesi, O.: 61-60, 62-63, 79, 90

Gingerich, K. A.: 65-50
Gittner-Mindermann, A.: 64-118
Giuffrè, L.: 60-135; 62-80
Giuliani, A. M.: 68-21
Gjems, O.: 53-18; 60-129
Gjessing, L.: 60-128
Gladysheva, K. F.: 67-42
Glauser, E. M.: 67-43
Glauser, S. C.: 67-43
Glavina, V. S.: 66-119
Glukhova, L. P.: 67-93
Godovvannaya, I. N.: 66-44
Goetz, C. A.: 50-7, 8; 55-74
Goffart, J.: 53-51
Gohrbrandt, E. C.: 54-12
Golby, R. L.: 58-93
Golden, H. J.: 67-81
Goldman, J. A.: 64-143; 68-22
Goldstein, D.: 59-151
Goldstein, G.: 62-32; 63-13
Goldstein, M.: 64-3
Golovina, A. P.: 56-65
Gomez, A. A.: 66-107
Görbe, E.: 67-26
Gorbenko, F. P.: 66-88, 89; 67-44
Gordon, L.: 56-10, 71; 57-75
Görner, F.: 64-33
Görög, S.: 60-193
Goryunova, N. N.: 65-28
Goryushina, V. G.: 57-66
Goto, K.: 67-147
Gotthardt, B.: 64-133
Gottschalk, G.: 60-127
Goyal, S. S.: 67-45
Goydish, B. L.: 62-52; 63-39; 66-8
Gramberg, K.: 64-119
Grankina, M. Y.: 64-2
Graske, A.: 61-143
Graue, G.: 53-52
Grdinic, V.: 67-40
Grecu, I.: 65-84
Green, H.: 55-45; 64-94
Greenblatt, I. J.: 51-17
Greene, Jr, A.: 57-69
Gregory, G. R. E. C.: 62-20
Griffenhagen, G. B.: 51-11
Grigor'ev, N. N.: 64-107
Grigor'eva, M. F.: 64-79
Gringras, L.: 57-24

Grintescu, P.: 58-111; 59-135
Groebel, W.: 55-48
Gromova, M. I.: 61-54, 176; 62-19;
  63-106
Grönquist, K. E.: 53-20
Grosskreutz, W.: 67-46
Grove, E. L.: 64-24
Grüner, K.: 58-75
Grünwald, A.: 53-49
Guerrieri, F.: 57-31
Guerrin, G.: 60-66, 67
Guerrini, F.: 53-50
Gupta, A. K.: 64-78
Gupta, M. P.: 61-116
Gupta, N.: 60-147
Gurevich, A. B.: 56-21
Gurevich, A. M.: 68-23
Gurkin, K. M.: 56-56
Gurn'ev, S. D.: 67-153
Gusenko, T. V.: 67-99
Gusev, S. I.: 62-113; 63-41; 64-53;
  66-16, 20, 49; 67-47, 48
Gut, R.: 54-27; 56-125
Gutman, F. W.: 56-51
Gutnikova, R. I.: 68-39
Gutzeit, G.: A-11
Gwilt, J. R.: 62-7
Gysling, H.: 49-4, 5

H

Haas, K.: 63-95; 65-22
Habercetl, M.: 62-139
Häberli, E.: 54-26
Hach, C. C.: 50-7
Hadjiioannou, T. P.: 58-110; 59-
  117, 118, 119, 120, 121; 60-78;
  66-66
Haerdi, W.: 66-75
Hagiwara, Z.: 55-11
Hahn, F. L.: 58-76; 60-124
Hahn, R. B.: 62-129
Hakkila, E. A.: 58-78
Halasz, A.: 61-64
Hall, J. L.: 54-32; 59-104; 64-147
Hall, L. C.: 61-35
Hall, L. H.: 68-24
Hamaguchi, H.: 61-172
Hamdy, M.: 59-125
Hamm, R.: 55-9

Hamm, R. E.: 62-36
Handa, R.: 68-49
Haniset, P.: 56-48
Hannema, V.: 67-49, 50
Hanok, A.: 51-13
Hanselman, R. B.: 60-123
Hanus, J.: 67-51
Hara, R.: 54-47; 55-41, 42, 43; 56-
  59
Hara, S.: 57-30; 61-61, 62
Harder, R.: 59-88
Harmelin, M.: 62-44
Harris, W. F.: 52-42; 54-49, 102;
  68-16
Hartman, S.: 51-17
Harzdorf, C.: 64-126
Hasegawa, M.: 67-145
Haslam, J.: 60-111
Hata, A.: 67-145
Hatanaka, T.: 64-98
Havír, J.: 56-93; 57-78; 58-46; 59-
  13; 66-13
Hayes, J. F.: 61-15
Hayes, T. J.: 53-33, 34, 35; 54-57
Haynes, B. J.: 67-114
Hazel, J. F.: 57-63
Headridge, J. B.: 58-22, 23; 60-83
Heeney, H. B.: 60-86
Hefly, A. J.: 68-25
Hegemann, F.: 61-118, 180
Heinle, P. J.: 62-28
Heller, J.: 51-1, 4, 6; 52-20
Helmboldt, O.: 67-15
Helmstädter, G.: 59-133, 134
Hennart, C.: 57-127, 129; 59-132;
  61-158, 159, 160, 161, 162; 62-69,
  70, 71; 65-76, 85, 86
Henry, S.: 56-48
Hensley, A. L.: 63-36
Hentrich, K.: 64-114; 66-57; 67-158
Herbinger, W.: 60-79
Herbolsheimer, R.: 59-51
Hermann, D.: 68-65
Hermans, M. E. A.: 65-51
Hernandez, H. R.: 50-9
Hernández-Cañavate, J.: 67-119
Herzog, K.: 54-85
Hess, G.: A-9
Hessler, W.: 64-19; 65-5

Heyer, F. C.: 65-32; 67-52
Heyl, H. J.: 66-32
Heyndrickx, P.: 55-5; 56-8; 59-40
Hibbs, L. E.: 58-42; 59-6, 8, 9, 11
Hickey, J. J.: 66-48
Hildebrand, G. P.: 57-123
Hill, A. G.: 65-52, 53
Hillebrand, G. E. F.: 53-56
Hindman, W. H.: 56-4, 14
Hinsvark, O. N.: 56-12
Hirata, H.: 67-91
Hirn, C. F.: 58-25; 59-89, 116; 60-131
Hisada, M.: 60-91
Hniličková, M.: 58-28; 60-20; 61-86
Hoard, J. L.: 61-177
Hodecker, J. H.: 58-24, 86
Hoelzl-Wallach, O. F.: 59-144
Hoffman, F. A.: 63-45
Hoffman, J. J.: 53-56
Hoffmann, H.: 63-93
Hofman, J.: 61-182
Hogiwara, Z.: 56-19
Hol, P. J.: 52-4; 54-64; 60-31
Holasek, A.: 52-23, 33, 34; 56-122; 57-86; 59-35, 36, 37; 60-2, 182, 183, 185; 61-33, 164, 167
Holbrook, M.: 52-7
Holeček, K.: 61-102; 64-92
Holloway, J. H.: 58-14; 60-198
Holt, R.: 54-54
Holtz, A. H.: 52-2
Holzer, S.: 59-95
Honda, M.: 60-107
Hoog de, P.: 55-12
Hopkinson, G.: 55-50
Horaček, J.: 60-120; 67-107, 108, 111, 142
Horakova, E.: 54-7
Horiuchi, T.: 66-90
Horiuchi, Y.: 66-91, 92; 67-53
Hoshikawa, G.: 55-23
Hoshino, Y.: 62-128
Hosokawa, Y.: 65-26
Houda, M.: 59-15
Hoyle, W.: 59-115
Hoyme, A.: 59-128
Hoyme, H.: 56-23; 57-27; 60-101
Hozumi, K.: 61-153; 64-104; 68-27

Hsu, T. M.: 59-84
Hsü, L.-Y.: 61-79
Huang, C. S.: 66-58
Huber, C. O.: 68-26
Hubicka, K.: 64-141
Hubmayer, W.: 60-79
Huditz, F.: 52-28, 29, 30, 31, 32
Huebner, K.: 67-15
Huitink, G. M.: 67-54
Huka, M.: 57-76
Hulanicki, A.: 64-58
Humlicek, D. E.: 62-28
Hung, S.-C.: 64-75; 65-44
Hunt, B. R.: 67-6
Hunter, J. A.: 56-25
Hunter, T. L.: 65-77
Hupfer, J.: 61-170

I

Ianescu, A. M.: 64-108
Ichijyo, O.: 66-91; 67-53
Idemori, M.: 64-56
Ideno, E.: 68-27
Ifkovits, E.: 67-43
Ikegami, A.: 57-115
Iimura, I.: 66-90
Imoto, H.: 60-112
Ince, A. D.: 60-179
Ingols, R. S.: 50-10
Ioanid, N.: 65-87
Iritani, N.: 56-18; 57-100; 58-117; 59-41; 60-84; 66-93; 67-55
Irving, H. M. N. H.: 54-55; 66-35
Isaac, R.: 66-14
Isachenko, A. V.: 61-54, 176
Isagai, K.: 55-22
Ishankodzayev, S. D.: 60-42
Ishibashi, N.: 66-70
Ishida, A.: 66-34
Ishidate, M.: 59-47, 66
Ishii, M.: 62-41, 42
Ishimori, T.: 65-71
Israel, Y.: 66-65
Ito, A.: 68-28
Ito, T.: 59-42
Ivanov, V. M.: 60-130; 61-132, 157; 62-92; 63-70
Ivanova, I.: 61-4, 56, 152; 67-13
Ivarsson, G.: 58-29

Iwamoto, T.: 61-66, 137; 62-43
Iwantscheff, G.: 58-116
Iwasaki, K.: 67-146
Iwase, A.: 59-52; 60-7
Iwata, S.: 60-19
Iyer, R. K.: 65-99

J

Jablonski, W. Z.: 60-30
Jackson, P. J.: 54-56
Jacobson, E.: 63-78
Jacobson, W. R.: 60-144
Jaimni, J. P. C.: 66-10, 11
Jain, D. S.: 67-39
James, D. B.: 60-100
Janácek, K.: 56-89
Jangida, B. L.: 64-135
Janiček, F.: 57-60
Jankovits, L.: 55-46
Jankovský, J.: 56-69; 57-80; 64-100
Janosi, A.: 61-64
Janoušek, I.: 58-101; 59-26
Jardin, Cl.: 59-68, 69
Jasinger, F.: 60-102
Jefferey, P. G.: 62-20
Jellinek, O.: 58-115
Jeničková, A.: 54-78, 81; 56-85, 88, 90
Jen, H.-T.: 65-44
Jenik, J.: 61-22; 63-119; 64-69
Jenness, R.: 53-54; 54-21
Jensen, B. S.: 60-18; 63-98
Jesenak, V.: 64-15
Ježhova, D.: 58-118
Jilek, A.: 56-93
Johannsen, W.: 56-106, 115; 57-4
Johannson, A.: 65-104
Johnson, C. A.: 55-72
Johnson, E. A.: 60-30
Johnson, M.: 55-108
Johnson, N. C.: 62-99
Johnston, M. B.: 60-17
Johri, K. N.: 67-56, 57
Jonckers, M. D. E.: 53-55
Jones, T. I.: 59-141
Jordan, D. E.: 67-58, 59
Jordan, J.: 57-97
Joussot-Dubien, J.: 64-57
Junek, H.: 62-47

Jurczyk, J.: 65-11; 68-29
Justus, N. L.: 56-50

K

Kabanov, B. N.: 56-6
Kabanova, O. I.: 61-18
Kabanova, O. L.: 60-10
Kachelska, O.: 64-115
Kaden, W.: 65-74
Kaimal, V. R. M.: 62-95
Kajiwara, T.: 65-62
Kajiyama, R.: 62-41, 42
Kakabadse, G.: 61-16
Kakác, B.: 58-48
Kaleis, O. Yu.: 57-98
Kalina-Zhikareva, V. I.: 56-21
Kalinichenko, I. I.: 58-16; 63-120
Kalland, G.: 63-78
Kalmykova, I. S.: 63-99; 66-124
Kalnzhnaya, G. A.: 61-41
Kamaeva, G.: 67-60
Kampitsch, E.: 45-1, 2; 46-1
Kanamori, K.: 62-43
Kaneniwa, N.: 56-17
Kang, K. H.: 66-96
Kanie, T.: 57-102; 59-43
Kanishi, S.: 64-101
Kao, H.-H.: 57-101
Kapitsa, N. V.: 67-151, 152
Kapp, F.: A-1
Karadakov, B.: 65-118, 119
Karakashov, A. V.: 60-81
Karanov, R. A.: 59-106; 62-152
Kardos, E.: 64-25; 65-89; 67-128
Karlicek, R.: 63-105
Karlik, M.: 57-34, 35; 65-67
Karolev, A.: 59-49; 62-152
Karolev, A. N.: 59-70, 106; 67-61
Karpov, O. N.: 67-62
Karraker, S. K.: 58-62
Karsten, P.: 54-43; 55-12
Kasaki, K.: 62-42
Kashikawa, K.: 60-91
Kashkovskaya, E. A.: 57-32; 58-68, 89
Kathen, H.: A-10
Katigawa, T.: 60-32
Katihara, S.: 57-30
Katiyar, S. S.: 64-47; 68-47

Kato, T.: 54-23; 55-11; 56-19; 62-54
Kats, A. L.: 58-41
Kawabe, A.: 67-63
Kawahata, M.: 62-41, 42
Kawassiades, C. Th.: 66-66
Keattch, C. J.: 63-44
Kékedy, L.: 62-158; 63-102; 64-28
Kellner, A.: 62-136; 65-88; 66-55; 68-59
Kemula, W.: 61-70, 71; 64-58; 66-94
Kennedy, J. H.: 61-169
Kenner, C. T.: 52-43; 54-8; 55-34
Kenny, A. D.: 54-53; 58-39
Keremedzhidi, L. N.: 63-113
Kerin, D.: 64-137
Khadeev, V. A.: 58-41; 60-42, 74, 75; 64-76; 67-64, 65, 66
Khalafallah, S.: 57-89
Khalifa, H.: 58-56, 57, 58, 59, 60; 59-124, 125; 60-163, 164; 61-13, 14; 62-75, 120; 64-99; 68-30
Khalinin, A. S.: 61-41
Khalizova, V. A.: 67-67
Khater, M. M.: 61-13, 14; 62-75, 120
Khersonskaya, L. M.: 55-51
Khlystova, K. B.: 66-95
Khristoforov, B. S.: 65-75
Kidama, M.: 67-68
Kies, H. L.: 54-43; 55-12; 58-44
Kim, J. H.: 62-154
Kim, H. R.: 66-96
Kimura, E.: 59-47, 66
Kimura, H.: 60-159
Kinnunen, J.: 52-6; 53-40, 41; 54-38, 39, 66, 67, 100, 101; 55-1, 2, 3, 4; 57-107, 137; 58-80, 81; 67-3
Kinoshita, S.: 64-104
Kirby, R. E.: 61-171; 63-12
Kirika, A. L.: 67-32, 69
Kirin, I. S.: 65-81
Kirkbright, G. F.: 62-151, 155
Kirtscheva, N.: 64-27
Kiseleva, L. V.: 58-108; 67-70
Kiss, T. A.: 62-138; 63-100, 101; 65-24; 66-97; 68-31, 32, 56
Kitagawa, H.: 62-45
Kitano, S.: 67-55

Kitazawa, C.: 64-98, 105
Kivalo, P.: 55-13
Klass, C. S.: 62-39
Klassova, N. S.: 67-71
Klein, P.: 61-45; 62-123; 63-52
Klemina, T. N.: 66-119
Kletenik, Yu. B.: 66-98
Klett, C. A.: 66-72
Klewska, A.: 63-124
Klygin, A. E.: 59-126; 61-89, 90, 91
Knappe, E.: 58-96
Knie, K.: 52-48
Knight, A. G.: 51-14
Knight, H. T.: 54-2, 3
Knight, W. S.: 59-73
Knizek, M.: 64-88
Knoblock, E. C.: 65-1
Knowles, C. J.: 64-110
Knyazeva, E. M.: 60-61; 62-61
Kobatake, T.: 67-131
Kobayashi, M.: 60-7
Kobyak, G. G.: 66-80, 81
Kochergina, S. A.: 67-65
Kodama, K.: 65-3
Kodama, M.: 56-94; 58-74
Koester, M.: 67-15
Kohare, H.: 66-70
Kohlschütter, H. W.: 63-53, 54; 64-130
Koizumi, T.: 58-74
Kojcev, M. K.: 59-70
Kolářik, Z.: 56-82; 57-99
Kolesnik, E. S.: 63-110
Kolobova, K. K.: 65-120
Kolthoff, I. M.: 47-3, 5; 52-11
Kolyada, N. S.: 61-89, 90, 91
Komar, N. P.: 59-65
Komaritskaya, I. D.: 61-17
Komashko, G. A.: 68-48
Komatsu, S.: 64-98, 105; 65-70; 66-53; 68-43
Kondela, Z.: 51-16
Konecny, J.: 66-1; 67-132
Konishi, S.: 62-143; 65-121, 122; 66-60, 99
Konkin, V. D.: 63-43, 115; 64-71, 81
Konkova, O. V.: 59-45; 66-114
Kononenko, L. I.: 62-60, 130

Konopik, N.: 67-8, 88; 68-33
Konrad, H.: 62-148
Kopanica, M.: 59-30, 31, 32; 60-48
Körbl, J.: 56-33, 39; 57-59, 60, 61,
  79, 83; 58-34, 47, 48, 49, 50, 51,
  52, 54; 59-15, 16, 18, 22, 27, 29;
  60-43, 44, 45, 51, 52; 61-85
Korenman, I. M.: 59-75; 60-191,
  195
Korkisch, J.: 58-91; 60-177, 178;
  61-36
Kornev, V. I.: 67-74
Koroki, C.: 60-138
Koroleva, V. I.: 64-2
Korondan, I.: 66-118; 67-127
Körös, E.: 53-32; 57-39, 135, 136;
  58-102, 103; 59-55, 58; 60-73; 61-
  21
Korosteleva, V. A.: 65-95
Korotkikh, A. E.: 64-30
Korotkova, O. I.: 62-62, 86
Kortüm, G.: 60-175
Koster, A. S.: 60-29
Kotouček, M.: 67-98
Kotrlý, S.: 57-77, 82; 58-46; 59-13;
  60-47; 66-100
Kotzian, H.: 62-9
Kovach, A.: 64-95
Kovárik, M.: 56-1
Kowalska, E.: 64-11
Kozhevnikova, I. A.: 62-113; 63-
  41; 67-47
Kozlicka, M.: 62-66
Kozlov, A. G.: 60-172
Kozlova, A. B.: 60-72
Kraft, G.: 62-38, 116; 68-34
Krajovan-Marjanovic, V.: 63-33
Kral, J.: 64-139
Krasyukova, I. G.: 67-67, 72
Kratochvil, M.: 61-92, 97
Kraus, E.: 57-60; 58-50
Kreingold, S. U.: 62-8, 149, 150;
  63-104; 67-130
Kriege, O. H.: 54-95; 55-59, 60;
  56-52; 66-23
Krijn, G. C.: 60-27, 28, 29; 62-37
Krishnamurthy, M.: 64-84
Krishnan, K.: 68-35
Kristiansen, H.: 59-67, 79; 61-44

Krivosheeva, V. I.: 63-130
Krleža, F.: 67-73
Krot, N. N.: 60-172; 62-81
Krowczynski, L.: 59-77
Krtil, J.: 66-33
Kruchkova, E. S.: 60-108; 61-5
Krumina, V. T.: 67-74
Kruysse, L.: 63-123
Krylova, A. N.: 57-114
Kuan, T.-Y.: 63-83; 65-97
Kudrunovsk-Pavlikova, E.: 63-24
Kula, R. J.: 66-38, 39
Kumar, S.: 68-54
Kumok, V. N.: 67-125
Kundu, P. C.: 61-1, 2
Kuperstein, M. A.: 64-30
Kurepa, G. A.: 67-48
Kurokawa, K.: 64-48
Kurpiel, I.: 61-163
Kurtz, T.: 52-46; 53-61
Kusakina, N. P.: 62-110; 63-94
Kushin, S. D.: 60-191, 195
Kustova, N. A.: 60-189; 64-79
Kuteinikov, A. F.: 62-89; 64-40;
  67-75, 76
Kutznetsova, E. T.: 57-110
Kutznetzov, V. I.: 57-109
Kutznetzova, L. A.: 64-30, 60
Kuzman-Anton, R.: 66-102
Kuz'menko, N. I.: 67-77
Kuznetsov, A. R.: 60-105; 61-75, 80
Kuznetsova, A. R.: 63-112
Kuznetzova, O. M.: 60-151
Kvapil, M.: 63-24
Kvashina, F. F.: 60-75
Kydd, P. H.: 52-38
Kyřš, M.: 63-51; 64-73; 67-78
Kysil, B.: 59-16

L
Lacourt, A.: 55-5; 56-8; 59-40
Lacy, J.: 63-46
Lada, Z.: 61-3
Laessig, R. H.: 65-106, 110; 66-64,
  101
Lagunova, N. L.: 64-146
Laitinen, H. A.: 54-88
Lakatos, A.: 66-87
Lambert, J. L.: 68-24

Lambert, R. S.: 65-90
Lamson, D. W.: 58-12
Lan, C. T.: 63-60
Landgren, O.: 51-10; 52-1
Lane, W. J.: 57-25, 122
Lang, K.: A-10
Langen de, D.: 63-71
Langer, H. G.: 62-100, 140; 64-123; 67-9
Langmyhr, F. J.: 59-67, 79
Lapitskaya, E. V.: 67-44; 68-63
Lapitskii, A. V.: 64-77
Lassner, E.: 57-21; 58-82, 83; 59-1, 2, 3; 60-3, 4, 5; 61-138, 139, 168; 62-50, 77; 63-37, 38; 64-70; 68-36
Lastovskii, R. P.: 67-79, 130
Laszlovsky, J.: 54-68; 57-119; 60-73
Latimer, G. W.: 66-30
Lauer, K. F.: 67-16
Laurent, B.: 60-68
Laurent, S.: 60-68
Laux, P. G.: 61-144
Lavrova, O. U.: 68-68
Lazar-Jucu, D.: 66-102
Lebedev, I. A.: 67-80; 68-37
Lebedeva, N. V.: 64-122
Leden, I.: A-4
Lee, Y. C.: 54-5
Leendertse, G. C.: 52-4
Lefevre, Y.: 61-160
Leftin, J. P.: 54-52
Leggieri, G.: 55-116
Legradi, L.: 61-140, 141
Lehmann, H.: 58-73
Leifer, E. I.: 59-75
Lembrez, Y.: 60-137
Leminger, O.: 54-16; 56-91, 92
Lengyel-Nadler, V.: 66-103
Leonard, M. A.: 58-2, 3; 60-194
Leonardova, L. A.: 65-75
L'Eplattenier, F.: 64-52
Leupin, K.: 64-118; 67-162
Levine, S. L.: 67-81
Lewandowska, K.: 65-69
Lewandowski, A.: 59-93
Lewis, L. L.: 60-33, 34; 61-68
Leyden, D. E.: 68-38
Lheureux, M.: 56-48
Li, H.-J.: 65-46

Lidin, R. A.: 63-77
Lieb, H.: 59-35; 60-185
Lieber, F.: 60-136
Lieberman, M. V.: 58-17
Lind, M.: 61-177
Lindgren, J. E.: 60-197
Lindley, G.: 56-64
Lindstrom, F.: 59-127; 60-134; 62-24; 66-14
Lingen van, R. L. M.: 62-145
Linko, E.: 53-46
Lisichenok, S. L.: 67-135
Liteanu, C.: 59-90; 60-132; 61-63; 63-116, 126, 127, 128; 65-58; 67-82, 83, 84
Liu, S.: 58-31
Lloyd, D. R.: 63-23
Longford, K. L.: 52-3; 54-11; 58-30
Loomis, T. C.: 50-8
de Lopidana: 66-104
Lorenzi de, F.: 52-10
Loriya, N. V.: 68-76
Lounamaa, N.: 63-131
Lott, P. F.: 56-60; 57-124; 59-102
Lovasi, J.: 61-67
Love, S. K.: 55-47
Lučanský, J.: 67-98
Lucchesi, C. A.: 58-25; 59-89, 116; 60-131
Lukachina, V. V.: 65-57
Lukáčová, O.: 64-121
Lukacs, I.: 61-63
Lukaszewski, G. M.: 57-84
Lukin, A. M.: 66-105, 106
Lukyanov, V. F.: 60-61, 62, 72; 62-61
Lupan, S.: 58-77
Lur'e, I. S.: 66-56
Lutchenko, N. N.: 67-153
Luzina, L. N.: 65-115
Lyashenko, T. V.: 65-96
Lydersen, D.: 53-18
Lyle, S. J.: 63-49, 50
Lysenko, S. A.: 64-40; 67-75, 76
Lystsova, G. G.: 61-109
Lyubomilova, G. V.: 65-45

M

Macek, K.: 55-28
MacNevin, M.: 58-78

MacNevin, W. M.: 54-95; 55-59, 60; 56-52
Madsen, N. P.: 60-143
Maekawa, S.: 67-85
Mahr, C.: 54-24
Majer, J.: 62-67; 63-105; 67-98
Majer, P.: 60-102
Majumdar, A. K.: 60-41
Majumdar, S. K.: 57-93
Makarova, N. I.: 63-103
Makarova, T. P.: 65-10; 67-126
Makashev, Yu. A.: 65-81
Makasheva, I. E.: 65-81
Makrickaja, E. K.: 56-3
Maksimova, A. M.: 67-80, 126
Malát, M.: 54-75, 77, 78, 81; 56-37, 38, 40, 41, 84, 85, 88, 90; 58-118; 59-14, 20; 61-101, 102, 185
Malevannyi, V. A.: 64-54
Malik, A. U.: 65-27
Malinek, M.: 56-31
Malissa, H.: 62-9
Malkus, Z.: 60-120
Malmstadt, H. V.: 54-12; 58-110; 59-117, 118, 119, 120; 60-78
Malowan, L. S.: 60-118
Mal'tseva, L. S.: 67-47
Manczyk, R.: 60-98
Mankota, M. S.: 60-99
Mann, J.: 67-37
Manning, D. L.: 62-32; 63-13
Manninger, D. L.: 58-70
Manns, Th. J.: 52-49
Manoliu, C.: 57-72
Marayama, T.: 58-74
Marchenko, P. V.: 64-125
Marcy, V. M.: 50-5, 6
Mareva, S.: 64-9
Margalet, A.: 64-96
Margerum, D. W.: 62-18
Margerum, S. L.: 54-46
Marincová, D.: 52-13
Marinescu, L.: 60-132
Marjanovic-Kratjovan, V.: 68-55
Marov, I. N.: 63-125
Marple, L. W.: 66-69
Martell, A. E.: 51-8; 54-9, 13; 63-25
Martens, G.: 57-70
Martin, A. E.: 59-148

Martiny, E.: 67-86
Martynenko, L. I.: 58-114
Masalova, L. S.: 67-66
Mashall, J.: 57-71; 58-5, 6
Mason, A. S.: 52-40
Matousek, J.: 65-37, 67
Matrasova, T. V.: 61-122
Matsui, Y.: 61-47
Matsuo, T.: 60-122
Matteucci, E.: 67-87
Mattocks, A. M.: 50-9
Matyska, B.: 51-15, 16
Maurice, M. J.: 56-79, 80
Maverick, E. F.: 54-96
Maxim, I.: 58-32; 59-60; 60-133
Mayer, J.: 63-105
Mazumdar, S. K.: 68-49
McBride, H. D.: 58-78
McCallum, J. R.: 56-54
McGlynn, J. A.: 64-29
McKend, J.: 60-69
McKenny, D. B.: 63-74
McNabb, W. M.: 57-63
Mechos, K. Z.: 68-39
Mee, J. E.: 61-78
Meissl, R. F.: 66-107
Meites, L.: 64-143; 67-10
Meites, T.: 67-10
Mekaota, T.: 64-149
Mekhanoshina, E. Y.: 63-120
Mekhanoshina, L. N.: 65-47
Melnick, L. M.: 60-33; 61-68
Melnikova, A. S.: 59-136; 61-40
Mendez-Bezzera, A. E.: 68-40
Menessy, I.: 59-76
Mengering, S.: 56-70
Menis, O.: 58-70
Merikanto, B.: 52-6; 54-39, 66; 55-1, 2, 3; 58-81
Merkulova, K. S.: 64-89; 66-82
Merlin, E.: 57-127, 129; 59-132; 61-158
Merrill, J. R.: 60-107
Mershad, E. A.: 63-6
Merz, K. W.: 58-73
Mesimer, W. J.: 58-69
Mesrob, B.: 66-47
Meszaros, P.: 67-88
Metcalfe, J.: 64-110

Metz, C. F.: 57-64
Michal, J.: 56-69
Michel, G.: 53-51; 54-29
Michizuki, H.: 62-41, 42
Michod, J.: 61-43
Mihalescu, M.: 67-89
Miketuková, V.: 59-21, 28
Mikhailov, V. A.: 57-109
Mikhaleva, Z. A.: 60-87; 63-120
Miklos, I.: 61-53
Milaev, S. M.: 65-96
Miles, M. J.: 58-69
Miles, T. D.: 63-45
Miller, C. C.: 56-25
Miller, R. L.: 62-28
Milner, G. W. C.: 54-91, 92; 55-79, 80, 81; 56-75, 123; 57-132, 133; 58-113; 59-110
Milner, O. I.: 65-14
Milner, O. J.: 54-31
Min, S. K.: 65-103
Minami, E.: 55-33; 60-119
Miner, F. J.: 65-15
Misumi, S.: 59-71, 87
Mitsui, T.: 61-178
Miyahara, T.: 67-55
Miyata, H.: 57-40
Miyazaki, M.: 63-59
Mizuno, K.: 61-153; 65-3, 80
Mlodečka, J.: 61-3
Moeller, T.: 62-101
Moiseeva, E. I.: 65-92
Moizhes, I. B.: 66-108
Mojejko, J.: 61-163
Molot, L. A.: 59-83; 66-109; 67-90
Momiki, K.: 64-46
Monk, R. G.: 62-25; 66-31
Monn, D. E.: 67-58, 59
Monnier, D.: 64-6, 7, 8; 66-75
Montelongo, G.: 68-18
Morachevskii, Yu. V.: 60-59, 60; 64-66
Morait, G.: 60-196; 62-56, 109
Mordovskaya, E. A.: 64-127
Morgan, L. O.: 56-50
Mori, I.: 65-33; 66-15
Mori, K.: 62-2, 3
Mori, M.: 55-69
Mori, S.: 67-91

Morimoto, Y.: 63-25
Morinaga, H.: 67-85
Morris, A. G. C.: 59-72; 60-113; 61-34
Morrison, G. H.: 57-141
Mosen, A. W.: 58-15
Moser, J. H.: 54-35
Moser, P.: 53-2
Moskvin, A. I.: 59-92
Mottala, H. A.: 67-92
Moučka, M.: 56-1
Moustafa, A. S.: 59-101; 60-188
Mtsunami, M.: 65-3
Múčka, V.: 61-101
Mukaida, M.: 65-71
Mukhadshina, R. A.: 65-93
Müller, F.: 59-142
Müller, K.: 66-71
Munemori, M.: 57-19; 59-122
Munger, J. R.: 50-10
Munnecke, D. L.: 66-45
Muraca, R. F.: 54-36
Murata, K.: 62-108
Murgu, G.: 60-132
Musha, S.: 57-19, 20; 59-122, 123; 60-109; 61-72; 64-80
Mushran, S. P.: 66-21
Musil, A.: 55-29, 30
Mustafin, I. S.: 57-32; 58-68, 89; 59-83; 60-108; 61-5; 64-16; 65-28; 67-90
Myshlyaeva, L. V.: 65-12

N
Nadezhdo, A. A.: 66-89
Nagamura, S.: 58-72
Nagashima, K.: 64-124
Nagayama, M.: 67-147
Nägeli, P.: 59-142
Nakagawa, G.: 64-31, 36, 45; 65-68, 111; 67-95, 96, 97; 68-70
Nakagowa, F.: 55-112, 117
Nakamoto, K.: 63-25
Nakasuka, N.: 66-110
Nakaya, S.: 68-64
Napadailo, I. N.: 61-165
Narang, A. S.: 61-115
Narasaraju, T. S. B.: 68-52
Nardozzi, M. J.: 61-68

Naruskevicius, L.: 62-111
Nassler, J.: 64-92
Nasu, S.: 64-105
Navrotskaya, V. A.: 64-109
Nazarchuk, T. N.: 64-103; 65-47
Nazarenko, V. A.: 64-122
Nebbia, L.: 53-50; 57-31
Nebergall, W. H.: 63-7, 40
Neis, P. J.: 55-52
Nemirovskay, A. F.: 63-113
Nemkova, N. K.: 67-90
Nemodruk, A. A.: 67-93
Neumann, J.: 68-41
Nevzorov, A. N.: 67-94
Ney, E.: 57-16; 58-97
Nezhnova, T. I.: 67-33
Nguyen, H. B.: 62-56
Nichols, P. N. R.: 61-104
Nickless, G.: 62-21; 68-2,3
Nicoara, N.: 62-109
Niculescu, I.: 64-108
Nielsch, W.: 55-109
Nielsen, H.: 52-16
Nigai, K. G.: 64-41, 42
Nigam, R. C.: 68-42
Nijst, L. J. H.: 55-52
Nikelly, J. G.: 56-22
Nikitina, E. I.: 59-85; 65-60
Nikolaeva, E. M.: 66-16, 49
Nikolaeva, E. R.: 64-90
Nikolskaya, N. A.: 59-126; 61-89
Nikol'skaya, N. N.: 67-154, 155, 156
Nippler, R. W.: 50-10
Nishimura, M.: 68-64
Nogami, H.: 55-112, 117
Noll, C. A.: 50-4
Nomura, T.: 65-111; 66-53; 67-95, 96, 97; 68-43
Nonova, D.: 64-140
Novák, J.: 63-108
Novak, R. J.: 66-7
Novak, V.: 67-98
Novakovskaya, E. G.: 62-10
Novozámský, I.: 62-122
Nunn, J. H.: 57-132, 133
Nydhal, F.: 60-110
Nygaard, O.: 55-53

O

Ochynska, J.: 59-77; 61-38
Odler, I.: 61-52
Oga, T.: 65-3
Ogawa, K.: 57-19, 20; 59-122, 123; 60-109; 64-80; 66-93
Ogino, H.: 64-93; 65-63, 66
O'Hara, F. J.: 56-12
Oishi, H.: 59-41
Oiwa, I. T.: 56-94
Okáč, A.: 61-142; 62-46
Okada, K.: 57-118
Okamoto, H.: 55-110
Okuno, H.: 65-71
Oliver, R. T.: 58-61
Olsen, E. D.: 64-106; 66-7; 67-1
Olsen, R. L.: 61-7
Olson, D. C.: 62-18
Omboly, V.: 62-132
Onisor, M.-I.: 67-26
Onosova, S. P.: 62-6
Onuki, S.: 62-142; 64-102, 124
Oprescu, C.-L.: 67-26
Orhanovic, M.: 67-159
Orlova, Yu. Ya.: 67-99
Orths, K.: 64-17
Osman, F. A.: 60-164
Osteryoung, K. A.: 59-73
Osyany, J. M.: 62-22
Ottendorfer, L. J.: 58-87; 59-86
Otterbein, H.: 54-24
Overbeck, C. J.: 66-48
Owens, E. G.: 60-106
Ozerskaya, F. A.: 67-36

P

Page, J. O.: 68-44
Palaček, M.: 62-121
Palatý, V.: 61-65
Paley, P. N.: 60-10, 57, 58; 61-79
Palmer, J. W.: 61-107
Pályi, G.: 67-100
Pande, C. S.: 60-14, 15, 16; 61-19
Pang, K.: 59-84
Paniti-Cetean, M.: 67-27
Pankratova, L. N.: 64-77
Pantani, F.: 66-51, 111
Parry, E. P.: 64-4

Partashnikova, M. Z.: 63-103; 65-25; 66-115
Parushev, M.: 63-58
Pasovkaya, G. B.: 57-28; 67-101; 68-45
Patrovský, V.: 53-38; 55-36; 57-76; 59-24; 65-30, 114; 67-102
Patton, J.: 56-26
Patzak, R.: 57-29; 58-57
Paul, R. M.: 56-28
Paul, S. D.: 57-121
Paulsen, A.: 56-29
Pavelkina, V. P.: 58-88
Pavlikova, E.: 56-69; 67-103
Pavlov, N. N.: 60-105; 61-75, 80; 63-112
Pavlova, I. V.: 67-137
Payne, M. A.: 61-112
Pawelczak, M.: 67-160, 161
Pease, B. F.: 59-82
Pećar, M.: 60-1, 185; 61-33, 167
Pechanec, V.: 62-103
Pechentovskaya, L. E.: 64-103
Peczok, R. L.: 54-96; 56-30
Pelikan, J.: 56-40
Pencheva, L.: 64-128, 136
Pensar, G.: 58-4
Pereshin, G. S.: 61-58
Peshkova, V. M.: 61-54, 176; 62-19; 63-106
Peták, P.: 67-140
Petrenko, A. G.: 58-109; 61-131; 64-30, 60; 65-75
Petrikova, G.: 66-109
Petrikova, M. N.: 59-64; 66-9
Petrova, G. S.: 60-199; 66-105
Petzold, Ingrid: 52-31
Pevzner, K. S.: 60-70
Pfeifer, S.: 64-114; 66-57; 67-158
Pfeiffer, M. M.: 67-28
Pfisterer, J. L.: 51-11
Phatak, G. M.: 68-46
Phennah, P. J.: 54-92
Phillipp, B.: 56-23; 57-27; 60-101
Phillips, H. O.: 54-32
Phillips, J. P.: 59-81; 62-1
Pickles, D.: 53-48
Pietrzyk, D. J.: 58-61; 59-96
Pilkington, E. S.: 65-49

Pilleri, R.: 57-68
Pilz, W.: 54-15; 58-112
Pinkston, J. L.: 55-34
Piriri, F.: 66-83
Piriri, I.: 66-83
Pison, R. E.: 67-105
Platanova, O. P.: 55-73
Plattner, E.: 53-5
Plattner, W.: 52-12
Pletneva, T. I.: 66-115
Pobedina, L. I.: 63-63
Poczok, I.: 58-103
Podany, V.: 60-102
Podgornova, V. S.: 67-13, 14
Podlaha, J.: 66-28, 112
Podlahova, J.: 65-54, 55, 56; 66-28, 112
Pokorný, J.: 54-98; 55-35
Polcin, J.: 57-22
Polky, J. R.: 60-139
Pollard, F. H.: 62-21
Polman, R. J.: 67-133
Pólos, L.: 57-117; 60-89
Polozhenskaya, L. P.: 68-23
Poluetkov, N. S.: 62-60, 130
Polvolotskaya, G. L.: 61-181
Polyak, L. Y.: 55-32; 56-6; 57-26
Popescu, F.: 64-108
Popescu, M.: 67-83
Popov, M. A.: 63-129
Popov, N. A.: 60-96
Popova, O. I.: 66-44
Portell, R.: 67-104
Porter, J. D.: 52-5
Porterfield, W. W.: 56-77
Portiglia, E.: 64-97
Potterat, M.: 54-93; 55-14
Pottie, R. F.: 56-118
Poulie, N. J.: 54-19
Povondra, P.: 60-49; 61-81, 84
Powell, J. E.: 60-100; 64-78
Poznanski, S.: 63-32
Preiss, M.: 63-53, 54
Přibil, R.: 51-15, 16, 19, 21; 52-13, 14, 36; 53-10, 11, 12, 13; 54-73, 74, 76, 79, 80, 82; 55-28, 89, 90, 114; 56-33, 39, 86, 87; 57-59, 60, 79, 83, 135, 136; 58-34, 49, 50, 52, 54; 59-15, 16, 18, 22, 23, 27, 30,

31, 32, 33, 34; 60-46, 48, 49; 61-21, 81, 84, 92, 93, 94, 95, 96, 98, 99, 100; 62-53, 87, 102; 63-17, 27, 61, 62, 91; 64-20, 21, 22, 23, 61, 62, 63, 72; 65-7, 8, 40, 41, 43; 66-2, 3, 24, 25, 26, 27, 54, 113; 67-2, 106, 107, 108, 109, 110, 111
Pribyl, J.: 54-98; 55-35
Priestley, P. T.: 63-66, 67
Prieto, A.: 62-98
Prince, R. H.: 63-23
Proctor, Z. G.: 62-4; 63-29
Protzer, H.: 67-15
Provaznik, J.: 64-88
Prudnikova, L. D.: 60-160
Prue, J. E.: 50-3
Prumbaum, R.: 64-17
Przyborowski, L.: 61-163
Pukhonto, A. N.: 65-92
Pungor, E.: 60-56; 61-187
Purdy, W. C.: 65-1
Purohit, D. N.: 66-10
Püschel, R.: 54-62; 55-66, 68; 56-47; 63-37; 64-70
Pushinov, Yu. V.: 65-48
Pustoshkina, M. P.: 64-26
Pyatosin, L. P.: 63-99

Q
Quattrone, C.: 64-97
Quentin, E.: 54-48

R
Rachkevich, V. F.: 66-114
Radhamma, D.: 63-38
Radko, V. A.: 60-76, 77; 61-77; 65-29; 67-112
Radović, G.: 62-133
Rady, C.: 62-63
Rády, G.: 56-74, 96; 61-60; 62-79, 90; 66-97
Raemaekers, R.: 56-117
Rahman, M. M.: 63-49, 50
Rais, J.: 64-73
Raisz, I.: 66-118; 67-127
Raklmilevich, N. M.: 64-85
Ralea, R.: 60-104
Ramaiah, N. A.: 57-13, 14; 60-103; 68-47

Rao, G. G.: 56-27; 57-11, 12
Rao, V. K.: 65-59; 66-37
Rao, V. N.: 56-27; 57-11
Rappart, B. R.: 68-74
Rathbun, J. C.: 59-63
Ray, K. L.: 56-58
Reber, L. A.: 55-55
Redfern, J. P.: 57-84
Reeder, W.: 56-26; 57-23
Rees, D. I.: 60-153; 62-155
Rehak, B.: 56-31; 60-51
Reichert, R.: 56-7
Reilley, C. N.: 55-75; 56-76, 77, 78, 124; 57-8, 9, 10, 123; 58-11, 12, 13, 14; 59-113, 114, 148; 60-67, 68, 198; 61-113; 62-33, 74
Reimers, H.: 57-44; 58-94
Reiser, P. L.: 59-1
Reishakhrit, L. S.: 57-33; 64-26
Reitz, M. T.: 54-36
Remport-Horvath, Z.: 57-39
Renault, J.: 57-126, 128
Renger, F.: 63-119; 64-69
Rentschler, H.: 57-95
Répás, P.: 53-44; 61-76
Reschowsky, M. U.: 52-49
Rethy, B.: 59-95
Reyder van der, A. J.: 62-145; 67-133
Rheinhart, R. C.: 55-10
Richard, M. J.: 58-62
Richards, C. S.: 64-59
Richards, P. J.: 64-94
Richardson, M. L.: 63-73
Riedel, K.: 59-74
Riemer, G.: 58-84
Rigg, B.: 61-186; 62-157
Rihová, J.: 60-35
Rik, V. M.: 64-82
Riley, J. P.: 59-112
Rim, H. K.: 62-154
Ringbom, A.: 53-45, 46, 57; 54-69, 70; 55-98; 56-73; 57-15; 58-4; 59-143; 62-118; 63-64; 65-109
Risch, A.: 64-130
Ristici, J.: 59-145; 62-134
Ritcey, G. M.: 61-106
Ritchie, J. A.: 55-84
Rittner, W.: 56-9, 119

Riva, B.: 58-85
Roberts, J. E.: 62-75
Roberts, M.: 57-36
Robertson, A.: 55-111
Robinson, H. M. C.: 59-63
Roche, M.: 67-113
Roester, U.: 53-53
Rogers, D. R.: 63-6
Rogers, J. R.: 63-16
Rogers, L. R.: 60-123
Rollins, O. W.: 67-114
Roman, J.: 62-73
Romanova, E. V.: 57-66
Rosenthal, M.: 57-86
Roskam, R. T.: 63-71
Rosselle, N.: 59-107
Rostromina, N. A.: 68-48
Roubal, Z.: 54-79
Roubalova, D.: 60-88
Rouéche, A.: 64-6, 7, 8
Roushdi, I. M.: 61-57, 156
Rowley, K.: 56-10
Rozhdestvenskaya, Z. B.: 63-48
Rozycki, C.: 67-115
Rubel, S.: 61-70
Ruckstuhl, P.: 49-7
Ruediger, H.: 67-38
Ruggli, P.: A-1
Rumler, F.: 59-51
Rush, R. M.: 54-90
Russ, J. J.: 57-23
Russeva, E.: 66-47
Russinova, R.: 64-128
Rusu, R.: 67-26
Ryabchikov, D. I.: 63-5, 125
Ryba, O.: 54-75, 77; 56-37, 38; 58-118
Rybinin, A. I.: 66-29
Ryskiewich, S. P.: 58-71

S

Sadek, F.: 56-42; 57-92, 125, 134; 58-36, 37
Sadek, F. S.: 57-8; 61-113
Saha, S. N.: 62-5; 64-39, 51; 68-49
Sahota, S. S.: 60-99; 61-115, 116
Saito, M.: 58-72
Sajó, I.: 53-43, 44; 54-41, 42; 55-96;

56-98, 99, 100; 58-66; 61-128, 129; 62-94; 63-1; 64-151; 66-5
Sakai, E.: 56-18
Sakai, T.: 64-120
Sakuma, Y.: 58-74; 59-80
Sallmann, R.: 52-19, 21; 53-1
Salmon, J. E.: 57-73, 84
Samoilova, O. I.: 67-118
Samsonova, N. N.: 66-109
Samuelson, O.: 58-29; 63-82
Sanchez, C.: 60-174; 62-14, 15
Sanchez-Pedrino, C.: 66-116
Sand, H. F.: 61-55
Sandås, P. E.: 53-45
Sandera, Jiri: 53-3
Sanderson, I. P.: 62-35
Sangal, G.: 66-21
Sangal, S. P.: 61-37, 73, 74; 64-44; 66-21
Sapek, A.: 67-116
Sára, J.: 58-106; 59-105; 60-148; 61-114
Šaršunóvá, M.: 55-6, 113; 56-121; 64-121
Sasaki, I.: 56-19
Sasongo: 65-91
Sastri, M. N.: 65-59; 66-37
Sato, G.: 55-33
Sato, T.: 57-115
Savariar, C. P.: 60-41
Savchenkova, V. P.: 68-17
Savinovskii, D. A.: 53-42; 54-71
Savos'kin, V. M.: 67-62
Savu, M.: 66-52
Sawicki, B.: 64-141
Sawyer, D. T.: 56-30; 63-26; 67-117
Sawyer, P.: 61-49; 62-27
Sawyer, W.: 61-15
Sayun, M. G.: 62-114
Scaife, D. B.: 63-47
Scavo, V.: 54-86, 87
Scharf, R.: 58-82, 83; 59-1, 2, 3; 60-3, 4, 5; 61-138, 139, 168; 62-50, 77
Schedle, H.: 68-36
Scheidhauer, G.: 61-8, 9
Scheppers, G. J.: 66-69
Scherzer, J.: 68-50
Schilling, D.: 65-83
Schlesinger, H.: 57-21

Schmid, R. W.: 56-76, 124; 57-10; 58-11, 12; 59-114; 62-29
Schmidlin, H. U.: 50-2
Schmidt, N. O.: 55-86; 58-33
Schmitz, B.: 54-83; 57-113; 60-173; 61-6
Schmuckler, G.: 61-113
Schneider, E.: 55-48
Schneider, F.: 51-18
Schneyder, J.: 56-101; 57-96
Schnorf, P.: A-8
Scholz, L.: 57-116
Schonebaum, R. C.: 64-145
Schöniger, W.: 56-20
Schöninger, W.: 51-20
Schott, D.: 62-82, 83
Schouwenburg van, J. Ch.: 60-8
Schraibar, M. S.: 62-12
Scribner, W. G.: 55-75; 56-78; 58-13; 59-111
Schultz, D.: 67-46
Schultz, F. A.: 67-117
Schwarz, C.: 62-34
Schwarz, K.: 57-70
Schwarzenbach, G.: A-6, 7, 8; 45-1, 2, 3; 46-1, 2; 47-1, 2; 48-1, 2, 3, 4, 5, 6, 7; 49-1, 2, 3, 4, 5, 6, 7; 50-3; 51-1, 2, 3, 4, 5, 6; 52-17, 18, 19, 20, 21, 50; 53-1, 2, 3, 4, 58; 54-27, 28; 55-118; 56-125; 57-138, 139; 59-140, 142; 63-28; 65-123
Schwarzenbach, K.: 63-28
Scribner, W. G.: 55-75; 56-78; 58-13; 59-111
Sebborn, W. S.: 63-67
Sedina, L. I.: 60-62
Sedláček, B. A. J.: 59-57; 60-39, 40; 61-133, 134
Seekles, L.: 52-2
Sekino, J.: 64-46
Selezneva, E. A.: 66-123
Seliverstova, I. A.: 67-79, 118
Selman, R. F.: 63-67
Selucký, P.: 67-78
Sen, A. B.: 62-137
Sen, B.: 57-108; 58-90, 99
Senn, H.: 57-139
Sen Sarma, R. N.: 55-115; 60-171
Serebrennikov, V. V.: 67-125

Sergeant, J. C.: 53-47; 54-89
Serikova, L. I.: 63-86
Séris, G.: 54-84
Sevy, R. W.: 67-43
Shafran, I. G.: 63-103; 65-25; 66-114, 115
Shakhanova, P. G.: 60-176
Shakhova, P. G.: 62-126; 66-43
Shakir, K.: 60-54, 55
Shalinets, A. B.: 67-80, 126; 68-37
Shapiro, L.: 55-87
Shapiro, M. Ya.: 59-39
Shardakova, M. A.: 66-81
Sharma, S. S.: 55-25; 58-95
Shatunova, T. G.: 65-13
Shchurova, L. M.: 64-53; 66-20
Shead, A. C.: 52-45
Sheka, Z. A.: 63-15
Sheldon, M. V.: 57-9; 60-67
Shesterova, I. P.: 67-14
Shevchenko, G. V.: 61-23
Shevchuk, I. A.: 67-154, 155, 156
Sheyanova, F. R.: 60-191, 195
Shibata, N.: 62-45
Shimozi, M.: 66-92
Shinozawa, S.: 55-11
Shipman, G. F.: 54-31
Shirai, K.: 68-61
Shirakashi, T.: 65-63
Shlenskaya, V. I.: 66-40
Shome, S. C.: 62-95; 66-18; 68-15
Shrivanck, V.: 61-45
Shteinma, E. A.: 64-127
Shu, S.-M.: 64-83
Shulgina, M. N.: 62-146
Shumina, V. A.: 64-54
Shumkova, N. G.: 66-40
Shvedova, N. V.: 68-9
Shvetsova, T. I.: 68-8
Sierra, F.: 60-174; 62-14, 15; 66-116; 67-119
Siggia, S.: 55-10
Siitonen, S.: 57-15
Sijderius, R.: 54-37
Silina, G. F.: 63-4
Sillén, L. G.: 57-138; 59-140
Silnichenko, V. G.: 64-65
da Silva, F. J. J. R.: 67-157; 68-51
Silver, G. L.: 67-120

Silverstone, N. M.: 62-16
Silverton, J. V.: 61-177
Simačkova, O.: 57-105
Simininc, E.: 60-104
Simoes, M. L. S.: 68-51
Simon, V.: 51-19; 52-14; 53-12, 13; 54-82
Simonescu, T.: 59-76
Simonova, T. N.: 67-154, 155, 156
Simova, L.: 61-4, 56
Singh, B.: 60-99; 61-115, 116
Singh, K.: 67-56, 57
Singh, N.: 66-17
Singh, R. P.: 68-52
Singhal, G. K.: 67-121, 122, 123; 68-53
Sinha, B. C.: 67-124; 68-54
Sinyavskay, E. I.: 63-15
Šir, Z.: 55-90; 56-86, 89
Sirina, A. M.: 64-146
Sjösted, G.: 57-24
Sjöström, E.: 56-9, 119
Skorik, N. A.: 67-125
Skrebkova, L. M.: 62-48, 115
Skrifvars, B.: 57-15
Skrivanek, V.: 61-45; 62-123; 63-52
Sloth, S. K.: 51-11
Smart, A.: 64-18
Smirnov, V. F.: 65-92
Smirnova, E. P.: 67-67
Smirnova, I. D.: 59-126
Smirnova, K. A.: 66-105
Smith, L. E.: 52-15
Smith, T. D.: 57-130; 61-39; 62-141
Smolenskaya, A. M.: 68-4
Sneshnova, L. P.: 62-40
Sobel, A. E.: 51-13; 58-20
Sochenvanova, M. M.: 59-139
Sochevanov, V. G.: 59-139; 60-9
Sochevanova, M. M.: 56-16; 59-139; 60-9
Socolovschi, R.: 62-55
Soderberg, J.: 59-144
Soe-Win: 68-19
Soffer, N.: 61-11
Sogani, N. C.: 66-10, 11
Sokolova, E. V.: 62-113
Sokolova, T. A.: 60-180; 65-61
Soliman, A.: 57-87, 88; 59-124, 125

Soljic, Z.: 68-55
Sollorz, J.: 64-11
Solntseva, L. F.: 68-23
Solodovnikov, P. P.: 63-56; 66-117
Solomin, G. A.: 61-10, 174
Solov'ev, E. A.: 63-103
Solov'eva, L. A.: 64-107
Solovykh, T. P.: 65-72; 66-41, 42; 67-35
Somidevamma, G.: 56-27; 57-12
Sommer, L.: 56-82; 57-99; 58-28; 60-20; 61-86; 67-18
Songina, O. A.: 63-48; 67-18
Soto, J.: 64-134
Soukup, M.: 58-75
Sousa de, A.: 53-31, 60; 54-40; 55-99; 60-21, 22, 23, 24, 25, 26; 61-25, 26, 27, 28, 29, 173, 179; 62-97, 105, 106; 63-30, 31, 68, 69
Southworth, B. C.: 58-24, 86
Spack, C.: 63-90
Spare, P. D.: 64-32
Spauszus, S.: 61-170; 62-34
Specht, F.: 56-11
Spedding, F. H.: 53-4
Speights, R.: 63-9
Spitsyn, V.I.: 63-4
Sporek, K. F.: 58-104, 105
Springer, V.: 62-67; 63-105
Squirrell, D. C. M.: 60-111
Srivastava, A. K.: 66-76
Srivastava, T. N.: 66-17
Srivastava, T. S.: 60-14, 15, 16; 61-19; 62-137
Srivastava, V. K.: 62-119, 147
Stambaugh, O. F.: 62-4; 63-29
Stankoviansky, S.: 60-102
Stanley, W. L.: 63-85
Starostin, V. V.: 63-4
Startseva, Z. P.: 63-97
Stearns, J. A.: 59-89
Steed, K. C.: 62-25
Stefanac, Z.: 62-58
Stefanovic, D.: 68-56
Steiner, R.: 45-1, 2; 46-1
Stengel, E.: 58-84
Stenger, V. A.: 47-3, 5
Stepanov, A. V.: 65-10; 67-80, 126
Stephen, W. I.: 60-153; 62-151, 155

Stephens, B. G.: 62-24
Stets, R. G.: 67-62
Stiles, R. E.: 66-45
Still, E.: 65-16, 109
Stoenner, R. W.: 56-10
Stoicescu, V.: 58-111; 59-135
Stolyarov, K. P.: 64-107; 68-57
Storck, J.: 57-52; 60-137
Strafelda, F.: 60-35; 62-17; 63-79; 65-4, 37, 67
Strafford, N.: 53-29; 55-61
Straub, W. A.: 60-34
Strebulaeva, E. N.: 67-36
Strecker, F. J.: 53-53
Street, H. V.: 58-98
Strelow, F. W. E.: 61-20; 66-50
Stresko, V.: 67-86
Strusivici, C.: 61-63
Strzeszewska, I.: 61-108
Študlar, K.: 58-101; 59-26; 60-92
Štulík, K.: 67-143; 68-58, 69
Styunkel, T. B.: 53-42; 54-71; 56-24; 58-9; 60-87; 63-120
Šubert, J.: 67-150
Sugawara, M.: 60-122
Sugisita, R.: 61-172
Sugiyama, T.: 57-118
Suh, Ch. C.: 60-12
Suk, J.: 56-88
Suk, V.: 54-75, 77, 78, 81; 56-37, 38, 40, 41, 84, 85, 88, 90; 58-118; 59-20, 21, 28; 68-41
Sukhobokova, N. S.: 57-33
Sukhorukova, N. V.: 68-10
Sulček, Z.: 55-36
Sun, P. J.: 60-13; 61-30, 31, 32; 62-30, 31
Sundra, D. S.: 65-59; 66-37
Suranyi, T. M.: 68-32
Surgenor, D. M.: 59-144
Suzuki, S.: 54-18, 103
Suzuki, T.: 64-68
Svasta, J.: 55-36
Sveinsson, S. L.: 55-53
Sverzhina, R. E.: 67-99
Svoboda, V.: 60-45, 50, 52; 61-85
Swann, M. H.: 55-44
Sweet, T. R.: 52-42; 54-49, 102; 55-85
Sweetser, P. B.: 53-37; 54-14

Sykes, A. G.: 68-62
Sympson, R. F.: 54-88
Szantho, C.: 64-131; 67-127
Szarvas, P.: 66-118; 67-127
Szegedi, R.: 61-53
Szekeres, G. L.: 64-25; 65-89; 67-128
Szekeres, L.: 62-136; 64-12, 25; 65-88, 89; 66-55; 67-128; 68-59
Szelag, M.: 62-66
Szonove, P. A.: 62-13

T

T'ao, T. C.: 65-101
Tachikawa, T.: 64-87; 65-112
Tackett, J. E.: 63-26
Tagliavini, G.: 60-201
Tajiri, M.: 63-55, 72, 122
Takagi, H.: 60-141
Takagi, T.: 60-112
Takamoto, S.: 55-100, 101, 102, 103; 57-41
Takao, H.: 61-72
Taketatsu, T.: 59-71, 87
Takeuchi, K.: 68-43
Talalaeva, O. D.: 57-110
Talipov, Sh. T.: 64-41, 42, 43; 65-23; 67-11, 32, 60, 69
Talipova, L. L.: 61-132, 157; 62-48, 49, 115; 63-70
Tallant, D. R.: 68-26, 60
Tamura, Z.: 63-59
Tanaka, H.: 64-38; 68-72
Tanaka, M.: 64-31; 66-34, 110; 68-61
Tanaka, N.: 56-94; 58-74; 59-80; 64-93; 65-63, 66
Tanaka, T.: 56-18; 57-100; 58-117; 59-41; 60-84; 66-93
Tananaev, I. V.: 61-23, 58; 62-91
Tănase, D.: 68-14
Tandon, J. P.: 67-45
Tandon, K. N.: 66-12; 67-121, 122, 123; 68-53
Tandon, S. N.: 64-47
Tanner, H.: 57-95
Tänzer, I.: 59-109
Tateshita, N.: 55-22
Taylor, D. H.: 63-85

Taylor, M. P.: 55-8
Teleshova, R. L.: 67-129
Temkina, V. Ya.: 67-130
Temple, C.: 56-78
Tenorová, M.: 59-14, 20
Teodorescu, Gr.: 64-108
Teodorescu, N.: 60-196; 62-109
Teodorovich, I. L.: 68-39
Terenteva, T. A.: 62-107
Tereshin, G. S.: 62-91
ter Haar, K.: 53-9, 22; 54-17, 104; 56-5
Terlain, B.: 60-137
Terpilowski, J.: 60-98
Terzeman, L. N.: 68-10
Teslinskii, Yu. K.: 66-89
Tettweiler, T.: 54-15
Thatcher, L. L.: 55-47
Theis, G.: 67-84
Theis, M.: 55-26, 27, 29, 30, 31
Theodore, M. L.: 66-23
Therattil, K. J.: 55-25; 58-95
Thikonov, A. S.: 57-110
Thomann, H.: 61-117, 118, 180
Thomas, R. S.: 63-65
Thompson, L. C.: 62-101
Thorneley, R. N. F.: 68-62
Tichomirova, C. T.: 57-105
Tikhanina, S. P.: 62-114
Tikhomirov, V. I.: 64-142
Tikhonov, V. N.: 62-88; 64-2, 16; 65-2, 12
Tikhonova, Z. I.: 64-26
Tiptsova, V. G.: 58-107; 60-167, 168, 180, 181; 61-130; 62-68, 92; 65-61
Tira, L.: 67-29
Tiwari, G. D.: 66-61; 68-47
Toei, K.: 57-40, 41, 50; 63-87; 65-26; 67-131
Toerien, F. v. S.: 66-50
Toft, R. J.: 63-74
Tölg, G.: 64-113
Tölgyessy, J.: 64-13, 14, 15; 66-1; 67-132
Tolmachev, C. N.: 57-104
Tomlinson, W. R.: 66-35
Tomori, T.: 62-125
Tonkoshurov, V. S.: 65-48

Tonooka, N.: 67-91
Toropova, V. F.: 62-64
Toschewa, D.: 62-152
Tournayan, L.: 68-66
Toverud, S. U.: 54-53
Tramm, R. S.: 60-70
Treindl, L.: 56-15
Trémillon, B.: 60-66
Trivedi, S. R.: 66-61; 68-47
Truta, L.: 63-126, 127, 128
Tsao, F.-Y.: 64-83
Tsareva, K. Kh.: 67-42
Tschan, D.: 67-162
Tse, Y.-H.: 61-103
Tselinsky, U. K.: 68-63
Tserkasevich, K. V.: 59-147
Tserkasevich, K. W.: 60-186
Tserkovnitsa, I. A.: 60-189
Tserkovnitskaya, I. A.: 63-109, 111; 64-79
Tsirul, V. A.: 63-84
Tsubota, H.: 63-92
Tsuchiva, Y.: 62-59
Tsuchiya, T.: 62-59
Tsukada, S.: 55-11
Tsunogai, S.: 68-64
Tsvetkova, E. V.: 61-121
Tsyvina, B. S.: 58-100; 59-45
Tucker, B. M.: 54-34; 55-24; 57-103
Tuckermann, M. M.: 58-24
Tukhonova, L. I.: 67-163
Türäkova, J.: 64-33
Turina, S.: 63-33
Tuyakov, N. B.: 64-76
Tympl, M.: 64-74
Tyukova, Z. V.: 62-78

U

Udaltsova, N. I.: 60-57
Uden, P. C.: 68-40
Udvardi, M.: 68-6
Ueda, S.: 62-108
Ueno, K.: 52-44; 57-65; 58-72; 63-14, 75, 76; 64-149; 67-146; 68-28
Uhlig, E.: 64-50; 68-65
Ukigaya, K.: 62-142
Umarova, M. M.: 63-118
Umeda, M.: 60-97
Umurzakov, I. A.: 63-114

Underwood, A. L.: 53-39; 54-50, 51; 55-97
Uno, T.: 68-72
Urbain, H.: 68-66
Usatenko, J. I.: 59-44; 60-95
Uvarova, E. I.: 64-82

V

Vainer, M. G.: 67-77
Vajgand, V.: 68-56
Vandael, C.: 62-135
Vanderdeelen, J.: 68-67
Van-der-Hende, A.: 68-67
Van der Plas, Th.: 65-51
Vándorffy, M. T.: 62-117
Vaníčková, E.: 56-34, 35; 57-56, 85; 60-44
Varde, M. S.: 64-135
Vasil'eva, V. F.: 68-68
Vasiliev, R.: 60-94; 64-129
Vassiliades, C.: 66-66
Vavoulis, A.: 59-113
Vazirani, V. A.: 67-134
Vdovenko, M. E.: 67-135
Večeřa, Z.: 58-65; 61-82, 87, 88
Veda, S.: 61-24
Venkatasubramanian, V.: 64-135
Verbeek, F.: 57-53
Vereph, J.: 62-131; 65-98
Verma, M. R.: 55-25; 57-120, 121; 58-95; 60-80; 61-59; 67-136
Veselý, V.: 61-92, 93, 94, 95, 96, 98, 99, 100; 62-53, 87, 102; 63-17, 27, 61, 62, 91; 64-20, 21, 22, 23, 61, 62, 63, 72; 65-7, 8, 40, 41, 43; 66-2, 3, 24, 25, 26, 27; 67-2, 109, 110, 111
Vestfried, T. Y.: 57-104
Vetter, G.: 63-34
Vibarova, L. I.: 64-122
Vičenova, E.: 52-36
Vichev, E. P.: 60-81
Vickery, R. C.: 54-65
Villanyi, K.: 61-64
Vinek, G.: 67-88
Vinogradov, A. V.: 67-137
Vinogradova, N. I.: 68-57
Vishnu: 57-13; 62-119, 147

Vitals'kaya, N. N.: 66-119
Vitkina, M. A.: 59-44; 60-95; 67-138
Vladimirov, L. V.: 62-146
Vladimirova, V. M.: 56-13; 58-100; 60-71
Vladimirtsev, I. F.: 63-96, 97; 65-29
Vlasof, L. G.: 64-77
Vobora, J.: 59-16
Vodák, Z.: 54-16; 56-91, 92
Volf, L. A.: 60-37, 38, 59, 60; 64-66
Volodarskaya, R. S.: 57-94; 60-93; 63-8
Vorišek, J.: 59-48
Vorlíček, J.: 64-64; 65-42, 64, 105; 66-120, 121, 122; 67-139, 140, 141
Vrchlabský, M.: 61-142; 62-46
Vřeštál, J.: 56-93; 57-77, 78; 58-46; 59-13; 60-47; 66-13, 59
Vromen, A.: 66-65
Vrzalová, D.: 62-124; 65-22
Vydrá, F.: 55-114; 56-87; 57-34, 35; 58-47, 52, 54; 59-22, 23, 27; 64-64; 65-42, 64, 105; 66-121, 122; 67-141, 142, 143; 68-58, 69

W

Wada, H.: 64-36, 45; 65-68; 68-70
Wada, T.: 57-50
Wakamatsu, S.: 57-106; 60-169, 170; 61-46, 155; 62-144; 65-65
Wakizaka, H.: 61-24
Waldvogel, P.: 59-150
Wallraf, M.: 57-17, 18; 61-42
Wang, E. K.: 65-100
Wänninen, E.: 54-70; 55-98; 58-4, 92; 60-53; 61-151; 65-104
Ward, G. M.: 60-86
Wardi, A. H.: 62-129
Warmuth, F. J.: 59-103
Washbrook, C. C.: 53-48
Washizuka, S.: 67-144
Watanabe, T.: 68-71
Watanake, M.: 62-42
Watanuki, K.: 60-119; 64-102, 124
Webb, M. M.: 52-8

Webster, H. L.: 55-111
Wehber, P.: 55-15, 16, 17; 56-106, 107, 108, 109, 110, 111, 112, 113, 114, 115; 57-2, 3, 4, 5, 6
Weiner, R.: 57-16; 58-97
Weiss, D.: 63-24; 67-103
Weitzel, G.: 53-53
Wendlandt, W. W.: 60-85
Weng, Y. K.: 63-60
Wennerstrand, B.: 53-40, 41; 54-38, 67, 100, 101; 55-4; 57-107, 137; 58-80; 67-3
West, P. W.: 54-47; 55-41, 42, 43; 56-59
West, T. S.: 54-1; 55-104, 105, 106; 57-42, 43; 58-1, 2, 3; 59-115; 60-154, 155, 156, 194; 62-35; 68-19
Wharton, H. W.: 64-5
Wheelwright, E. J.: 53-4
Whidby, J. F.: 68-38
White, J. C.: 60-152
White, J. C. D.: 62-96
Wilhite, R. N.: 55-97
Wilke, K. T.: 67-46
Wilkins, D. H.: 58-42, 43; 59-4, 5, 6, 7, 8, 9, 10, 11, 12, 149; 60-63, 64, 65
Wilkins, R. G.: 67-159
Wilkinson, J. V.: 61-110
Wilkinson, P. R.: 54-32; 59-104; 64-147
Willard, H. H.: 58-15
Willi, A.: A-6; 47-1; 51-5
Williams, A. E.: 54-55
Williams, H. P.: 59-112
Williams, M. B.: 54-35; 59-82
Williams, R. J. P.: 53-30; 54-55
Williams, Jr, T. R.: 55-20
Wills, V.: 62-4; 63-29
Wilsen, A. E.: 50-11
Wilson, D. W.: 65-50
Wilson, H. J.: 61-16
Wilson, W.: 65-49
Winsauer, K.: 59-35
Wise, W. S.: 55-86
Witkowski, H.: 59-93
Wittmann, H.: 62-47
Wolf, G.: 59-51

Wolfsberg, K.: 62-127
Wood, D. F.: 62-85
Woodhead, J. L.: 54-91; 55-81; 56-75
Wronski, M.: 59-46
Wu, C. H.: 59-84
Wünsch, L.: 55-91; 57-111; 60-6; 61-12
Wünscher, H. J.: 63-93

Y

Yabro, C. L.: 58-93
Yaguchi, H.: 65-62
Yakimets, E. M.: 53-42; 54-71; 56-2, 24; 58-9; 60-76, 77, 161, 162; 61-77; 62-110; 65-29; 67-77
Yakolev, N.: 68-37
Yamada, T.: 63-42
Yamaguchi, H.: 67-146
Yamaguchi, K.: 63-14, 75, 76; 64-149; 67-146
Yamaguchi, N.: 67-145
Yamaguchi, O.: 68-72
Yamaguchi, T.: 67-147
Yamamoto, Y.: 61-24; 62-108
Yamamura, S. S.: 64-86
Yamauchi, O.: 64-38
Yano, K.: 67-55
Yao, K. M.: 63-5, 125
Yao, M. S.: 66-58
Yaroshenko, G. F.: 67-130
Yashina, N. I.: 63-113
Yashunskii, V. G.: 67-118; 68-68
Yatsimirskii, K. B.: 55-83
Yatsyk, I. Ya.: 68-73
Yeh, J. C.: 63-109, 111
Yih, I. M. L.: 62-11
Ymauchi, T.: 60-159
Yoe, J. H.: 54-90; 60-106
Yofé, J.: 68-74
Yoneyama, Y.: 67-85
Yoshino, Y.: 64-102, 124
Yoshiwaka, K.: 61-178
Yotsuyanagi, T.: 67-147
Young, A.: 55-85
Yu, V.: 64-66
Yun, W.: 57-101
Yurasova, G. M.: 65-9

Yurist, I. M.: 60-176; 62-62, 78, 86, 126; 64-111; 65-94; 66-62; 67-148, 149; 68-75

Z

Zabolotskii, T. V.: 64-109
Zahradníček, M.: 67-150
Zaikovskii, F. L.: 60-11
Zaikovsky, F. V.: 59-138
Zaitsev, A. A.: 57-112
Zak, B.: 56-4, 14
Zaki, M. R.: 54-59; 60-54, 55
Zapp, E. E.: 60-56
Zarinskii, V. A.: 63-5
Zavarikhina, G. B.: 66-106
Zavrazhnova, D. M.: 61-89, 90
Zelichenok, S. L.: 63-103
Zemlyanski, N. I.: 61-111
Zenin, A. A.: 55-82
Zhavoronkova, A. Ya.: 65-92

Zhdanov, A. K.: 56-3; 58-41; 60-42; 63-114, 118; 67-151, 152
Zhigalkina, T. S.: 68-1
Zhikareva, V. I.: 63-43, 115; 64-71, 81
Zhirnova, N. M.: 65-39, 82
Zhivopitsev, V. P.: 63-99; 66-123, 124
Zieger, M.: 67-38
Zimmer, P.: 54-85
Zimmermann, A.: A-1
Zimmermann, Th.: 65-123
Zinov'eva, L. D.: 65-9
Zittel, H. E.: 62-32; 63-13
Zminkowska, T.: 67-116
Zöhler, A.: 53-52
Zosin, Z.: 61-108
Zsigrai, J.: 68-32
Zuse, M.: 65-102
Zýka, J.: 53-12, 13; 54-82; 55-88; 62-122; 64-67

Subject Index

ABSORPTIOMETRIC determination of end-points, 90–8
Absorption spectrum, of Erio T, 45
of metalphthalein, 41
of murexide, 61
of pyrocatechol violet, 38
A-cations, 9
Accuracy of titrations (*see also* Titration error), 113–15, 117, 121
Acetate buffers, 21, 152
Acetato-complexes, 9, 20, 192
Acetylene, 285
Acid Alizarin Black, 49, 73, 162
SE, 49, 73
SN, 49, 73
Acid chlorides of oxalic acid, 285
Acid Chrome Blue K, 47, 73, 161
Acid Chrome Dark Blue, 47, 73, 161
Acid Chrome Violet B, 47, 73, 161
Acid, sensitivity to, 7
A.C. polarization, 103
Actinides, 10
Activity coefficients, 11
Albumen, 290
Alcohols, 286
Aldehydes, determination of, 289
Alizarin, 58, 73
Alizarin Chrome Black R, 46, 73
Alizarin red S, 59, 73, 89, 96
Alizarincomplexone, 58, 59, 73, 97
Alkalimetric titrations, 9, 14, 127, 128
Alkaline earths, 22, 22–30, 80, 81, 85, 86, 159–83
Alkaloids, 287, 291
Alkoxylsulphonates, 287
Alpha coefficients, xii, xxi, 15, 16, 17, 19, 20, 23
Aluminate, 133, 188
Aluminium, 184–92
Aluminon, 36, 39, 162
Amalgam methods, 124–7, 258, 290
Amalgam reduction, apparatus for, 126

Amine bases, 286
Aminoacids, 13, 290
β-Aminoethylmercaptan, 134
Aminopolycarboxylic acids, 8, 130
Aminopyrine, 286, 288
Ammine complexes, 5, 18, 20, 24, 56, 92, 134, 259
stability constants, 5
Ammoniacal buffers, 153
Ammonium bases, quaternary, 288
Amperometric titrations, 106–9, 137
Analergen, 287
Analysis, indirect, 141
Aneurine, 287, 288
Anthraquinone, 59
Antihistamines, 287
Antimony, 308, 309
Anti-oxidants, 286, 289
Antipyrine, 291
Argenticyanide ion, 124, 259
Arsenates, 126, 306
Arsenazo, 47, 73, 96, 97
Arsenic, 306–8
Arsonium bases, 287
Ascorbic acid, 135, 137, 154, 155, 230, 272
Ascorbin palmitate, 289
Automated photometric titrations, 98
Automation, 90, 98, 106, 111
Auxiliary complexing agents, 19, 21–3, 25, 118, 129, 132
Auxochrome, 34, 43
Auxochromic oxygen, 36, 37, 40, 41, 55
Azides, 32
Azo dyes, 43–57, 72
Azoxine, 52, 56, 96

Back titration, 25, 103, 115–17, 182
BAL (British Anti-Lewisite), 134, 154, 234, 263, 270, 285
Barbital, 286
Barbituric acid, derivatives of, 286, 289

Barium, procedures for, 179–83
titration of, 28, 29, 181–3
Barium–Erio T, transition curve, 176
Barium–metalphthalein, transition curve, 86, 178
Barium and magnesium, simultaneous titration of, 28
Benzaurin, 35, 73
Benzidine, as redox indicator, 71, 73
derivatives of, 66, 67, 71
Benzhydroxamic acid, 32, 33, 73
Beryllium, 137, 158, 159
Beryllon II, 48, 73
Bichromates, 222
Bimetallic complexes, 18, 38, 224, 227, 292, 310
Bindschedler's green, 68, 69, 73
Biological fluids, analysis of, 170
Bis-(aminoethyl)-glycolether-N,N,N′,N′-tetra-acetic acid (*see* EGTA)
Bismuth, 309–13
Bismuthiiodide ion, as indicator, 34, 287, 309
as precipitant, 158, 287
Blocking (of indicators), 90, 136, 137, 154, 156, 160, 210, 225, 255
Blood, 170
B-metals, 9
Boron, determination of, 184
Brilliant Congo Blue, 50, 73
Bromates, determination of, 323
Bromides, determination of, 323
Bromocresol green, 128
Bromoform, 286
Bromophenol blue, 36
Brompyrogallol red, 42, 73
Brucine, 70, 287, 288
Buffers and buffer materials, 19, 152–3

Cacotheline, 69
Cadmium, demasking of cyanide complexes, 137
determination of, 268–71
masking with cyanide, 134

masking with iodide, 122, 134
titration in presence of zinc, 264, 265
Caesium, 156–8
Caffeine, 288
Calcein, 42, 43, 64, 65, 72, 73, 76, 96, 163
as fluorescence indicator, 64, 163
Calcichrome, 50, 72, 73, 163, 251
Calcion, 50, 73
Calcium, 22, 27–30, 81, 85, 86, 159, 162–5, 170–2, 174–9
direct titration with metalphthalein, 81, 177
direct titration with murexide, 81, 85, 175–7
substitution titration with Erio T, 27, 174
Calcium–Erio T, transition curves, 85, 173
Calcium and magnesium, simultaneous titration, 27–30, 173, 174
mixtures of, 165–71
Calcium–metalphthalein, transition curves, 81, 86, 178
Calcium–murexide, transition curves, 81, 176
Calcon, 48, 55, 162, 167, 178
indicator constants, 82
Calmagite, 46, 55, 89, 94, 96, 161
indicator constants, 82
Cal-Red, 48, 55, 162, 167
Captax, 285
Carbonates, 281
Carbon compounds, 281–91
Carbonic acid, 281
Carboxyl groups, determination of, 290
Carminic acid, 58, 59
Chelate effect, 4, 5
Chelate ring, 4, 8, 34, 55, 62
Chelidonine, 287
Chelon wave, 180
Chloral as demasking agent, 137, 154, 266
Chloranilic acid, 58, 59, 73
Chlorates, determination of, 323
Chlorides, determination of, 322
Chloroform, 288

Chloropromazine, 288
Chromates, determination of, 123, 126, 221–3
Chromeazurol S, 73, 96, 97, 161, 162, 241
Chrome Bordeaux B, 46, 73
Chrome Brown RR, 46, 73
Chrome Fast Blue RR, 46, 73
Chromium, 218–23
Chromophore, 31, 34, 43
Chromotrope 2C, 46, 73
Chromotrope 2R, 46, 73
Chromoxane Green GG, 39, 73, 161
Chromoxane Pure Blue BLD, 39, 73
Chromoxane Violet 5B, 39, 73
Chronopotentiometric   titration, 110
Cinnamohydroxamic acid, 32
Citrate complexes, 20, 25, 133
Cobalt, 242–5
Cochineal, 58
Codeine, 287, 291
Coefficient α, xii, xxi, 15–17, 19, 20, 23
Collective titrations, 27–30, 119–22
Colour indicators, 26–76
    theory of, 77–89
    transition interval, 84, 85, 173, 176
    visibility of colour change, 86–9
Common names (of indicators), 71–6
Complementary colours, 87–9
Complex equilibria, effect of pH on, 15, 16
Complex formation, 3, 4, 9, 10, 11–21
    and masking, 131, 132–5
    and redox potentials, 99–101
    stepwise, 4
    stoicheiometry of, 4
Complexes, bimetallic, 18, 39, 224, 227, 292, 310
    inert, 3, 116, 117, 219, 220, 247
    labile, 3
    1:2, 18
    sensitivity to acid, 7

2:1, 18, 38, 224, 227, 292, 310
Complexone wave, 108, 138, 261, 269
Complexones, 8, 9, 150
Concentration constants, 11
Conditional stability constants, vii, xxii, 11, 15
Conductimetric titrations, 104, 105
Consecutive titrations, 25, 27, 103, 106, 111, 130, 137–41, 169
Constants, conditional, xxii, 11, 15
    effective, xxii, 15, 19, 21, 22, 24, 29, 115, 118, 120–2, 129, 137, 214, 236
    individual, 5
    stoicheiometric, 11
    thermodynamic, 11
Co-ordination sphere, 18
Copper, 252–8
Copper   diethyldithiocarbamate, 118
Copper–EDTA complex, titration curves, 23, 26
Copper–murexide,   transition curves, 85, 248
Copper–PAN complex, 55, 96, 119, 215, 253, 257
Coulometric titrations, 109, 110
Cumarones, 65
Cyanides, 134, 282, 283
Cyanide complexes, as a basis for titrations, 3, 124, 134, 258, 259
Cyanide masking, for large amounts of iron, 135, 240
Cyanide methods, 123, 124, 233, 258, 259
Cyanopalladate ion, 124, 250
Cystein, 134, 135

DCTA (DCyTA), xix, 9, 10, 16, 104, 139, 150, 189, 202, 220, 226, 233, 247, 296
Dead stop end-point, 107
Demasking, 125, 136, 137, 139, 140
Diacetyl, 286
Diamine Green, 89
Diaminocyclohexane-N,N,N′,N′-
    tetra-acetic acid (*see* DCTA)

Di-(2-aminoethyl)ethyleneglycol-
N,N,N′,N′-tetra-acetic acid
(*see* EGTA)
Dianisidine complexone, 66
Diethylenetriaminepenta-acetic acid
(*see* DTPA)
o,o′-Dihydroxyazo dyestuffs, 44, 55,
57, 71, 162, 231
Dimethylglyoxime, 249
3,3′-Dimethylnaphthidine, 71, 73,
267
Dinitrosochromotropic acid, 60
Dioximes, 286
Dioximes, vicinal, 286
Diphenylcarbazide, 70, 74, 271
Diphenylcarbazide–vanadium sys-
tem, 70, 214
Diphenylcarbazone, 70
Diphenylformazan, 57, 83
Dissociation constants, 12
Distribution coefficients, xii, xxi,
20, 21, 84, 118, 120–2, 132
Dithizone, 57, 58, 74, 152, 155, 187,
192, 243, 296
Divascol, 287
Double bonds, determination of,
290
Drawn-out, end-points, 87
Dropping-mercury electrode, 107
DTPA, xix, 9, 10, 11, 98, 150, 263
Dyestuffs, complexes of, 31, 34,
77–84

EDTA, xix, 8, 9, 10, 146
Effective stability constants, vii,
xxii, 15, 19, 21, 22, 24, 29, 115,
118, 120–2, 129, 137, 214, 236
EGTA, xix, 9, 11, 98, 121, 122, 130,
138, 150, 165, 168, 263
End-points, detection by instrumental methods, 90–112, 113,
115, 121
indication of, 31
sharpness of, 87, 115, 117, 119
pM at, 114, 115, 117, 120, 121
Energy, free, 3, 15, 18
Enthalpy titration, 110
Entropy effect, 5
Equilibrium constants, xxi, xxii

Equivalence point, 25, 88, 113, 114,
117, 120, 121
Erio A, 48, 55, 74
indicator constants, 82
Erio B, 48, 74
indicator constants, 82
Erio SE, 47, 55, 162
indicator constants, 82
Erio T, 44, 48, 55, 71, 72, 74, 96,
151, 173, 264, 265, 272, 281,
299
absorption spectrum, 45
indicator constants, 82
Erio T, transition curves,
with barium, 176
with calcium, 80, 85, 173
with magnesium, 79, 80, 85, 173,
176
with strontium, 176
with zinc, 85, 248
Eriochrome Azurol S, 39
indicator constants, 82
Eriochrome Black PV, 47, 74
Eriochrome Black T, vii, 44, 72, 74
Eriochrome Blue SE, 47, 74
Eriochrome Cyanine R, 37, 74
Eriochrome Green H, 47, 74, 89
Eriochrome Navy Blue BRL, 47, 74,
162
Eriochrome Red B, 54, 56, 74
Eriochrome Violet, 47, 74
Error in titrations, 114, 115, 117,
118, 121
Ethylenediamine-N,N,N′,N′-tetra-
acetic acid (*see* EDTA)
as primary standard, 146
dissociation constants, 10, 13
keeping qualities, 146, 147
purity of, 146
salts of, 146
standard solutions of, 146
structure of, 9
Euphylline, 287
Exchange reaction, 117
Extinction coefficient, 33
Extraction procedures, 131
Extractive end-points, 89, 236
cobalt titration, 242
theory, 89

Fast sulphon black F, 49, 74
Ferri-ferrocyanide in redox indication systems, 70, 101, 267, 273, 297
Ferron, 64, 74
Flexadil, 288
Fluorescein, 43, 64, 74
Fluorescein complexone, 74, 76, 98, 162
Fluorescence indicators, 31, 63–7
transition, 63
Fluorexone, 43, 64, 65
Fluorides, determination of, 319–22
masking by, 131–7
Fluorine, 319–22
Formaldehyde as demasking agent, 125, 137, 154, 266
Formation constants, xxii, 5, 12
individual (stepwise), 5
overall (cumulative), 5
Formazans, 57
Functional groups, determination of, 290

Gallates, 286, 287, 289
Gallion, 47, 74
Gallium, 273–5
Gallocyanine, 42, 74, 274
Gentian violet, 89
Germanium, 292
Glass containers, 147
Glycine, as buffer constituent, 152
Glycine naphthol violet, 46, 74
Glycine thymol blue, 39, 74
Glyoxal-bis-(2-hydroxyanil), 62, 162
Gold, 124, 260
Graphite electrodes, 107
Grignard reagents, determination of, 291

Haematoxylin, 59, 74
Hafnium, 203–7
Halogens, 124, 319–23
Hardness of water, titration of, 96, 105, 171–3, 174
Hexamethyleneimine, 286
Hexamethylenetetramine, 153

Hexamine, 4, 5
HHSNN, 72, 74
High-frequency titrations, 105, 106
Hydration shell, 3
Hydrazine, 97, 135, 224, 289, 300, 304
Hydrazine derivatives, determination of, 300
Hydrogen peroxide, 135, 137, 154, 164, 200, 201, 243
Hydron (I), 74, 163
Hydron (II), 89
Hydroxamate complexes, 32, 33
Hydroxamic acid, 32, 33, 97
Hydroxide precipitates, 6, 19, 131, 166, 230, 274
Hydroxyanthraquinone, 58
o-Hydroxy-o′-carboxyazo dyes, 55
Hydroxy complexes, 18, 21, 159, 186, 192, 236, 273, 292
Hydroxylamine, 154, 230, 289, 300
3-Hydroxy-2-naphthoic acid, 32, 64, 74
Hydroxynaphthoquinone, 58
Hydroxyquinoline, determination of, 288
8-Hydroxyquinoline, 56, 63
as fluorescence indicator, 63
8-Hydroxyquinoline-5-sulphonic acid, 64, 74
Hydroxyquinone, 58
Hypophosphites, 305

Imidazole, as buffer, 153
Iminoacetate groups, 71
Indicator constants, 33, 78, 81–4, 119
Indicator electrodes, 100, 101
Indicator mixtures (solid), 151
Indicators, amperometric, 106–8
colour theory of, 86–8
list of synonyms, 73–6
Indicators, metallochromic, 33, 34–63, 77–90
colour changes of, 77–90
effect of pH on colour changes, 40, 44, 77, 78–81, 128
Indicators, 'metalfluorechromic', 63

'metal fluorescent', 63
monocolour, 40, 86
photometric, 93
purity of, 71
solutions of, 151, 152
weak colour, 31
Indigo carmine, 89
Indirect analyses, 141
Indirect analysis of Zr–Hf mixtures, 207
Indium, 275–7
Inert dyestuff, 87
Infrared spectra of complexes, 12
Iodides, as indicators, 34, 74
masking by, 122, 133, 134, 263, 272
Iodine, determination of, 259, 260, 323
Ion-exchangers, 130, 155, 157, 161, 197, 205, 212, 230, 275, 291, 316
Ionization stages, 12, 19
Iridium, 250
Iron, 235–41
Iron–EDTA, redox potentials, 99
titration curves, 23, 25, 26
Iron and manganese, simultaneous masking, 233, 240
Isonicotinic acid hydrazide, 288, 300
Isopropylphenazone, 286

Kinetic masking, 135, 136, 139, 220

Lacquer scarlet C, 41, 74, 161, 162
Lanthanides, 10, 129, 194–7
Larusan, 288
Lead, 295–300
Leptazol, 285
Ligand, 4
Liquid–liquid extraction, 131, 207, 230, 231, 267, 269, 275, 277
Lithium, 154

Magnesium, 22, 27–30, 159–62, 170–3
standard solutions of, 150
Magnesium and barium, total titration of, 28

Magnesium complexonate, 174, 175, 181
Magnesium–Erio T, transition curves, 85, 173
Magnesium and strontium, total titration of, 28
Magneson, 46, 72, 74
Mandelic-azochromotropic acid, 47
Manganese, 230–4
Manganese and iron, simultaneous masking of, 233, 240
Masking, 106, 122, 131–6, 139, 140, 155, 164, 232, 269
agents, 103, 130, 132–4, 154
by complex formation, 132, 133, 137, 232, 233
by precipitation, 106, 131
kinetic, 135, 136, 139, 220
Mercaptopropionic acid, 134
Mercury, determination of, 158, 271–3
drop electrode, 101–3, 107, 109, 271
Mesomeric systems, 35, 36, 37, 60, 62
Metal complexonates, preparation of standard solutions of, 150
Metal hydroxides, precipitation of, 6, 19, 131, 166, 230, 274
Metallochrome violet A, 39, 74
Metallochromic indicators, 33, 34–63, 71, 77, 78
Metalphthaleins, 39–42, 81, 82, 86, 89, 152, 177, 183
absorption spectra of, 41
indicator constants, 82
transition curve of barium complex, 86, 178
transition curve of calcium complex, 81, 86, 178
Metals, standard solutions of, 150, 151
Metaphosphates, 304
Metasystox-i, determination of, 291
4-Methoxy-4'-aminodiphenylamine, 69
Methylanabasinoazoresorcinol, 51, 74
Methylcalcein, 42, 43, 65, 74

Methylcalcein blue, 65, 66, 74
Methylene blue, 89
Methyleneiminodiacetate groups, 36
Methyl red, 89, 128
Methylthymol blue, 39, 74, 163, 296
  indicator constants, 82
Mixed indicators, 88, 89, 128, 162
Molybdenum, 123, 223–6
Mono-*o*-hydroxyazo dyestuffs, 55
Monoximes, 286
Mordant dyestuffs, 72
Morin, 59, 63, 74, 274, 277
Morphine, 288
Murexide, vii, 60, 61, 74, 79, 89, 95, 151, 162, 175, 244, 248, 256
  absorption spectra, 61
  indicator constants, 83
  transition curves with calcium, 81, 86, 176
  transition curves with copper, 85, 248
  transition curves with nickel, 85, 248

Naphthidine, 71
Naphthol green B (in mixed colour indicators), 89
Naphthol purple, 59, 74, 96
Naphthol violet, 46, 74
Naphthol yellow S, 74, 89
Naphthyl azoxine, 52, 74
Naphthyl azoxine S, 52, 74
Naphthyl azoxine 2S, 52, 74
Narcotine, 287
NAS, 52
Neutralization, stepwise, 3
Nickel, determination of, 245–50
  titration curve with EDTA, 23
  titration curve with ammonia, 6
Nickel cyanide (*see* Tetracyano nickelate)
Nickel–murexide, transition curve, 85, 248
Nicotinamide, 286
Ninhydrin dyestuff, 62, 74
Niobium, 216–18
Nitrilotriacetic acid (NTA), vii, xix, 8, 9, 11

Nitro compounds, 125, 126
Nitrogen derivatives, 300
Nitro-groups, determination of, 290
Nitrosochromotropic acid, 60
Nitroso compounds, 125
Nitroso-groups, determination of, 290
Nitrosonaphthol, 60
Nitrosophenols, 60
Nitroso-R-salt, 60
Nitroso-SNADNS, 48, 74
NMR measurements on complexes, 12
Nordihydroguajaretic acid, 289
NTA, xix, 9, 11, 128, 150, 217

Omega chrome fast blue, 47, 74
Omega chrome garnet, 46, 74
Organic compounds, 123, 125, 283–91
  amalgam reduction of, 290
Organo-tin compounds, 291
Overall (gross) stability constants, xii, xxii, 5
Oxalates, determination of, 123, 285
Oxalic acid, esters of, 285
Oxygen, 343
Oxines, 48, 63, 74

PACA, 51, 75
PADNS, 51, 75
PAHA, 51, 75
Palatine chrome violet, 46, 82, 75
Palladium, 124, 250, 251, 252
PAN, 51, 56, 75, 119, 152, 257, 295
  indicator constants, 83
PAO, 51, 75
Papaverine, 287, 288
PAR, 51, 56, 75, 83, 96, 300
  indicator constants, 83
Pectin, 123, 285
Penten, xix, 4–8
Peptides, determination of, 290
Perchlorate, determination of, 323
Peroxy-complexes, of niobium and tantalum, 202, 216, 217
  of titanium, 200, 201
Peroxydisulphates, 318

Pharmaceuticals, 123, 272, 283–90
pH, dependency, 17, 18
  effect, 13, 14, 15, 130
  of indicators, 40, 44, 77, 78–81, 128
pHg method, 102
Phenanthroline, 134, 187, 189
Phenetidine complexone, 66, 75
Phenolphthalein, 36, 40
Phenol reaction, 32
Phenol red, 36
Phosphates, 123, 161, 301–6
Phosphonium bases, 287
Phosphorus, 301–6
Photometric indicator, 93
Photometric titrations, 90–8, 138
  applications of, 95–8
  automation of, 98
  selectivity of, 97
  theoretical considerations, 94, 95
Phthalein complexone, 75, 96, 162
Phthaleins, 34, 36, 40
Picric acid, 89
Piperazine salts, 287
Plastic vessels, 147, 183
Plateau potential, 107
Platinum, 250–2
Platinum electrode, vibrating, 107
Platinum electrodes, 100, 101, 107, 109
Plumbon, 63, 75
Plutonium, 198–9
pM indicators, 40, 77
  curves of, 22, 23, 28, 29, 86
pM, jump in, during titrations, 21, 24, 25, 87, 99, 113–15, 120
pM values, 21, 113, 114, 116
Poirrier's blue, 89
Polarography, 106, 107, 108
Polyamine complexes, 130
Polyamines, 4, 103, 130, 134, 263
Polyphenols, 4
Polythene vessels, 147, 183
Ponceau 3R, 48, 75
Potassium, 156
Potentiometric titrations, 98-103
Precipitation reactions, 123, 130, 131, 132, 155, 156, 158, 159, 166, 184, 223, 227, 228, 249, 267, 278, 282, 284–9, 291, 292, 298, 301, 302, 303, 306, 307, 314, 316
Primary standards, 96, 105, 148, 149
Proton complexes, 12
Purpureate ion, 61
Purpuric acid, 60, 75
Pyramidone, 287, 288
Pyridine, 288
Pyridine-azo-chromotropic acid, 51
Pyridine-azo-dihydroxynaphthol sulphonic acid, 51, 75
Pyridine-azo-H acid, 51, 75
Pyridine-azo-naphthol, 51, 75
Pyridylazo-orcinol, 51, 75
Pyridylazoresorcinol, 51, 75
Pyridylazoxine, 53, 75
Pyrocatechol, 55, 75, 289
Pyrocatechol violet, 37, 38, 39, 42, 75, 82, 96, 97, 152, 162, 212, 294, 310
  absorption spectrum of, 38
  indicator constants, 82
  metal complexes of, 38, 39
Pyrogallol red, 42, 75
Pyrophosphate, determination of, 123, 304

Quaternary ammonium bases, 288
Quercitin, 59, 75, 98
Quinine, 285, 287, 288
Quinizarin, 59, 75
Quinizarin sulphonic acid, 59, 75

Radiometric titrations, 111, 112
Rare-earths, 194–7
Redox behaviour, 67–71, 99–101
Redox indicators, 31, 67–71
  masking, 135
Redox potentials, 67, 68, 69, 98
  and complex formation, 98, 99
Redox titrations, 99–103
Rhenium, 234
Rhodium, 251
Rhodizonic acid, 58, 75
Rubeanic acid, 62
Rubidium, 156
Ruthenium, 250

Salicylate complexes, 33
Salicylic acid, as indicator, 32, 33, 34, 75, 97, 191, 208, 241
Scandium, 193–4
Screening, of indicators, 88
Selective titrations, 119–22, 129
Selectivity, 25, 97, 106, 129–41
Selenium, 319
Self-indicating systems, 92, 93, 97
Semicarbazide, 300
Semixylenol orange, 39, 75
Separation procedures, 130
Serum, 84, 138, 155, 157, 158, 170
Shade (of an indicator), 87
Silicate analysis, 96
Silicic acid, 291, 292
Silicon, 291–2
Silver, 124, 258–60
Silver halides, 124
Simultaneous titrations, 28, 29
    of magnesium and calcium, 29
    of zinc and barium, 29
Slope indication, 91, 97
Slope indicator, 93
SNADNS, 48, 75, 96
SNADNS-5, 48, 75
SNAZOXS, 52, 75
Soap analysis, 285
Sodium, 123, 154–6
Solochrome Black WDFA (or T), 48, 75
Solochrome Fast Violet B, 47, 75
Solubility (of EDTA), 13
SPADNS, 47, 75
Spinal fluid, 170
Square-wave polarography, 106, 138
Stability constants, xii, 10, 11, 12, 82, 83, 121, 122, 158, 159, 203, 214, 228, 229, 250, 266, 277, 278, 309
    conditional, 11, 15
    effective, 15, 19, 21, 22, 24, 26, 29, 115, 118, 120, 121
    stoicheiometric, 11
    thermodynamic, 11
Stability constants of complexes of
    DCTA, 10
    DTPA, 11
    EDTA, 10
    EGTA, 11
    NTA, 11
Stages, 3, 19
Standard solutions, preparation of, 146–51
    storage of, 147
Standardization of solutions, 147, 148
Stepwise indication, 91
Stepwise processes, 19, 120
Stilbazo, 50, 75
Stilbene derivatives, 67
Stoicheiometry, 4
Strontium, 28, 179–83
Strontium and magnesium, simultaneous titration, 28
Strychnine, 287
Substitution titrations, 25, 27–9
Sugars, complexometric determination of, 289
Sulpharsazen, 63, 75
Sulphates, determination of, 123, 314–18
Sulphide, determination of, 123, 291, 314
Sulphites, determination of, 318
Sulphonamide, 286
Sulphonaphthylazoresorcinol, 46, 75
Sulphonphthaleins, 36, 40
Sulphosalicylic acid, 32, 33, 75, 96, 133, 241
Sulphur, 313–19
    determination in organic compounds, 291, 317
Symbols, used in text, xix–xxii

TAC, 53, 75
TAM, 53, 75
TAN, 54, 75
TAN-6-S, 54, 56, 75
Tantalum, 216–18
Tantalum electrode, 107
TAO, 53, 75
TAR, 53, 75
Tartrate complexes, 20, 25, 133, 167, 230
Temperature dependence (of complexity constants), 12

Tetracyanonickelate ion, 96, 124, 154, 247, 250, 251, 258, 259, 283, 286
Tetraethylenepentamine, 103, 134, 138, 263
Tetrahydroxyfluoranes, 42
Tetramines, 4
Tetraphenylborate method, 158
Thallium, 277–81
Theobromine, 286
Theophyllin, 285, 287
Thermometric titrations, 110, 111, 138
Thiazolylazocresol, 53, 75
Thiazolylazonaphthol, 54, 75
Thiazolylazonaphthol sulphonic acid, 54, 75
Thiazolylazo-orcinol, 53, 75
Thiazolylazophenols, 56, 75
Thiazolylazoresorcinol, 53, 75
Thioacetamide, 123, 130
Thiocyanate complexes, 32
Thiocyanates, 31, 32, 124, 319, 323
Thioglycollic acid, 134
Thiosemicarbazide, 134
Thiosulphates, 134, 137, 314
Thiourea, 34, 75, 89, 134, 137, 240
    determination of, 291
Thorin, 46, 72, 75
Thorium, 209–13
Thymol blue, 36, 86
Thymolphthalein, 36, 40
Thymolphthalexone, 39, 75, 86
Tin, 292–5
Tin compounds, organic, 291
Tiron, 32, 33, 133, 152, 154, 235, 241
Titanium, 199–202
Titanium–hydrogen peroxide complex, 200, 201
Titration, accuracy of, 113–15, 117, 121
    alkalimetric–complexometric, 127, 128
    amperometric, 106–9
    back-, 115–17
    chronopotentiometric, 110
    conductimetric, 104, 105

consecutive, 25, 27, 103, 106, 111, 130, 137–41, 169
    coulometric, 109, 110
    curves, 6, 21–5
    degree of, 22, 26
    different types of, 113–28
    direct, 113–15
    error, 114, 115, 117, 118, 120–2
    high frequency, 105, 106
    photometric, 90–8, 138
    potentiometric, 98–103
    radiometric, 111, 112
    selective, 119–22, 129
    substitution, 117–19
    thermometric, 110
Titration error, 114, 115, 117, 118, 121
    in collective titrations, 121, 122
Transition curves, 84–6
Transition interval, 79, 84, 86, 87
Transition pM, 122
Tren, xix, 4, 8, 122
Trien, 97
Triethanolamine, 133, 134, 136, 153, 154, 232
Triethylenetetramine, 97, 103, 122, 138, 263
Triethylenetetraminepenta-acetic acid, 18, 139
Trimethylphenylarsonium chloride, as indicator, 220, 243
Triphenylmethane dyes, 35, 41
Tripolyphosphates, determination of, 304
Tristimulus colorimetry, 88
Tristimulus procedure, 88
Trivial names (of indicators), 72, 76
Tungsten, 123, 226, 227

Ultramicro scale, 93, 171, 244, 245, 260
Umbellicomplexone, 65, 66, 75
Unithiol, 134, 181, 263, 270, 308
Uranium, 155, 227–30
Urea, 156
Urine, 170
Urotropine, 97, 153, 157
    buffers, 153
    determination of, 288

Vanadium, 213–16
Variamine blue, 69, 75, 241
Visibility of colour change, 86–9

Water, hardness of, vii, 27, 171–4

Xanthates, 286
Xanthones, 65
Xylene cyanole FF, 89
Xylenol orange, 37, 39, 40, 75, 82, 89, 96, 97, 152, 155, 208, 238, 299
indicator constants, 82

Yttrium, 194–7

Zinc, 260–8
demasking of CN complex, 125, 137, 261, 262, 266
masking with CN, 122
standard solution of, 263
Zinc complexonate, 182
Zinc–EDTA, titration curves, 23
Zinc–Erio T, transition curves, 85, 248
Zinc uranyl acetate solution, precipitation with, 123, 154, 155
Zincon, 57, 72, 75, 264
indicator constants, 83
Zirconium, 203–7
Zwitterion, 37, 40